Concepts and Techniques of Inorganic Chemistry

Concepts and Techniques of Inorganic Chemistry

Editor: Bernard Wilde

NY RESEARCH
P R E S S

New York

Published by NY Research Press
118-35 Queens Blvd., Suite 400,
Forest Hills, NY 11375, USA
www.nyresearchpress.com

Concepts and Techniques of Inorganic Chemistry
Edited by Bernard Wilde

International Standard Book Number: 978-1-63238-565-9 (Hardback)

Cataloging-in-Publication Data

Concepts and techniques of inorganic chemistry / edited by Bernard Wilde.
p. cm.
Includes bibliographical references and index.
ISBN 978-1-63238-565-9
1. Chemistry, Inorganic. 2. Chemistry. I. Wilde, Bernard.
QD151.3 .C65 2018
546--dc23

Contents

Preface

I am honored to present to you this unique book which encompasses the most up-to-date data in the field. I was extremely pleased to get this opportunity of editing the work of experts from across the globe. I have also written papers in this field and researched the various aspects revolving around the progress of the discipline. I have tried to unify my knowledge along with that of stalwarts from every corner of the world, to produce a text which not only benefits the readers but also facilitates the growth of the field.

The branch of chemistry, which studies compounds that contain no carbon, is known as inorganic chemistry. Topics related to the field of inorganic chemistry are combination reactions, single displacement reactions, decomposition reactions, double displacement reactions, etc. This book unravels the recent studies in the field of inorganic chemistry. It is a vital tool for all researching and studying this field. Scientists and students actively engaged in this field will find this book full of crucial and unexplored concepts. It will help the readers in keeping pace with the rapid changes in this field.

Finally, I would like to thank all the contributing authors for their valuable time and contributions. This book would not have been possible without their efforts. I would also like to thank my friends and family for their constant support.

Editor

An iron-based green approach to 1-h production of single-layer graphene oxide

Li Peng[1], Zhen Xu[1], Zheng Liu[1], Yangyang Wei[1], Haiyan Sun[1], Zheng Li[1], Xiaoli Zhao[1] & Chao Gao[1]

As a reliable and scalable precursor of graphene, graphene oxide (GO) is of great importance. However, the environmentally hazardous heavy metals and poisonous gases, explosion risk and long reaction times involved in the current synthesis methods of GO increase the production costs and hinder its real applications. Here we report an iron-based green strategy for the production of single-layer GO in 1 h. Using the strong oxidant K_2FeO_4, our approach not only avoids the introduction of polluting heavy metals and toxic gases in preparation and products but also enables the recycling of sulphuric acid, eliminating pollution. Our dried GO powder is highly soluble in water, in which it forms liquid crystals capable of being processed into macroscopic graphene fibres, films and aerogels. This green, safe, highly efficient and ultralow-cost approach paves the way to large-scale commercial applications of graphene.

[1] MOE Key Laboratory of Macromolecular Synthesis and Functionalization, Department of Polymer Science and Engineering, Zhejiang University, Polymer Building, 38 Zheda Road, Hangzhou 310027, P.R. China. Correspondence and requests for materials should be addressed to C.G. (email: chaogao@zju.edu.cn).

Graphene has been the focus of significant attention for its potential across a broad spectrum of applications due to its unrivalled mechanical, electrical and thermal properties[1-3]. Thus far, two main strategies have been developed for the production of graphene from graphite: mechanical exfoliation (including solvent and ultrasonic-assisted methods)[4-6] and chemical oxidation–reduction[7-18]. Mechanically exfoliated graphene possesses few or no defects[6,19], but suffers from poor solubility ($<0.1\,\mathrm{mg\,ml}^{-1}$)[4] and extremely low productivity (for example, $\sim 2.0 \times 10^{-3}\,\mathrm{g\,h}^{-1}$)[20]. In addition, because of strong π–π stacking, such graphene is prone to irreversible aggregation after concentration and drying.

A recently applied process of high rate-shear exfoliation in N-methyl-2-pyrrolidone provides notable increases in productivity ($\sim 5.3\,\mathrm{g\,h}^{-1}$ (ref. 19), still far too low for commercial needs); however, the addition of polymer surfactants is necessary, otherwise the pristine graphene would aggregate and precipitate. Such graphene sheets are a mixture of different layers, limiting experimental reproducibility and inhibiting its use in fine applications. By comparison, preparation by chemical oxidation yields highly soluble single-layer graphene oxide (slGO; solubility $>110\,\mathrm{mg\,ml}^{-1}$)[21] in large-scales (up to tons scale), enabling easy processing of slGO into high performance composites and macroscopic materials such as fibres[22,23], films/papers[24] and aerogels[25] by solution-based polymer-type techniques. Although slGO is often denounced for containing defects[16] that may influence its properties, such defects can be easily repaired through simple chemical reduction[26]. Thermal treatment has the capacity to restore the chemically converted graphene (CCG) back to a material with ultrahigh electrical conductivity ($1.83 \times 10^5\,\mathrm{S\,m}^{-1}$) and thermal conductivity ($1434\,\mathrm{W\,m}^{-1}\,\mathrm{K}^{-1}$)[27]. These values are far higher than those of mechanically exfoliated defect-free graphene ($2.2 \times 10^4\,\mathrm{S\,m}^{-1}$, $313\,\mathrm{W\,m}^{-1}\,\mathrm{K}^{-1}$)[28]. For these reasons, the slGO–CCG route is the more attractive of the two for the industrial production of graphene.

Generally, GO is prepared by the ultrasonic exfoliation of graphite oxide[29]. The preparation methods of GO can be classified by the oxidant employed as either the $KClO_3$-based Brodie–Staudenmaier[8-10] method or the $KMnO_4$-based Hummers method[11-14]. The $KClO_3$-based method was first introduced by Brodie[8] in 1859, modified by Staudenmaier[9] in 1898 and again modified by Hofmann[10] in 1937. The reaction medium for this process is nitric acid, which presents the inherent disadvantages of explosion risk, release of hazardous gases (for example, NO_X and ClO_2) and the generation of carcinogenic ClO^-. The Hummers method was first reported in 1958 (ref. 11). Although the change of oxidant circumvented a number of $KClO_3$-based issues, it is plagued by the necessity of polluting heavy metal ions (Mn^{2+}) and the explosion risk that accompanies the unstable Mn_2O_7 intermediates[30]. Various modifications involving minor optimization of the Hummers method have been employed for the synthesis of GO; however, no significant improvements have been made despite the intensive interest in this material[1,7,31]. In addition, the two methodologies used to obtain slGO require long reaction times (6 h–5 days), relatively high temperatures ($>50\,^{\circ}\mathrm{C}$) and often additional intercalation and ultrasonication processes. These shortcomings result in a costly process in terms of time and energy, a complicated fabrication procedure and carry high costs related to waste treatment. Hence, a green (free of toxic gases and polluting heavy metals), safe (no explosive risk), ultrafast and low cost methodology is eagerly sought.

Herein, we propose a strong yet green oxidant, K_2FeO_4, and establish an ultrafast, safe and non-toxic methodology for the scalable production of slGO. The entire fabrication process requires only 1 h, and the as-prepared large GO sheets are nearly 100% single layered without any ultrasonic treatment. Our slGO has a similar chemical structure and solubility to materials prepared using the conventional long-time modified Hummers method. Furthermore, the GO powder obtained by drying slGO solutions can be re-dissolved in water or organic solvents to form stable liquid crystals (LC) and subsequently assembled into macroscopic materials such as one-dimensional (1D) fibres, 2D films and 3D aerogels. In addition, sulphuric acid is recycled in our protocol. Through the refreshing of oxidant, our approach dramatically reduces the effluent and lowers the operating cost. This method paves the way for cheap, eco-friendly, large-scale production of slGO and its macroscopic materials.

Results

Selection of oxidant. Oxidant is the most important controlling factor in the preparation of GO. The Brodie–Staudenmaier[8-10] method and Hummers[11-14] method essentially differ in their choice of oxidant. The prevailing oxidants, predominantly $KClO_3$ and $KMnO_4$, provide high oxygen content to the resultant GO materials; however, their byproducts are highly polluting and intermediates in the processes carry a high risk of explosion. For example, $KClO_3$ is a key ingredient in blasting caps and is prone to explode when mixed with combustible materials. It is also frequently used in explosives and fireworks, and is thus strictly controlled in China. In the synthesis of GO with $KClO_3$, the toxic and explosive gas ClO_2 is generated in the concentrated sulphuric acid solvent. In addition, $KMnO_4$ is easily converted into Mn_2O_7, which is prone to explode above $55\,^{\circ}\mathrm{C}$ in an acidic environment[30]. The use of $KMnO_4$ generates massive amounts of the heavy metal pollutant Mn^{2+}, which can cause great damage to human and plant life in an ecosystem. The various modifications of these two methodologies over the past decade have not been able to remedy the substantial inherent environmental and safety issues related to the production of toxic gases and heavy metal pollutants or the risk of explosion.

To resolve the problems posed by the conventional methods, an alternative oxidant for GO production is sought. The new oxidant must satisfy the following prerequisites: (1) high oxidation efficiency, (2) no risk of explosion and (3) no toxic or polluting byproducts. After numerous experiments, we identified K_2FeO_4 as the novel oxidant of choice. K_2FeO_4 is an eco-friendly and highly efficient oxidant with harmless byproducts. Currently, it is widely used in the fields of environmental protection and water treatment[32,33]. K_2FeO_4 has an electrode potential of 2.2 V, which is much higher than that of $KMnO_4$ (1.36 V) in acid environments, and should thereby considerably decrease the required reaction time. As opposed to $KMnO_4$, K_2FeO_4 can be safely used at temperatures as high as $100\,^{\circ}\mathrm{C}$, due to the absence of explosive intermediates. In addition, as a commonly used water treatment agent, K_2FeO_4 is inexpensive and commercially available. Therefore, K_2FeO_4 is attractive as a new-generation oxidant for the preparation of GO in the desired eco-friendly and highly efficient manner.

Preparation and characterization of GO. Typically, concentrated sulphuric acid, K_2FeO_4 and flake graphite were loaded into a reactor and stirred for 1 h at room temperature. The dark green suspension gradually became a grey viscous fluid. After recycling the H_2SO_4 reaction medium by centrifugation, the precipitate was purified by repeated centrifugation and water-washing to obtain highly water soluble slGO (solubility $>27\,\mathrm{mg\,ml}^{-1}$), coined as GO^{Fe}. Because the reaction process is extremely simple and requires no energy transfer (either heating or cooling), it is straightforward to scale up.

For instance, we successfully used a 20-l reactor to prepare 750 g of GO^{Fe} in one pot (Supplementary Fig. 1), corresponding to a 75 l GO^{Fe} aqueous solution with a concentration of $10\,mg\,ml^{-1}$ (Fig. 1a).

The composition of GO^{Fe} was analyzed via combustion analysis, quantitative X-ray photoelectron spectroscopy (XPS) and inductively coupled plasma mass spectrometry (ICP-MS). Combustion analysis showed that GO^{Fe} has a relative composition of $CO_{0.51}H_{0.22}S_{0.028}$. The XPS spectrum confirmed the composition of GO^{Fe} as follows (at.%): C (68.51%), O (31.14%), S (0.30%), Si (0.03%), N (0.01%), P (0.01%). ICP-MS measurements demonstrated the existence of negligible metal ion content: Fe (0.13 p.p.m.), Mn (0.025 p.p.m.), Co (0.073 p.p.m.), Cu (0.017 p.p.m.), Pb (0.033 p.p.m.) and Ni (0.014 p.p.m.). Notably, despite the high concentration of K_2FeO_4 in the reaction, the negligible iron content in the final GO^{Fe} after purification by the centrifugation/water-washing protocol indicates that no insoluble byproducts, such as Fe_2O_3, are generated in the fabrication and post-treatment processes.

The single-layered nature of the GO^{Fe} dispersion was demonstrated via scanning electron microscopy (SEM), transmission electron microscopy (TEM) and atomic force microscopy (AFM; Fig. 1). Under SEM inspection (Fig. 1c), the GO^{Fe} sheets show typical wrinkles, implying fine flexibility in the slGO sheets. According to the statistics from the SEM images, the GO^{Fe} sheets have a number-average width of $\sim 10\,\mu m$ and 53% of the relative size distribution (σ_w; Fig. 1d). TEM image also shows an abundance of wrinkles (Fig. 1e), and the selected area electron diffraction patterns (the insert) indicate its single-layer character[34]. The thickness of the GO^{Fe}, as measured by AFM, is $\sim 0.9\,nm$ (Fig. 1f), which confirms the single-layered state and the presence of oxygen-containing functional groups on the basal plane[17].

Raman spectra, X-ray diffraction and ultraviolet–visible spectra show that the GO^{Fe} has a similar structure to GO prepared by the modified Hummers method[24] using $KMnO_4$ as the oxidant (GO^{Mn}; average lateral size $= 8\,\mu m$; $\sigma_w = 79\%$, Supplementary Fig. 3). The Raman spectrum (Fig. 2a) of GO^{Fe} shows the typical D peak ($1,353\,cm^{-1}$), G peak ($1,600\,cm^{-1}$), 2D peak ($2,698\,cm^{-1}$) and D + G peak ($2,945\,cm^{-1}$) with an I_D/I_G intensity ratio of 0.93, confirming lattice distortions induced by oxidation[16]. The XRD curve of the vacuum-assisted filtration paper indicates that the interlayer spacing of GO^{Fe} is $\sim 9.0\,\mathring{A}$ (Fig. 2b), which is similar to that of GO^{Mn} ($8.7\,\mathring{A}$). The ultraviolet/Vis spectra of both GO^{Fe} and GO^{Mn} present a strong absorption peak at 230 nm ($\pi \to \pi^*$ transitions of the conjugation domains) and a weak shoulder peak at $\sim 300\,nm$ ($n \to \pi^*$ transitions of the carbonyl groups; Fig. 2c), revealing their similar domain structures[35].

The thermogravimetric analysis (TGA) profiles of both GO^{Fe} and GO^{Mn} show similar weight loss plots (48–50% mass loss at 800 °C, Fig. 2d). The Fourier transform infrared spectra identify the same functional groups in GO^{Fe} as GO^{Mn} (Fig. 2e): O–H stretching vibrations ($3,412\,cm^{-1}$), C $=$ O stretching vibration ($1,726\,cm^{-1}$), C $=$ C from sp^2 bonds ($1,624\,cm^{-1}$), O–C–O vibrations ($1,260\,cm^{-1}$) and C–O vibration ($1,087\,cm^{-1}$). As shown in Fig. 2f–h, the XPS spectra confirm the presence of similar chemical bonds in both GO^{Fe} and GO^{Mn}: C $=$ C (284.86 eV), epoxy/hydroxyls (C–O, 287.0 eV), C $=$ O (288.0 eV) and O–C $=$ O (289.2 eV) (ref. 21).

The oxygen-rich functional groups impart a high zeta potential to GO^{Fe} ($-58\,mV$) and excellent solubility in both water and polar organic solvents, as is the case for GO^{Mn} (Fig. 1b). The GO^{Fe} solution retains a homogenously dispersive constitution, without any precipitate, even after storage for 1 year at a concentration of $3\,mg\,ml^{-1}$ in water or N,N-dimethylformamide (Fig. 1g). The excellent solubility of the highly asymmetrical GO sheets may enable the formation of a lyotropic LC[23,36] phase, which is a criterion used to evaluate the 'true' solubility of graphene derivatives. Our GO^{Fe} aqueous dispersions display the vivid textures typical of nematic LCs between crossed polarizers (Fig. 1h).

Recycling and post-treatment of sulphuric acid. In addition to the problems of polluting heavy metals, toxic gases and tedious reaction times associated with the conventional methods, another

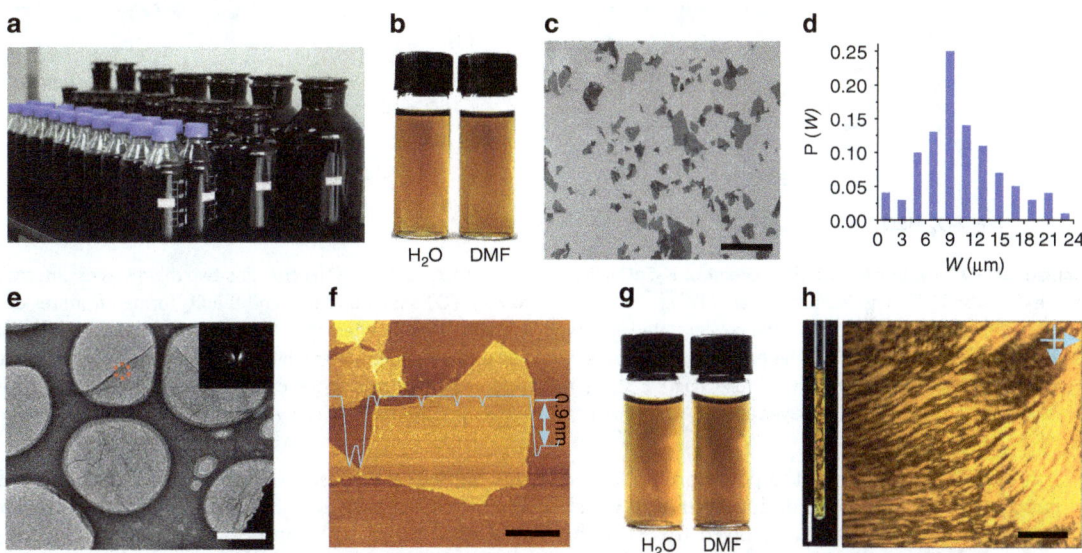

Figure 1 | Large-scale synthesis of single-layer GO^{Fe} via K$_2$FeO$_4$-based methodology. (a) Seventy-five litre GO^{Fe} aqueous solution with a concentration of $10\,mg\,ml^{-1}$. **(b)** GO^{Fe} solution in H$_2$O and N,N-dimethylformamide (DMF) with a concentration of $3\,mg\,ml^{-1}$. **(c)** SEM image of GO^{Fe} on Si/SiO$_2$ substrate. **(d)** The size distribution of the GO^{Fe} sheets, counted and calculated from **c** and Supplementary Fig. 2. **(e)** TEM image of GO^{Fe} and its SAED diffraction patterns (inset). **(f)** Tapping mode AFM image and height profile of GO^{Fe}. **(g)** GO^{Fe} solution of H$_2$O and DMF with a concentration of $3\,mg\,ml^{-1}$ after storage for 1 year. **(h)** Image of aqueous LCs in a quartz tube between crossed polarisers and POM image between crossed polarisers in planar cells of aqueous GO^{Fe} LCs at a concentration of $3\,mg\,ml^{-1}$. Scale bars, $20\,\mu m$ (**c**), $2\,\mu m$ (**e**), $4\,\mu m$ (**f**) and $5\,mm$ (**h**, left), $1\,\mu m$ (**h**, right).

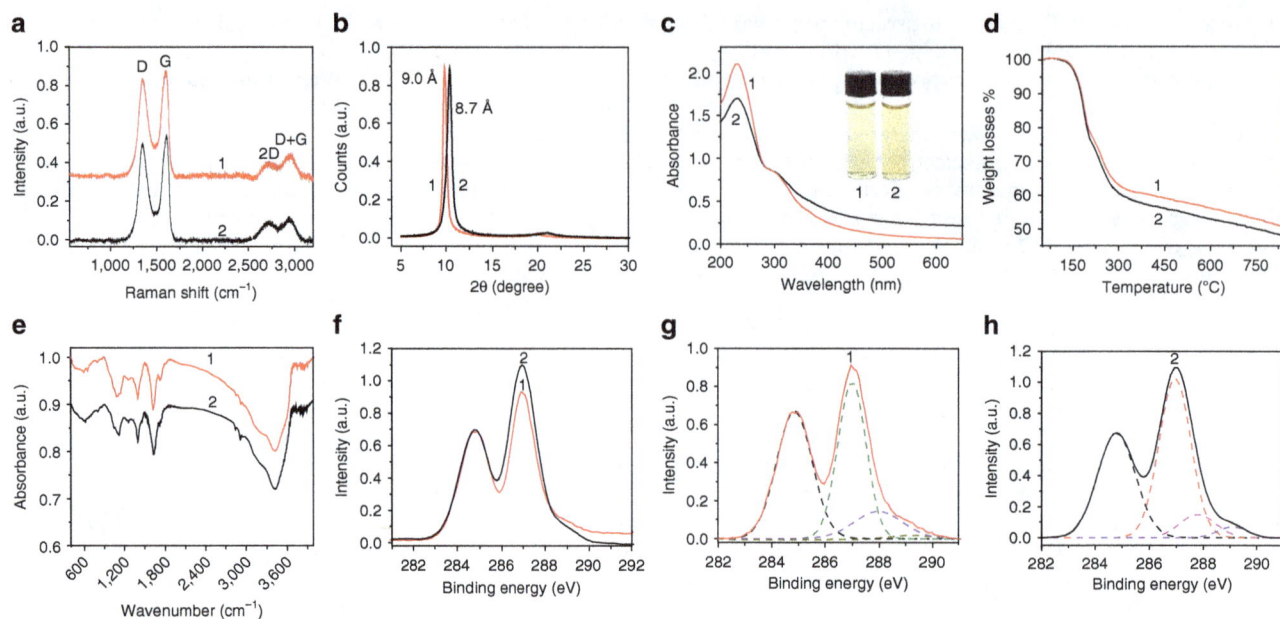

Figure 2 | Comparison of GO^Fe and GO^Mn. (**a**) Raman spectra recorded using 514 nm laser excitation, (**b**) XRD spectra, (**c**) ultraviolet–visible spectra recorded in aqueous solution at 0.05 mg ml^{-1}, (**d**) TGA plots, (**e**) Fourier transform infrared spectra and (**f–h**) XPS spectra and its C1s XPS spectra of GO^Fe and GO^Mn. **1** and **2** denote GO^Fe and GO^Mn, respectively. All of these data show that GO^Fe and GO^Mn have similar composition and structures.

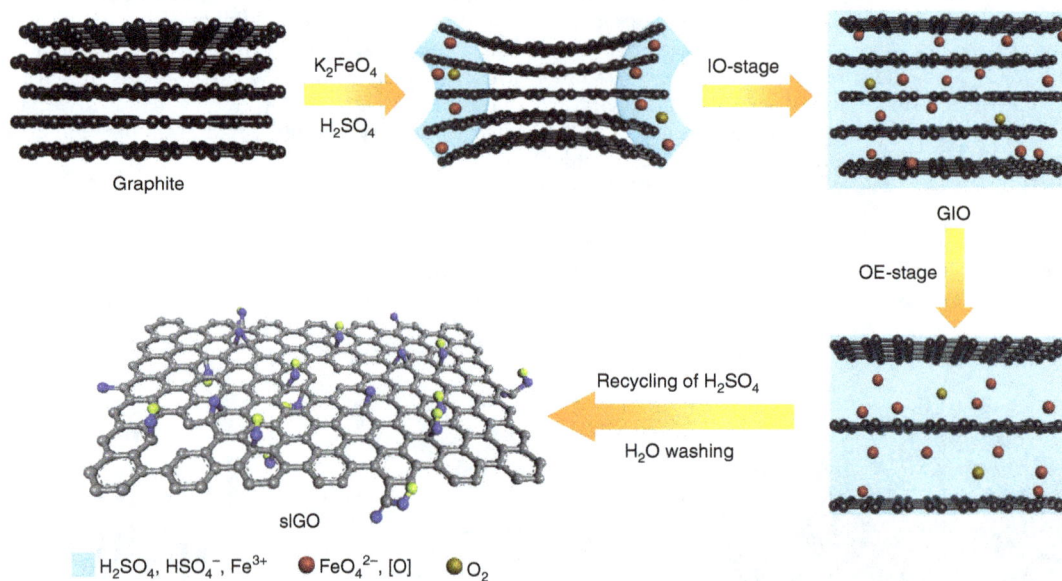

Figure 3 | Mechanism of GO^Fe synthesis with the oxidant of K$_2$FeO$_4$. The whole synthetic process (1 h) contains two main stages: intercalation–oxidation (IO) and oxidation–exfoliation (OE). The *in situ* generated FeO$_4^{2-}$ and atomic oxygen [O] act as oxidants and the O$_2$ formed from residual [O] provides mild and durative gas exfoliation. In the IO stage, the concentrated sulphuric acid and oxidants intercalate into the layers of graphite to form intercalated graphite oxide (GIO). During the intercalation, the oxidants break the π-π conjugated structures of graphite, generating negatively charged functional groups, and increasing the interlayer spacing. In the following OE stage, the oxidants further oxidize the carbon basal planes of GIO, giving rise to more functional groups and enlarging the interlayer space. After recycling of sulphuric acid and washing with water, 100% slGO is achieved.

persistent criticism of GO production is the pollution associated with the use of concentrated sulphuric acid, the disposal of which significantly adds to the costs of GO. We resolved this issue by recycling concentrated sulphuric acid, a process which was enabled by the strong oxidation ability of K$_2$FeO$_4$. We recycled the concentrated sulphuric acid at least 10 times without change to the fabrication efficiency (1 h) or the GO quality. Notably, even if the collected sulphuric acid was not immediately reused, its removal proved greatly beneficial to the subsequent GO purification by either centrifugation or sieving/filtration, as well as to

the subsequent waste treatment steps. The small amount of H$_2$SO$_4$ complexed to K$_2$SO$_4$ and Fe$_2$(SO$_4$)$_3$ in the washing water was neutralized with ammonia, forming mixtures consisting of (NH$_4$)$_2$SO$_4$, K$_2$SO$_4$ and Fe$_2$(SO$_4$)$_3$, which are used as fertilisers in agriculture. This protocol significantly decreases the cost of slGO.

Discussion

To deeply understand the fast oxidation–exfoliation process in our K$_2$FeO$_4$-based system, we investigated the effects of oxidation

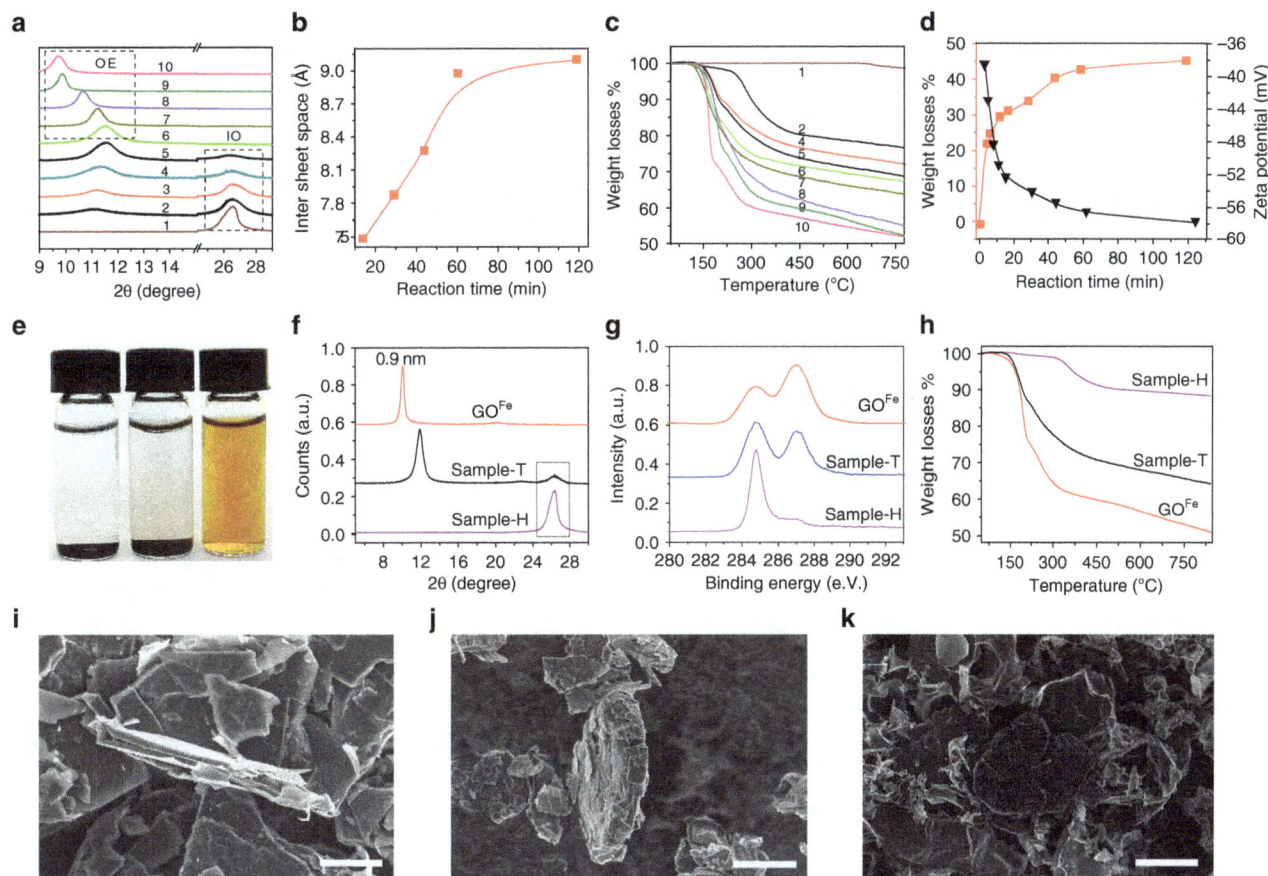

Figure 4 | Kinetics of the synthesis of GO^Fe. (**a**) XRD spectra of the samples taken from the synthesis process at the reaction times = 0 min, 3 min, 5 min, 8 min, 11 min, 15 min 30 min, 45 min, 1 h and 2 h (**1-10**), respectively. (**b**) Interlayer spacing of selected samples at the OE stage versus reaction time. (**c**) TGA plots of the same samples as shown in **a**. (**d**) Weight loss of GO^Fe at 600 °C (left, red) and corresponding zeta potential (right, blue) as a function of reaction time. The kinetics of GO^Fe confirms that the whole reaction process completes in 1 h, including ∼15 min of intercalation–oxidation and 45 min of oxidization–exfoliation. (**e**) Sample-H, Sample-T and GO^Fe (2 mg ml^{-1}) placed in water, indicating that only GO^Fe is well-soluble. (**f**) XRD spectra, (**g**) C1s XPS spectra and (**h**) TGA plots of GO^Fe, Sample-H and Sample-T with the reaction time 1 h. (**i-k**) SEM images of graphite, Sample-T and Sample-H, showing that the conventional Hummers methods with the oxidant of KMnO$_4$ can only result in thick graphite-like particles rather than slGO in 1 h of reaction time. Scale bar, 20 μm (**i-k**).

Table 1 | A comparison of our K$_2$FeO$_4$-based methodology with KClO$_3$- and KMnO$_4$-based methodologies.

Method (Year)	KClO$_3$ based			KMnO$_4$ based				K$_2$FeO$_4$ based
	Brodie[8]	Staudenmaier[9]	Hofmann[10]	Hummers[11]	Modified-1 (1999)[12]	Modified-2 (2004)[13]	Modified-3 (2010)[14]	Our work (2014)
Reaction time	10 h	1–10 days	4 days	2–10 h	8 h	5 days	12 h	1 h
Interlayer spacing	5.95 Å	6.23 Å	—	6.67 Å	6.9 Å	8.3 Å	9.3 Å	9.0 Å
C/O ratio	2.16	—	—	2.25	2.3	1.8	—	2.2
Toxic gas	ClO$_2$	ClO$_2$, NO$_X$	ClO$_2$, NO$_X$	NO$_X$	—	NO$_X$	—	No
Exploder	KClO$_3$	KClO$_3$	KClO$_3$	Mn$_2$O$_7$	Mn$_2$O$_7$	Mn$_2$O$_7$	Mn$_2$O$_7$	No
Heavy metal in GO (p.p.m.)	—	—	—	97 (Mn^{2+})	—	—	87 (Mn^{2+})	0.025 (Mn^{2+}) 0.13 (Fe^{3+})
Mn^{2+} generated (for 1 ton graphite)	—	—	—	1 ton	1 ton	1.5 ton	2 ton	0

C, carbon; GO, graphene oxide; O, oxygen.

on the dispersive state in water for samples collected at different reaction times. Supplementary Figure 4 shows the dispersion states of the materials after standing for 24 h. Only the solution observed at 1 h of reaction time has no precipitate, implying that the functional group density is high enough to overwhelm the aggregation tendency. Furthermore, the colour of the solutions becomes lighter with increasing oxidation time due to the gradual destruction of π–π conjugate structures by the formation of functional groups. More subtle analyses by XRD, TGA and zeta potential demonstrated that the entire reaction process (1 h) can be divided into two stages: intercalation–oxidation (IO) and oxidization–exfoliation (OE; Fig. 3)[37].

Figure 5 | Spray-dried GO^{Fe} powder for re-dissolving. (**a**) Fresh GO^{Fe} LC solution of H_2O with a concentration of 6 mg ml^{-1}. (**b**) Macroscopic photograph of spray-dried GO^{Fe} powders with a density of 224 mg cm^{-3}. (**c,d**) SEM images of GO^{Fe} powders, showing that the GO^{Fe} individual particles have a peony-like morphology. The insert of **d** is a peony. (**e**) Re-dissolved GO^{Fe} solutions of H_2O and N,N-dimethylformamide with a concentration of 4 mg ml^{-1}. (**f**) SEM image of re-dissolved single-layered GO^{Fe} sheets on Si/SiO$_2$ substrate. (**g**) Tapping mode AFM image and height profile of re-dissolved GO^{Fe}. (**h**) POM images of re-dissolved GO^{Fe} aqueous LCs in a quartz tube and a planar cell between crossed polarisers at a concentration of 4 mg ml^{-1}. Scale bars, 3 μm (**c**), 500 nm (**d**), 10 μm (**f**), 2 μm (**g**) and 5 mm (**h**, left), 1 μm (**h**, right).

In the first IO stage, concentrated sulphuric acid and K_2FeO_4 intercalate into the interlayer spacing of graphite. The oxidant then breaks the π–π conjugated structures at the edges and defects of the graphite, weakening the conjugate forces between pristine graphitic lamellae. In the corresponding XRD patterns (Fig. 4a), the appearance of a new peak at $2\theta = 11.4°$ accompanies the gradual fading of the 002 peak at $2\theta = 26.5°$ with increasing reaction time. At ~15 min, the diffraction peak of graphite at 26.5° disappears completely, indicating the completion of the IO stage and the formation of intercalated and partially oxidized graphite (GIO). An increase of the d-spacing of GIO is observed from 0.34 to 0.75 nm due to intercalation and oxidation. Intercalation and oxidation of graphite occur simultaneously, as confirmed by the dramatic mass loss from 0 to 30 wt % at 15 min in the corresponding TGA curves (Fig. 4c,d). The zeta potential also decreases rapidly to -52 mV (Fig. 4d), demonstrating the generation of negatively charged functional groups.

In the following OE stage, the oxidant further oxidized the carbon basal planes of GIO, giving rise to more functional groups and enlarging the d-spacing from 0.75 to 0.91 nm (Fig. 4a,b). In the TGA curves, the weight loss at 600 °C further increases from 30% at 15 min to 43% at 60 min (Fig. 4c). Notably, 100% slGO was achieved by 1 h, verifying the ultrafast OE process of our protocol. In fact, further extending the reaction time to 2 h gave little changes in the d-spacing, weight loss or zeta potential.

The entire reaction process is proposed by the following two steps:

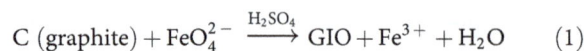

$$C \text{ (graphite)} + FeO_4^{2-} \xrightarrow{H_2SO_4} GIO + Fe^{3+} + H_2O \quad (1)$$

$$GIO + FeO_4^{2-} \xrightarrow{H_2SO_4} GO + Fe^{3+} + H_2O \quad (2)$$

In addition, FeO_4^{2-} reacts with H^+ or water to produce atomic oxygen [O] that also effectively oxidizes carbon[38]. FeO_4^{2-} and [O] work synergistically to efficiently yield slGO. The residual [O] forms oxygen gas, making both intercalation and exfoliation much more powerful and ultrafast[39]. Accordingly, all the reactions can be listed as follows:

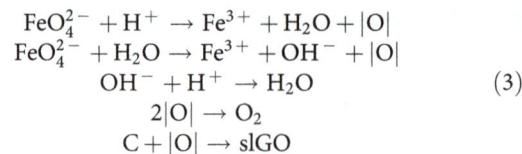

$$\begin{aligned} FeO_4^{2-} + H^+ &\rightarrow Fe^{3+} + H_2O + |O| \\ FeO_4^{2-} + H_2O &\rightarrow Fe^{3+} + OH^- + |O| \\ OH^- + H^+ &\rightarrow H_2O \\ 2|O| &\rightarrow O_2 \\ C + |O| &\rightarrow slGO \end{aligned} \quad (3)$$

This unique reaction mechanism results in an ultrafast oxidation and exfoliation rates, providing slGO without additional ultrasonic treatment.

To analyze the oxidation efficiency of K_2FeO_4, we quantified the oxygen yield during GO^{Fe} production (Supplementary Methods). The results show that 70.2% of the K_2FeO_4 is consumed in the oxidation of graphite, 17.3% is decomposed into oxygen and 12.5% remains in the reaction suspension. This indicates that ~80% of the reacted K_2FeO_4 is converted into the oxygen-containing moieties of GO, confirming the extremely high oxidation efficiency of K_2FeO_4.

For comparison, we also studied samples oxidized for 1 h by two popular modified Hummers methods: Tour's method[14] (Sample-T) and Hirata's method[13] (Sample-H). Figure 4e shows that the two samples precipitated completely after 1 h of sonication after storage for 12 h, showing almost no solubility. XRD profile of Sample-H shows a strong graphite peak at $2\theta = 26.5°$ without the characteristic peak of graphite oxide. The Sample-T exhibits an obvious graphite peak and a graphite oxide peak at $2\theta = 11.9°$ (Fig. 4f), indicating strong oxidation but poor exfoliation. The XPS spectra of Sample-T and Sample-H reveal C/O ratios of 7.3 and 18.5, which are much higher than those found in GO^{Fe} (2.2; Fig. 4g). From the TGA plots of the two samples, ~32 and 10% mass losses are found, which are much lower than those found in GO^{Fe} (45%; Fig. 4h). A sample oxidized by $KClO_3$ (Sample-B) for 1 h was also prepared and tested, indicating no solubility in water, a small degree of oxidation (C/O = 13.4, 25 % wt loss) and poor exfoliation (with a

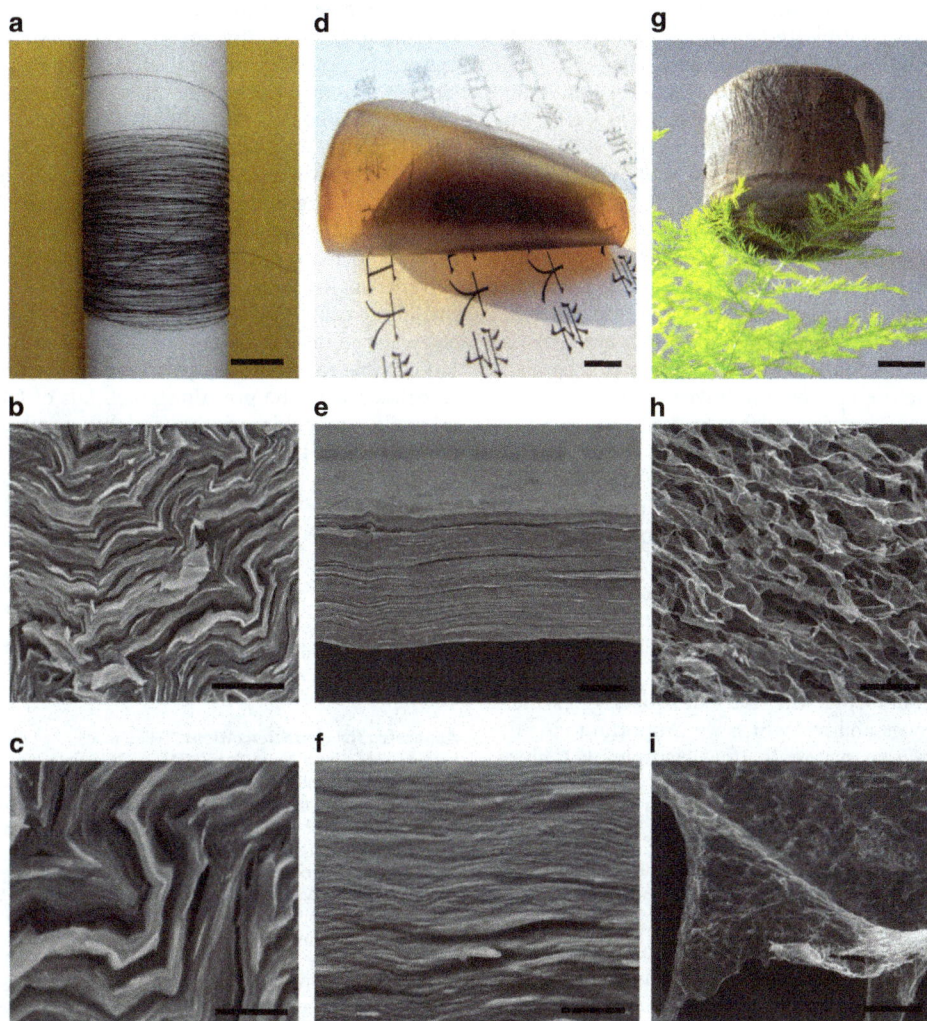

Figure 6 | Macroscopic assembled materials of re-dissolved GO^{Fe}. (**a**–**c**) A wet-spun 14-m long continuous fibre with diameter 10 μm and its SEM images at the cross-section of fibre. (**d**–**f**) A film made by the filtration method and its SEM image of a section. (**g**–**i**) Ultralight weight GO^{Fe} aerogel with a density of 2 mg cm^{-3} and its SEM images showing CNT-coated graphene morphology. Scale bars, 3 cm (**a**), 1 μm (**b**), 500 nm (**c**), 1 cm (**d**), 3 μm (**e**), 400 nm (**f**), 2 cm (**g**), 30 μm (**h**) and 2 μm (**i**).

strong graphite peak at $2\theta = 26.5°$ and graphite oxide peak at $2\theta = 12.5°$; Supplementary Fig. 5). SEM images show that Sample-H has a similar thickness (~ 0.6 μm) as that of raw graphite and that a portion of Sample-T has a similar appearance to that of raw graphite, which confirm their multilayered state (Fig. 4i–k). These results demonstrate that our K$_2$FeO$_4$-based methodology, capable of both highly efficient oxidation and ultrafast exfoliation, is superior to the conventional methods.

Table 1 lists the comprehensive comparison of our K$_2$FeO$_4$-based methodology with the conventional methods. Generally, the new method possesses the following merits: ultrafast reaction rate, safe and environmentally friendly processing, no heavy metal pollution and ultralow cost. In our new method, 1 h is sufficient to obtain slGO without any additional post treatments such as ultrasonication or H$_2$O$_2$ washing, which are normally required in the Hummers methods. By comparison, the conventional methods require ~ 6 h–5 days of reaction time, as described in the 57 most cited studies on GO preparation (Supplementary Table 1). All of the conventional methods based on the KClO$_3$ and KMnO$_4$ oxidants as well as their optimized modifications produce toxic gases (ClO$_2$, NO$_X$) and explosive intermediates (for example, Mn$_2$O$_7$). In addition, for the KMnO$_4$-based methodology, consumption of 1 ton of graphite would

result in 1–5 ton of neat Mn^{2+} and 40–120 ton of sulphuric acid waste, leading to pollution, tedious post treatments and high costs. The high concentration of manganese in the system also stains GO with a Mn content of up to 97 p.p.m., which may cause significant injury to the body in cases where GO is used as a vehicle for drugs[40,41]. On the contrary, our K$_2$FeO$_4$-based approach has no safety or pollution issues, and the Mn content in GO^{Fe} is negligible (~ 0.025 p.p.m.). Moreover, the GO^{Fe} contains almost no iron (0.13 p.p.m.) despite the use of an iron-based oxidant, to the benefit of the eventual applications of GO and CCG.

Even though the fabrication is ultrafast at room temperature, the resulting GO^{Fe} is highly soluble in water and polar organic solvents and has both a composition and morphology comparable to GO^{Mn}. As such, GO^{Fe} can be directly used in fields where GO^{Mn} has been demonstrated to be effective.

The preparation of GO powders is another very important issue that greatly affects the practical use of GO and its transport. Freeze-drying is commonly used to obtain solid GO. As shown in Supplementary Fig. 6, commercial GO powders apparently precipitate in minutes at 2 mg ml^{-1} even after 12 h of ultrasonic agitation. GO sheets laminate together as a result of π–π conjugation in the process of solvent removal. These aggregates are difficult to disrupt by the re-addition of solvents. We adopt a spray-drying method to control the morphology of the GO^{Fe} sheets and obtain only soluble GO powders (Fig. 5b). The dried GO powders can be completely dissolved in water and N,N-dimethylformamide (Fig. 5e) to form lyotropic LCs (Fig. 5h), identical to the fresh GO solutions before drying (Fig. 1b,h). The GO sheets are all dispersed in a single-layered state, as confirmed by SEM and AFM measurements (Fig. 5f,g).

As shown in Fig. 5c,d, the surfaces of the dried GO sub-microspheres are full of folds because the GO sheets shrink inwardly, forming peony-like 3D crumpled structures under the surface tension experienced in the spray-drying process. Such 3D crumpled sub-microsphere morphologies effectively prevent GO stacking, favouring the unfolding of sub-microspheres into plane sheet morphologies when re-dissolved in solvents. The GO^{Fe} sub-microsphere powder has a specific surface area of 1,467 m^2 g^{-1}, indicating 1–2 atomic layer structures (Supplementary Fig. 7). The dried GO^{Fe} powders are highly soluble in water and polar organic solvents. Significantly, our GO^{Fe} powder has a very high density (>224 mg cm^{-3}), which facilitates its storage, transport and application. By comparison, despite a very low density (<30 mg cm^{-3}) resulting from the freeze-drying process, the undissolved commercial GO powders have a very low specific surface area (<10 m^2 g^{-1}) due to their multilayer structure (Supplementary Fig. 8).

The excellent dispersibility of the GO^{Fe} powders gives them superior solution processability, which is important in the fabrication of macroscopic materials (for example, 1D fibres, 2D films and 3D frameworks). A re-dissolved aqueous GO^{Fe} solution (~ 6 mg ml^{-1}) shows a colourful optical texture typical of a nematic LC mesophase, identical to the appearance of fresh aqueous GO^{Fe} solutions (Fig. 5a and Supplementary Fig. 9)[36]. In a macroscopic quartz tube, the birefringence Schlieren texture between crossed polarisers can be seen with the naked eye across the entire solution (Fig. 5h). Such a LC suspension establishes the foundation to fabricate GO fibres, which has been demonstrated by our group and other independent researchers[22,42]. We subsequently obtained a continuous fibre by wet-spinning of the LC dope (Fig. 6a–c and Supplementary Fig. 10). It shows a highly compact and ordered structure, similar to previous GO fibres made directly from undried GO suspensions.

A film was made from the re-dissolved GO^{Fe} solution (Fig. 6d–f and Supplementary Fig. 11a) by the vacuum-assisted filtration method, which shows a well-aligned lamellar structure and comparable mechanical performance to GO^{Mn} papers[43]. After reduction with HI, our graphene film exhibits an electrical conductivity of 374 S cm^{-1} (Supplementary Fig. 11c), close to that (400 S cm^{-1}) of defect-free graphene made by a high-shear exfoliation method[19]. A 3D aerogel prepared by a synergistic assembly of GO^{Fe} and carbon nanotubes (CNTs, 50 wt %) shows the same appearance and internal structure (Fig. 6g–i and Supplementary Fig. 12a) as an assembly prepared from GO^{Mn} and CNTs reported by our group previously[25]. After reduction with N$_2$H$_4$, an aerogel with a density of 2.0 mg cm^{-3} shows complete recovery even after 1,000 cycles of 87% compression. Significantly, the aerogel still remains elastic and intact after being compressed by a weight 5,000 times its own (Supplementary Fig. 12b). These results demonstrate the 'true' solution state of our re-dissolved GO^{Fe} and suggest the wide application of GO and CCG.

In conclusion, we established an industrially viable one-pot method for the production of slGO in 1 h at room temperature with ultralow cost based on the use of the novel oxidant of K$_2$FeO$_4$. The reaction process includes ~ 15 min of intercalation-oxidation and ~ 45 min of oxidization–exfoliation. The excellent oxidation capabilities of both K$_2$FeO$_4$ and the *in situ* generated

atomic oxygen, accompanied by the exfoliation capacity of oxygen gas, make the intercalation, oxidation and exfoliation extremely powerful and ultrafast. The as-prepared slGO has a similar composition, chemical structure and solubility to materials prepared by the conventional Hummers method. Significantly, our dried slGO powders maintain excellent solubility in water and polar organic solvents and readily form stable LCs. Therefore, they retain the capacity to assemble into macroscopic materials such as continuous fibres, films and aerogels displayed by fresh GO solutions. The sulphuric acid solvent can be recycled in our protocol due to the ultrastrong oxidation capability of K$_2$FeO$_4$, which dramatically reduces the effluent and lowers the cost of GO. Our fast, eco-friendly and safe K$_2$FeO$_4$-based methodology circumvents the intrinsic problems associated with the prevailing methods of GO production, and it is easily amenable to the scalable production and industrial application of GO and CCG.

Methods

Synthesis of GO^{Fe}. K$_2$FeO$_4$ (60 g, 6 wt equiv.) was added to concentrated H$_2$SO$_4$ (93%, 400 ml) at room temperature. Graphite (10 g, 1 wt equiv., 40 μm) was then added and the mixture was kept at room temperature for 1 h (note: the flask was not sealed due to the release of oxygen during the reaction). The mixture was centrifuged (10,000 r.p.m. for 3 min) to recycle the concentrated sulphuric acid. The paste-like product was collected by repeated centrifugation and washing with 1 l of water until the pH of the supernatant solution approached 7.

Apparatus for characterizations. AFM images of GO sheets were taken in the tapping mode on a Nano Scope IIIA, with samples prepared by spin-coating diluted aqueous solutions onto freshly exfoliated mica substrates at 1,000 r.p.m.. SEM images were taken on a Hitachi S4800 field-emission SEM system. TEM was performed on a JEM-1200EX with an accelerating voltage of 120 kV. Zeta potential measurements were performed on a ZET-3000HS apparatus. Fourier transform infrared spectra were recorded on a PE Paragon 1000 spectrometer (film or KBr disk). Ultraviolet–visible spectra were obtained using a Varian Cary 300 Bio UV-visible spectrophotometer. Tensile tests were carried out on a HS-3200C at a loading rate of 1 mm min^{-1}. XPS was performed using a PHI 5000C ESCA system operated at 14.0 kV. All binding energies were referenced to the C1s neutral carbon peak at 284.8 eV. TGA was carried out using a thermogravimetric analyser (PerkinElmer Pyris 1) from room temperature to 850 °C at 10 °C min^{-1} heating rate under air atmosphere. XRD data were collected with an X'Pert Pro (PANalytical) diffractometer using monochromatic Cu Kα1 radiation ($\lambda = 1.5406$ Å) at 40 kV. Raman spectra were recorded on a Labram HRUV spectrometer operating at 632.8 nm. Mechanical property tests were carried out on a HS-3002C at a loading rate of 10% per minute. Elemental analyses were performed using an Agilent model 7700 × ICP-MS. BET surface area measurements were performed by nitrogen adsorption on a Quantachrome NOVA 2000 surface analyzer. POM observations were performed with a Nikon E600POL, and the liquid samples were loaded into the planar cells for observations. Combustion analysis was performed on an elemental analyzer (Vario Micro).

References

1. Geim, A. K. & Novoselov, K. S. The rise of graphene. *Nat. Mater.* **6**, 183–191 (2007).
2. Tung, V. C., Allen, M. J., Yang, Y. & Kaner, R. B. High-throughput solution processing of large-scale graphene. *Nat. Nanotechnol.* **4**, 25–29 (2009).
3. Novoselov, K. S. *et al.* Electric field effect in atomically thin carbon films. *Science* **306**, 666–669 (2004).
4. Hernandez, Y. *et al.* High-yield production of graphene by liquid-phase exfoliation of graphite. *Nat. Nanotechnol.* **3**, 563–568 (2008).
5. Coleman, J. N. *et al.* Two-dimensional nanosheets produced by liquid exfoliation of layered materials. *Science* **331**, 568–571 (2011).
6. Smith, R. J. *et al.* Large-scale exfoliation of inorganic layered compounds in aqueous surfactant solutions. *Adv. Mater.* **23**, 3944–3948 (2011).
7. Dreyer, D. R., Park, S., Bielawski, C. W. & Ruoff, R. S. The chemistry of graphene oxide. *Chem. Soc. Rev.* **39**, 228–240 (2010).
8. Brodie, B. C. On the atomic weight of graphite. *Phil. Trans. R. Soc. Lond.* **149**, 249–259 (1859).
9. Staudenmaier, L. Verfahren zur darstellung der graphitsäure. *Ber. Dtsch. Chem. Ges.* **31**, 1481–1487 (1898).
10. Hofmann, U. & König, E. Untersuchungen über graphitoxyd. *Z. Anorg. Allg. Chem.* **234**, 311–336 (1937).
11. Hummers, W. S. & Offeman, R. E. Preparation of graphitic oxide. *J. Am. Chem. Soc.* **80**, 1339–1339 (1958).

12. Kovtyukhova, N. I. *et al.* Layer-by-layer assembly of ultrathin composite films from micron-sized graphite oxide sheets and polycations. *Chem. Mater.* **11**, 771–778 (1999).

13. Hirata, M., Gotou, T., Horiuchi, S., Fujiwara, M. & Ohba, M. Thin-film particles of graphite oxide 1. *Carbon* **42**, 2929–2937 (2004).

14. Marcano, D. C. *et al.* Improved synthesis of graphene oxide. *ACS Nano* **4**, 4806–4814 (2010).

15. Pei, S., Zhao, J., Du, J., Ren, W. & Cheng, H. M. Direct reduction of graphene oxide films into highly conductive and flexible graphene films by hydrohalic acids. *Carbon* **48**, 4466–4474 (2010).

16. Stankovich, S. *et al.* Synthesis of graphene-based nanosheets via chemical reduction of exfoliated graphite oxide. *Carbon* **45**, 1558–1565 (2007).

17. Becerril, H. A. *et al.* Evaluation of solution-processed reduced graphene oxide films as transparent conductors. *ACS Nano* **2**, 463–470 (2008).

18. Moon, I. K., Lee, J., Ruoff, R. S. & Lee, H. Reduced graphene oxide by chemical graphitization. *Nat. Commun.* **1**, 73 (2010).

19. Paton, K. R. *et al.* Scalable production of large quantities of defect-free few-layer graphene by shear exfoliation in liquids. *Nat. Mater.* **13**, 624–630 (2014).

20. Hamilton, C. E., Lomeda, J. R., Sun, Z., Tour, J. M. & Barron, A. R. High-yield organic dispersions of unfunctionalized graphene. *Nano Lett.* **9**, 3460–3462 (2009).

21. Xu, Z., Zhang, Y., Li, P. & Gao, C. Strong, conductive, lightweight, neat graphene aerogel fibers with aligned pores. *ACS Nano* **6**, 7103–7113 (2012).

22. Xu, Z. & Gao, C. Graphene chiral liquid crystals and macroscopic assembled fibres. *Nat. Commun.* **2**, 571 (2011).

23. Xu, Z. & Gao, C. Graphene in macroscopic order: liquid crystals and wet-spun fibers. *Acc. Chem. Res.* **47**, 1267–1276 (2014).

24. Han, Y., Xu, Z. & Gao, C. Ultrathin graphene nanofiltration membrane for water purification. *Adv. Funct. Mater.* **23**, 3693–3700 (2013).

25. Sun, H., Xu, Z. & Gao, C. Multifunctional, ultra-flyweight, synergistically assembled carbon aerogels. *Adv. Mater.* **25**, 2554–2560 (2013).

26. Bagri, A. *et al.* Structural evolution during the reduction of chemically derived graphene oxide. *Nat. Chem.* **2**, 581–587 (2010).

27. Xin, G. *et al.* Large-area freestanding graphene paper for superior thermal management. *Adv. Mater.* **26**, 4521–4526 (2014).

28. Wu, H. & Drzal, L. T. Graphene nanoplatelet paper as a light-weight composite with excellent electrical and thermal conductivity and good gas barrier properties. *Carbon* **50**, 1135–1145 (2012).

29. Liang, Y. *et al.* Co_3O_4 nanocrystals on graphene as a synergistic catalyst for oxygen reduction reaction. *Nat. Mater.* **10**, 780–786 (2011).

30. Koch, K. R. & Krause, P. F. Oxidation by Mn_2O_7—an impressive demonstration of the powerful oxidizing property of dimanganeseheptoxide. *J. Chem. Educ.* **59**, 973–974 (1982).

31. Dreyer, D. R., Ruoff, R. S. & Bielawski, C. W. From conception to realization: an historial account of graphene and some perspectives for its future. *Angew. Chem. Int. Ed.* **49**, 9336–9344 (2010).

32. Hoppe, M. L., Schlemper, E. O. & Murmann, R. K. Structure of dipotassium ferrate(VI). *Acta Crystallogr.* **B38**, 2237–2239 (1982).

33. Audette, R. J., Smith, P. J. & Quail, J. W. Oxidation of substituted benzyl alcohols with ferrate (Vi) ion. *J. Chem. Soc. Chem. Commun.* 38–39 (1972).

34. Park, S. & Ruoff, R. S. Chemical methods for the production of graphenes. *Nat. Nanotechnol.* **5**, 309 (2010).

35. Gao, X., Jang, J. & Nagase, S. Hydrazine and thermal reduction of graphene oxide: reaction mechanisms, product structures, and reaction design. *J. Phys. Chem. C* **114**, 832–842 (2010).

36. Xu, Z. & Gao, C. Aqueous liquid crystals of graphene oxide. *ACS Nano* **5**, 2908–2915 (2011).

37. Dimiev, A. M. & Tour, J. M. Mechanism of graphene oxide formation. *ACS Nano* **8**, 3060–3068 (2014).

38. Rush, J. D. & Bielski, B. H. J. Kinetics of ferrate (V) decay in aqueous solution. A pulse-radiolysis study. *Inorg. Chem.* **28**, 3947–3951 (1989).

39. Geng, X. *et al.* Interlayer catalytic exfoliation realizing scalable production of large-size pristine few-layer graphene. *Sci. Rep.* **3**, 1134 (2013).

40. Liu, Z., Robinson, J. T., Sun, X. M. & Dai, H. PEGylated nanographene oxide for delivery of water-insoluble cancer drugs. *J. Am. Chem. Soc.* **130**, 10876–10877 (2008).

41. Sun, X. *et al.* Nano-graphene oxide for cellular imaging and drug delivery. *Nano Res.* **1**, 203–212 (2008).

42. Cong, H., Ren, X., Wang, P. & Yu, S. Wet-spinning assembly of continuous, neat, and macroscopic graphene fibers. *Sci. Rep.* **2**, 613 (2012).

43. Dikin, D. A. *et al.* Preparation and characterization of graphene oxide paper. *Nature* **448**, 457–460 (2007).

Acknowledgements

This work is supported by the National Natural Science Foundation of China (no. 21325417) and Fundamental Research Funds for the Central Universities (no. 2013XZZX003).

Author contributions

C.G., L.P. and Z.X. conceived and designed the research; L.P. conducted the experiments and analyzed the data; Z. Liu, Z. Li, H.S., X.Z. and Y.W. discussed the data and provided some useful suggestions; C.G. supervised and directed the project; all of the authors read and revised the paper.

Additional information

2

Single-site trinuclear copper oxygen clusters in mordenite for selective conversion of methane to methanol

Sebastian Grundner[1], Monica A.C. Markovits[1], Guanna Li[2], Moniek Tromp[3], Evgeny A. Pidko[2,4], Emiel J.M. Hensen[2], Andreas Jentys[1], Maricruz Sanchez-Sanchez[1] & Johannes A. Lercher[1,5]

Copper-exchanged zeolites with mordenite structure mimic the nuclearity and reactivity of active sites in particulate methane monooxygenase, which are enzymes able to selectively oxidize methane to methanol. Here we show that the mordenite micropores provide a perfect confined environment for the highly selective stabilization of trinuclear copper-oxo clusters that exhibit a high reactivity towards activation of carbon–hydrogen bonds in methane and its subsequent transformation to methanol. The similarity with the enzymatic systems is also implied from the similarity of the reversible rearrangements of the trinuclear clusters occurring during the selective transformations of methane along the reaction path towards methanol, in both the enzyme system and copper-exchanged mordenite.

[1] Department of Chemistry and Catalysis Research Center, Technische Universität München, Lichtenbergstrasse 4, Garching 85748, Germany. [2] Schuit Institute of Catalysis, Inorganic Materials Chemistry Group, Department of Chemical Engineering and Chemistry, Eindhoven University of Technology, PO Box 513, Eindhoven 5600 MB, The Netherlands. [3] Van't Hoff Institute for Molecular Sciences, University of Amsterdam, PO Box 94215, Amsterdam 1090GE, The Netherlands. [4] Institute for Complex Molecular Systems, Eindhoven University of Technology, PO Box 513, Eindhoven 5600 MB, The Netherlands. [5] Institute for Integrated Catalysis, Pacific Northwest National Laboratory, PO Box 999, Richland, Washington 99352, USA. Correspondence and requests for materials should be addressed to J.A.L. (email: johannes.lercher@ch.tum.de).

The recent marked increase in the availability of methane as well as its global dispersion requires novel chemistry to convert it into easily condensable energy carriers that can be readily integrated into the existing chemical infrastructure[1-3]. This has triggered a worldwide quest for new processes, allowing economical small-scale operations at remote locations[4]. Such boundary conditions rule out the current dominant method to first convert methane with H_2O, O_2 or CO_2 to synthesis gas (a mixture of CO and H_2), followed by synthesis of methanol and the subsequent conversion of methanol to hydrocarbons with zeolite catalysts[5,6] or via direct hydrocarbon synthesis via the Fischer–Tropsch process. While oxidative coupling of methane to ethane and ethylene[7], dehydroaromatization[8] and the direct partial oxidation of methane to methanol are conceptually promising for a direct conversion of natural gas, only the latter permits lower operating temperatures.

Nature has found a way to convert methane in a single step to methanol via a biocatalytic transformation under aerobic conditions with methane monooxygenase (MMO) as catalysts using Cu and Fe as potential active metals. Two forms of MMOs at different cellular locations are known, a cytoplasmic MMO (soluble MMO) and a membrane-bound MMO (particulate, pMMO). In soluble MMOs, the active site of the hydroxylase contains a bis(μ-oxo)diiron core[9], while in pMMOs the active site is represented by a Cu cluster that catalyses the insertion of oxygen into the methane C–H bond with a very high rate (rate normalized to the concentration of active sites, that is, the turnover frequency of about $1\,s^{-1}$)[10].

Aerobic and anaerobic handling during purification led to drastically different concentrations of Cu in the enzyme without inducing marked changes in the enzyme structure, and in turn, differences in the conclusions about the nature of the active site[11]. Rosenzweig et al.have attributed the catalytic activity to a Cu dimer[12-14], while Chan et al. suggested a cluster of three Cu atoms to form the active site[11,15-17]. Following these leads, homogeneous[18-20] and heterogeneous[21,22] Cu-based catalysts have been explored. In particular, studies of the heterogeneous Cu catalysts led to limited and conflicting insights despite the excellent spectroscopic work, because in nearly all cases the active site has been concluded to represent only a minority of the Cu species in the catalyst.

Using leads, especially from the work on zeolite catalysis for the conversion of methane to methanol[23-28], we develop an

approach to prepare uniform Cu-oxo species in zeolites able to activate and convert methane. Here, we demonstrate that by using this approach, a selective stabilization of single-site Cu-oxo clusters in mordenite micropores is possible. Zeolite mordenite (MOR) has non-intersecting 12-membered ring (12-MR) channels with 8-membered ring (8-MR) pockets. In mordenite, not only the uniformity of Cu species but also a high concentration of these sites is achieved, resulting in an unprecedented activity of the material. During methane oxidation, these sites dynamically disintegrate and reform upon reoxidation, a rearrangement that is needed to close the (catalytic) cycle.

Results

Formation of an active copper single site in mordenite. For preparation of a single type of cluster on a solid support via ion exchange sites, two main requirements have to be met, that is, (i) the exchanging species must be well defined in the aqueous solution, and precipitation of the metal cations by changes in the pH during the exchange procedure has to be avoided; (ii) the exchanged cations need to form a well-defined thermo-dynamically stable configuration upon coordination to the zeolite lattice. We have selected mordenite as a matrix for the synthesis of Cu clusters, because it is known that this framework allows a preferential exchange of the sites located in the more constrained side pockets (SP)[29]. The ion exchange was carried out using Cu-acetate as a precursor and under conditions (pH = 5.7) to maximize the concentration of partially hydrolysed $Cu(OH)^+$ ions to avoid their further hydrolysis and precipitation as $Cu(OH)_2$. The concentration of alkali cations that could compete for the exchange sites in the zeolite support was also minimized. Using this approach, a series of ion-exchanged Cu-MOR catalyst precursors with varying Cu concentrations were prepared. Subsequent calcination in flowing O_2 at 450 °C converted these precursors to active materials. The high-temperature activation is necessary for the dehydration of the materials and the induced migration of the exchanged Cu ions towards formation of the oxide clusters. In the oxidized and dry state, it was surprisingly observed that only two lattice aluminium ions were involved in binding of three Cu cations (Fig. 1a). A blank experiment showed the preparation procedure eliminates ~5% of the total Brønsted acid sites (BASs), ca. $70\,\mu mol\,g^{-1}$, by dealumination.

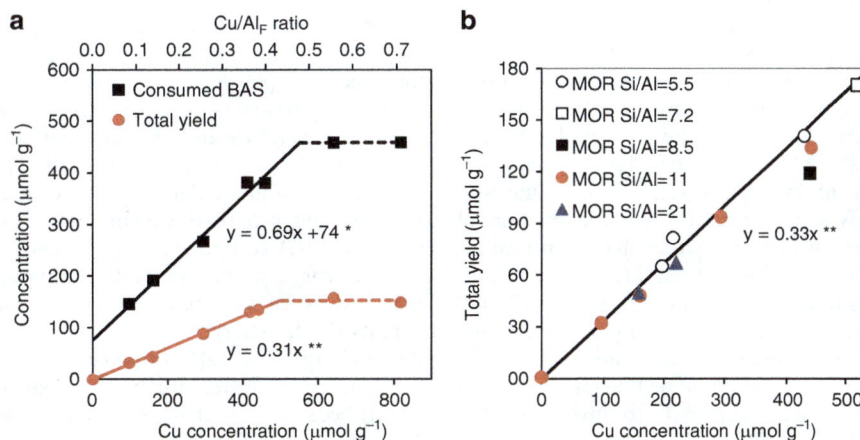

Figure 1 | Activity and framework aluminium coordination upon copper loading. (**a**) The concentration of tetrahedrally coordinated aluminium acting as an ion exchange site for Cu^{2+} with total yield for Cu-MOR with Si/Al = 11 and (**b**) total yield of methane oxidation as a function of Cu concentration in Cu-MOR for various Si/Al ratios. *The slope of 0.69 indicates an exchange stoichiometry of 2/3 meaning that two H^+ are substituted by three Cu^{2+}. The offset of $74\,\mu mol\,g^{-1}$ shows slight dealumination of framework Al (~5%) during Cu exchange. **The slopes of 0.31 and 0.33, respectively, indicate that three Cu centres are involved in the oxidation of one methane molecule.

Table 1 | Acidity of Cu-MOR.

Cu conc. [μmol g^{-1}]	BAS$_{main\ channel}$[†] [μmol g^{-1}]	BAS$_{SP\ bottom}$[‡] [μmol g^{-1}]	BAS$_{SP\ pore\ mouth}$[§] [μmol g^{-1}]	Total BAS [μmol g^{-1}]
0	400	310	380	1,090
100	430	270	330	1,030
160	420	270	290	980
290	410	320	160	890
440	440	370	20	830

BAS, Brønsted acid site; MOR, mordenite; SP, side pocket.
Quantification of acid sites for a series of Cu-MOR* (Si/Al = 11, Cu/Al ≤ 0.4).
*Total concentration of BAS in H-MOR (Si/Al = 11) was determined by Na$^+$ exchange. For Cu-exchanged MOR, the normalized integral of the O-H vibration of BAS was used for deconvolution and quantification.
†Obtained by quantification of the band at 3,612 cm^{-1} (after deconvolution of the band at 3,605 cm^{-1} into 3,612, 3,590 and 3,500 cm^{-1}; see Supplementary Fig. 2).
‡Calculated by the difference of BAS concentration determined as in * and BAS concentration determined by pyridine.
§Calculated by the difference between BAS concentration quantified after n-hexane adsorption (band at 3,590 cm^{-1}) and BAS concentration in the SP bottom (‡); an offset of 70 μmol g^{-1} due to dealumination during Cu exchange was substracted for H-MOR.

The activity of these clusters was evaluated by exposing the activated catalyst to methane at 200 °C followed by purging the zeolite with water to release the formed products. Approximately 80% of the methane converted by the materials was desorbed as methanol or dimethyl ether in the purge step. The total yield of methane oxidation products per gram of zeolite catalyst was an order of magnitude higher than the maximum methanol yields reported in the recent literature for this catalyst class (160 versus 13 μmol g^{-1})[27,30]. The productivity of the active materials scaled linearly with the Cu concentration pointing to a stoichiometry of three Cu cations needed to convert one methane molecule (Fig. 1b). The linear dependence of the activity on the Cu^{2+} concentration strongly suggests that only one type of active site has been formed on a large series of samples with different Cu loading. The stoichiometry of methane activated per Cu, together with the observed Cu/consumed BAS ratio valid for different Si/Al ratios (see Supplementary Fig. 1 and Supplementary Table 1) is the first evidence of an active site involving three Cu atoms anchored to two Al framework sites. With these findings in hand, we redirected our attention to the nature of these sites and the origin of their high reactivity towards methane.

Siting of copper-oxo species. The first key question to be addressed is the structure and location of the active Cu-oxo cluster within the zeolite micropores. The use of different probe molecules has allowed a precise determination of the acid site distribution in H-MOR, which can be controlled by varying the zeolite synthesis conditions. The diameter of MOR side pockets (SP,8-MR) is substantially smaller than the aperture of the large straight channels composed of 12 Si(Al) atoms. Pyridine, owing to its basicity and spherical shape, can access both BASs located in the main channel, as well as in the pore mouth of the side pockets[31]. On the contrary, the more bulky elongated n-hexane molecule interacts only with BASs in the large 12-MR channel[32]. By analysing the results of infrared spectra of pyridine and n-hexane adsorption, it was demonstrated that the fraction of BASs in MOR side pockets, which are inaccessible for n-hexane, constitute about 65% in the H-MOR used in this study (see Supplementary Fig. 2). From this amount, 55% of BASs are accessible to pyridine, meaning that these protons are in the side pockets but located close to the 8-MR mouth pore.

To obtain insight into the location of extra-framework Cu clusters, we employed the same approach to investigate the distribution of residual BAS in Cu-MOR materials after their activation in O$_2$. By direct comparison of the concentration of OH groups associated with BAS in the activated Cu-MOR with the concentration of these groups of the parent H-MOR, the location of the Cu-oxo clusters was inferred. It should be emphasized that deconvolution of the SiO-HAl vibration band

associated with BAS did not show a change of concentration of BAS in the main channel (3,612 cm^{-1}) with Cu loading. Only the band associated with BAS in the side pockets (3,590 cm^{-1}) decreased with increasing Cu concentration. Perturbation of the main channel BAS with n-hexane confirmed these findings. In turn, this allows the conclusion that Cu^{2+} exchanges selectively for H$^+$ in the side pockets. It is hypothesized that the relatively high concentration of framework Al (65% of the total) in the side pockets is stabilizing the Cu-oxo ions. A linear decrease of BAS concentration probed by pyridine adsorption with increasing Cu loading indicates location of the Cu ions balancing Al atoms near the pore mouth of the MOR side pocket (see Table 1 and Supplementary Fig. 3).

In addition, the concentration of Al sites paired in the MOR samples was probed by Co^{2+} exchange, following the method described in ref. 33. It was observed that a majority of Al atoms (66% for Si/Al 11; 60% for Si/Al 21) are separated from another by only one or two Si units. Combining this information, it is concluded that the Cu clusters are balancing the charge of two Al sites located in the 8-MR of the side pockets.

Structure determination of the copper-oxo active site. The information obtained about the nuclearity and location of the Cu-oxo active sites was used to propose several model structures for trinuclear Cu clusters in MOR to be studied by density functional theory (DFT) calculations. Their stability was evaluated and compared with dicopper clusters, as those are typically described to be the active species for Cu-ZSM-5 in the literature[34]. Calculated reaction energies for interconversion of different Cu species indicated a high intrinsic stability of binuclear complexes at 0 K. However, ab initio thermodynamic analysis in terms of reaction Gibbs free energy depending on system temperature and pressure showed that the [Cu$_3$(μ-O)$_3$]$^{2+}$ complex (Fig. 2) is the most stable species at 700 K in O$_2$ atmosphere and under dry conditions (see Supplementary Fig. 4). The proposed stoichiometry suggests a formal mixed Cu(II)/Cu(III) composition of the cluster. However, analysis of the electronic properties of the computed structures indicates that, because of the substantial anion-radical nature of the oxygen ligands (Bader charge −0.76 e and −0.63 e compared with the Bader charge of −1.09 e on oxygens in bulk CuO), all Cu sites in the trinuclear cluster are more adequately described as being Cu(II) because their charges are very close to those computed for Cu centres in bulk CuO (+1.09 e). The results of Bader charge and spin-polarized charge density analysis are summarized in Supplementary Fig. 5.

To verify the nature of the Cu-oxo cluster predicted by DFT calculations, we analysed the Cu species of an activated Cu-MOR sample by X-ray absorption spectroscopy (XAS). To date,

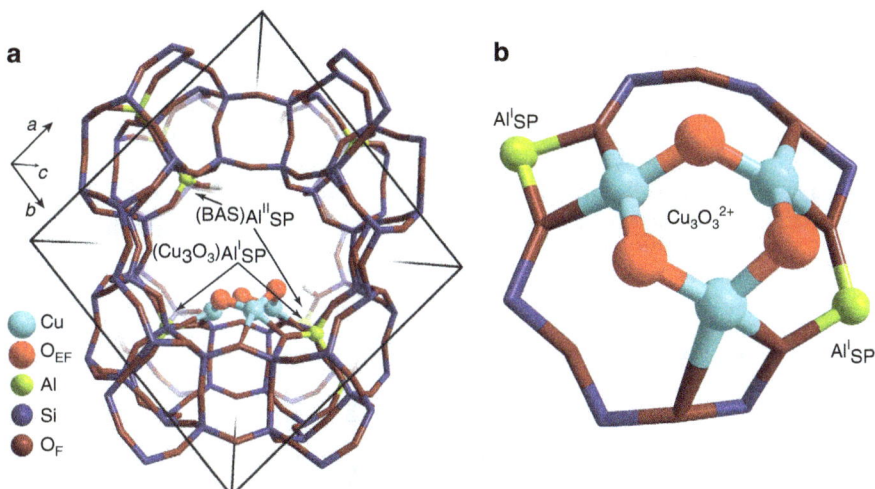

Figure 2 | Structure and location of [Cu₃(μ-O)₃]²⁺ cluster in mordenite predicted by DFT. The zeolite model contained paired (type I) and isolated (type II) Al atoms located at the pore mouth of the side pocket. The cluster is stabilized by two anionic centres due to Al^I_{SP} lattice sites at the entrance of the MOR side pocket (**b**) so that the extra-framework oxygens responsible for the initial C–H activation are pointing towards the main channel of MOR (**a**). The charge due to the remaining Al^{II}_{SP} is compensated by acidic protons resulting in BAS formation.

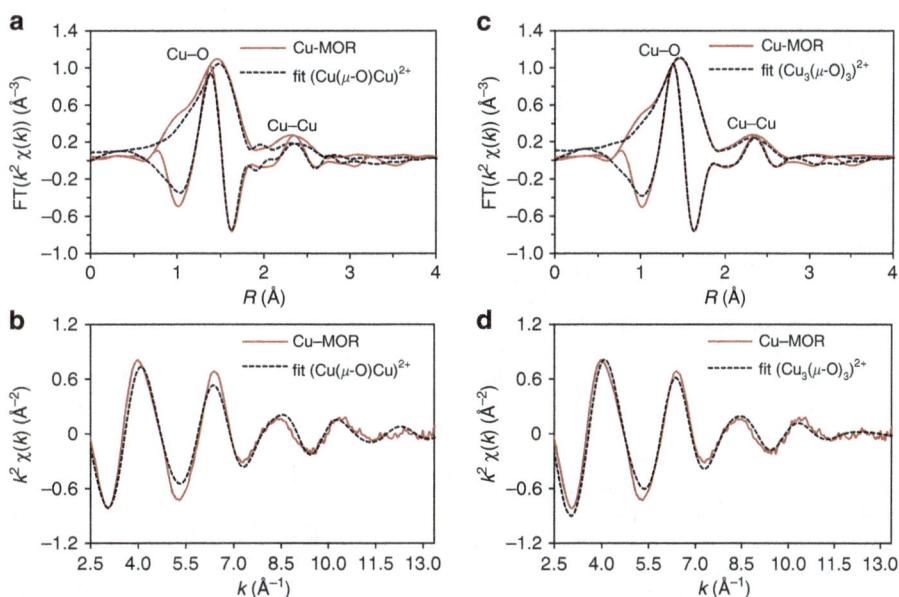

Figure 3 | Copper EXAFS data and fitting for Cu-MOR. Comparison of the k^2-weighted Fourier transformed EXAFS at the Cu K-edge of the Cu-MOR zeolite activated in O₂ at 450 °C with EXAFS simulation of an intrazeolite (**a**) binuclear [Cu(μ-O)Cu]²⁺, (**c**) trinuclear [Cu₃(μ-O)₃]²⁺ complexes and (**b,d**) the corresponding k^2-weighted experimental EXAFS oscillations and their simulation using the DFT-computed model. Colour key: measured spectra (red lines), simulated spectra (black lines).

XAS analysis has not provided unambiguous information on the nature of the active Cu species in Cu-zeolites, due to the heterogeneity of the Cu species in conventional materials. However, the reactivity data discussed above point to a uniform nature of the Cu sites in the Cu-MOR materials prepared by the optimized procedure presented here and, therefore, Cu K-edge spectra may yield direct information of the structure of the active clusters.

Figure 3a,b compares the k^2-weighted and Fourier transformed EXAFS (Extended X-ray Absorption Fine Structure) data measured at the Cu K-edge with the simulated EXAFS, using the DFT-optimized [Cu(μ-O)Cu]²⁺ cluster in the MOR unit cell with a structure previously proposed as the active site for methane activation in Cu-ZSM-5 (ref. 35). The multiple

k-weighted analysis is crucial for a reliable analysis to recognize and properly analyse both light and heavy scatterers and, in addition, to consider the anti-phase behaviour of the different Cu–Cu and Cu–O shells, with constructive and destructive interferences in different parts of the EXAFS range[36]. Therefore, the full EXAFS data were analysed in k- and R-space using a combined k^1-k^2-k^3-fitting procedure. A fit of the fully refined cluster was only acceptable, if the fit was of high quality in all k-weightings.

The most prominent peak can be observed below 2 Å in R-space and is associated with backscattering from the oxygen atoms to which Cu is directly bonded. The complex line shape suggests at least two distinct types of O-donor ligands with significantly different Cu–O distances due to framework (O_F) and extra-framework oxygen (O_{EF}). Features above 2.0 Å in R-space arise from Cu–Cu and second shell Cu–O single-scattering paths. The experimental data significantly deviates from the simulated spectrum of a binuclear $[Cu(\mu\text{-}O)Cu]^{2+}$ complex. A particularly strong deviation can be seen at large interatomic distances ($R \sim 2.25$ Å—not phase corrected), corresponding to the second coordination shell of Cu. Multiple scattering paths must be visible for a dimeric structure with a O–Cu–O structure, similar to typical CuO spectra[37]. The absence of this feature in the experimental spectra suggests the existence of >1 Cu–Cu path and, therefore, the presence of a Cu cluster with a nuclearity higher than 2. Conversely, a good fit of the experimental EXAFS data is achieved by the simulated EXAFS based on the DFT-optimized structure of the $[Cu_3(\mu\text{-}O)_3]^{2+}$/MOR cluster model (Fig. 3c,d and Supplementary Fig. 6 for comparison of k^1, k^2 and k^3 weighted plots). Results of the fitting are summarized in Table 2. The DFT-optimized geometric parameters of the proposed trinuclear Cu cluster show a low symmetry of this species (see Supplementary Table 2). Therefore, Cu atoms in the cluster are not equivalent and two different shells need to be included. Coordination numbers (CNs) are averaged for the three Cu scatterers. The average CNs derived from the DFT-optimized geometric parameters of the most stable trinuclear cluster are applied for the fit to take all Cu–O and Cu–Cu contributions into account (see Supplementary Tables 2 and 3). When the EXAFS fitting was started from an alternative trinuclear model, similar parameters were obtained. This indicates that the values shown in Table 2 are a true minimum. Details on the EXAFS fitting for different model clusters are available in Supplementary Tables 4 and 5.

***In situ* monitoring of active copper species.** Having shown that the single site in the present Cu-MOR catalysts is a $[Cu_3(\mu\text{-}O)_3]^{2+}$ cluster, the next step was to monitor the formation of the cluster during activation under O_2 and its interaction with CH_4 under reaction conditions. For this purpose, an *in situ* study was performed by XAS and ultraviolet–visible (UV-vis) spectroscopy. Figure 4 shows the XANES (X-ray Absorption Near Edge Structure) and EXAFS of Cu-MOR at different stages of the catalytic cycle. Dehydration of the fresh catalysts led to a change from the octahedral coordination sphere Cu^{2+} in hexaquo-complexes to tetrahedral Cu^{2+} species. At temperatures above 200 °C, the progressive formation of a feature at $R > 2$ Å in EXAFS was observed (see Supplementary Fig. 7), which can be attributed to the new Cu–Cu path and therefore to the formation of Cu-oxo clusters with nuclearity ≥ 2.

In situ ultraviolet–visible spectroscopy of the activation of Cu-MOR in O_2 showed the development of a very broad band

centred at ca. 31,000 cm^{-1} (see Supplementary Fig. 8), while a band at 22,700 cm^{-1}, assigned to extra-framework $O \rightarrow Cu(II)$ charge transfer for the active species $[Cu(\mu\text{-}O)Cu]^{2+}$ in Cu-ZSM-5 (ref. 35), was not observed in any of the tested conditions. This fact further supports the conclusion that the active Cu-oxo clusters reported here have a structure different from those described for conventionally prepared Cu-ZSM-5.

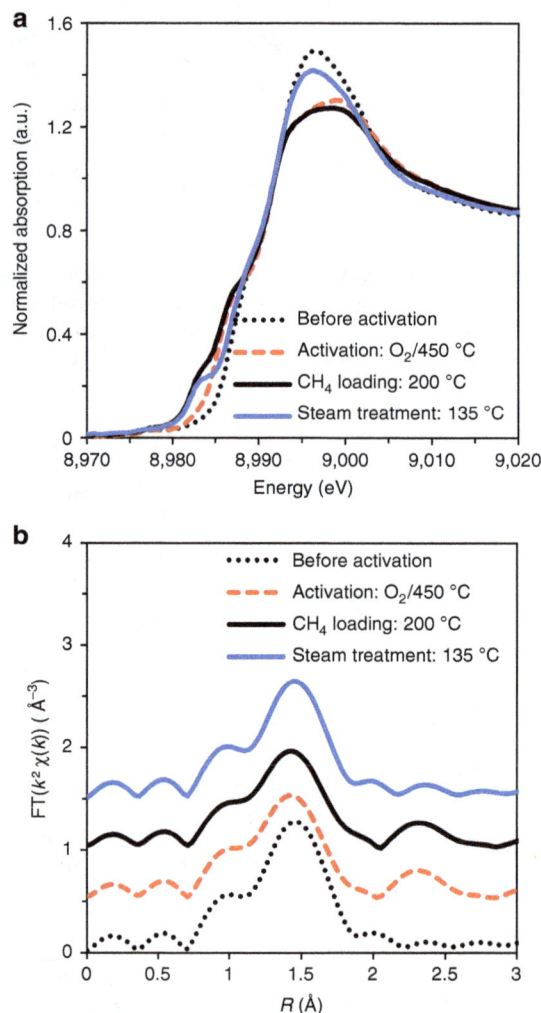

Figure 4 | *In situ* X-ray absorption spectroscopy. (a) *In situ* XANES and **(b)** Fourier transformed EXAFS during a full cycle of selective partial oxidation of methane to methanol.

Table 2 | Copper EXAFS fitting results.

Backscatterer	Coordination numbers N^{DFT}	Coordination numbers N^{EXAFS}	Distance R^{DFT} (Å)	Distance R^{EXAFS} (Å)	Debye–Waller factor[†] $\Delta\sigma^2$ (Å^2)
Cu-O_{EF}	2	2.2 (± 0.8)	1.80	1.91 (± 0.03)	0.003 (± 0.003)
Cu-O_F	1.66	1.6 (± 0.5)	2.02	2.04 (± 0.07)	0.004 (± 0.005)
Cu-O_F	0.33	0.4 (± 0.5)	2.63	2.35 (± 0.05)	0.003 (± 0.010)
Cu-Cu	0.66	0.7 (± 0.4)	2.74	2.86 (± 0.04)	0.005 (± 0.005)
Cu-Cu	1.33	1.5 (± 0.7)	3.04	3.02 (± 0.05)	0.010 (± 0.006)
Cu-O_{EF}	1	1.3 (± 1.1)	3.23	3.50 (± 0.09)	0.008 (± 0.025)

EXAFS, Extended X-ray Absorption Fine Structure; DFT, density functional theory; MOR, mordenite.
Comparison of Cu K-edge EXAFS fit results* for O_2-activated Cu-MOR zeolite with DFT-optimized geometric parameters of $[Cu_3(\mu\text{-}O)_3]^{2+}$ in Cu-MOR.
*Combined k^1, k^2 and k^3-weighted fit, $2.4 < k < 12.0$ Å, $1 < R < 3.6$, $E_0 = -1$ (3), R-factor = 0.003, S_0^2 (fixed) = 0.9, statistical errors in brackets.
[†]Debye–Waller factors were fixed (to the values obtained in the best fit with set coordination numbers) during EXAFS fit to reduce the number of fitting parameters. The values predicted by DFT calculations are averaged over three Cu scatterers. See also Supplementary Tables 2-5.

Upon reaction with CH_4, the development of a new strong feature at 8,983 eV has been observed in XANES, which indicates a reduction of a fraction of intrazeolite Cu^{2+} to Cu^+ (Fig. 4a). This is in line with the thermal autoreduction reported in the literature[37,38], and also in good agreement with different mechanisms proposed for the donation of an oxygen atom from the metal oxide cluster to the methyl moiety[39]. On the other hand, significant changes could not be noted in the Fourier transformed EXAFS upon the treatment with CH_4, indicating that the trinuclear structure of the active site is preserved at this step, presumably because the oxygenated products remained strongly attached to the cluster. In the UV–vis spectra, the broad band centred at 31,000 cm^{-1}, which is stable at 200 °C under O_2 or N_2, disappears only after 20–30 min contact with CH_4 flow at 200 °C (Fig. 5). The low rate of disappearance of this feature is in good agreement with the low rate of reaction predicted for the catalyst (at least 30 min in contact with CH_4 was necessary to measure significant amounts of methanol).

Mechanism of C–H bond activation on $[Cu_3(\mu\text{-}O)_3]^{2+}$ cluster. To understand the elementary energetics of methane to methanol oxidation on the proposed trinuclear complex $[Cu_3(\mu\text{-}O)_3]^{2+}$ on a more quantitative mechanistic basis, DFT calculations were performed. The methane activation proceeds via a homolytic C–H bond cleavage followed by a direct radical rebound reaction mechanism[34], resulting in methanol adsorbed to a reduced Cu cluster. The C–H bond activation over the extra-framework $Cu_3O_3^{2+}$ cluster is facilitated by the interaction with a formally radical-anionic extra-framework oxygen centre. The interaction of the respective single-occupied molecular orbital of $Cu_3O_3^{2+}$ with the antibonding CH orbital ($\sigma^*(CH)$) of methane results in the cleavage of the CH bond, resulting in CH_3 and an OH group bound to the cluster (see Supplementary Fig. 9). The role of Cu centres and, accordingly, the orbital interactions with the metal ions, is to stabilize the electronic configuration of the extra-framework cluster with the anion-radical character of the oxygen ligands necessary for the facile C–H activation[40]. The C–H bond activation barrier of 74 kJ mol^{-1} (see Supplementary Fig. 10) is comparable to that of 78 kJ mol^{-1} on binuclear $[Cu(\mu\text{-}O)Cu]^{2+}$ in ZSM-5 (ref. 34). The CH activation is followed by a barrierless recombination of the CH_3 radical with the Cu-bound OH group in the cluster. The so-formed methanol molecule is preferentially coordinated to two neighbouring Cu centres of the trinuclear species (see Supplementary Fig. 10), in good agreement with the conservation of the cluster structure upon methane activation observed by XAS.

As has been outlined above, desorption of methanol is only accomplished by steam treatment of the catalysts at 135 °C, which led to a substantial decrease of the Cu–Cu path in the EXAFS (Fig. 4b). We conclude from these data that the trinuclear Cu cluster is hydrolysed by contact with water. Interestingly, the renewed activation of the material in O_2 at 500 °C completely restored the activity, even when the procedure was repeated up to eight cycles (see Supplementary Fig. 11). Identical XAS and UV–vis spectra were obtained for samples after a second activation, confirming that the $[Cu_3(\mu\text{-}O)_3]^{2+}$ cluster is highly stable and re-forms by self-organization under dry oxidation conditions in mordenite.

Discussion
The choice of a MOR with high concentration of Al in the side pockets, together with an optimized synthetic approach for copper exchange has yielded Cu-MOR materials, showing an outstanding activity in methane activation, which is at least one order of magnitude higher than those reported in the literature for analogous systems. In contrast to other metal-exchanged zeolites where a mixture of cationic species with different structures and reactivities is detected[41,42], the stoichiometry of converted methane to Cu for these Cu-MOR materials has shown that it is possible to develop a Cu zeolite with only one type of active site.

In situ XAS demonstrated that the homogeneous single sites in activated Cu-MOR are the trinuclear Cu-oxo clusters, namely $[Cu_3(\mu\text{-}O)_3]^{2+}$, anchored to two framework Al atoms located at the pore mouth of the 8-MR side pockets. Furthermore, this active $[Cu_3(\mu\text{-}O)_3]^{2+}$ species has been found to be highly stable under dry conditions, in agreement with *ab initio* thermodynamic analysis based on DFT results. Even though the reaction with methane followed by steam treatment led to the hydrolysis of the cluster, it can be re-formed without loss of activity by reactivation in O_2.

The present results show conclusively that trimeric Cu-oxo clusters are active and selective for partial methane oxidation. It should be noted in this context that Chan *et al.*, recently proposed such a cluster to be active in pMMO and reported that it was possible to use tri-copper complexes for selective methane oxidation to methanol, albeit with H_2O_2 as an oxidant[17,43]. The multiple Al framework atoms in the 8-MR side pockets of H-MOR provide the conditions to stabilize $[Cu_3(\mu\text{-}O)_3]^{2+}$ clusters. We hypothesize, in addition, that the 8-MR side pockets in MOR enhance the activity of the clusters by providing similar steric constraints as found for the hydrophobic cavity formed by the pmoA and pmoC subunits of pMMO[15,16].

The material presented here is one of the few examples of catalysts with well-defined active sites evenly distributed in the zeolite framework, a truly single-site heterogeneous catalyst. This not only allows for much higher efficiencies in conversion of methane to methanol than previously reported, it also enables the unequivocal linking of the structure of the sites with their catalytic activity. Understanding why Cu clusters form reversibly, in varying reaction environments, while similar clusters remain stable in the enzyme, despite unfavourable conditions, is one of the big hurdles to achieve similar activities and selectivities in heterogeneous catalysts, as are usually found in enzymatic systems alone. The presented system is therefore a more than promising basis to tackle this challenge.

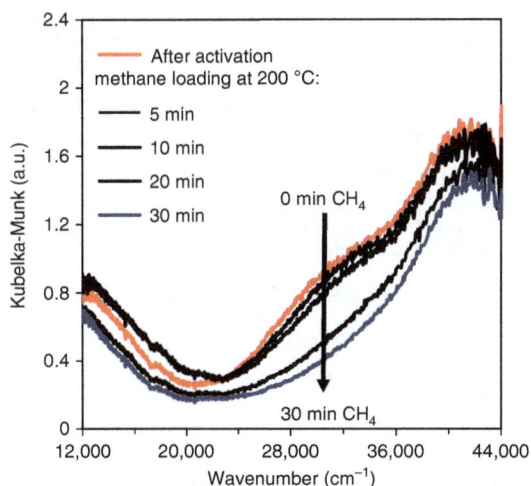

Figure 5 | Ultraviolet–visible spectroscopy of Cu-MOR. *In situ* UV–vis spectra of Cu-MOR after activation in oxygen at 450 °C and subsequent methane loading at 200 °C.

Methods

Preparation of Cu-exchanged zeolites. H-MOR was obtained by calcination of commercial zeolite NH_4-MOR (Clariant, Si/Al 11, 21) in synthetic air at 500 °C for 8 h. Cu-MOR with different Cu/Al ratios was prepared by aqueous ion exchange of H-MOR with Cu^{2+}. The Cu^{2+} exchange was carried out at ambient temperature by contacting 5 g zeolite with 300 ml of aqueous $Cu(CH_3COO)_2$ (Sigma-Aldrich, 99.99%) solution. The reaction time and the molarity of the solution was varied between 0.0025 and 0.01 M $Cu(CH_3COO)_2$ to obtain Cu/Al ratios between 0.1 and 0.4. A series of several subsequent cycles of ion exchange with intermediate rinsing was performed in an attempt to increase the Cu/Al ratio (0.4–0.6). The pH of the solution was 5.5–6.0 during exchange. A typical exchange time was 20 h. After the last exchange step, the samples were rinsed four times with doubly deionized water (50 ml g^{-1} MOR each time) with an intervening centrifugation step between each rinse. These rinse cycles were performed to ensure that the pores did not contain further non-exchanged Cu ions, which would form large CuO clusters during activation. Samples were then dried in static ambient air at 110 °C for 24 h. The Si, Al, Na and Cu contents were measured by atomic absorption spectroscopy on a UNICAM 939 AA spectrometer after dissolution in boiling hydrofluoric acid. Brunnauer-Emmet-Teller (BET) surface area was measured on a PMI automated Sorptomatic 1990 after activation at 350 °C. Co^{2+} exchange was prepared by aqueous ion exchange of Na-MOR in 0.05 M $Cu(NO_3)_2$ solution at room temperature, following the procedure described in ref. 33. Na-MOR was prepared by Na^+ exchange of freshly calcined H-MOR with 0.5 M $NaNO_3$ solution for 24 h at 60 °C.

Testing of activity for selective oxidation of methane. Cu-MOR samples were tested for their activity towards methane oxidation in an atmospheric pressure stainless steel plug flow reactor with a 4-mm inner diameter. The reaction included three consecutive steps: (i) activation, (ii) CH_4 loading and (iii) steam-assisted CH_3OH desorption. In a typical experiment, 0.1 g of Cu-MOR (250–400 μm) was calcined in an O_2 flow (16 ml min^{-1}) at 450 °C for 1 h. The activated catalyst was cooled to 200 °C in O_2 and flushed in He. In the subsequent CH_4-loading step, 90% CH_4 in He (16 ml min^{-1}) was passed over the sample for 4 h. The temperature was then decreased in He to 135 °C. A steam-assisted CH_3OH desorption step was carried out by passing an equimolar mixture of H_2O steam and He (20 ml min^{-1}) through the reactor bed for 30 min. The reaction products were identified and quantified by online mass spectroscopy by monitoring the time-dependent evolution of signals at m/e 28, 31, 44 and 46 characteristic for CO, CH_3OH, CO_2 and $(CH_3)_2O$, respectively. The He signal ($m/e = 4$) was used as an internal standard. Productivity was calculated as the product of the effluent flow rate and the integral of the product concentrations as a function of time. The product $(CH_3)_2O$ was assumed to be formed via condensation of two partially oxidized CH_4 molecules corresponding to two CH_3OH equivalents. The sum of all detected products is referred to as total yield.

Infrared spectroscopy. The samples for infrared spectroscopy were prepared as self-supporting wafers with a density of ca. 10 mg cm^{-2}. Samples were first activated in vacuum (1.0×10^{-7} mbar) at 450 °C for 1 h with a heating rate of 10 °C min^{-1}. Infrared spectra of adsorbed n-hexane were recorded on a Vertex 70 spectrometer from Bruker Optics at a resolution of 4 cm^{-1}. After pretreatment, the activated samples were cooled to 30 °C, n-hexane (0.5–5 mbar) was adsorbed and equilibrated for at least 30 min. All spectra were recorded at 30 °C. Infrared spectra of adsorbed pyridine were measured on Thermo Nicolet 5,700 FT-IR spectrometer with a resolution of 4 cm^{-1}. After activation, the total concentration of BAS was determined at 150 °C after adsorption of 0.1 mbar pyridine and subsequent evacuation for 30 min at the same temperature. All spectra were recorded at 150 °C.

DFT calculation. All periodic DFT calculations were performed using VASP software with a generalized gradient-approximated PBE exchange-correlation functional[44,45]. Projected augmented wave method and plane wave basis set with a cutoff of 400 eV were employed. Brillouin zone-sampling was restricted to the Γ point[46]. A supercell of all-silica MOR, constructed by a doubling monoclinic primitive cell along the c axis with lattice parameters of $a = b = 13.648$, $c = 15.015$ Å and $\gamma = 97.2°$ as optimized by DFT, was used as an initial model[47]. To compensate for the positive charge of the extra-framework cationic Cu complexes, two framework Si^{4+} ions in the MOR supercell were substituted by two Al^{3+} at the side-pocket position of Al^I_{SP}. The other two $[AlO_2]^-$ units at the side-pocket position of Al^{II}_{SP} were charge compensated by two BASs[48]. The resulting MOR model had a Si/Al ratio of 11. The nudged elastic band method[49] was used to determine the minimum energy path and to locate the transition-state structure for the methane oxidation to methanol reaction. The maximum energy geometry along the reaction path obtained by the nudged elastic band method was further optimized using a quasi-Newton algorithm. In this step, only the extra-framework atoms were relaxed. Spin-polarized calculations were performed throughout this study. The calculated reaction paths following the spin PESs (potential energy surfaces) of ground electronic states with $S = 1/2$ and $S = 3/2$ were very close in energy both for intermediates and transition states of methane activation. Vibrational frequencies were calculated using the finite-difference

method, as implemented in VASP. Small displacements (0.02 Å) were used to estimate the numerical Hessian matrix. The transition state was confirmed by the presence of a single imaginary frequency corresponding to the reaction path. Electron density analysis was carried out using VESTA[50].

For molecular orbital analysis, single-point calculations at the PBE/6-31 + G(d,p) level of theory were carried out using the Gaussian 09 program[51] on a 8-MR cluster model (see Supplementary Fig. 9) directly cut from the periodic DFT-optimized Cu_3O_3/MOR structure. The dangling Si–O bonds at the periphery of the cluster model were saturated by hydrogen atoms oriented in the direction of the next T-sites of the zeolite lattice. Both doublet and quartet ground spin states ($S = 1/2$ and $S = 3/2$) were considered.

Ab initio thermodynamic analysis. To account for the effect of temperature as well as the presence of H_2O and O_2 upon the catalysts activation on the stability of different extra-framework Cu complexes in Cu/MOR, ab initio thermodynamic analysis was employed. In this study, the thermodynamic analysis is performed with a reference to bulk copper oxide as the most plausible alternative to extra-framework Cu species formed in the zeolite. The following reversible reactions were considered to compare equilibria among species with different chemical compositions:

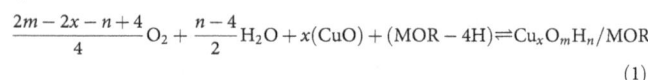

$$\frac{2m-2x-n+4}{4}O_2 + \frac{n-4}{2}H_2O + x(CuO) + (MOR-4H) \rightleftharpoons Cu_xO_mH_n/MOR$$

(1)

The reaction Gibbs free energy ΔG for equilibrium (1) is:

$$\Delta G(T,p) = G^s_{Cu_xO_mH_n/MOR} - G^s_{MOR-4H} - xG^s_{CuO} - \frac{2m-2x-n+4}{2}\mu^g_O - \frac{n-4}{2}\mu^g_{H_2O}$$

(2)

The vibrational and pressure-volume contributions of solids are neglected, and the Gibbs free energies of zeolite and bulk copper oxide are approximated as their respective electronic energies directly computed by DFT. The chemical potentials of the gas phase O (O_2) and H_2O depend on T and p. We assume that the surrounding O_2 atmosphere forms an ideal gas-like reservoir, and we chose the reference state of $\mu_{O_2}(T,p)$ to be the total electronic energy of an isolated O_2 molecule (E_{O_2}). In other words, the chemical potential of oxygen at the reference state at 0 K is $\mu_O(0K) = 1/2E_{O_2}$ in which E_{O_2} is the DFT-calculated total energy of O_2. Then the chemical potential of oxygen at arbitrary T and p can be written as:

$$\mu_O(T,p) = 1/2E_{O_2} + \Delta\mu_O(T,p)$$

(3)

Where

$$\begin{aligned}
\Delta\mu_O(T,p) &= \Delta\mu_O(T,p^0) + \frac{1}{2}RT\ln\left(p_{O_2}/p^0_{O_2}\right)\\
&= \frac{1}{2}\left[\Delta\mu_{O_2}(T,p^0) + RT\ln\left(p_{O_2}/p^0_{O_2}\right)\right]\\
&= \frac{1}{2}\left[H(T,p^0,O_2) - H(0K,p^0,O_2) - T\left(S(T,p^0,O_2)\right.\right.\\
&\quad \left.\left. - S(0K,p^0,O_2)\right) + RT\ln\left(p_{O_2}/p^0_{O_2}\right)\right]
\end{aligned}$$

(4)

The chemical potential change ($\Delta\mu_O(T,p)$) defined in such a manner includes all temperature- and pressure-dependent free-energy contributions. The temperature and pressure dependency of the chemical potential is obtained from the differences in the enthalpy and entropy of an O_2, as well as H_2O molecules with respect to the reference state at the 0-K limit. For standard pressure (1 atm), the values tabulated in thermodynamic tables were employed[52]. This approach has been previously proven to provide rather accurate results for ab initio thermodynamic calculations[53,54].

The chemical potential of H_2O as well as the chemical potential change were calculated exactly in the same way as described in the above:

$$\mu_{H_2O}(T,p) = E_{H_2O} + \Delta\mu_{H_2O}(T,p)$$

(5)

Bringing equations (3) and (5) into equation (2) and considering the Gibbs free energies of zeolite and bulk copper oxide are approximated as their respective DFT-computed electronic energies, we arrive at:

$$\Delta G(T,p) = \Delta E - \frac{2m-2x-n+4}{2}\Delta\mu_O - \frac{n-4}{2}\Delta\mu_{H_2O}$$

(6)

where

$$\Delta E = E_{Cu_xO_mH_n/MOR} - \frac{2m-2x-n+4}{4}E_{O_2} - \frac{n-4}{2}E_{H_2O} - xE_{CuO} - E_{MOR-4H}$$

(7)

$E_{Cu_xO_mH_n/MOR}$ is the total electronic energy of a given Cu-containing MOR model, E_{MOR-4H} is the energy of the H-form of MOR with four Al^{3+} substituted at Al^I_{SP} and Al^{II}_{SP}, E_{CuO}, E_{O_2} and E_{H_2O} correspond to the electronic energies of bulk CuO, gaseous H_2O and O_2, respectively. The factor x denotes the number of Cu atoms in the unit cell of $Cu_xO_mH_n/MOR$. Depending on the structure of the Cu complex, x can be 1, 2 or 3.

X-ray absorption spectroscopy. X-ray absorption spectra were recorded at Diamond Light Source in Oxford Shire, UK, on beamline B18. The electron energy was 3 GeV with a beam current of 300 mA. The beam size at the sample was $200 \times 250\,\mu m$. Samples were prepared as self-supporting wafers (60–80 mg) and placed into an *in situ* XAS cell. The X-ray absorption spectra were collected *in situ* at the Cu K-edge (8,979 eV) during activation in oxygen at 450 °C, during exposure of the sample to CH_4 and after steam treatment. To avoid condensation, all lines of the set-up were thermostated at 110 °C. The samples were activated in an O_2 flow of 30 ml min^{-1} at 450 °C for 1 h (heating ramp 10 °C min^{-1}) and afterwards cooled to 200 °C. After a short flush with He, CH_4 was loaded for 4 h at 220 °C (flow 30 ml min^{-1}). The temperature was then decreased under He flow to 135 °C and an equal molar mixture of water steam/He (50 ml min^{-1}) was passed for 2 h through the cell. The Cu K-edge XANES data processing and EXAFS analysis were performed using IFEFFIT version 1.2.11d with the Horae package (Athena and Artemis)[55,56]. The amplitude reduction factor, that is, $S_0{}^2$, was experimentally derived to be 0.9, from EXAFS analysis of Cu reference compounds with known structures, that is, $Cu(OAc)_2$ and $Cu(OH)_2$ (refs 57,58). Fitting was done in k- and R-space and in multiple weightings of k^1, k^2 and k^3, simultaneously. A fit was only concluded to be good, if all fits in all weightings, as well in k- as in R-space were good and all included contributions were determined to be significant, tested by refinement of CNs. Refinement of CNs gave values with a deviation of $<10\%$ from the values predicted by the DFT model for all refined paths. Fits were performed using the optimized geometrical parameters for Cu-MOR obtained from the periodic DFT calculations as an input model.

EXAFS fitting and analysis. For the fitting of EXAFS spectra, the amplitude ($S_0{}^2$), determined from reference materials, as well as the CNs derived from the DFT models were fixed to reduce the number of fitting parameters. In a second step, Debye–Waller factors were fixed to the values obtained in the best fit with set CNs, and thus CNs and bond distances were refined. Only if the refined values of the distances and the corresponding CNs were in good agreement with the DFT model, a fit was considered as good. Although the R-factor (0.009, see Supplementary Table 4) is low, a significant deviation between the experimental data and the DFT model of a binuclear $[Cu(\mu\text{-O})Cu]^{2+}$ complex can be observed and is particularly pronounced at larger interatomic distances (R) (see Fig. 3a,b). Not just the intensity of the fitted model, but also the imaginary part of the Fourier transform, does not fit well in the Cu–Cu region. Moreover, large Debye–Waller factors with large statistical errors are obtained for both the Cu–Cu and Cu–O$_F$ contributions (Supplementary Tables 4 and 5). Possible multiple scattering paths, likely for a dimeric structure with fairly linear O–Cu–O structure motifs and, for example, clearly pronounced in CuO[37], cannot be observed or analysed. This suggests the presence of >1 Cu–Cu path and therefore the presence of a Cu species with nuclearity higher than 2. Comparison of various Cu trinuclear models proposed by DFT (Supplementary Fig. 4) showed that a good fit of the experimental EXAFS data is achieved by a trinuclear $[Cu_3(\mu\text{-O})_3]^{2+}$ cluster (Fig. 3c,d), in good agreement with our calculations predicting this cluster to be the most stable under activation conditions. Results of the fitting are summarized in Table 2. It should be noted that the three Cu atoms in the trinuclear $[Cu_3(\mu\text{-O})_3]^{2+}$ cluster have the same CNs but slightly different bond lengths, and therefore EXAFS fitting requires averaging over three scatterers (Supplementary Tables 2 and 3). EXAFS fitting starting from alternative trinuclear models resulted in similar parameters, indicating that the values shown in Table 2 are a true minimum. In addition, full refinement of all parameters including CNs did not result in significant deviation of the parameters as obtained. A R-factor of 0.004 indicates a good fit. Naturally, the more complex structure of the trinuclear Cu cluster leads to more first- and second-shell scattering paths compared with the simpler binuclear structure. The quality of the data, in addition to fixing the CNs and amplitudes, allows reliable refinement of all other parameters (number of independent data points is 16, with 13 parameters fitted). The Cu–Cu contributions are significant, and realistic distances and low Debye–Waller factors with low errors are obtained. The Cu-O$_{EF}$ contribution is shifted significantly compared with the model, with large errors, especially in the Debye–Waller factor. This is due to the small contribution of this shell at long distance, and hence this shell cannot be determined with high accuracy. Different four-coordinate Cu clusters, for example, cubane or ring-type structures, were also tested as a starting model for the EXAFS analysis. A full refinement of experimental data with the Cu$_4$ model results in exactly the same analysis as presented in the manuscript, which corresponds to the trimeric model rather than a Cu$_4$ cubane-type structure. This is another proof that we have refined a true minimum.

Ultraviolet–visible spectroscopy. Ultraviolet–visible measurements of the Cu-exchanged H-MOR samples were performed with an Avantes Avaspec 2,048 spectrometer in the diffuse reflectance mode. The samples were measured as powders and placed in a quartz flow reactor (6 mm inner diameter) with square optical-grade quartz windows. The reactor was placed horizontally in a lab-made heating chamber with an 8-mm diameter hole on top, through which a high-temperature optical fibre (Avantes FCR-7UV400 – 2ME-HTX UV-vis reflection probe) could be vertically directed to the reactor. The temperature was measured by a thermocouple located on the bottom of the quartz reactor. In a typical experiment, the UV-vis spectra were collected during treatment in oxygen,

nitrogen or methane. The intensity of the diffuse reflectance UV-vis spectra is presented in the form of the Kubelka–Munk function, defined as $F(R) = (1 - R)^2/(2 \times R)$ with $R = R_s/R_r$, with R_s—the reflectance of the sample and R_r—the reflectance of the H-MOR parent material used as a reference. The samples were first treated at 450 °C for 1 h in He (flow 16 ml min^{-1}), heating at a rate of 10 °C min^{-1}. Subsequently, the sample was cooled to ambient temperature, He was replaced with O_2 and the sample was heated to 450 °C in O_2 flow (flow 16 ml min^{-1}). After cooling of the activated sample to 200 °C, the sample was contacted with CH_4 (flow 16 ml min^{-1}).

References

1. Olah, G. A. Beyond oil and gas: the methanol economy. *Angew. Chem. Int. Ed.* **44**, 2636–2639 (2005).
2. Kerr, R. A. Natural gas from shale bursts onto the scene. *Science* **328**, 1624–1626 (2010).
3. Malakoff, D. The gas surge. *Science* **344**, 1464–1467 (2014).
4. Jensen, K. F Microreaction engineering - is small better? *Chem. Eng. Sci.* **56**, 293–303 (2001).
5. Henriciolive, G. & Olive, S. Fischer-Tropsch synthesis: molecular-weight distribution of primary products and reaction-mechanism. *Angew. Chem. Int. Ed.* **15**, 136–141 (1976).
6. Olsbye, U. *et al.* Conversion of methanol to hydrocarbons: how zeolite cavity and pore size controls product selectivity. *Angew. Chem. Int. Ed.* **51**, 5810–5831 (2012).
7. Keil, F. J. Methane activation oxidation goes soft. *Nat. Chem.* **5**, 91–92 (2013).
8. Guo, X. G. *et al.* Direct, nonoxidative conversion of methane to ethylene, aromatics, and hydrogen. *Science* **344**, 616–619 (2014).
9. Rosenzweig, A. C., Frederick, C. A., Lippard, S. J. & Nordlund, P. Crystal-structure of a bacterial nonheme iron hydroxylase that catalyzes the biological oxidation of methane. *Nature* **366**, 537–543 (1993).
10. Bordeaux, M., Galarneau, A. & Drone, J. Catalytic, mild, and selective oxyfunctionalization of linear alkanes: current challenges. *Angew. Chem. Int. Ed.* **51**, 10712–10723 (2012).
11. Chan, S. I., Chen, K. H. C., Yu, S. S. F., Chen, C. L. & Kuo, S. S. J. Toward delineating the structure and function of the particulate methane monooxygenase from methanotrophic bacteria. *Biochemistry* **43**, 4421–4430 (2004).
12. Himes, R. A., Barnese, K. & Karlin, K. D. One is lonely and three is a crowd: two coppers are for methane oxidation. *Angew. Chem. Int. Ed.* **49**, 6714–6716 (2010).
13. Lieberman, R. L. & Rosenzweig, A. C. Crystal structure of a membrane-bound metalloenzyme that catalyses the biological oxidation of methane. *Nature* **434**, 177–182 (2005).
14. Balasubramanian, R. *et al.* Oxidation of methane by a biological dicopper centre. *Nature* **465**, 115–U131 (2010).
15. Ng, K. Y., Tu, L. C., Wang, Y. S., Chan, S. I. & Yu, S. S. F. Probing the hydrophobic pocket of the active site in the particulate methane monooxygenase (pMMO) from Methylococcus capsulatus (bath) by variable stereoselective alkane hydroxylation and olefin epoxidation. *ChemBioChem* **9**, 1116–1123 (2008).
16. Chan, S. I. *et al.* Redox potentiometry studies of particulate methane monooxygenase: Support for a trinuclear copper cluster active site. *Angew. Chem. Int. Ed.* **46**, 1992–1994 (2007).
17. Chan, S. I. *et al.* Efficient oxidation of methane to methanol by dioxygen mediated by tricopper clusters. *Angew. Chem. Int. Ed.* **52**, 3731–3735 (2013).
18. Labinger, J. A. & Bercaw, J. E. Understanding and exploiting C-H bond activation. *Nature* **417**, 507–514 (2002).
19. Periana, R. A. *et al.* A mercury-catalyzed, high-yield system for the oxidation of methane to methanol. *Science* **259**, 340–343 (1993).
20. Periana, R. A. *et al.* Platinum catalysts for the high-yield oxidation of methane to a methanol derivative. *Science* **280**, 560–564 (1998).
21. Palkovits, R., Antonietti, M., Kuhn, P., Thomas, A. & Schuth, F. Solid catalysts for the selective low-temperature oxidation of methane to methanol. *Angew. Chem. Int. Ed.* **48**, 6909–6912 (2009).
22. Soorholtz, M. *et al.* Direct methane oxidation over Pt-modified nitrogen-doped carbons. *Chem. Commun.* **49**, 240–242 (2013).
23. Smeets, P. J., Woertink, J. S., Sels, B. F., Solomon, E. I. & Schoonheydt, R. A. Transition-metal ions in zeolites: coordination and activation of oxygen. *Inorg. Chem.* **49**, 3573–3583 (2010).
24. Starokon, E. V., Parfenov, M. V., Pirutko, L. V., Abornev, S. I. & Panov, G. I. Room-temperature oxidation of methane by alpha-oxygen and extraction of products from the FeZSM-5 surface. *J. Phys. Chem. C* **115**, 2155–2161 (2011).
25. Vanelderen, P. *et al.* Spectroscopy and redox chemistry of copper in mordenite. *ChemPhysChem* **15**, 91–99 (2014).
26. Alayon, E. M., Nachtegaal, M., Ranocchiari, M. & van Bokhoven, J. A. Catalytic conversion of methane to methanol over Cu-mordenite. *Chem. Commun.* **48**, 404–406 (2012).
27. Alayon, E. M. C., Nachtegaal, M., Kleymenov, E. & van Bokhoven, J. A. Determination of the electronic and geometric structure of Cu sites during

methane conversion over Cu-MOR with X-ray absorption spectroscopy. *Microporous Mesoporous Mater.* **166,** 131–136 (2013).

28. Smeets, P. J., Groothaert, M. H. & Schoonheydt, R. A. Cu based zeolites: a UV-vis study of the active site in the selective methane oxidation at low temperatures. *Catal. Today* **110,** 303–309 (2005).

29. Veefkind, V. A., Smidt, M. L. & Lercher, J. A. On the role of strength and location of Bronsted acid sites for ethylamine synthesis on mordenite catalysts. *Appl. Catal. A* **194,** 319–332 (2000).

30. Groothaert, M. H., Smeets, P. J., Sels, B. F., Jacobs, P. A. & Schoonheydt, R. A. Selective oxidation of methane by the bis(μ-oxo)dicopper core stabilized on ZSM-5 and mordenite zeolites. *J. Am. Chem. Soc.* **127,** 1394–1395 (2005).

31. Moreau, F. *et al.* Influence of Na exchange on the acidic and catalytic properties of an HMOR zeolite. *Microporous Mesoporous Mater.* **51,** 211–221 (2002).

32. Eder, F., Stockenhuber, M. & Lercher, J. A. Bronsted acid site and pore controlled siting of alkane sorption in acidic molecular sieves. *J. Phys. Chem. B* **101,** 5414–5419 (1997).

33. Dedecek, J., Kaucky, D., Wichterlova, B. & Gonsiorova, O. Co^{2+} ions as probes of Al distribution in the framework of zeolites. ZSM-5 study. *Phys. Chem. Chem. Phys.* **4,** 5406–5413 (2002).

34. Woertink, J. S. *et al.* A [Cu$_2$O]$^{2+}$ core in Cu-ZSM-5, the active site in the oxidation of methane to methanol. *Proc. Natl Acad. Sci. USA* **106,** 18908–18913 (2009).

35. Groothaert, M. H., Lievens, K., Leeman, H., Weckhuysen, B. M. & Schoonheydt, R. A. An operando optical fiber UV-vis spectroscopic study of the catalytic decomposition of NO and N$_2$O over Cu-ZSM-5. *J. Catal.* **220,** 500–512 (2003).

36. Tromp, M. *et al.* Cu K-Edge EXAFS characterisation of copper(I) arenethiolate complexes in both the solid and liquid state: detection of Cu-Cu coordination. *Chemistry* **8,** 5667–5678 (2002).

37. Neylon, M. K., Marshall, C. L. & Kropf, A. J. In situ EXAFS analysis of the temperature-programmed reduction of Cu-ZSM-5. *J. Am. Chem. Soc.* **124,** 5457–5465 (2002).

38. Alayon, E. M. C., Nachtegaal, M., Bodi, A. & van Bokhoven, J. A. Reaction conditions of methane-to-methanol conversion affect the structure of active copper sites. *ACS Catal.* **4,** 16–22 (2014).

39. Vanelderen, P., Vancauwenbergh, J., Sels, B. F. & Schoonheydt, R. A. Coordination chemistry and reactivity of copper in zeolites. *Coord. Chem. Rev.* **257,** 483–494 (2013).

40. Dietl, N., Schlangen, M. & Schwarz, H. Thermal hydrogen-atom transfer from methane: The role of radicals and spin states in oxo-cluster chemistry. *Angew. Chem. Int. Ed.* **51,** 5544–5555 (2012).

41. Beznis, N. V., Weckhuysen, B. M. & Bitter, J. H. Partial oxidation of methane over Co-ZSM-5: tuning the oxygenate selectivity by altering the preparation route. *Catal. Lett.* **136,** 52–56 (2010).

42. Beznis, N. V., van Laak, A. N. C., Weckhuysen, B. M. & Bitter, J. H. Oxidation of methane to methanol and formaldehyde over Co-ZSM-5 molecular sieves: tuning the reactivity and selectivity by alkaline and acid treatments of the zeolite ZSM-5 agglomerates. *Microporous Mesoporous Mater.* **138,** 176–183 (2011).

43. Maji, S. *et al.* Dioxygen activation of a trinuclear CuICuICuI cluster capable of mediating facile oxidation of organic substrates: competition between O-atom transfer and abortive intercomplex reduction. *Chem. Eur. J.* **18,** 3955–3968 (2012).

44. Kresse, G. & Hafner, J. Ab-initio molecular-dynamics for open-shell transition-metals. *Phys. Rev. B* **48,** 13115–13118 (1993).

45. Perdew, J. P., Burke, K. & Ernzerhof, M. Generalized gradient approximation made simple. *Phys. Rev. Lett.* **77,** 3865–3868 (1996).

46. Vankoningsveld, H., Jansen, J. C. & Vanbekkum, H. The monoclinic framework structure of zeolite H-ZSM-5. Comparison with the orthorhombic framework of as-synthesized ZSM-5. *Zeolites* **10,** 235–242 (1990).

47. Pidko, E. A., van Santen, R. A. & Hensen, E. J. M. Multinuclear gallium-oxide cations in high-silica zeolites. *Phys. Chem. Chem. Phys.* **11,** 2893–2902 (2009).

48. Pidko, E. A. & van Santen, R. A. Structure-reactivity relationship for catalytic activity of gallium oxide and sulfide clusters in zeolite. *J. Phys. Chem. C* **113,** 4246–4249 (2009).

49. Mills, G., Jonsson, H. & Schenter, G. K. Reversible work transition-state theory: application to dissociative adsorption of hydrogen. *Surf. Sci.* **324,** 305–337 (1995).

50. Momma, K. & Izumi, F. VESTA 3 for three-dimensional visualization of crystal, volumetric and morphology data. *J. Appl. Crystallogr.* **44,** 1272–1276 (2011).

51. Frisch, M. J. *et al. Gaussian 09, Revision D.01* (Gaussian, Inc., Wallingford, CT, 2009).

52. Stull, D. R. & Prophet, H. in *JANAF Thermochemical Tables* (U.S. National Bureau of Standards, U.S. EPO, 1971).

53. Guhl, H., Miller, W. & Reuter, K. Water adsorption and dissociation on SrTiO$_3$(001) revisited: a density functional theory study. *Phys. Rev. B* **81,** 155455 (2010).

54. Reuter, K. & Scheffler, M. Composition and structure of the RuO$_2$(110) surface in an O$_2$ and CO environment: Implications for the catalytic formation of CO$_2$. *Phys. Rev. B* **68,** 045407 (2003).

55. Newville, M. IFEFFIT: interactive XAFS analysis and FEFF fitting. *J. Synchrotron Rad.* **8,** 322–324 (2001).

56. Ravel, B. & Newville, M. ATHENA, ARTEMIS, HEPHAESTUS: data analysis for X-ray absorption spectroscopy using IFEFFIT. *J. Synchrotron Rad.* **12,** 537–541 (2005).

57. Brown, G. M. & Chidamba, R. Dinuclear copper(II) acetate monohydrate: a redetermination of structure by neutron-diffraction analysis. *Acta Crystallogr. B Struct. Sci.* **29,** 2393–2403 (1973).

58. Oswald, H. R., Reller, A., Schmalle, H. W. & Dubler, E. Structure of copper(II) hydroxide, Cu(OH)$_2$. *Acta Crystallogr. C Cryst. Struct. Commun.* **46,** 2279–2284 (1990).

Acknowledgements

The research was partly supported by the Inorganometallic Catalyst Design Center (ICDC), an Energy Frontier Research Center funded by the U.S. Department of Energy, Office of Basic Energy Sciences, Division of Chemical Sciences under Award DE-SC0012702. It was also partly supported by the EU NEXT-GTL (Innovative Catalytic Technologies & Materials for Next Gas to Liquid Processes) project. The XAS measurements were carried out with the support of the Diamond Light Source on Beamline B18 under Proposal SP8508. SurfSARA and NWO (The Netherlands Organisation for Scientific Research) are acknowledged for providing access to supercomputer resources. We also thank Gary L. Haller for fruitful discussions.

Author contributions

S.G. and M.A.C.M. performed experiments. G.L. performed the DFT calculations. S.G., M.T., E.A.P., E.J.M.H., A.J., M.S.-S. and J.A.L. contributed to experimental design, data analysis and manuscript preparation.

Additional information

3

Nano-socketed nickel particles with enhanced coking resistance grown *in situ* by redox exsolution

Dragos Neagu[1,*], Tae-Sik Oh[2,*], David N. Miller[1], Hervé Ménard[3], Syed M. Bukhari[1], Stephen R. Gamble[1], Raymond J. Gorte[2], John M. Vohs[2] & John T.S. Irvine[1]

Metal particles supported on oxide surfaces are used as catalysts for a wide variety of processes in the chemical and energy conversion industries. For catalytic applications, metal particles are generally formed on an oxide support by physical or chemical deposition, or less commonly by exsolution from it. Although fundamentally different, both methods might be assumed to produce morphologically and functionally similar particles. Here we show that unlike nickel particles deposited on perovskite oxides, exsolved analogues are socketed into the parent perovskite, leading to enhanced stability and a significant decrease in the propensity for hydrocarbon coking, indicative of a stronger metal–oxide interface. In addition, we reveal key surface effects and defect interactions critical for future design of exsolution-based perovskite materials for catalytic and other functionalities. This study provides a new dimension for tailoring particle–substrate interactions in the context of increasing interest for emergent interfacial phenomena.

[1]School of Chemistry, University of St Andrews, St Andrews, KY16 9ST Scotland, UK. [2]Department of Chemical and Biomolecular Engineering, University of Pennsylvania, Philadelphia, Pennsylvania 19104, USA. [3]Sasol Technology (UK) Ltd., St Andrews, KY16 9ST Scotland, UK. *These authors contributed equally to this work. Correspondence and requests for materials should be addressed to D.N. (email: dn67@st-andrews.ac.uk) or to J.M.V. (email: vohs@seas.upenn.edu) or to J.T.S.I. (email: jtsi@st-andrews.ac.uk).

Several particle characteristics, including size and morphology, but most importantly interaction with the oxide support determine the activity, selectivity and stability of supported metal catalysts; thus, controlling these aspects is essential for both fundamental and applicative reasons[1-3]. The vast majority of supported particles are prepared by deposition methods (for example, infiltration, Supplementary Fig. 1a), which although widely applicable, provide limited control over particle interaction with the support, during deposition and over time[4,5]. This leads to deactivation by agglomeration[5] or by coking (carbon accumulation on the metal in hydrocarbon environment) in industrially critical processes such as syngas production by methane steam reforming[6,7]. Several post-particle growth procedures have been developed to delay agglomeration, by partly embedding or fully encapsulating the particles in thin oxide layers[8,9], while coking may be diminished by mild conditioning or alloying[10], although these intricate solutions may be temporary or compromise activity.

Previous studies demonstrated that catalytically active transition metals can be substituted on the B-site of perovskite oxides (ABO_3), in oxidizing conditions, and released (exsolved) on the surface as metal particles following reduction (Supplementary Fig. 1b), with applications in catalysis ranging from automotive emission control to solid oxide fuel/electrolysis cells[11-17]. Interestingly, several reports find exsolved particles to be more resilient to agglomeration and coking as compared to deposited analogues, although the origin of this stability is unclear[11,18].

Here we reveal that this stability is due to exsolved particles being partially embedded in the surface of a parent perovskite and thus exsolution may be regarded as an elegant one-step environmentally friendly method to grow pinned, coking-resistant, socketed particles. We also provide critical insights into surface effects and defect interactions relevant for the future development of exsolution process but also for perovskite bulk or surface related applications.

Results

Surface effects controlling exsolution. In this work, we employ compositions derived from $SrTiO_3$, an archetype oxide of considerable interest for applications ranging from solid oxide fuel cells to complex oxide electronics[19-22]. We introduce A-site deficiency, $La_xSr_{1-3x/2}TiO_3$ ($x = 0.4$), which we have shown to promote B-site redox exsolution[14,17]. We use modest Ni^{2+} doping levels, $La_{0.4}Sr_{0.4}Ni_yTi_{1-y}O_{3-y}$ ($y = 0.03$) or $La_{0.4+2x}Sr_{0.4-2x}Ni_xTi_{1-x}O_3$ ($x = 0.06$), to improve relevance to other systems where low loadings are desirable due to cost and/or low solubility in the host lattice (for example, noble metals).

To form the oxide supports with given stoichiometry into microstructures relevant for applications (usually in flat dense, porous or powder form), the oxides have to be exposed to synthetic- or processing-specific conditions that may cause the surface stoichiometry to deviate as compared with the nominal bulk (that is, 'native surface'). For example, for porous $La_{0.4}Sr_{0.4}Ni_{0.03}Ti_{0.97}O_{3-\gamma}$, while the bulk displayed a quasi-cubic perovskite structure (Supplementary Fig. 2) with an A/B ratio close to the nominal value of 0.8 ($La_{0.4}Sr_{0.4}Ni_{0.03}Ti_{0.97}O_{3-\gamma} \equiv$ '$A_{0.8}BO_{3-\gamma}$', see Fig. 1c), the native surface showed a high A/B value of ~0.98 (Fig. 1a). Perovskite oxides are known to develop A-site overstoichiometric surfaces ('$A_{1+\alpha}BO_{3+\gamma}$'), probably as a result of cation size mismatch and charge compensation[23], with detrimental effects on certain catalytic processes, such as the oxygen reduction reaction[24]. Surface A-cation enrichment is also expected to hinder or supress exsolution since we have found previously that A-site stoichiometric formulations are less prone to exsolve from the B-site as compared with A-site-deficient

ones[14]. The surface excess is thought to be accommodated as perovskite-type Ruddlesden-Popper structures or AO islands[25,26]. Here the A/B ratio is just below unity, suggesting the bulk perovskite structure persists up to the surface, consistent with transmission electron microscopy (TEM) observations, but the surface itself has higher A-site occupancy as compared with the bulk (Supplementary Fig. 4). This implies that surface A-site vacancies are unfavourable and will be naturally filled given sufficient temperature, but also that A-site-deficient formulations may be used to minimize surface A-site cation segregation where this is regarded to be detrimental.

An important feature is that several perovskites, and possibly other oxides used as supports, may develop faceted surfaces, as exemplified in Fig. 1a, inset, suggesting that surfaces can be spatially inhomogeneous as well. This is revealed unquestionably through reduction, which triggers particle growth preferentially on certain facets (Fig. 1b). By contrast, a surface of nominal stoichiometry, which can be exposed by polishing away the native surface or by cleaving through the bulk, grows particles much more uniformly (Fig. 1d; Supplementary Fig. 5). Atomic force microscopy (AFM) surface reconstruction carried out on a representative faceted surface with exsolved particles, allowed us to measure the angles between different facets (Supplementary Fig. 6), and show that they correspond to different perovskite orientations, as highlighted in Fig. 1e. On the basis of this, TEM of near surface region (Supplementary Fig. 4), and the surface composition derived from X-ray photoelectron spectroscopy (XPS), the facets with smooth appearance in Fig. 1a,b, corresponds to A-site terminated (100) and (111) orientations, while the ones with rough morphology correspond to ABO^{4+}/O_2^{4-} terminated (110) orientations, as illustrated in Fig. 1f. On the basis of our previous observations, exsolution occurs when the oxygen vacancy concentration induced through reduction reaches a sufficiently high concentration (δ_{lim}) in the presence of A-site vacancies (α) such that the perovskite lattice becomes destabilized by high deficiency on two of its three primitive sites and will spontaneously exsolve from the B-site to re-establish stoichiometry[14]:

$$A_{1-\alpha}BO_{3-\delta_{lim}} \xrightarrow{\text{Exsolution}} (1-\alpha)ABO_{3-\delta} + \alpha B. \quad (1)$$

Therefore, the systematic presence of particles on (110) terminations is unlikely to be incidental, and may be related to the fact that this is the only orientation in this quasi-cubic crystal in which all primitive perovskite sites holding the key defects required for exsolution are coplanar (Figs 1f,2) and thus potentially in favourable proximity to nucleate particles, on a surface that is overall quasi-stoichiometric and thus not prone to exsolve (Fig. 1b). Nucleation on (110) surfaces may also be facilitated by their rough morphology as compared with (100) and (111) (see Fig. 1a,b), since the nucleation barrier is generally lowered by the presence of crystal defects.

Insights into bulk processes from surface effects. It should also be noted that the preferential exsolution of particles on (110) terminations could also represent a reflection of bulk processes. According to De Souza et al.[27], B-cation migration in perovskites is likely to occur between adjacent B-sites along a curved trajectory in the (110) planes, and is substantially facilitated by the presence of A-site vacancies due to lowering of migration repulsions. Figure 2a illustrates a view of the perovskite lattice in the direction of migration, down the cubic [001] direction (additional projections are given in Supplementary Fig. 7). Thus, it seems that (110) planes are particularly suitable for B-site cation diffusion and the abundance of A-site vacancies in our systems may well facilitate the process, supplying exsolvable

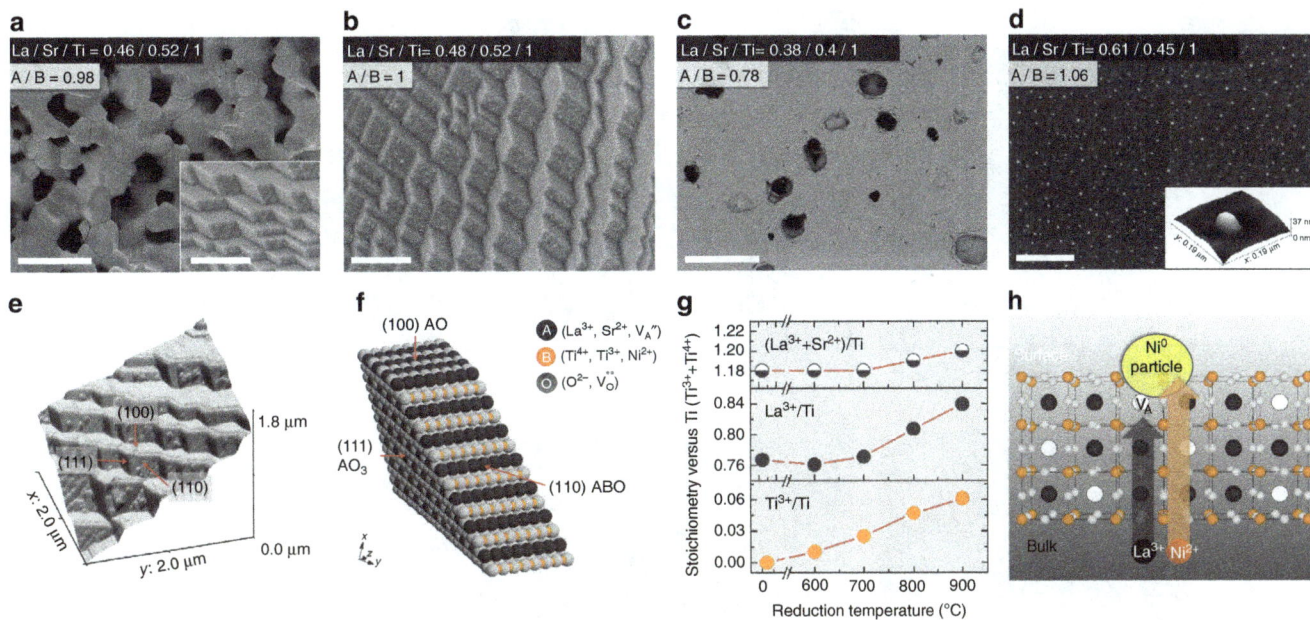

Figure 1 | Surface effects controlling exsolution. SEM micrographs of the native surface of a porous $La_{0.4}Sr_{0.4}Ni_{0.03}Ti_{0.97}O_{3-\gamma}$ sample (**a**) before (scale bars, 50 µm (overview); 1 µm (detail)) and (**b**) after reduction (5% H_2/Ar, 930 °C, 20 h); scale bar, 1 µm. SEM micrographs of the polished surface of a 94% dense $La_{0.4}Sr_{0.4}Ni_{0.03}Ti_{0.97}O_{3-\gamma}$ pellet (**c**) before (scale bar, 50 µm) and (**d**) after reduction (5% H_2/Ar, 900 °C, 20 h); the inset depicts a three-dimensional (3D) AFM image of a particle; scale bar, 1 µm. (**e**) 3D AFM reconstruction of a native surface similar to **a** and **b** highlighting the calculated orientations of the facets (see Supplementary Fig. 6 for details); (**f**) atomic scale model highlighting the orientation and probable termination layers of the terraces found in samples **a**, **b** and **e**. (**g**) Surface composition versus reduction temperature by *in situ* XPS, carried out on a sample with nominal composition $La_{0.52}Sr_{0.28}Ni_{0.06}Ti_{0.94}O_3$, in 5% H_2/Ar, using 0.5 h isotherms (see Supplementary Fig. 8 and Supplementary Table 1 for the corresponding XPS spectra and analysis, respectively). (**h**) Schematic of the key processes occurring during the reduction of an A-site-deficient surface such as **c**, highlighting that Ni^{2+} and La^{3+} diffuse in parallel from the bulk to the surface, forming Ni particles, and filling available A-site vacancies, respectively. The ratios in (**a**–**d**) indicate surface (2–10 nm) stoichiometry from XPS (error ± 0.01 versus Ti; the corresponding spectra is given in Supplementary Fig. 3).

species to the surface and providing additional reasons as to why A-site-deficient perovskites were found to be much more effective towards exsolution as compared with their stoichiometric analogues[14].

Further insight into the diffusion processes occurring during exsolution may be extracted from the XPS data in Fig. 1. While native surfaces are largely unchanged following exsolution reflecting their stable, yet rigid nature (Fig. 1a versus b), surfaces of nominal (A-site deficient) stoichiometry show an enrichment of A-site cations, in particular La^{3+} (Fig. 1c versus d), perhaps shedding light on the transport processes occurring inside the bulk. Similar cation transport to the surface was observed by *in situ* XPS (Fig. 1g; Supplementary Fig. 8; Supplementary Table 1), which shows that as the reduction temperature increased, the perovskite was reduced to higher extents (indicated by an increase in Ti^{3+}), and in parallel the A/B and La/Ti ratios also increased. The exsolution of Ni particles was observed by XPS and scanning electron microscope (SEM), but quantification was not included in Fig. 1g, due to significant errors associated with low Ni doping level and overlap with the La peak. Thus, during exsolution on bulk-like surfaces nickel diffuses from the bulk to the surface to form particles, while in parallel La^{3+} (and Sr^{2+} to less extent) gradually fills surface A-site vacancies (Fig. 1h). The filling of surface vacancies, eventually leading to a newly formed 'rigid' 'native surface', is expected to gradually limit lanthanum and nickel diffusion from the bulk to the surface, and thus exsolution, and may possibly contribute to locking the particles into place.

By mapping Ni particles in Fig. 1d to estimate the total amount exsolved, it becomes apparent that the Ni must have diffused

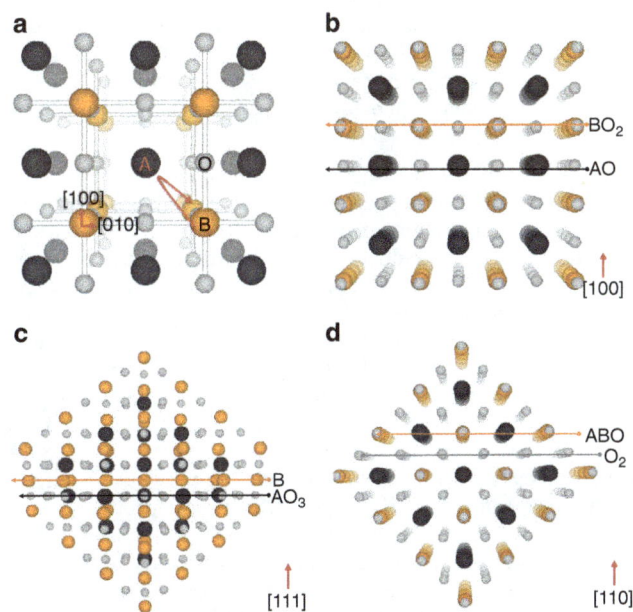

Figure 2 | Possible orientations and termination layers for a cubic perovskite crystal ABO₃. Depending on the direction from which it is viewed, the perovskite crystal structure (**a**) may be described as alternating planes of (**b**) AO and BO_2 in the [100] direction; (**c**) B and AO_3 planes in the [111] direction; (**d**) ABO and O_2 planes in the [110] direction. The curved arrow in **a** illustrates the diffusion trajectory of B-site cations down the [001] direction, according to De Souza *et al*[27].

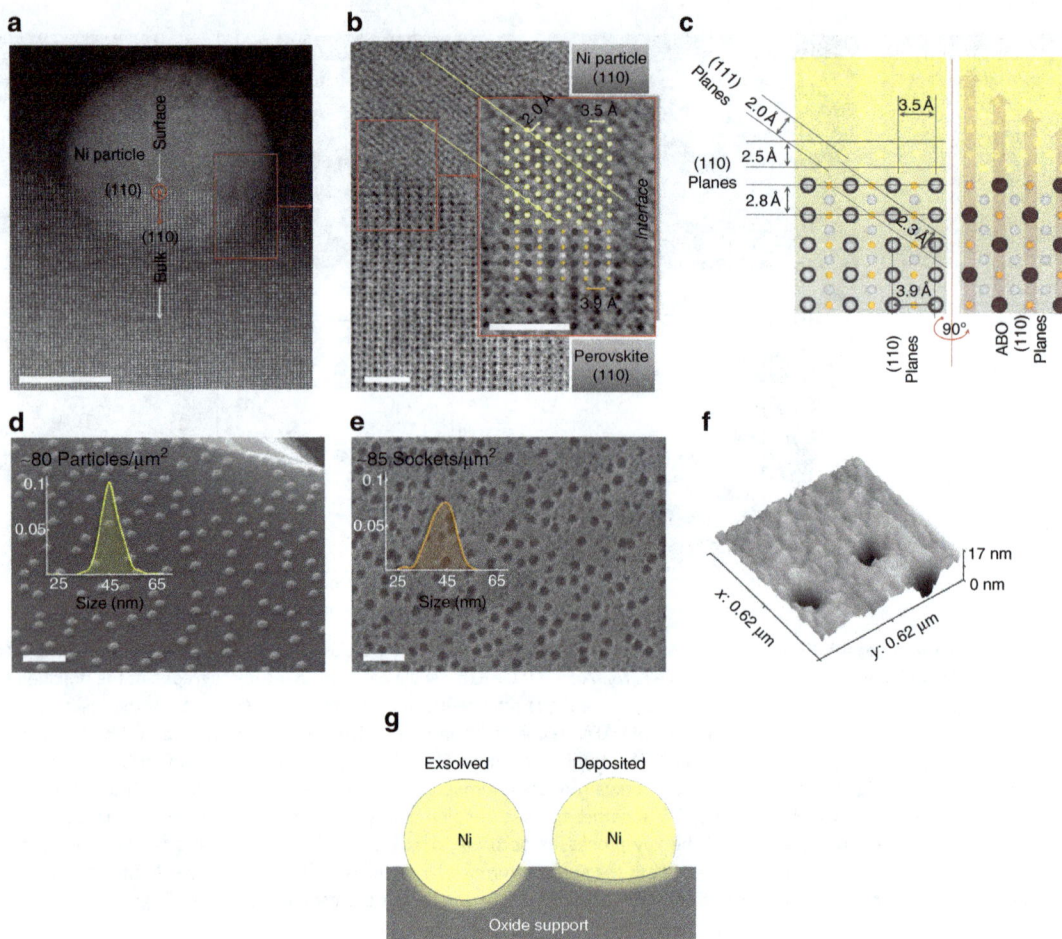

Figure 3 | Exsolved particle–substrate interface. (a) TEM micrograph (dark field) of a Ni particle exsolved on (110) native surface facet (see Fig. 1b) after ageing ($\sim$3% H_2O/5% H_2/Ar, 930 °C, 60 h); scale bar, 10 nm. **(b)** TEM micrograph detail (bright field) of the metal–perovskite interface highlighting the corresponding atomic planes and orientations; scale bars, 1 nm. **(c)** Schematic atomic model of the metal–perovskite orientation relationship based on **b**. SEM micrographs of exsolved Ni particles ($La_{0.52}Sr_{0.28}Ni_{0.06}Ti_{0.94}O_3$, 5% H_2/Ar, 920 °C, 12 h) **(d)** before and **(e)** after etching particles in HNO_3; the insets show the size histogram of the particles and sockets, respectively, as determined through image analysis; scale bars, 200 nm. **(f)** Three-dimensional AFM of sockets similar to those in **e**. **(g)** Schematic illustration of the particle–substrate interface for deposited and exsolved nickel particles.

from $\sim$100 nm deep in the bulk, confirming that B-site transport to the surface is critical for exsolution. Moreover, this Ni amount is in similar quantity to the La^{3+} that enriched the surface during reduction, as determined by XPS (see Supplementary Fig. 9 and Supplementary Note 1 for particle analysis and quantification, respectively), suggesting that B-site and A-site diffusion are probably correlated, possibly as tungsten bronze ephemeral units (Supplementary Fig. 10). Notably, the different mobility of A-site cations during reduction shows that their nature may directly influence B-site diffusion and hence exsolution, providing another dimension to tailor it. Indeed, La facilitating exsolution is consistent with our previous findings by which compositions with high La/Sr ratios produced more numerous and better distributed particles[14], which may provide a conceptual approach for overcoming the rigidity of the native surface in exsolution, as exemplified in Supplementary Fig. 11 for $La_{0.46}Sr_{0.34}Ni_{0.03}Ti_{0.97}O_3$, which displayed improved native surface particle coverage compared with $La_{0.4}Sr_{0.4}Ni_{0.03}Ti_{0.97}O_{3-\gamma}$ (Fig. 1b).

Particle–substrate interface. TEM imaging of the particles exsolved on (110) terminations revealed that when the bulk is also aligned to [110], this corresponds to a profile view of the

interface, the particle consists of a slightly oblate spheroid, $\sim$30% submerged into the parent oxide surface (Fig. 3a; Supplementary Fig. 12). The fact that particles are indeed universally socketed in the surface was confirmed by etching them in concentrated HNO_3, which left behind pits with similar number density and size distribution as the particles themselves (Fig. 3d,e). An AFM detail of an etched surface is shown in Fig. 3f demonstrating they are indeed embedded to a considerable depth in the parent oxide support. Figure 3a,b shows that the bulk 'cubic' perovskite structure appears to be retained all the way to the surface and to the metal–oxide interface, which is smooth and continuous. Particles are metallic in nature and epitaxial with respect to the parent oxide as determined by electron diffraction (Supplementary Fig. 13). The epitaxic relationship is highlighted in Fig. 3b which depicts the perovskite (111) planes aligned to the (111) planes of the Ni lattice. A more detailed analysis of the proposed (ideal) orientation relationship is given in Fig. 3c, although it should be noted that dislocations are likely to occur in the metal lattice (possibly visible in Fig. 3b) to accommodate the lattice mismatch (the cell parameters for the perovskite and nickel lattices are $\sim$3.9 and $\sim$3.5 Å, respectively). In addition, the fact the metal lattice is growing from the oxide lattice might naturally facilitate interdiffusion between the two, which has been shown to significantly increase adhesion between metal and oxide phases

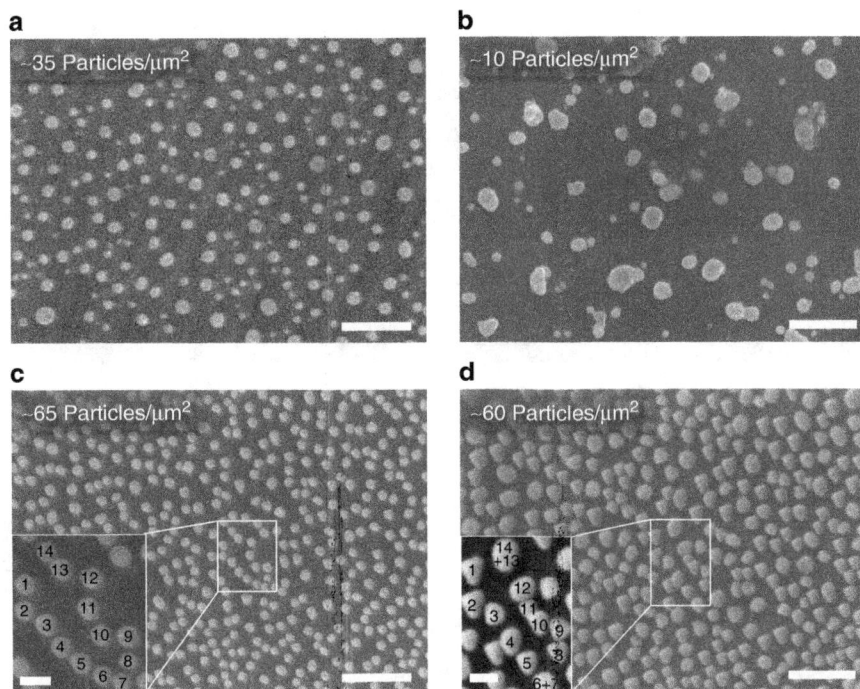

Figure 4 | Thermal stability of deposited and exsolved Ni particles. SEM micrographs of vapour-deposited Ni particles on $La_{0.4}Sr_{0.4}TiO_3$ (**a**) before and (**b**) after ageing (H_2, 650 °C, 24 h and 800 °C, 6 h). SEM micrographs of Ni particles exsolved from cleaved bulk surface of $La_{0.52}Sr_{0.28}Ni_{0.06}Ti_{0.94}O_3$ (5% H_2/Ar, 900 °C, 12 h) (**c**) before and (**d**) after ageing (5% H_2/Ar, 900 °C, 70 h). Scale bars, 500 nm (overview), 100 nm (detail).

even when occurring over 1–2 unit cell thin interfaces[28]. Overall, these features appear to contribute synergistically towards considerably improving the anchorage of exsolved particles, explaining both our current observations and those described previously[18]. By contrast, little to no particle embedding could be observed by AFM in samples prepared by conventional deposition on similar A-site-deficient perovskites (see Supplementary Fig. 14), indicating a much smaller degree of interaction occurred during growth between the deposited metal and oxide phases.

Implications of particle–substrate interactions. The profound morphological differences that distinguish exsolved particles from the deposited analogues, highlighted above and summarized in Fig. 3g, are clearly reflected by their stability and functionality. Exsolved Ni particles display considerably lower tendency to agglomerate and coke as opposed to deposited ones, as illustrated below for various scenarios. Figure 4 compares the thermal stability of deposited and exsolved particles of similar size showing that while the former coalesce rapidly at 800 °C or below (Fig. 4a,b), the latter are reasonably stable over tens of hours at 900 °C, in spite of having almost double initial particle loading (Fig. 4d). Examination of Fig. 4b,d reveals that the size of exsolved particles does increase over time possibly because of additional exsolution from the bulk and to less extent due to coalescence since ~90% of the particles are preserved throughout the ageing test. Coalescence seems to occur predominantly when particles were initially tangent or in close proximity to each other (see inset of Fig. 4d). Nonetheless, exsolved particles generally behave as if pinned to their original location, showing a level of stability beyond metal–oxide interfaces produced through conventional means on a similar A-site-deficient perovskite.

The second consequence of particle–substrate interaction studied here is in relation to the tendency of Ni to grow carbon fibres in a hydrocarbon environment, which is detrimental in

various applications, as explained before. Ni metal particles are well-known to catalyse carbon fibre formation in a wide range of particle sizes and hydrocarbon-containing environments. To test coking stability, samples were exposed to 20% CH_4/H_2, at 800 °C, for 4 h. Consistent with previous reports, relatively small Ni particles (~20 nm) prepared by infiltration on $La_{0.4}Sr_{0.4}TiO_3$ (Supplementary Fig. 15a) coked severely, showing well developed fibres, as illustrated in Fig. 5a (refs 6,7). Similarly, Ni particles prepared by vapour deposition on $La_{0.4}Sr_{0.4}TiO_3$ with particle size within 30–100 nm range (see Supplementary Fig. 15b for initial microstructure and size distribution) also produced large amounts of micron-long carbon fibres, as shown in Fig. 5b. Notably, exsolved Ni metal particles of comparable sizes, ~25, 60 and 80 nm, displayed considerably less carbon fibre growth, as illustrated in Fig. 5d,g and Supplementary Fig. 15d, respectively. Carbon fibre growth on Ni particles has been generally shown to occur in a characteristic manner, by a so-called 'tip-growth'[7] mechanism illustrated in Fig. 5c (left). In this mechanism, exhibited by the deposited Ni particles (see Fig. 5a,b), carbon initially dissolves into the Ni lattice, while the fibre grows at the metal particle–oxide support interface, resulting in particle uplifting from its original location. Most likely, the remarkable decrease in the tendency to grow carbon fibres observed in the exsolved Ni systems is due to the strong interaction between the exsolved socketed particle and the parent oxide support which prevents particle uplifting and subsequent fibre growth. Limited carbon fibre formation may also be found occasionally in the exsolved Ni samples (see Fig. 5d–h), although it should be noted that in this case the fibres were considerably shorter than those found in the deposited Ni samples (compare Fig. 5b with Fig. 5g,h). It should also be noted that for the ~25 nm exsolved Ni particles most of the short carbon fibres that form appear to be laying on the oxide support (see Fig. 5e) rather than standing perpendicular on it as expected for a tip-growth mechanism. This type of carbon fibre growth is reminiscent of a so-called 'base growth' mechanism where the particle remains attached to the

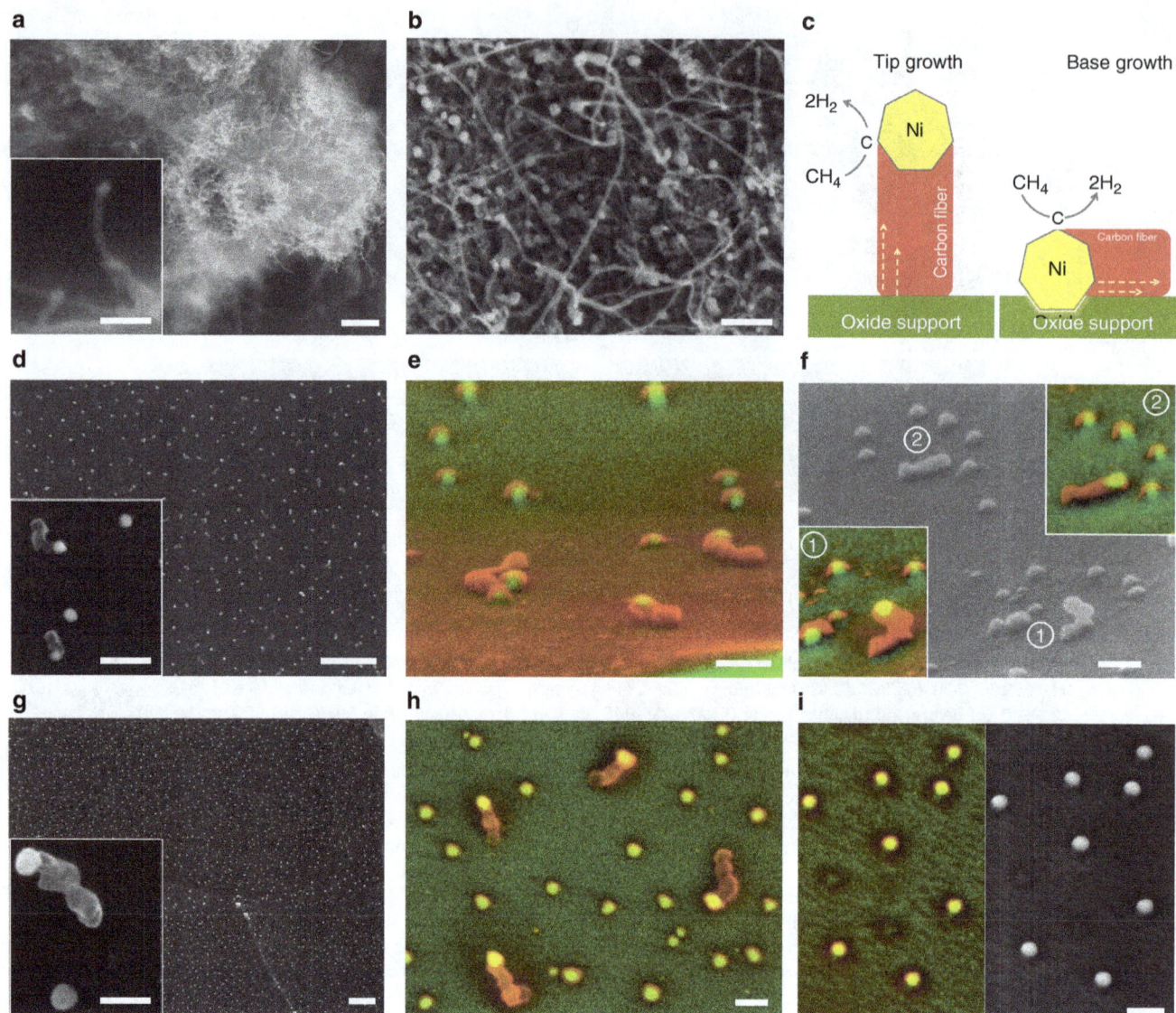

Figure 5 | Emergent anti-coking trait of exsolved particles. (**a**) Approximately 20 nm Ni particles formed by infiltration on $La_{0.4}Sr_{0.4}TiO_3$ (see Supplementary Fig. 15a for initial microstructure) after coking test, showing significant carbon fibre growth; scale bars, 1 μm (overview); 100 nm (detail). (**b**) Ni particles (30–100 nm) prepared by vapour deposition on $La_{0.4}Sr_{0.4}TiO_3$ (see Supplementary Fig. 15b for initial microstructure) showing considerable carbon fibre growth; scale bar, 0.5 μm. (**c**) Schematic of possible carbon fibre growth mechanisms based on refs 6, 7, 29. (**d**) Approximately 25 nm Ni particles formed by exsolution from $La_{0.52}Sr_{0.28}Ni_{0.06}Ti_{0.94}O_3$ (5% H_2, 880 °C, 6 h), after coking test, showing limited carbon fibre growth; scale bars, 0.5 μm (overview); 100 nm (detail). (**e**) False colour micrograph depicting a side view detail of sample (**d**); scale bar, 100 nm. (**f**) Side view micrograph and false colour micrograph detail insets of different region in sample **d**; scale bar, 100 nm. (**g**) Approximately 60 nm Ni particles formed by exsolution from $La_{0.52}Sr_{0.28}Ni_{0.06}Ti_{0.94}O_3$ (5% H_2/Ar, 1,000 °C, 6 h) after coking showing limited coking; scale bars, 1 μm (overview); 100 nm (detail). (**h**) False colour top view micrograph of sample **g**; scale bar, 100 nm. (**i**) False colour and corresponding secondary electron micrographs of a different region in sample **g** showing particles alongside empty sockets; scale bar, 100 nm. In all cases, the coking test was carried out in 20% CH_4/H_2, at 800 °C, for 4 h. In the false colour micrographs, green was used for the perovskite, red for carbon and yellow for Ni metal.

substrate while the fibre grows on top of it (see Fig. 5c, right). Interestingly, base growth is generally thought to occur particularly when particles adhere strongly to the support[29], which seems to provide additional proof towards the argument that exsolved particles possess superior adhesion to the parent substrate, which in turn prevents particle uplifting. Nonetheless, close examination of this sample reveals that particle uplifting may still occur occasionally (area 1 in Fig. 5f) alongside seemingly base growth (area 2 in Fig. 5f). Similar behaviour was found for the ~60 nm exsolved Ni particle system, which also displayed remarkably low extent of coking with limited formation of

relatively short carbon fibres, as shown in Fig. 5g,h. Interestingly, in this sample, a few areas displaying empty sockets alongside metal particles were also observed (see Fig. 5i), indicating that probably for this particle size range carbon growth by particle uplifting may start to dominate. Carbon growth by particle uplifting seems to occur exclusively for the ~80 nm exsolved particle also (Supplementary Fig. 15d), perhaps indicating that the interaction and adhesion between exsolved and parent oxide diminish with increasing particle size.

It is important to highlight that this improved coking resistance does not come at the cost of reduced activity for

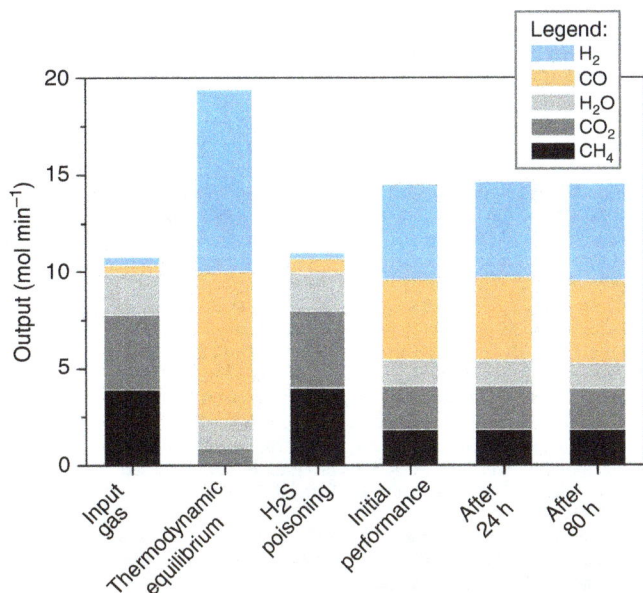

Figure 6 | Reforming test on a La$_{0.52}$Ca$_{0.28}$Ni$_{0.06}$Ti$_{0.94}$O$_3$ perovskite powder with exsolved Ni particles. The powder ($\sim$1 m^2 g^{-1} total surface area) was reduced in the testing setup at 900 °C, for 4 h in 5% H$_2$/Ar before the reforming test, to exsolve Ni particles on the surface. The reforming test was carried out at 900 °C, and the input gas consisted of: 4% H$_2$, 4% CO, 36% CH$_4$, 36% CO$_2$ and 20% H$_2$O. The performance of the catalyst following poisoning with 4 p.p.m. of H$_2$S is shown to illustrate reforming does not occur in absence of an adequate catalyst.

desirable reactions, since these materials show stable catalytic activity throughout long-term reforming test (see Fig. 6) during which deposited Ni particles coke significantly. Indeed, the Ni-based catalyst prepared by exsolution exhibits a level of activity and sensitivity to H$_2$S comparable to deposited Ni analogues prepared by conventional deposition methods, but with considerably improved coking resistance[30].

Discussion

In this work, we reveal the unique interface developing between the exsolved particles and parent support that leads to at least two emergent effects of great utility: particle pinning and anti-coking behaviour. Thus, redox exsolution may provide an attractive route for tailoring particle–substrate interactions and their subsequent functionality, beyond the capabilities of conventional deposition methods. In particular, the exsolution process may enable the production of socketed metal particles through a single-step reduction treatment. In addition, redox exsolution could be used to deliver such functionality in a selective manner, at a desired location in intricate devices, otherwise inaccessible to deposition methods, by adjusting initial composition at a localized region such as an electrode/electrolyte interface.

We also illustrate some of the key factors that govern exsolution, which may be used for future tailoring of this process, including surface structuring and transport of B-site cations to the surface while highlighting the possible links between them. In particular, we reveal the unexpected role of A-site cations and A-site vacancies in the bulk transport of B-site cations in the perovskite lattice, and hence towards B-site exsolution. Last, the surface structuring occurring in certain perovskites and described here in detail is critical for understanding and interpreting not only exsolution itself, but also the general surface activity and further functionalization of related oxides. A surface consisting of

multiple facets and orientations (see, for example, Figs 1b,f and 2) would not only exhibit different catalytic activity from facet to facet depending on the local configuration of atoms, but also particles deposited on such an inhomogeneous surface could exhibit different morphology, distribution and anchorage depending on the polarity and composition of the underlying oxide termination.

Methods

Sample preparation and processing. The perovskite oxides were prepared by a modified solid-state synthesis[31]. High purity precursors including La$_2$O$_3$ (Pi-Kem, >99.99%), TiO$_2$ (Alfa Aesar, >99.6%), SrCO$_3$ (Aldrich, >99.9%) and Ni(NO$_3$)$_2$*6H$_2$O (Acros, >99%) were used in the appropriate stoichiometric ratios. Certain oxides and carbonates were dried at different temperatures (TiO$_2$ and SrCO$_3$—300 °C and La$_2$O$_3$—800 °C) and weighed while warm. The mixture, including nickel nitrate, was quantitatively transferred to a beaker and mixed with acetone and $\sim$0.05 wt% Hypermer KD1 dispersant. An ultrasonic Hielscher UP200S probe was used into break down agglomerates and homogenize the mixture into a fine, stable dispersion. The acetone was then evaporated at room temperature under continuous stirring and the content of the beaker was quantitatively transferred to a crucible and calcined at 1,000 °C for 12 h to decompose the carbonates and start forming the perovskite phase. The calcined powder was then pressed into pellets and fired at 1,400 °C for 12 h to form the perovskite phase. Porous ceramics, such as the one illustrated in Fig. 1a, were prepared by firing mixtures of as-prepared oxide in powder form with glassy carbon serving as pore-former[31]. Polishing (for example, Fig. 1c) was carried out on a dense ($\sim$94%, polycrystalline) pellet, with a Metaserv 2000 polisher, using initially MetPrep P1200 polishing paper, followed by cloth polishing with MetPrep 6, 3 and 1 µm diamond paste, respectively. Cleaving was carried out on porous pellets, by fracturing the samples with a pestle in a mortar.

To exsolve particles on the surface reduction was carried out in a controlled atmosphere furnace, under continuous flow of 5% H$_2$/Ar or pure H$_2$, at the temperatures indicated in the main text, with heating and cooling rates of 7 °C min^{-1}.

Vapour-deposited Ni (Fig. 5e) was formed using a thermal evaporator (Kurt J. Lesker PRO Line PVD 75), under base pressure of 1.2×10^{-6} torr and deposition rate of 0.5 Å s^{-1}, followed by annealing in H$_2$, at 800 °C, for 4 h.

Sample in Fig. 5a was prepared by infiltration through the following procedure. La$_{0.4}$Sr$_{0.4}$TiO$_3$ powder was infiltrated with nickel nitrate aqueous solution at room temperature. After heating in air at 450 °C for 15 min to decompose nickel nitrate, the powder mixture was reduced in dry H$_2$ at 500 °C for 5 h. Estimated nickel loading was 2 wt.%.

Sample characterization. The crystal structure of the prepared perovskites was analysed using powder X-ray diffraction. X-ray diffraction patters were collected at room temperature on a PANalytical Empyrean Diffractometer operated in reflection mode. Selected data were analysed and refined using FullProf software to confirm their crystal structure and determine unit cell parameters (Supplementary Fig. 2). The diffraction peaks were fitted using a pseudo-Voigt profile. Refined parameters include: scale factor, background polynomial parameters or linear interpolation between a set of background points of refinable heights, unit cell parameters, peak profile parameters u, v, w and η (Lorentzian/Gaussian distribution), zero shift, atomic positions and site occupancies. A general thermal factor was initially set for the whole pattern and in latter stages refined and eventually converted to atomic isotropic displacement factors for individual atomic sites. The structural information was then used to construct the crystal structure by using Crystal Maker for Windows software (Supplementary Fig. 2b,c).

Nano-E microscope (Pacific Nanotechnology) was used to collect AFM images in tapping mode with silicon tips (Aspire CT300R) from NanoScience Instrument. All AFM image analyses were carried out using Gwyddion software package.

A JEOL JSM-6700 field emission SEM equipped with secondary and backscattered electron detector was used for investigating the surface morphology and phase homogeneity, respectively. All SEM samples were sufficiently conductive; no gold or carbon deposition was required to prevent specimen charging. High-magnification secondary and backscattered electron images were obtained using a FEI Scios electron microscope. Selected micrographs (for example, Fig. 5) were converted to false colour micrographs by colouring the secondary electron image in green, the backscattered analogue in red and then blending them with the help of Mathematica 10 for Windows.

Preparation of a specimen for TEM was carried out using a JEOL JIB-4501 multibeam focused ion beam-scanning electron microscope system. The nickel particles and surface were preserved by the deposition of a protective carbon layer, before milling and thinning using the gallium focused ion beam. High-resolution scanning transmission electron microscopy was carried out on a JEOL ARM 200F. Further electron diffraction characterization was carried out on a JEOL JEM-2010 TEM.

XPS was carried out on a Kratos Axis Ultra-DLD photoelectron spectrometer equipped with an Al monochromatic X-ray source, and the data were analysed using CasaXPS software. The spectra were calibrated based on the C $1s$ peak from adventitious carbon. Quantification was performed based on the area of peaks of interest (La $3d_{5/2}$, Ti $2p_{3/2,1/2}$, Sr $3d_{5/2,3/2}$) after a linear type background subtraction, and the main results summarized in Fig. 1g and Supplementary Table 1. Initially, additional peaks were used for quantification for each chemical species to ensure that the calculated stoichiometry does not depend significantly on different core levels. Ti^{3+} quantification was carried out by fitting the Ti $2p_{3/2}$ peak with individual components.

Particles on flat oxide surfaces were analysed using ImageJ software according to the following procedure. SEM images of adequate magnifications were selected and their contrast and sharpness slightly increased. These were then imported into ImageJ software where particles were outlined based on contrast and distances (in pixels) were calibrated based on the corresponding SEM image scale (in nm). Built-in ImageJ functions were then used to calculate the area of particles based on the number of pixels contained within their set boundaries. This area was assumed to correspond to the average area of the equatorial circle of the spheroid-like particles and thus was used to estimate their average size and further particle size distribution.

Coking and reforming experiments. The coking test was carried out at 800 °C by flowing 20% CH_4/H_2 without humidification. The two gases were mixed right before the reactor inlet.

For the reforming test, the catalyst bed was 3 mm deep, and had an internal diameter of 20 mm. Gas hourly space velocity was $19,000\,h^{-1}$. The gas contained $16.5\,ml\,min^{-1}$ Ar as an internal standard, while the total input gas flow was $300\,ml\,min^{-1}$.

References

1. Farmer, J. A. & Campbell, C. T. Ceria Maintains Smaller Metal Catalyst Particles by Strong Metal-Support Bonding. *Science* **329**, 933–936 (2010).
2. Yamada, Y. *et al.* Nanocrystal bilayer for tandem catalysis. *Nat. Chem.* **3**, 372–376 (2011).
3. Cargnello, M. *et al.* Control of Metal Nanocrystal Size Reveals Metal-Support Interface Role for Ceria Catalysts. *Science* **341**, 771–773 (2013).
4. Wallace, W. T., Min, B. K. & Goodman, D. W. The nucleation, growth, and stability of oxide-supported metal clusters. *Top. Catal.* **34**, 17–30 (2005).
5. Jiang, S. P. Nanoscale and nano-structured electrodes of solid oxide fuel cells by infiltration: Advances and challenges. *Int. J. Hydrogen Energ.* **37**, 449–470 (2012).
6. Helveg, S. *et al.* Atomic-scale imaging of carbon nanofibre growth. *Nature* **427**, 426–429 (2004).
7. Abild-Pedersen, F., Nørskov, J. K., Rostrup-Nielsen, J. R., Sehested, J. & Helveg, S. Mechanisms for catalytic carbon nanofiber growth studied by ab initio density functional theory calculations. *Phys. Rev. B* **73**, 115419 (2006).
8. De Rogatis, L. *et al.* Embedded phases: a way to active and stable catalysts. *ChemSusChem* **3**, 24–42 (2010).
9. Lu, J. *et al.* Coking- and sintering-resistant palladium catalysts achieved through atomic layer deposition. *Science* **335**, 1205–1208 (2012).
10. Rostrup-Nielsen, J. R. & Alstrup, I. Innovation and science in the process industry: Steam reforming and hydrogenolysis. *Catal. Today* **53**, 311–316 (1999).
11. Shiozaki, R. *et al.* Partial oxidation of methane over a Ni/BaTiO3 catalyst prepared by solid phasecrystallization. *J. Chem. Soc. Faraday Trans.* **93**, 3235–3242 (1997).
12. Nishihata, Y. *et al.* Self-regeneration of a Pd-perovskite catalyst for automotive emissions control. *Nature* **418**, 164–167 (2002).
13. Madsen, B. D., Kobsiriphat, W., Wang, Y., Marks, L. D. & Barnett, S. SOFC anode performance enhancement through precipitation of nanoscale catalysts. *ECS Trans.* **7**, 1339–1348 (2007).
14. Neagu, D., Tsekouras, G., Miller, D. N., Ménard, H. & Irvine, J. T. S. In situ growth of nanoparticles through control of non-stoichiometry. *Nat. Chem.* **5**, 916–923 (2013).
15. Jardiel, T. *et al.* New SOFC electrode materials: The Ni-substituted LSCM-based compounds ($La_{0.75}Sr_{0.25}$)($Cr_{0.5}Mn_{0.5-x}Ni_x$)$O_{3-\delta}$ and ($La_{0.75}Sr_{0.25}$)($Cr_{0.5-x}Ni_xMn_{0.5}$)$O_{3-\delta}$. *Solid State Ionics* **181**, 894–901 (2010).
16. Katz, M. B. *et al.* Reversible precipitation/dissolution of precious-metal clusters in perovskite-based catalyst materials: Bulk versus surface re-dispersion. *J. Catal.* **293**, 145–148 (2012).
17. Tsekouras, G., Neagu, D. & Irvine, J. T. S. Step-change in high temperature steam electrolysis performance of perovskite oxide cathodes with exsolution of B-site dopants. *Energy Environ. Sci.* **6**, 256–266 (2013).
18. Kobsiriphat, W., Madsen, B. D., Wang, Y., Marks, L. D. & Barnett, S. A. $La_{0.8}Sr_{0.2}Cr_{1-x}Ru_xO_{3-\delta}$–$Gd_{0.1}Ce_{0.9}O_{1.95}$ solid oxide fuel cell anodes: Ru precipitation and electrochemical performance. *Solid State Ionics* **180**, 257–264 (2009).
19. Ohtomo, A. & Hwang, H. Y. A high-mobility electron gas at the $LaAlO_3/SrTiO_3$ heterointerface. *Nature* **427**, 423–426 (2004).
20. Hwang, H. Y. *et al.* Emergent phenomena at oxide interfaces. *Nat. Mater.* **11**, 103–113 (2012).
21. Szot, K., Speier, W., Bihlmayer, G. & Waser, R. Switching the electrical resistance of individual dislocations in single-crystalline $SrTiO_3$. *Nat. Mater.* **5**, 312–320 (2006).
22. Ruiz-Morales, J. C. *et al.* Disruption of extended defects in solid oxide fuel cell anodes for methane oxidation. *Nature* **439**, 568–571 (2006).
23. Lee, W., Han, J. W., Chen, Y., Cai, Z. & Yildiz, B. Cation size mismatch and charge interactions drive dopant segregation at the surfaces of manganite perovskites. *J. Am. Chem. Soc.* **135**, 7909–7925 (2013).
24. Chen, Y. *et al.* Impact of Sr segregation on the electronic structure and oxygen reduction activity of $SrTi_{1-x}Fe_xO_3$ surfaces. *Energy Environ. Sci.* **5**, 7979–7988 (2012).
25. Szot, K. & Speier, W. Surfaces of reduced and oxidized $SrTiO_3$ from atomic force microscopy. *Phys. Rev. B* **60**, 5909–5926 (1999).
26. Bonnell, D. A. & Garra, J. Scanning probe microscopy of oxide surfaces: atomic structure and properties. *Rep. Prog. Phys.* **71**, 044501 (2008).
27. De Souza, R. A., Islam, M. S. & Ivers-Tiffée, E. Formation and migration of cation defects in the perovskite oxide LaMnO3. *J. Mater. Chem.* **9**, 1621–1627 (1999).
28. Chambers, S. A. *et al.* Ultralow contact resistance at an epitaxial metal/oxide heterojunction through interstitial site doping. *Adv. Mater.* **25**, 4001–4005 (2013).
29. Gohier, A., Ewels, C. P., Minea, T. M. & Djouadi, M. A. Carbon nanotube growth mechanism switches from tip- to base-growth with decreasing catalyst particle size. *Carbon* **46**, 1331–1338 (2008).
30. Bøgild Hansen, J. & Rostrup-Nielsen, J. in *Handbook of Fuel Cells: Fundamentals Technology and Applications* 957–969 (John Wiley & Sons, 2009).
31. Neagu, D. & Irvine, J. T. S. Enhancing electronic conductivity in strontium titanates through correlated A and B-site doping. *Chem. Mater.* **23**, 1607–1617 (2011).

Acknowledgements

D.N. thanks the European Project METSAPP (FCH JU-GA 278257) for funding. We also thank NSF and EPSRC for Materials World Network funding refs EP/J018414/1 and DMR-1210388, EPSRC Platform Grant EP/K015540/1, EPSRC Capital Equipment Grant EP/L017008/1, Royal Society Wolfson Merit Award WRMA 2012/R2. We thank JEOL Japan for TEM sample preparation and imaging and SASOL St Andrews for access to the XPS facility.

Author contributions

D.N. carried out sample preparation (exsolution), SEM, micrograph and XPS analysis. T.-S.O. carried out sample preparation (deposition), AFM, SEM and coking experiments. D.N.M. collected and analysed TEM data and assisted with SEM image collection. H.M. performed XPS and assisted in data analysis. S.M.B. contributed to sample preparation and data acquisition. S.R.G. carried out catalysis tests. R.J.G., J.M.V. and J.T.S.I. coordinated and supervised the project. D.N. drafted the manuscript and all authors commented on it.

Additional information

Accession codes: The research data supporting this publication can be accessed at http://dx.doi.org/10.17630/2CA88024-4A83-4690-B66F-B63B98A052D1,

Competing financial interests: The authors declare no competing financial interests.

4

Microporous metal–organic framework with dual functionalities for highly efficient removal of acetylene from ethylene/acetylene mixtures

Tong-Liang Hu[1,2], Hailong Wang[2], Bin Li[2], Rajamani Krishna[3], Hui Wu[4], Wei Zhou[4], Yunfeng Zhao[5], Yu Han[5], Xue Wang[6], Weidong Zhu[6], Zizhu Yao[7], Shengchang Xiang[7] & Banglin Chen[2]

The removal of acetylene from ethylene/acetylene mixtures containing 1% acetylene is a technologically very important, but highly challenging task. Current removal approaches include the partial hydrogenation over a noble metal catalyst and the solvent extraction of cracked olefins, both of which are cost and energy consumptive. Here we report a microporous metal–organic framework in which the suitable pore/cage spaces preferentially take up much more acetylene than ethylene while the functional amine groups on the pore/cage surfaces further enforce their interactions with acetylene molecules, leading to its superior performance for this separation. The single X-ray diffraction studies, temperature dependent gas sorption isotherms, simulated and experimental column breakthrough curves and molecular simulation studies collaboratively support the claim, underlying the potential of this material for the industrial usage of the removal of acetylene from ethylene/acetylene mixtures containing 1% acetylene at room temperature through the cost- and energy-efficient adsorption separation process.

[1] Department of Chemistry, TKL of Metal- and Molecule-Based Material Chemistry, Collaborative Innovation Center of Chemical Science and Engineering (Tianjin), Nankai University, Tianjin 300071, China. [2] Department of Chemistry, University of Texas at San Antonio, One UTSA Circle, San Antonio, Texas 78249-0698, USA. [3] Van 't Hoff Institute for Molecular Sciences, University of Amsterdam, Science Park 904, Amsterdam 1098 XH, The Netherlands. [4] NIST Center for Neutron Research, Gaithersburg, Maryland 20899-6102, USA. [5] Advanced Membranes and Porous Materials Center, Physical Sciences and Engineering Division, King Abdullah University of Science and Technology, Thuwal 23955-6900, Saudi Arabia. [6] Key Laboratory of the Ministry of Education for Advanced Catalysis Materials, Institute of Physical Chemistry, Zhejiang Normal University, Jinhua 321004, China. [7] College of Chemistry and Chemical Engineering, Fujian Provincial Key Laboratory of Polymer Materials, Fujian Normal University, Fuzhou 350007, China. Correspondence and requests for materials should be addressed to B.C. (email: Banglin.Chen@utsa.edu).

Ethylene is one of the most essential raw chemicals and widely used to produce polymers and other useful chemicals[1]. During the production of ethylene through the cracking of ethane, propane and heavier hydrocarbons, a small amount of acetylene as an impurity of about 1% is also generated. It is imperative that acetylene in the ethylene feed should be reduced to an acceptable level because acetylene has a deleterious effect on end products of ethylene: acetylene can cause a catalyst poison during ethylene polymerization and thus significantly affect the quality of the resulting polyethylene. Furthermore, acetylene can form solid metal acetylides, which can block the fluid stream and lead to explosion[2].

Extensive efforts have been pursued to remove acetylene from ethylene/acetylene mixtures[3,4]. In the petrochemical industry, current commercial approaches include partial hydrogenation of acetylene into ethylene over a noble metal catalyst such as a supported Pd catalyst and solvent extraction of cracked olefins using an organic solvent such as DMF and acetone. Both of which have some drawbacks: the former process suffers from the need of noble metal catalyst and the loss of olefins due to the over hydrogenation to paraffins, while the latter wastes a significant amount of solvents. Porous materials through selective adsorption separation of acetylene over ethylene might provide an alternative cost- and energy-efficient approach for this industrially very important while quite challenging task, though this has not been fully explored.

Among diverse porous materials, the emerging microporous metal–organic frameworks (MOFs) are of particular interest and are important for gas separation and thus removal of acetylene from ethylene/acetylene mixtures. This is because the pores within microporous MOFs can be straightforwardly and rationally tuned to enforce their size selective sieving effects while their pore surfaces can be readily functionalized to induce preferential interactions with specific gas molecules[5–35]. Although such a potential has been speculated, MOFs for the separation of C_2H_2/C_2H_4 have not been fully explored. We realized the first microporous MOF for this challenging separation in 2011[36]. We have been further able to tune the micropores through the interplay of metalloligands and organic linkers, and thus to optimize separation selectivities. Although the separation selectivities of these MOFs for the separation of C_2H_2/C_2H_4 are quite high because of their extraordinarily high sieving effects; their extremely narrow pores have also limited their acetylene uptake, which eventually affects their overall performance for separation of C_2H_2/C_2H_4, as clearly demonstrated in the simulated breakthrough curves[37]. Further development led to the discovery of MOF-74 series for C_2H_2/C_2H_4 separation in 2012[38,39]. This series of MOFs has high densities of open metal sites, which can significantly enforce their high acetylene uptake[40,41], but their pores are too large to introduce size-sieving effects. Furthermore, the open metal sites have quite strong interactions with ethylene molecules, so MOF-74 series have systematically quite low selectivities for C_2H_2/C_2H_4 separation. The ideal MOFs for C_2H_2/C_2H_4 separation are those with high C_2H_2/C_2H_4 sieving effects but without sacrificing acetylene uptake. From the MOF structural point of view, these MOFs should still have comparatively narrow pores to enforce their high sieving effects, but these narrow pores should still take up moderate amount of acetylene molecules. Furthermore, some additional intercrossed cage spaces will be required to exclusively bind acetylene molecules and thus to maximize acetylene uptake. It should be theoretically feasible for us to design and realize MOFs to meet these criteria; however, in the reality, it is still a daunting challenge. Recently, there has been some progress on microporous MOFs for C_2H_2/C_2H_4 separation; however, their pore structures still

cannot meet the above mentioned criteria and their separation performances are comparable to established ones[42,43]. Here we report a microporous MOF [Cu(ATBDC)] · G (UTSA-100; $H_2ATBDC = 5$-(5-Amino-1H-tetrazol-1-yl)-1,3-benzenedicarboxylic acid; G = guest molecules), based on our extensive research endeavours on microporous MOFs for C_2H_2/C_2H_4 separation, which can indeed meet those mentioned criteria. UTSA-100 is thus superior to other MOFs, exhibiting highly efficient removal of acetylene from ethylene/acetylene mixtures containing 1% acetylene.

Results

Preparation and characterization of UTSA-100. The amino derivative of tetrazol-1,3-benzenedicarboxylic acid (5-(5-Amino-1H-tetrazol-1-yl)-1,3-benzenedicarboxylic acid, H_2ATBDC) was prepared based on alkaline decomposition of the tetrazole ring and heterocyclization of the resulting N-arylcyanamides on interaction with ammonium azide generated *in situ*. Reaction of $CuCl_2 · 2H_2O$ with H_2ATBDC in the solvothermal condition at 353 K formed UTSA-100 as green block single crystals. It was formulated as [Cu(ATBDC)] · G (UTSA-100) by single-crystal X-ray diffraction (SXRD) studies, and the phase purity of the bulk material was independently confirmed by powder X-ray diffraction (PXRD) (Supplementary Figs 4 and 5). The desolvated [Cu(ATBDC)] (UTSA-100a) for the adsorption studies was prepared from the acetone-exchanged samples followed by the activation under ultrahigh vacuum at room temperature (296 K) for one day, and then at 358 K for another 3 days. The PXRD profile of desolvated UTSA-100a indicates that it maintains the crystalline framework structure (Supplementary Fig. 5).

X-ray single-crystal structure reveals that UTSA-100 has a three-dimensional framework with rhombic open zigzag nano-channels with amino and tetrazole functionalized wall running in the c-direction (Fig. 1; Supplementary Table 1; Supplementary

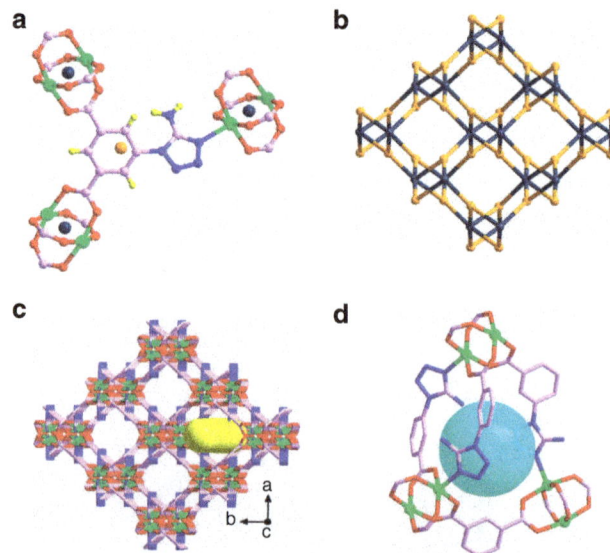

Figure 1 | X-ray crystal structure of UTSA-100. (a) The coordination environment of organic ligand $ATBDC^{2-}$ and Cu(II), and dinuclear copper(II) unit as a 6-connected node (purple balls) and $ATBDC^{2-}$ as a 3-connected node (orange ball). **(b)** The framework topology of *apo*-type (3,6)-connected network with Schläfli symbol $\{4.6^2\}_2\{4^2.6^9.8^4\}$. **(c)** The 3D structure viewed along the c axis showing the 1D rhombic channels of about 4.3 Å in diameter. **(d)** The cage with a diameter of about 4.0 Å between 1D channels with window openings of 3.3 Å. Solvent molecules were omitted for clarity. Colour scheme: Cu, bright green; O, red; N, blue; C, orchid; and H, yellow.

Data 1). There are 6-connected binuclear $Cu_2(COO)_4$ units, which are bridged by 3-connected $ATBDC^{2-}$ anions to form a (3,6)-connected *apo*-type network with Schläfli symbol $\{4.6^2\}_2\{4^2.6^9.8^4\}$ (Fig. 1b). The one-dimensional (1D) open zigzag channels with a diameter of about 4.3 Å are filled with the disordered solvent molecules (DMF and CH_3OH), and there are small cages with the diameter of about 4.0 Å between the 1D channels with window openings of 3.3 Å (Fig. 1d). The calculated solvent accessible volume of UTSA-100a is 51.0%, estimated using the PLATON program[44].

Microporous nature of UTSA-100. To assess the permanent porosity, the acetone-exchanged UTSA-100 was further activated under high vacuum to obtain the desolvated UTSA-100a. The porosity of UTSA-100a was evaluated by N_2-gas sorption at 77 K. The type I isotherm shows a very sharp uptake at $P/P_0 < 0.1$, which clearly indicates its microporous nature (Supplementary Fig. 7a). The nitrogen physisorption of UTSA-100a reaches a plateau at around $P/P_0 = 0.1$, the saturation uptake is $257.6 \, cm^3 \, g^{-1}$ and the corresponding specific pore volume is $0.399 \, cm^3 \, g^{-1}$. The Langmuir (Brunauer, Emmett and Teller) surface area based on the N_2 adsorption isotherm at 77 K is 1,098 (970) $m^2 \, g^{-1}$ for UTSA-100a, within the pressure range of $0.05 < P/P_0 < 0.3$ (Supplementary Fig. 7b–d). The experimental determined Brunauer, Emmett and Teller surface of $970 \, m^2 \, g^{-1}$ matches well with the simulated one ($913 \, m^2 \, g^{-1}$) from its crystal structure data.

Sorption of acetylene and ethylene within UTSA-100a. The unique pore structure encouraged us to examine the capacities of UTSA-100a for the selective separation of C_2H_2/C_2H_4. The low-pressure sorption isotherms of acetylene and ethylene were collected at 273 and 296 K, respectively. At 296 K and 1 atm, the acetylene and ethylene uptake amounts of UTSA-100a are 95.6 and $37.2 \, cm^3 \, g^{-1}$, respectively (Fig. 2). This is really encouraging: the acetylene uptake of UTSA-100a is moderately high, while ethylene uptake is much lower. As shown in Table 1, the C_2H_2/C_2H_4 uptake ratio of 2.57 in UTSA-100a is systematically higher than the examined MOFs except M'MOF-3a with very narrow pores, indicating its bright promise for the C_2H_2/C_2H_4 separation.

The measured pure component isotherm data for acetylene and ethylene on UTSA-100a were fitted with the dual-Langmuir–Freundlich isotherm model. The fitted parameter values are presented in Supplementary Table 2. As illustration of the goodness of the fits, Supplementary Fig. 8 presents a comparison of component loadings for acetylene and ethylene at 296 K in UTSA-100a with the isotherm fits. The fits are excellent for both components over the entire pressure range.

The binding energy of acetylene is reflected in the isosteric heat of adsorption, Q_{st}, defined as

$$Q_{st} = RT^2 \left(\frac{\partial \ln p}{\partial T} \right)_q \qquad (1)$$

Supplementary Figure 10 presents a comparison of the heats of adsorption of acetylene in UTSA-100a with five other representative MOFs (M'MOF-3a, MgMOF-74, CoMOF-74, FeMOF-74 and NOTT-300); the calculations are based on the use of the Clausius–Clapeyron equation. We note that values of Q_{st} in UTSA-100a and M'MOF-3a are significantly lower than that of MOFs with coordinately unsaturated metal atoms FeMOF-74, CoMOF-74, and MgMOF-74. The value of Q_{st} in UTSA-100a is also lower than for NOTT-300. This implies that the regeneration energy requirement of UTSA-100a will be lower than that of FeMOF-74, CoMOF-74, MgMOF-74 and NOTT-300.

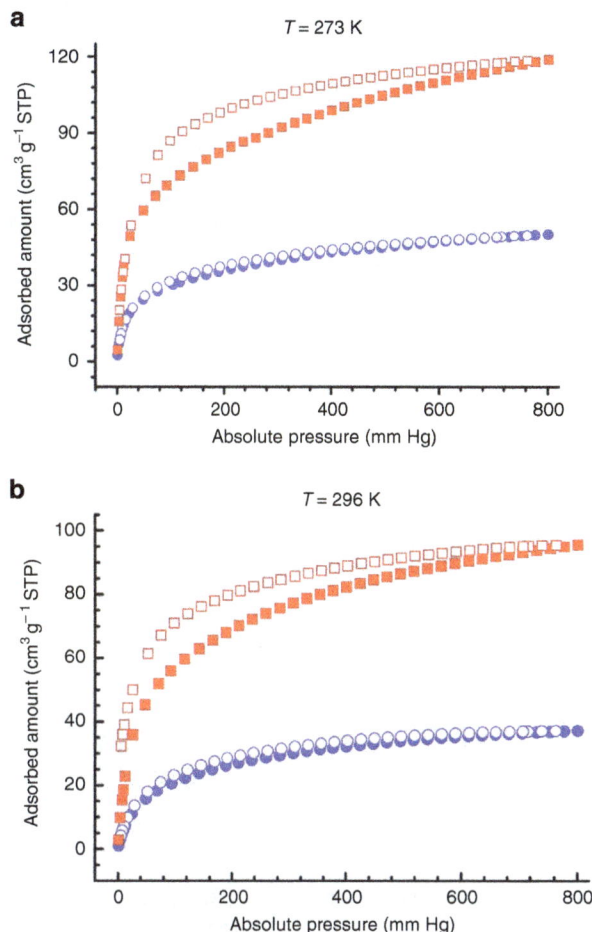

Figure 2 | Gas sorption isotherms on the activated UTSA-100a.
(**a**) Acetylene (red) and ethylene (blue) sorption at 273 K. (**b**) Acetylene (red) and ethylene (blue) sorption at 296 K. Adsorption and desorption branches are shown with closed and open symbols, respectively.

IAST calculations of adsorption selectivities. Analysis of the pure component isotherms at 296 K via ideal adsorbed solution theory (IAST)[45] was carried out to estimate the selectivity between acetylene and ethylene. We consider the separation of binary ethylene/acetylene mixtures containing 1%, that is, 10,000 p.p.m., of acetylene, in the mixture; this composition is typical of industrial mixtures. Figure 3a presents the IAST calculations of the ethylene/acetylene adsorption selectivity, defined by

$$S_{ads} = \frac{q_1/q_2}{p_1/p_2}. \qquad (2)$$

The highest adsorption selectivity is with M'MOF-3a, and this is followed by UTSA-100a. MOFs with coordinately unsaturated metal atoms MgMOF-74, CoMOF-74 and FeMOF-74 have selectivities that are in the range of 1.6 to 2.2. NOTT-300 has selectivities in the range 1.8–2.1.

In the paper by Yang *et al.*[42], the ethylene/acetylene mixtures for NOTT-300 are compared with FeMOF-74 for varying mole fraction of acetylene in the gas phase, keeping the total gas phase pressure constant at 100 kPa. It must be remarked that in industrial practice, the compositions of acetylene in the gas phase are < 1%. Also, acetylene is selectively adsorbed from ethylene/acetylene mixtures, and the bulk gas phase compositions will vary from 1% at the inlet to the desired 40 p.p.m. limit at the outlet of the fixed-bed absorber. Nevertheless, in order to compare our

Table 1 | Adsorption data.

	M'MOF-3a	MgMOF-74	CoMOF-74	FeMOF-74	NOTT-300	UTSA-100a
Surface area ($m^2 g^{-1}$; BET)	110	927	1,018	1,350	1,370	970
Pore volume ($cm^3 g^{-1}$)	0.165	0.607	0.515	0.626	0.433	0.399
Framework density($kg\,m^{-3}$)	1,023	909	1,169	1,126	1,062	1,146
Size of pore window (Å)	3.4×4.8	11×11	11×11	11×11	6.5×6.5	4.3×4.3
C_2H_2 uptake at 1.0 bar ($mmol\,g^{-1}$)	1.90	8.37	8.17	6.80*	6.34[†]	4.27
C_2H_4 uptake at 1.0 bar ($mmol\,g^{-1}$)	0.40	7.45	7.02	6.10*	4.28[†]	1.66
C_2H_2/C_2H_4 uptake ratio	4.75	1.12	1.16	1.11	1.48	2.57
Selectivity for C_2H_2/C_2H_4[‡]	24.03	2.18	1.70	2.08	2.17	10.72
Q_{st} (C_2H_2, $kJ\,mol^{-1}$)[§]	25	41	45	46	32	22

BET, Brunauer, Emmett and Teller.
Summary of the adsorption data for acetylene and ethylene in representative MOFs at 296 K[38,39,42].
*At temperature of 318 K.
[†]At temperature of 293 K.
[‡]IAST analysis for ethylene/acetylene mixtures containing 1% acetylene at 100 kPa.
[§]Q_{st} values at low surface coverage.

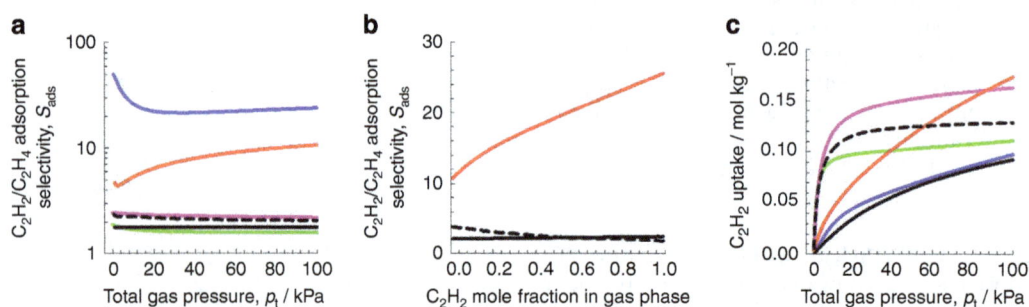

Figure 3 | IAST calculations for binary C_2H_2/C_2H_4 mixture on the MOFs. (**a**) C_2H_2/C_2H_4 adsorption selectivity, and (**c**) uptake capacity of C_2H_2 for adsorption from C_2H_2/C_2H_4 mixture containing 1% C_2H_2. The total bulk gas phase is at 296 K and 100 kPa. The partial pressures of C_2H_2, and C_2H_4 are, respectively, $p1 = 1$ kPa, $p2 = 99$ kPa. (**b**) IAST calculations of the C_2H_2/C_2H_4 adsorption selectivity for FeMOF-74, NOTT-300, and UTSA-100a as a function of the mole fraction of C_2H_2 in the gas phase. The total gas pressure is constant at 100 kPa. The data for FeMOF-74 is at the temperature of 318 K; this is the lowest temperature used in the isotherm measurements of Bloch et al.[38]. The data for NOTT-300 is at 293 K. M'MOF-3a (blue), MgMOF-74 (pink), CoMOF-74 (green), FeMOF-74 (black dash), NOTT-300 (black) and UTSA-100a (red).

IAST calculation methodology with those of Yang et al., we carried out similar comparison, also including UTSA-100a; the results are shown in Fig. 3b. Our IAST selectivity calculations for NOTT-300 and FeMOF-74 agree reasonably well with those of Yang et al., We note that UTSA-100a has significantly higher selectivity than both NOTT-300 and FeMOF-74.

As shown in Fig. 3c, the gravimetric uptake capacity of acetylene in UTSA-100a for adsorption from C_2H_2/C_2H_4 mixtures containing 1% C_2H_2 is compared with other five MOFs (M'MOF-3a, MgMOF-74, CoMOF-74, FeMOF-74 and NOTT-300). At a total gas phase pressure of 100 kPa, the hierarchy of uptake capacities for acetylene is UTSA-100a > MgMOF-74 > FeMOF-74 > CoMOF-74 > M'MOF-3a ≈ NOTT-300.

Ethylene/acetylene breakthrough simulations and experiments. We carried out breakthrough simulations for C_2H_2/C_2H_4 (1:99, v/v) mixture, whose composition is typical of industrial mixtures, in a fixed-bed to demonstrate the feasibility of purification of ethylene in a Pressure Swing Adsorption operation. The transient breakthrough simulations show the concentrations of C_2H_2/C_2H_4 exiting the adsorber packed with UTSA-100a as a function of the dimensionless time, τ (Fig. 4a). Analogous breakthrough simulations were performed for M'MOF-3a, MgMOF-74, CoMOF-74, FeMOF-74 and NOTT-300. On the basis of the gas phase concentrations, we can calculate the impurity level of acetylene in the

gas mixture exiting the fixed-bed packed with six different MOFs. Figure 4b shows the p.p.m. C_2H_2 in the outlet gas mixture exiting an adsorber packed with M'MOF-3a, MgMOF-74, CoMOF-74, FeMOF-74, NOTT-300 and UTSA-100a. At a certain time, τ_{break}, the impurity level will exceed the desired purity level of 40 p.p.m. (indicated by the dashed line), that corresponds to the purity requirement of the feed to the polymerization reactor. The adsorption cycle needs to be terminated at that time τ_{break} and the regeneration process needs to be initiated. From a material balance on the adsorber, the amount of acetylene captured during the time interval $0-\tau_{break}$ can be determined. The amount of acetylene captured in UTSA-100a during the time $0-\tau_{break}$ is 137.6 mmol l^{-1}, which is the best one among the compared six MOFs and approximately twice that of NOTT-300 (Supplementary Table 5). A plot of the amount of acetylene captured plotted as a function of the time interval τ_{break} is presented in Fig. 4c. The hierarchy of acetylene capture capacities is UTSA-100a > MgMOF-74 > FeMOF-74 > CoMOF-74 > M'MOF-3a > NOTT-300. The significantly superior performance of UTSA-100a is attributable to a combination of high adsorption selectivity and high uptake capacity. M'MOF-3a has the highest selectivity but the lowest uptake capacity; this results in poorer performance in the industrial fixed-bed adsorber. It needs to be pointed out that the separations in fixed-bed adsorbers are determined by a combination of adsorption selectivity and uptake capacity. Transient breakthrough simulations can provide us some guidance on the potentials of materials for

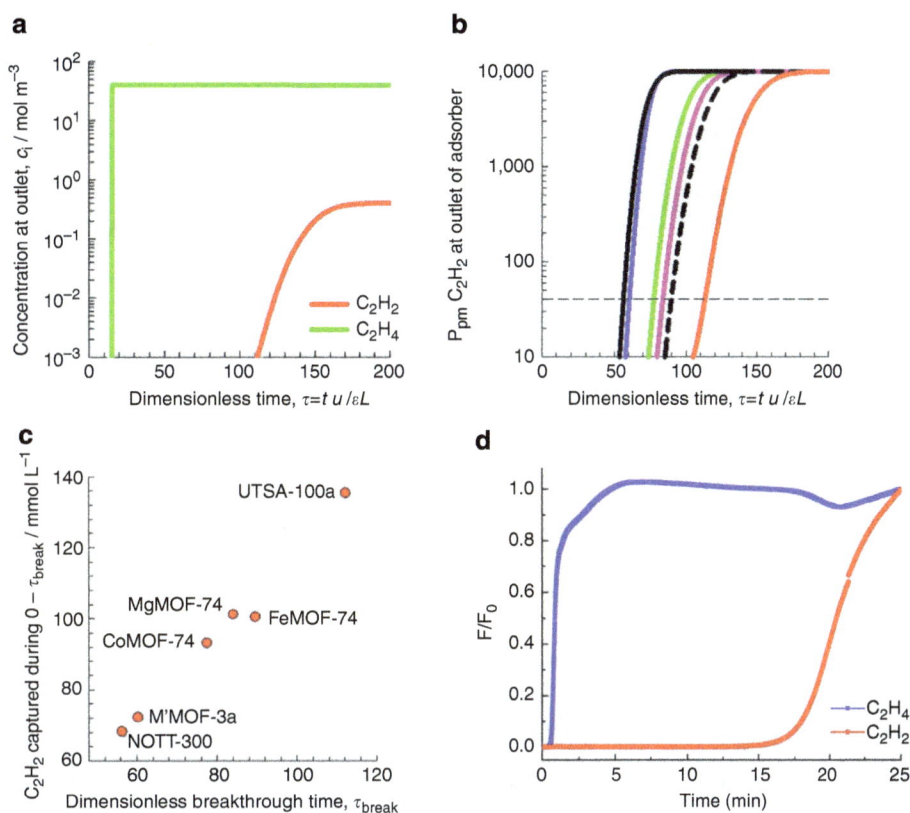

Figure 4 | Simulative and experimental column breakthrough experiments. (**a**) Transient breakthrough curve of C_2H_2/C_2H_4 mixture containing 1% C_2H_2 in an adsorber bed packed with UTSA-100a. The partial pressures of C_2H_2, and C_2H_4 in the inlet feed gas mixture are, respectively, $p_1 = 1$ kPa, $p_2 = 99$ kPa. For the breakthrough simulations, the following parameter values were used, $L = 0.12$ m; $\varepsilon = 0.75$; $u = 0.00225$ m s^{-1}. (**b**) Ppm C_2H_2 in the outlet gas of an adsorber bed packed with UTSA-100a and various MOFs. M'MOF-3a (blue), MgMOF-74 (pink), CoMOF-74 (green), FeMOF-74 (black dash), NOTT-300 (black), and UTSA-100a (red). (**c**) Plot of C_2H_2 captured per L of adsorbent (< 40 ppm C_2H_2 in outlet gas), during the time interval $0-\tau_{break}$, plotted as a function of the time interval τ_{break}. (**d**) Experimental column breakthrough curve for C_2H_2/C_2H_4 mixed gas containing 1% C_2H_2 over UTSA-100a. The experiment temperatures are 296 K except FeMOF-74 (318 K) and NOTT-300 (293 K).

their real industrial applications. They can also offer more information than IAST calculations, and help us to make a clear comparison with other MOFs.

To evaluate the performance of UTSA-100a in the actual adsorption-based separation and purification processes, breakthrough experiments were performed in which an C_2H_2/C_2H_4 (1:99, v/v) mixture was flowed over a packed bed of UTSA-100a solid with a total flow of 2 ml min^{-1} at 296 K. As shown in Fig. 4d, the separation of C_2H_2/C_2H_4 (1:99, v/v) mixture through the column packed bed of UTSA-100a solid can be efficiently achieved. To the best of our knowledge, this is the first example of porous materials whose separation for C_2H_2/C_2H_4 (1:99, v/v) has been clearly established by experimental breakthrough, enabling UTSA-100a to be a potential high-performance material for real industrial ethylene purification application.

Pore structure analysis and first-principles calculations. To correlate the excellent C_2H_2/C_2H_4 separation performance of UTSA-100a to its structure, we carried out more detailed pore structure analysis. As mentioned earlier, the overall MOF pore size is centred at 4.3 Å, with rather narrow distribution (Fig. 5c). Gas adsorption and diffusion would thus take place predominantly within individual channel pores, with interchannel diffusion limited by the narrow window opening between adjacent channels and the size of gas molecules. The pore size variation along the pore channel (in crystallography c axis direction) is shown in

Fig. 5d. Clearly, the pore limiting size (for the access of guest gas molecule into the MOF crystal) is 3.96 Å, while the largest cavity within the channel is 4.6 Å in diameter. Note that the empirical kinetic diameters of C_2H_2/C_2H_4 are $\sim 3.3/4.2$ Å ($3.32 \times 3.34 \times 5.70$ Å^3 for C_2H_2 versus $3.28 \times 4.18 \times 4.84$ Å^3 for C_2H_4), respectively[46–49]. For ethylene, its kinetic diameter is slightly larger than the channel pore opening of UTSA-100a, and much larger than the interchannel window size. These two factors may significantly hinder/block the ethylene adsorption and diffusion in the MOF structure, leading to the much lower ethylene uptakes than acetylene ones.

To further understand the acetylene adsorption in UTSA-100a, we performed detailed computational investigations (Supplementary Methods). We first optimized the bare UTSA-100a structure by first-principles DFT-D (dispersion-corrected density-functional theory) calculations, where van der Waals interactions were corrected by empirical r^{-6} terms[50]. The optimized structure is fairly close to the experimentally determined structure. We then introduced acetylene molecules to various locations of the channel pore, and further optimized the 'UTSA-100a + C_2H_2' structures using DFT-D. Interestingly, the guest acetylene molecules all get relaxed to a particular adsorption sites. In Fig. 5b, we plot this preferred acetylene adsorption location. The acetylene sits right at the small cage connecting two adjacent channel pores. The relatively strong binding clearly comes from multiple-point interactions of the molecule with framework (particularly, the metal center O, the linker tetrazole rings and –

a

b

c

d

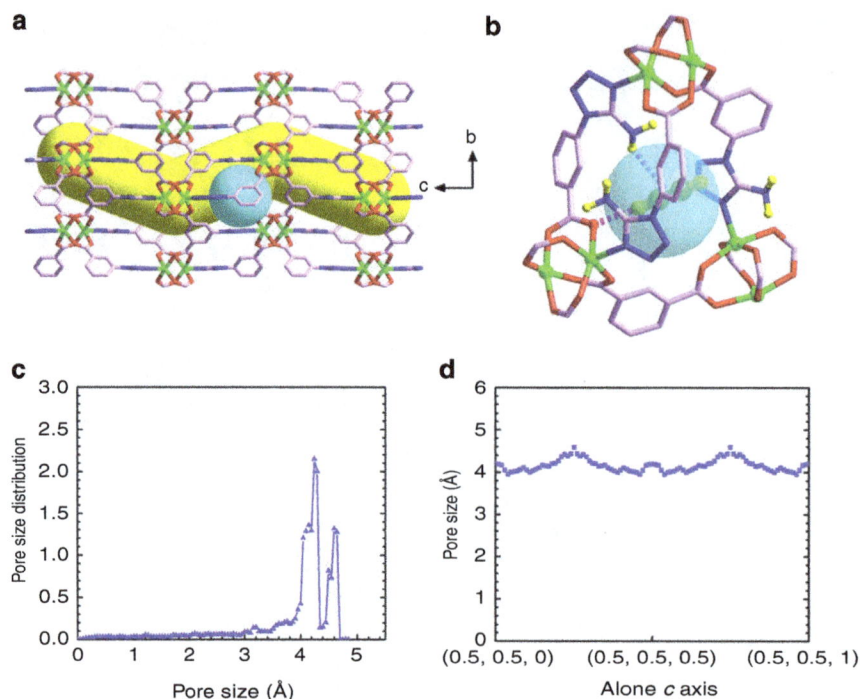

Figure 5 | The pore structure of UTSA-100 and the C$_2$H$_2$ binding site. (a) The pore structure showing the zigzag channels along the c axis and the cage with a diameter of about 4.0 Å in the pore wall with window openings of 3.3 Å. **(b)** The acetylene sits right at the small cage connecting two adjacent channel pores. (multiple-point interactions of the acetylene molecule with framework: d [O(-CO$_2$)$\cdots$H(C$_2$H$_2$)] = 2.252 Å, d [H(-NH$_2$)$\cdots$(C$_2$H$_2$)] = 2.856 Å). **(c)** Pore size distribution (PSD) of UTSA-100a. PSD was calculated using the well-known method by Gubbins et al.[58]. The van der Waals diameters of the framework atoms were adopted from the Cambridge Crystallographic Center. **(d)** Pore size variation along the pore channel (in c axis direction), within the crystal unit cell of UTSA-100a. Colour scheme: Cu, bright green; O, red; N, blue; C, orchid (green in acetylene); and H, yellow.

NH$_2$ groups, Fig. 5b). These binding sites are typically classified as specific and/or strong sites for gas recognitions[29]. The interactions between the metal center O with C$_2$H$_2$ can be assigned as hydrogen bonding interactions. Given the fact that aromatic –NH$_2$ has weak basicity, while C$_2$H$_2$ has weak acidity (pKa = 25)[51], there might exist weak acid–base interactions between –NH$_2$ groups and C$_2$H$_2$ molecules. Because C$_2$H$_2$ is more acidic than C$_2$H$_4$ (pKa = 44)[51], the –NH$_2$ groups have stronger interactions with C$_2$H$_2$ than C$_2$H$_4$, which enforces the selective binding of UTSA-100a for C$_2$H$_2$ than C$_2$H$_4$ as well. It is suggested that the relatively narrower pore size allows and reinforces one acetylene molecule to interact mutually with –NH$_2$ group and one metal center O atom. We note that the adsorbed acetylene is slightly distorted, with an induced dipole moment. The H–C–C bond angle of acetylene is 178.8°, comparable to that of acetylene adsorbed on the open-Cu site in HKUST-1 ($\sim$178°)[40]. The static acetylene binding energy, derived from the DFT-D calculation, is $\sim$31.3 kJ mol^{-1}. This value (without considering the thermal correction, which is typically a few kJ mol^{-1}) is somewhat larger than the experimental Q_{st} value but still reasonable, considering the accuracy limitation of the DFT-D approach.

Discussion

Removal of acetylene from ethylene/acetylene mixtures containing 1% acetylene is a very important but challenging industrial separation task. It has been speculated that adsorption-based porous materials could be the good alternative for the highly efficient removal of acetylene from ethylene steam, but have not been fully fulfilled. To evaluate a porous material for acetylene removal from ethylene, adsorption selectivity and saturation uptake capacity have been deemed as two most important criteria,

and high values for both of them are needed to achieve high effectiveness and high efficiency for acetylene removal. However, which factor plays the dominant role depends on the composition of gas mixtures. For the ethylene/acetylene mixtures containing 1% acetylene, adsorption selectivity will be enough more important than acetylene adsorption capacity. Comparing with other five well-known MOFs (M'MOF-3a, MgMOF-74, CoMOF-74, FeMOF-74 and NOTT-300), the significantly superior performance of UTSA-100a in removing acetylene from ethylene/acetylene mixtures containing 1% acetylene is attributable to the collaboration of high adsorption selectivity and high uptake capacity at ambient conditions. From the structure point of view, UTSA-100a has suitable pores and opening windows to enforce its high sieving effects and thus high adsorption selectivities, while the suitable cages and immobilized functional sites such –NH$_2$ further maximize the acetylene uptakes. It is speculated that weak acid–base interactions between –NH$_2$ and C$_2$H$_2$ molecules also play the important roles for the preferential binding of UTSA-100a with C$_2$H$_2$ over C$_2$H$_4$. Incorporation of stronger basic sites such as alkaneamines into porous MOFs might significantly differentiate their interactions with C$_2$H$_2$ over C$_2$H$_4$, leading to even more efficient MOF materials for C$_2$H$_2$/C$_2$H$_4$ separations in the future.

It is to be noted that our comparisons of different MOFs are for ethylene/acetylene mixtures containing 1% acetylene that is representative of compositions encountered in industry. If we were to compare the performance of UTSA-100a, NOTT-300 and FeMOF-74 for 50/50 ethylene/acetylene mixtures, the conclusions are different because in this case, capacity considerations would be very important. Supplementary Figure 11 presents a comparison of breakthroughs for UTSA-100a, NOTT-300, and FeMOF-74 for 50/50 ethylene/acetylene mixtures. In this case the performance of NOTT-300 and UTSA-100a are nearly the same. The best separations are achieved with FeMOF-74 that has the highest capacity to adsorb acetylene. These results also underscore the need for a proper evaluation of MOFs using transient breakthroughs. Comparisons based pure on selectivities may lead to wrong conclusions.

In conclusion, we have prepared a microporous MOF (UTSA-100) for highly efficient removal of acetylene from ethylene/acetylene mixtures containing 1% acetylene. Experimental and computational simulation results demonstrate the high efficiency of UTSA-100a in the removal of acetylene from ethylene/acetylene mixtures containing 1% acetylene, which is a very important but challenging industrial separation task. The results of this research have important significance on the practical design and preparation of porous materials for light hydrocarbon separations. It will also provide some guidance on the design and synthesis of microporous MOFs for other gas separations.

Methods

Materials and measurements. Commercially available reagents were purchased in high purity and used without further purification. 5-(5-amino-1H-tetrazol-1-yl)-1,3-benzenedicarboxylic acid (H$_2$ATBDC) was synthesized according to the literature method[52] (Supplementary Fig. 1). ^{1}H NMR and ^{13}C NMR spectra were obtained using a Varian INOVA 500 MHz spectrometer at room temperature. FTIR spectra were performed on a Bruker Vector 22 spectrometer at room temperature. Thermal gravimetric analysis (TGA) was performed under a nitrogen atmosphere with a heating rate of 3 °C min^{-1} using a Shimadzu TGA-50 thermogravimetric analyzer. PXRD patterns were measured by a Rigaku Ultima IV diffractometer operated at 40 kV and 44 mA with a scan rate of 1.0 deg min^{-1}.

Synthesis of 5-(1H-tetrazol-1-yl)-1,3-benzenedicarboxylic acid. Glacial acetic acid (10.0 ml) was added with stirring to a suspension of 5-aminoisophthalic acid (4.54 g, 0.025 mol), and sodium azide (1.79 g, 0.0275 mol) in triethyl orthoformate (14.2 ml, 0.075 mol), and the mixture was stirred at 80–90 °C for 6 h. The reaction mixture was cooled, and concentrated hydrochloric acid (4.2 ml, 0.05 mol) and water (12.5 ml) were added. The precipitated solid was separated by filtration, washed with water and dried. The obtained raw product was recrystallized from DMF. Yield: 82% (4.79 g). ^{1}H NMR (500 MHz, d^6-DMSO, p.p.m.): δ 13.78 (s, 2H, –CO$_2$H), 10.33 (s, 1H, –N$_4$CH), 8.65 (s, 2H, –C$_6$H$_3$), 8.55 (s, 1H, –C$_6$H$_3$).

Synthesis of 5-(cyanoamino)-1,3-benzenedicarboxylic acid. DMSO (20.0 ml) was added dropwise with constant stirring to a suspension of 5-(1H-tetrazol-1-yl)-1,3-benzenedicarboxylic acid (4.68 g, 0.02 mol) in 22% aqueous KOH solution (12.0 ml). Gas evolution was observed, accompanied by self-heating of the reaction mixture. Stirring of the reaction mixture was continued for 2 days. The mixture was then diluted to 160 ml with water, acidified with concentrated hydrochloric acid to pH 3–4 and stored at 5–10 °C until precipitation of solid. The obtained product was filtered off and dried in vacuum. Another portion of product was reprecipitated from the filtrate through salting out method using sodium salt. Yield: 93% (3.83 g). ^{1}H NMR (500 MHz, d^6-DMSO, p.p.m.): δ = 7.67 (s, 1H, –C$_6$H$_3$), 7.40 (s, 2H, –C$_6$H$_3$).

Synthesis of 5-(5-amino-1H-tetrazol-1-yl)-1,3-benzenedicarboxylic acid. A suspension of 5-(cyanoamino)-1,3-benzenedicarboxylic acid (2.06 g, 0.01 mol), sodium azide (0.98 g, 0.015 mol) and ammonium chloride (1.07 g, 0.02 mol) in DMF (25 ml) was stirred at 70–80 °C for 6 h, after which water (100 ml) was added to the reaction mixture. The white solid was precipitated through salting out method using sodium salt. The obtained product was filtered off, and dried in vacuum. Yield: 82% (2.04 g). ^{1}H NMR (500 MHz, D$_2$O, p.p.m.): δ = 8.47 (s, 1H, –C$_6$H$_3$), 8.14 (s, 2H, –C$_6$H$_3$) (Supplementary Fig. 2). ^{13}C NMR (D$_2$O, p.p.m.): δ = 173.00, 138.49, 132.35, 130.43, 126.94 (Supplementary Fig. 3).

Synthesis of UTSA-100. A mixture of CuCl$_2$·2H$_2$O (34 mg, 0.2 mmol) and the organic linker H$_2$ATBDC (50 mg, 0.2 mmol) was dispersed into an 8 ml mixed solvent (DMF/MeOH, 5/3, v/v) in a screw-capped vial (20 ml). And five drops of HBF$_4$ (48% w/w aqueous solution) were added. The suspension was sonicated until homogenous. The vial was capped and heated in an oven at 80 °C for 24 h. Green block crystals were obtained by filtration and washed with DMF several times to afford UTSA-100. IR (neat, cm^{-1}): 1,739w; 1,630s; 1,596m; 1,490w; 1,457m; 1,377s; 1,255m; 1,128w; 1,097m; 1,062w; 908m; 780m; 725s; 677m; 661m.

Gas sorption studies. A Micromeritics ASAP 2020 surface area analyzer was used to measure gas adsorption isotherms. To remove all the guest solvents in the

framework, the fresh sample of UTSA-100 was guest exchanged with dry acetone at least 10 times, filtered and degassed at room temperature (296 K) for one day, and then at 358 K for another 3 days until the outgas rate was 5 μmHg min^{-1} before measurements. A sample of activated UTSA-100a (100–150 mg) was used for the sorption measurement and was maintained at 77 K with liquid nitrogen, at 273 K with an ice-water bath. As the center-controlled air conditioner was set up at 23 °C, a water bath was used for adsorption isotherms at 296 K.

Fitting of pure component isotherms. Experimental data on pure component isotherms for acetylene and ethylene in UTSA-100a were measured at temperatures of 273 and 296 K. The pure component isotherm data for acetylene and ethylene were fitted with the dual-Langmuir–Freundlich isotherm model

$$q = q_{A,\text{sat}} \frac{b_A p^{\nu_A}}{1 + b_A p^{\nu_A}} + q_{B,\text{sat}} \frac{b_B p^{\nu_B}}{1 + b_B p^{\nu_B}} \qquad (3)$$

with T-dependent parameters b_A and b_B

$$b_A = b_{A0} \exp\left(\frac{E_A}{RT}\right); \quad b_B = b_{B0} \exp\left(\frac{E_B}{RT}\right)$$

The fitted parameter values are presented in Supplementary Table 2.

For FeMOF-74, the dual-site Langmuir–Freundlich parameters are from Bloch et al.[38]; for convenience, the parameters are summarized in Supplementary Table 3. For NOTT-300, the isotherm data at 293 K were fitted with a single-site Langmuir isotherm model; the fit parameters are specified in Supplementary Table 4. Supplementary Figure 9 presents a comparison of component loadings for acetylene and ethylene at 293 K in NOTT-300 with 1-site Langmuir isotherm fits. The Langmuir fits are of good accuracy. For all other MOFs, the isotherm data are from He et al.[39].

Transient breakthrough of ethylene/acetylene mixtures in fixed-bed adsorbers. The performance of industrial fixed-bed adsorbers is dictated by a combination of adsorption selectivity and uptake capacity. For a proper comparison of various MOFs, we perform transient breakthrough simulations using the simulation methodology described in the literature[53–55]. For the breakthrough simulations, the following parameter values were used: framework density, ρ (1,146 kg m^{-3}); length of packed bed, L (0.12 m); voidage of packed bed, ε (0.75); superficial gas velocity at inlet, u (0.00225 m s^{-1}). For breakthrough simulations with NOTT-300, we use calculated the framework density from the crystal structure information provided in the paper of Yang et al.[42]; the resultant value is ρ (1,062 kg m^{-3}). The framework densities for all other MOFs are available in the papers by Bloch et al.[38] and He et al.[39]. The transient breakthrough simulation results are presented in terms of a dimensionless time, τ, defined by dividing the actual time, t, by the characteristic time, $\frac{L\varepsilon}{u}$.

Column breakthrough tests. The breakthrough separation experiments were conducted on a home-made apparatus (Supplementary Fig. 12) with a set-up similar to what was described in Yaghi's paper[56]. In a typical experiment, 1.00 g of UTSA-100a powders were thoroughly mixed and packed into a quartz column (5.8 mm inner diameter × 150 mm) with quartz wool filling the void space. The sample was in situ activated under vacuum (6.5 × 10^{-4} Pa) at 353 K to remove adsorbed molecules and make the active sites accessible. The sample was then purged with He flow (2.0 ml min^{-1}) for 1 h while the temperature of the column was decreased to 296 K. The mix gas (C$_2$H$_2$:C$_2$H$_4$ = 1: 99 by volume) flow was then introduced at 2.0 ml min^{-1}. Effluent from the column was monitored using a mass spectrometer.

Caution: Because of its wide flammability limits and a potential for explosive decomposition[57], acetylene should be handled with care.

References

1. Sundaram, K. M., Shreehan, M. M. & Olszewski, E. F. Kirk-Othmer Encyclopedia of Chemical Technology 4th edn 877–915 (Wiley, 1995).
2. Molero, H., Bartlett, B. F. & Tysoe, W. T. The hydrogenation of acetylene catalyzed by palladium: hydrogen pressure dependence. J. Catal. 181, 49–56 (1999).
3. Studt, F. et al. Identification of non-precious metal alloy catalysts for selective hydrogenation of acetylene. Science 320, 1320–1322 (2008).
4. Lewis, J. D. Separation of acetylene from ethylene-bearing gases. US Patent 3,837,144 (1974).
5. Matsuda, R. et al. Highly controlled acetylene accommodation in a metal-organic microporous material. Nature 436, 238–241 (2005).
6. Furukawa, H., Cordova, K. E., O'Keeffe, M. & Yaghi, O. M. The chemistry and applications of metal-organic frameworks. Science 341, 1230444 (2013).
7. Sato, H. et al. Self-accelerating CO sorption in a soft nanoporous crystal. Science 343, 167–170 (2014).
8. Zhao, X. et al. Selective anion exchange with nanogated isoreticular positive metal-organic frameworks. Nat. Commun. 4, 2344 (2013).
9. An, J. et al. Metal-adeninate vertices for the construction of an exceptionally porous metal-organic framework. Nat. Commun. 3, 604 (2012).

10. Xiang, S. *et al.* Microporous metal-organic framework with potential for carbon dioxide capture at ambient conditions. *Nat. Commun.* **3,** 954 (2012).

11. Li, J.-R. *et al.* Porous materials with pre-designed single-molecule traps for CO_2 selective adsorption. *Nat. Commun.* **4,** 1538 (2013).

12. Vaidhyanathan, R. *et al.* Direct observation and quantification of CO_2 binding within an amine-functionalized nanoporous solid. *Science* **330,** 650–653 (2010).

13. Chen, B., Xiang, S. & Qian, G. Metal-organic frameworks with functional pores for recognition of small molecules. *Acc. Chem. Res.* **43,** 1115–1124 (2010).

14. Lin, Q., Wu, T., Zheng, S.-T., Bu, X. & Feng, P. Single-walled polytetrazolate metal-organic channels with high density of open nitrogen-donor sites and gas. *J. Am. Chem. Soc.* **134,** 784–787 (2012).

15. Férey, G. *et al.* A chromium terephthalate–based solid with unusually large pore volumes and surface area. *Science* **309,** 2040–2042 (2005).

16. Zhao, X. *et al.* Hysteretic adsorption and desorption of hydrogen by nanoporous metal-organic frameworks. *Science* **306,** 1012–1015 (2004).

17. Farha, O. K. *et al.* De novo synthesis of a metal-organic framework material featuring ultrahigh surface area and gas storage capacities. *Nat. Chem.* **2,** 944–948 (2010).

18. Mohideen, M. I. H. *et al.* Protecting group and switchable pore-discriminating adsorption properties of a hydrophilic-hydrophobic metal-organic framework. *Nat. Chem.* **3,** 304–310 (2011).

19. Zhang, Y.-B. *et al.* Geometry analysis and systematic synthesis of highly porous isoreticular frameworks with a unique topology. *Nat. Commun.* **3,** 642 (2012).

20. Zhang, J.-P. & Chen, X.-M. Optimized acetylene/carbon dioxide sorption in a dynamic porous crystal. *J. Am. Chem. Soc.* **131,** 5516–5521 (2009).

21. Li, B. *et al.* Enhanced binding affinity, remarkable selectivity, and high capacity of CO_2 by dual functionalization of a rht-type metal–organic framework. *Angew. Chem. Int. Ed.* **51,** 1412–1415 (2012).

22. Burd, S. D. *et al.* Highly selective carbon dioxide uptake by [Cu(bpy-n)$_2$(SiF$_6$)] (bpy-1 = 4,4'-bipyridine; bpy-2 = 1,2-bis(4-pyridyl)ethene). *J. Am. Chem. Soc.* **134,** 3663–3666 (2012).

23. Nugent, P. *et al.* Porous materials with optimal adsorption thermodynamics and kinetics for CO_2 separation. *Nature* **495,** 80–84 (2013).

24. Shekhah, O. *et al.* Made-to-order metal-organic frameworks for trace carbon dioxide removal and air capture. *Nat. Commun.* **5,** 4228 (2014).

25. Li, B. *et al.* Metal-cation-directed *de novo* assembly of a functionalized guest molecule in the nanospace of a metal – organic framework. *J. Am. Chem. Soc.* **136,** 1202–1205 (2014).

26. Ma, L., Mihalcik, D. J. & Lin, W. Highly porous and robust 4,8-connected metal – organic frameworks for hydrogen storage. *J. Am. Chem. Soc.* **131,** 4610–4612 (2009).

27. Lan, Y. Q., Jiang, H. L., Li, S. L. & Xu, Q. Mesoporous metal-organic frameworks with size-tunable cages: selective CO_2 uptake, encapsulation of Ln^{3+} cations for luminescence, and column-chromatographic dye separation. *Adv. Mater.* **23,** 5015–5020 (2011).

28. Motkuri, R. K. *et al.* Fluorocarbon adsorption in hierarchical porous frameworks. *Nat. Commun.* **5,** 4368 (2014).

29. Li, B., Wen, H. M., Zhou, W. & Chen, B. Porous metal – organic frameworks for gas storage and separation: what, how, and why? *J. Phys. Chem. Lett.* **5,** 3468–3479 (2014).

30. Li, B. *et al.* A porous metal – organic framework with dynamic pyrimidine groups exhibiting record high methane storage working capacity. *J. Am. Chem. Soc.* **136,** 6207–6210 (2014).

31. Chen, B. *et al.* A microporous metal–organic framework for gas-chromatographic separation of alkanes. *Angew. Chem. Int. Ed.* **45,** 1390–1393 (2006).

32. Guo, Z. *et al.* A metal–organic framework with optimized open metal sites and pore spaces for high methane storage at room temperature. *Angew. Chem. Int. Ed.* **50,** 3178–3181 (2011).

33. He, Y., Zhou, W., Krishna, R. & Chen, B. Microporous metal–organic frameworks for storage and separation of small hydrocarbons. *Chem. Commun.* **48,** 11813–11831 (2012).

34. Farha, O. K. *et al.* Metal – organic framework materials with ultrahigh surface areas: Is the sky the limit? *J. Am. Chem. Soc.* **134,** 15016–15021 (2012).

35. DeCoste, J. B. *et al.* Metal-organic frameworks for oxygen storage. *Angew. Chem. Int. Ed.* **53,** 14092–14095 (2014).

36. Xiang, S. *et al.* Rationally tuned micropores within enantiopure metal–organic frameworks for highly selective separation of acetylene and ethylene. *Nat. Commun.* **2,** 204 (2011).

37. Das, M. Y. *et al.* Interplay of metalloligand and organic ligand to tune micropores within isostructural mixed–metal organic frameworks (M'MOFs) for their highly selective separation of chiral and achiral small molecules. *J. Am. Chem. Soc.* **134,** 8703–8710 (2012).

38. Bloch, E. D. *et al.* Hydrocarbon separations in a metal-organic framework with open iron(II) coordination sites. *Science* **335,** 1606–1610 (2012).

39. He, Y., Krishna, R. & Chen, B. Metal–organic frameworks with potential for energy–efficient adsorptive separation of light hydrocarbons. *Energy Environ. Sci.* **5,** 9107–9120 (2012).

40. Xiang, S., Zhou, W., Gallegos, J. M., Liu, Y. & Chen, B. Exceptionally high acetylene uptake in a microporous metal-organic framework with open metal sites. *J. Am. Chem. Soc.* **131,** 12415–12419 (2009).

41. Xiang, S. *et al.* Open metal sites within isostructural metal–organic frameworks for differential recognition of acetylene and extraordinarily high acetylene storage capacity at room temperature. *Angew Chem. Int. Ed.* **49,** 4615–4618 (2010).

42. Yang, S. *et al.* Supramolecular binding and separation of hydrocarbons within a functionalized porous metal–organic framework. *Nat. Chem.* **7,** 121–129 (2015).

43. Wen, H. M. *et al.* A microporous metal–organic framework with rare lvt topology for highly selective C_2H_2/C_2H_4 separation at room temperature. *Chem. Commun.* **51,** 5610–5613 (2015).

44. Spek, A. L. *PLATON, A Multipurpose Crystallographic Tool* (Utrecht Univ., 2001).

45. Myers, A. L. & Prausnitz, J. M. Thermodynamics of mixed gas adsorption. *AIChE. J.* **11,** 121–130 (1965).

46. Breck, D. W. *Zeolite Molecular Sieves: Structure, Chemistry and Use* (John Wiley and Sons, Inc., 1974).

47. Sircar, S. & Myers, A. L. Gas Separation by Zeolites. In *Handbook of Zeolite Science and Technology.* (eds Anesbach, S. M., Carrado, K. A. & Dutta, P. K.) (Marcel Dekker Inc., 2003).

48. Li, J. R., Kuppler, R. J. & Zhou, H. C. Selective gas adsorption and separation in metal–organic frameworks. *Chem. Soc. Rev.* **38,** 1477–1504 (2009).

49. Aguado, S., Bergeret, G., Daniel, C. & Farrusseng, D. Absolute molecular sieve separation of ethylene/ethane mixtures with silver zeolite A. *J. Am. Chem. Soc.* **134,** 14635–14637 (2012).

50. Giannozzi, P. *et al.* QUANTUM ESPRESSO: A modular and open-source software project for quantum simulations of materials. *J. Phys. Condens. Matter* **21,** 395502 (2009).

51. Smith, M. B. & March, J. *March's Advanced Organic Chemistry* 6th edn 363–364 (Wiley, 2007).

52. Voitekhovich, S. V., Vorobév, A. N., Gaponik, P. N. & Ivashkevich, O. A. Synthesis of new functionally substituted 1-R-tetrazoles and their 5-amino derivatives. *Chem. Heterocycl. Compd* **41,** 999–1004 (2005).

53. Krishna, R. & Long, J. R. Screening metal – organic frameworks by analysis of transient breakthrough of gas mixtures in a fixed bed adsorber. *J. Phys. Chem. C* **115,** 12941–12950 (2011).

54. Krishna, R. The maxwell – stefan description of mixture diffusion in nanoporous crystalline materials. *Microporous Mesoporous Mater* **185,** 30–50 (2014).

55. Krishna, R. Separating mixtures by exploiting molecular packing effects in microporous materials. *Phys. Chem. Chem. Phys.* **17,** 39–59 (2015).

56. Britt, D., Furukawa, H., Wang, B., Glover, T. G. & Yaghi, O. M. Highly efficient separation of carbon dioxide by a metal-organic framework replete with open metal sites. *Proc. Natl Acad. Sci. USA* **106,** 20637–20640 (2009).

57. Hilden, D. L. & Stebar, R. F. Evaluation of acetylene as a spark ignition engine fuel. *Energy Res.* **3,** 59–71 (1979).

58. Bhattacharya, S. & Gubbins, K. E. Fast method for computing pore size distributions of model materials. *Langmuir.* **22,** 7726–7731 (2006).

Acknowledgements

This work was supported by the Welch Foundation AX-1730 (B.C.) and the NSFC (grant no. 21371102; T.L.H.). Y.H. thanks the support from the KAUST Office of Competitive Research Funds (OCRF, Awards no. URF/1/1672-01-01).

Author contributions

T.L.H. and B.C. conceived and designed the experiments and co-wrote the paper. T.L.H. performed most of the experiments and analysed the data. H.W. and B.L. performed a part of characterization of MOF material. R.K. contributed to the IAST calculations and breakthrough simulations. H.W. and W.Z. worked on all computational investigations. Y.Z., Y.H., XW and W.Z. performed the column breakthrough tests. Z.Y. and S.X assisted the analysis of crystal structure. All authors discussed the results and commented on the manuscript.

Additional information

Accession codes: The X-ray crystallographic coordinates for structures reported in this Article have been deposited at the Cambridge Crystallographic Data Centre (CCDC), under deposition number CCDC 1044083. These data can be obtained free of charge from The Cambridge Crystallographic Data Centre via www.ccdc.cam.ac.uk/data_request/cif.

Competing financial interests: The authors declare no competing financial interests.

5

Bio-inspired electron-delivering system for reductive activation of dioxygen at metal centres towards artificial flavoenzymes

Yoann Roux[1], Rémy Ricoux[1], Frédéric Avenier[1] & Jean-Pierre Mahy[1]

Development of artificial systems, capable of delivering electrons to metal-based catalysts for the reductive activation of dioxygen, has been proven very difficult for decades, constituting a major scientific lock for the elaboration of environmentally friendly oxidation processes. Here we demonstrate that the incorporation of a flavin mononucleotide (FMN) in a water-soluble polymer, bearing a locally hydrophobic microenvironment, allows the efficient reduction of the FMN by NADH. This supramolecular entity is then capable of catalysing a very fast single-electron reduction of manganese(III) porphyrin by splitting the electron pair issued from NADH. This is fully reminiscent of the activity of natural reductases such as the cytochrome P450 reductases with kinetic parameters, which are three orders of magnitude faster compared with other artificial systems. Finally, we show as a proof of concept that the reduced manganese porphyrin activates dioxygen and catalyses the oxidation of organic substrates in water.

[1] Laboratoire de Chimie Bioorganique et Bioinorganique, Institut de Chimie Moléculaire et des Matériaux d'Orsay (UMR 8182), Univ Paris Sud, Université Paris Saclay, rue du doyen Georges Poitou, 91405 Orsay, France. Correspondence and requests for materials should be addressed to F.A. (email: frederic.avenier@u-psud.fr) or to J.-P.M. (email: jean-pierre.mahy@u-psud.fr).

Selective catalytic oxidations are crucial reactions for the chemical industry, and new environment-friendly processes are highly needed to meet new standards for a sustainable growth. Yet, Nature has figured out an elegant manner to perform such reactions by the reductive activation of dioxygen at metal centres of multicomponent monooxygenases[1,2]. For example, in liver cells, cytochromes (P450) catalyse the oxidation of various xenobiotics and metabolites by the reductive activation of dioxygen at their haem cofactors, via the formation of high-valent iron-oxo intermediates, capable of transferring an oxygen atom into the C–H bond of substrates[2]. In bacteria, another protein complex, methane monooxygenase, catalyses the selective oxidation of methane into methanol via the reductive activation of dioxygen and the formation of a di-iron(IV) di-oxo intermediate[1]. Other metallic monooxygenases also activate dioxygen to perform oxidation reactions, and in all cases the two electrons needed at each catalytic cycle are provided to the metal centres by reductase proteins capable of harvesting electrons from NAD(P)H owing to their flavin cofactors FMN or FAD[3,4]. These cofactors collect electron pairs as hydride ions from NAD(P)H and ensure a stepwise electron transfer towards the catalytic metal centres of monooxygenases. The protein environment surrounding these flavin cofactors in the reductase proteins plays two key roles in the hydride transfer. It provides a binding site for NAD(P)H and it finely tunes the redox potential of flavin cofactors by creating a local specific microenvironment[5,6].

These selective electron transfers constitute the main scientific lock in the perspective of performing a bio-inspired oxidation catalysis using dioxygen with metal-based catalysts. It has, indeed, been proven very difficult to provide electrons for the reductive activation of dioxygen, without quenching the active species formed on reaction with dioxygen. Hence, most of the bio-inspired catalysts developed so far had to shunt this reduction process by the use of hydrogen peroxide (see Fig. 1), and only a handful examples of catalysts have been described that are capable of using dioxygen directly[7].

Nolte and co-workers have incorporated hydrophobic manganese porphyrins into vesicles in which electrons were provided by the oxidation of either hydrogen with colloidal Platinum[8] or formate at rhodium complexes[9,10]. Other systems also used manganese porphyrins in combination with either nicotinamide and benzoic anhydride[11], or pyruvate oxidase and flavin cofactors[12]. More recently, reductase proteins were substituted by ruthenium-based photosentisizers capable of gathering electrons on light irradiation and transferring them to cytochrome (P450) enzymes[13–15]. A similar strategy was also successful in the reduction of di-iron(III) complexes as functional mimics of methane monooxygenase[16]. However, despite the few examples cited above, efficient water-soluble systems capable of selectively driving electrons to metallic centres for the reductive activation of dioxygen are still highly needed.

In this perspective, multibranched polyethyleneimine (PEI), which are water-soluble polymers, can easily be modified with various chemical groups and they generate supramolecular entities bearing a specific local microenvironment, reminiscent to the active site of enzymes. These so-called 'synzymes' (synthetic enzymes) have been shown to mimic various biological activities in the presence, or in the absence, of natural or handmade cofactors[17–24]. In terms of oxidation catalysis, the incorporation of polyoxometalate allowed, for example, the epoxidation of styrene in good yield in the presence of hydrogen peroxide[25].

Herein, we describe the preparation and study of an artificial reductase on the basis of modified polyethyleneimine derivatized with guanidinium and octyl groups to bind the phosphate group of the FMN and bring this cofactor into a locally hydrophobic microenvironment. Once incorporated into the polymer, the FMN is able to efficiently collect electron pairs from NADH and insure a stepwise single-electron delivery to manganese(III) porphyrins with no use of other heavy metal, enzyme or additional chemical. Finally, we demonstrate that the obtained Mn(II) porphyrin can then activate dioxygen from the air and perform catalytic oxidation in water at room temperature.

Results

Construction of the artificial flavoenzyme. Since the flavin mononucleotide (FMN) cofactor (Fig. 2) usually binds to proteins by simple electrostatic interactions[26], we decided to design bio-inspired artificial systems based on the same concept. To do so, a modified PEI bearing guanidinium and octyl groups was prepared as previously described, by reaction of the commercial PEI (25 kDa, multibranched) with praxadine and iodooctane in DMF[23,24]. Its purification was then performed by extensive dialysis in aqueous solutions (Fig. 3a,b). In this system, guanidinium groups were chosen for their specific affinity towards phosphate groups and their stability over a large pH range, whereas octyl groups were added for the creation of a locally hydrophobic microenvironment.

Flavin cofactors, such as FMN and riboflavin (Fig. 2), absorb light with two specific bands at 370 and 470 nm in aqueous solution. The former band at 370 nm is known to be sensitive to its environment, with a hypsochromic shift of ~10 nm when solubilized in hydrophobic environments[27,28]. Figure 3d shows the ultraviolet-visible absorption spectra of FMN in pure water and in the presence of the modified PEI (guanidinium groups 0.2 equiv. per monomer/octyl groups 0.4 equiv. per monomer). As one can observe, the addition of the modified PEI to the aqueous solution of FMN induces a 12-nm shift of the band at 370 nm. When the same experiment was realized with riboflavin, that bears no phosphate group in its structure, no shift could be observed (Supplementary Fig. 1). This clearly indicates that the FMN is brought into an hydrophobic environment on the addition of the polymer (Fig. 3c), suggesting that its phosphate group interacts

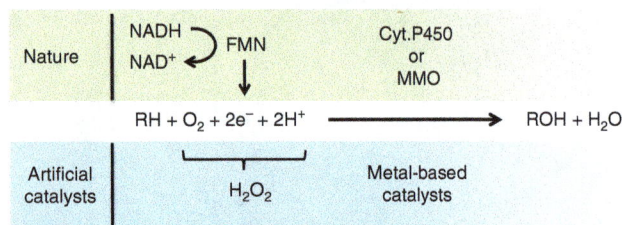

Figure 1 | Natural and bio-inspired catalytic oxidations. Catalytic oxidation performed at the active site of monooxygenase enzymes by reductive activation of dioxygen, and the corresponding peroxide shunt used by most of the handmade catalysts developed so far.

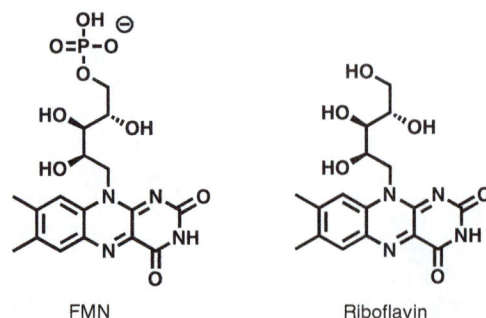

Figure 2 | Flavin cofactors. FMN and riboflavin.

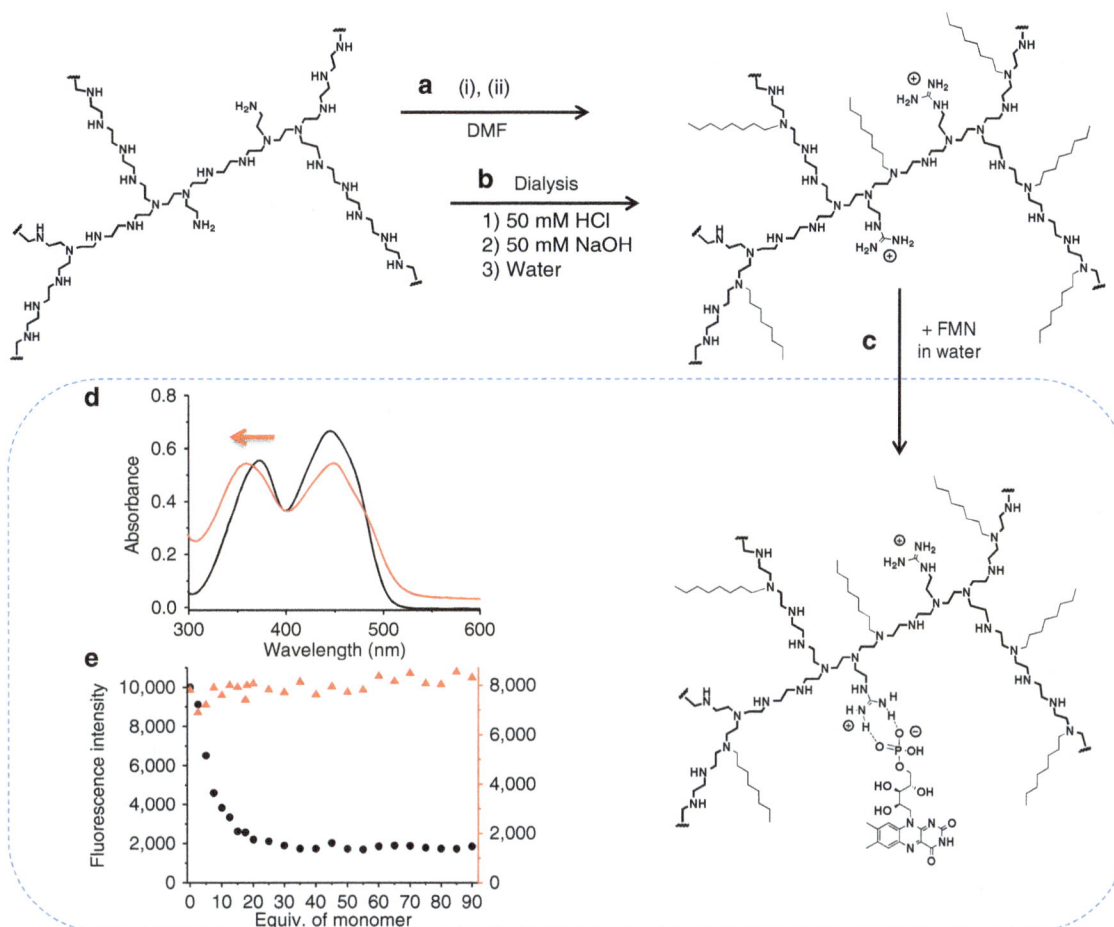

Figure 3 | Preparation of the artificial flavoenzyme. (**a**) Synthesis of the modified PEI by reaction of a commercial 25-kDa multibranched PEI with (i) praxadine (1*H*-pyrazole-1-carboxamidine hydrochloride) and (ii) iodooctane in DMF at room temperature, (**b**) purification by extensive dialysis in 20% EtOH in 50 mM HCl; 50 mM HCl; water; 50 mM NaOH and water, (**c**) incorporation of FMN into the polymer, (**d**) ultraviolet-visible spectra of a 50-μM solution FMN in water (black trace) and in the presence of the modified PEI (4 mM in monomer; red trace). (**e**) Fluorescence intensity of 1 μM solutions of FMN (black circles) and riboflavin (red triangles) at 530 nm on excitation at 450 nm in the presence of an increasing amount of the modified PEI in water.

with the guanidinium groups of the polymer, whereas the riboflavin remains in solution. To confirm this incorporation, fluorescence spectra of both FMN and riboflavin were measured on the addition of increasing amount of the modified polymer. Figure 3e (and Supplementary Fig. 2) shows that only the fluorescence of the FMN is quenched by the presence of the polymer, indicating that riboflavin remains in aqueous solution, whereas the FMN interacts closely enough with the polymer to have its fluorescence quenched by its direct environment[29].

Reactivity of the artificial flavoenzyme with NADH. With the FMN buried into the locally hydrophobic microenvironment of the modified PEI, we endeavoured to study its reactivity with NADH in water. As one could expect, when a 200-μM (or even a 500 μM) water solution of NADH was added to a 50-μM FMN water solution, the hydride transfer between the two molecules did not occur, and no reduction of the FMN was observed (Fig. 4a). Conversely, when the FMN (50 μM) was incorporated into the modified PEI, the addition of NADH (200 μM) to the solution induced the decrease in the absorbance at 470 nm, indicating a reduction of the FMN (Fig. 4b)[30]. This reduction process was optimal for the addition of 4 equiv. of NADH with respect to the FMN, and did not improve for higher concentrations.

The initial rate for the NADH oxidation by the FMN was then measured by following the disappearance of NADH at 340 nm in the presence and in the absence of the modified PEI. The values obtained are, respectively, 4.00 ($\pm$ 0.10) $\times$ 10^{-6} and 2.00 ($\pm$ 0.20) $\times$ 10^{-8} mol l^{-1} min^{-1}, giving a rate enhancement of 200 when the FMN was buried into the polymer. To compare this polymeric system to other NADH/FMN oxidoreductases, we also calculated the second-order rate constant of the reaction $k = 500\,\mathrm{M}^{-1}\,\mathrm{s}^{-1}$, which is, respectively, only three or four orders of magnitude slower than the natural NADH-specific FMN oxidoreductases from *Beneckea harveyi*[31] and from *Pseudomonas putida*[32], but is sevenfold faster compared with the semi-synthetic enzyme, flavopapain developed by Kaiser and co-workers[33,34].

Mechanistic implications on the hydride transfer. Since the FMN is not reduced by NADH in solution, one may suggest two effects of the modified PEI for this reduction. First, the polymer may favour the interaction between the two molecules into its locally hydrophobic microenvironment, by interaction of its guanidinium groups with the phosphate groups of the FMN and NADH. Alternatively, the incorporation of the FMN into the microenvironment of the polymer may change its redox potential and thermodynamically allow its reduction by NADH. To answer this question, we performed electrochemical analysis with the

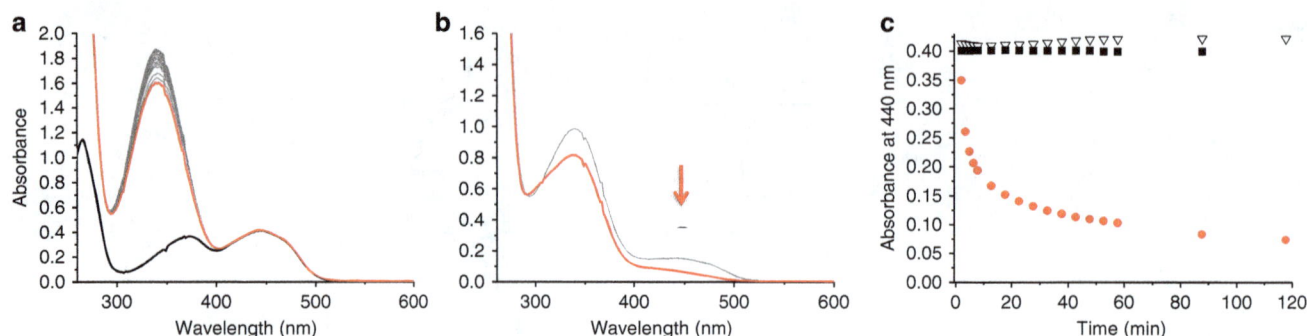

Figure 4 | Influence of the modified PEI on the reduction of FMN by NADH. (**a**) Ultraviolet-visible spectra of a 50-μM deoxygenated aqueous solution of FMN before (black trace) and after addition of NADH (500 μM) followed for 2 h with the final curve in red. (**b**) Ultraviolet-visible spectra of a deoxygenated aqueous solution of FMN (50 μM) incorporated in the modified PEI (1.25 mM in monomer) in the presence of NADH (200 μM; (black traces) followed for 2 h with the final curve in red. (**c**) Time courses for the reduction of FMN (50 μM) by NADH (500 μM) in deoxygenated water only (reverse triangle), by NADH (200 μM) in presence of the modified polymer (1.25 mM in monomer; red circle), without NADH but in the presence of the modified polymer (1.25 mM; black square).

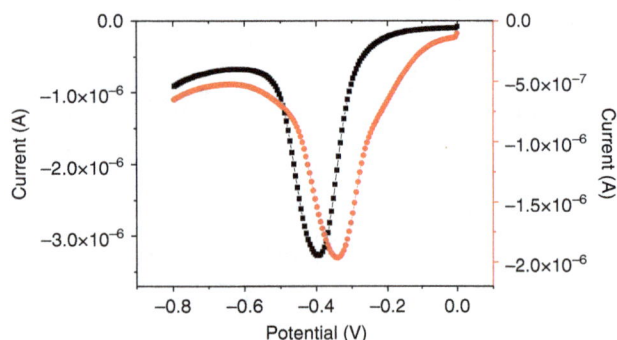

Figure 5 | Electrochemistry. Square wave voltamogram of FMN (0.1 mM) in HEPES buffer (10 mM; pH = 7) with KCl 10 mM (black squares) and square wave voltamogram of FMN (0.1 mM) in HEPES buffer (10 mM; pH = 7) in the presence of the modified PEI (2,5 mM in monomer; red circles). Step potential 0.005 V, amplitude − 0.020 V, frequency 25 Hz.

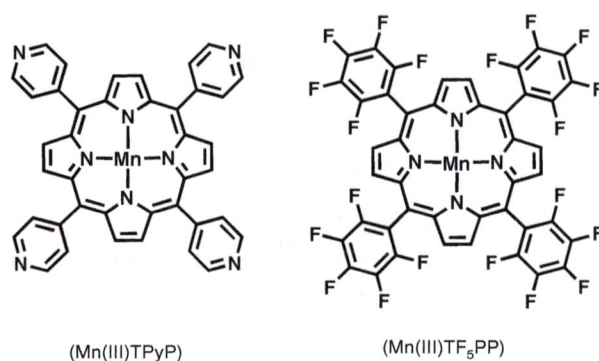

(Mn(III)TPyP)   (Mn(III)TF₅PP)

Figure 6 | Manganese porphyrins used in this work. Manganese(III)-meso-tetrapyridinyl-porphyrin (Mn(III)TPyP) and manganese(III)-meso-tetra(pentafluorophenyl)porphyrin (Mn(III)TF$_5$PP).

FMN free in solution, and with the FMN incorporated into the polymer (Fig. 5). The reduction potential measured by square wave voltammetry experiments for the FMN in aqueous solution was − 395 mV versus Ag/AgCl (− 203 mV versus Normal Hydrogen Electrode (NHE)), whereas in the presence of polymer the reduction potential was shifted to − 330 mV versus Ag/AgCl (− 138 mV versus NHE). This shift of 65 mV clearly indicates that the hydrophobic environment, created by the modified PEI, has a rather large influence on the reduction potential of the FMN. For comparison, the same experiment realized with the riboflavin shows only a 10-mV shift, which correlates with its none or poor incorporation into the polymer (Supplementary Fig. 3).

The thermodynamic redox potential for the NAD$^+$/NADH couple was previously estimated based on equilibrium measurements for some enzyme-catalysed reactions to be − 315 mV versus NHE[35]. This potential is thermodynamically low enough to reduce both the FMN in solution and the FMN buried into the polymer, implying that the 65-mV shift observed is not responsible for the generation of the reductase activity within the polymer. One can then conclude that the main role of the polymer is to bring together the two protagonists of the reaction in the same local hydrophobic microenvironment, and to facilitate the hydride transfer by lowering the energy barrier of the transition state. This is supported by results obtained with

supramolecular flavin/NADH models, which clearly demonstrate the importance of an optimal relative disposition of the redox partners for the hydride transfer[36]. Finally, it is also worth to note that once reduced by the addition of NADH, the FMN was easily (re-)oxidized on exposure to dioxygen and then reduced again by the addition of NADH, making this system capable of multiple turn over reduction catalysis.

Reduction of a water-soluble manganese porphyrin. Since this artificial enzyme (modified PEI + FMN) was demonstrated to efficiently gather electrons from NADH, we also attended to study the possible transfer of those electrons towards a redox partner, such as a manganese(III)-meso-tetrapyridiniumyl-porphyrin (Mn(III)TPyP; Fig. 6). In this case, manganese porphyrins were chosen for the following two reasons: (1) Mn(III) and Mn(II) porphyrins have distinctive absorption bands at 470 and 440 nm, respectively, and (2) their absorption molar coefficients are high enough to be observed even in the presence of FMN. In a typical experiment, a stoichiometric amount of Mn(III)TPyP was added to a 10-μM aqueous solution of FMN before adding 4 equiv. of NADH, and the reaction was followed by monitoring the ultraviolet-Vis spectrum of the resulting solution as a function of time (Fig. 7). When the experiment was realized in the absence of the modified polymer, the characteristic band of the Mn(III) porphyrin at 470 nm did not change on addition of NADH, as well as for the following 60 min (Fig. 7a).

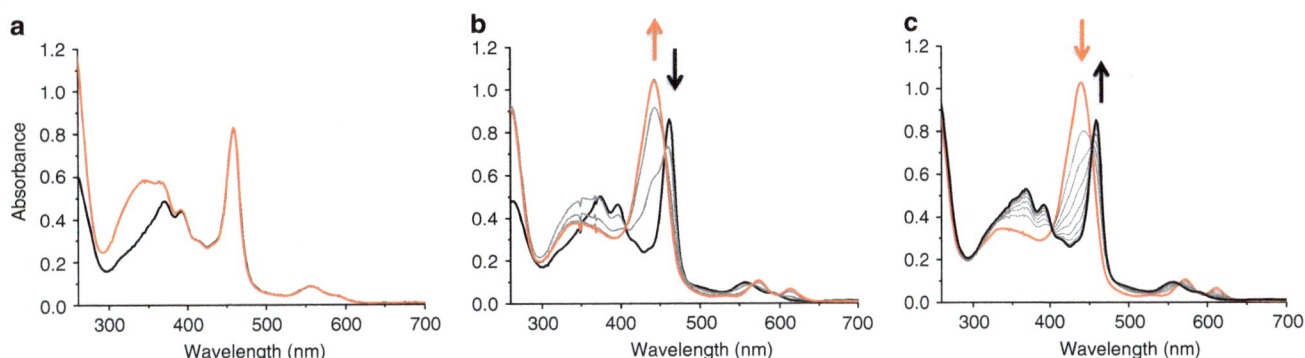

Figure 7 | Reduction of a water-soluble manganese porphyrins. (**a**) Ultraviolet-visible spectrum of a deoxygenated aqueous solution of Mn(III)TPyP (10 μM) and FMN (10 μM; black trace) and its evolution for 1 h after addition of NADH (4 equiv.) at 20 °C (red trace). (**b**) Ultraviolet-visible spectrum of a deoxygenated aqueous solution of Mn(III)TPyP (10 μM) and FMN (10 μM) in the presence of the modified polymer (250 μM in monomer; black trace) and its evolution for 2 min after the addition of NADH (4 equiv.) at 20 °C (red trace). (**c**) Ultraviolet-visible spectra of a deoxygenated 10-μM aqueous solution of the reduced Mn(II)TPyP (10 μM) with FMN (10 μM; red trace) and its evolution for 6 min on exposure to O₂ at 20 °C (black traces).

Conversely, when the same experiment was realized in the presence of the modified polymer, the absorption band at 470 nm was rapidly replaced by a new band at 440 nm, characteristic of the reduced Mn(II)TPyP (Fig. 7b)[37]. The appearance of clear isosbestic points at 405, 460 and 570 nm during this fast process also demonstrated the direct one electron reduction of the manganese(III) porphyrin, without the observation of any other intermediate (Fig. 7b). Finally, once exposed to dioxygen, this Mn(II)TPyP was rapidly re-oxidized to Mn(III)TPyP with the same isosbestic points previously observed during the reduction process (Fig. 7c). It is worth noting that this one electron reduction of the manganese(III) porphyrin implies the splitting of the initial electron pair collected from NADH as an hydride ion.

Splitting electron pairs from NADH, to perform single-electron transfer, is one of the key features involved in important biological processes. Hence, we decided to titrate the reduction process of the Mn(III)TPyP by successive addition of NADH in the presence of the modified polymer to better understand this electron transfer. Figure 8 shows the time course for this reduction by following the formation of the Mn(II)TPyP at 440 nm. As one can observe, below 0.5 equiv. of NADH added, the reduction of the Mn porphyrin was not completed. However, the addition of 0.5 equiv. of NADH was enough to fully reduce the Mn porphyrin, while the addition of larger amount of the reducing agent only changed the kinetics of the reduction. This titration demonstrates that one electron pair, issued from one NADH molecule, is enough to reduce two Mn porphyrin molecules via the FMN cofactor, implying the splitting of the electron pair during the process. This is fully reminiscent of the biological activity of cytochrome P450 reductases and will be of great interest in the perspective of the development of new bio-inspired reduction systems, especially for the development of environment-friendly oxidation catalysts.

In terms of kinetics, Fig. 9 shows that the reduction of the Mn(III) porphyrin, in the presence of the modified polymer, is completed within 2 min after the addition of NADH (4 equiv.), whereas almost no reduction is observed in pure water. The initial rates of the reactions were measured and gave $3.90\,(\pm 0.60) \times 10^{-6}\,M\,min^{-1}$ in the presence of the modified polymer and $9.65\,(\pm 0.57) \times 10^{-10}$ $M\,min^{-1}$ in water only. In these conditions, second-order rate constant calculated from the equation $v = k[MnTPyP][NADH]$ gave $k_2 = 163\,M^{-1}\,s^{-1}$ and $k_2 = 0.04\,M^{-1}\,s^{-1}$, which correspond to a rate enhancement of 4×10^3. If the commercial polymer is used instead of the 'synzyme', the reaction does occur, but much more slowly compared with the modified polymer (Fig. 9a).

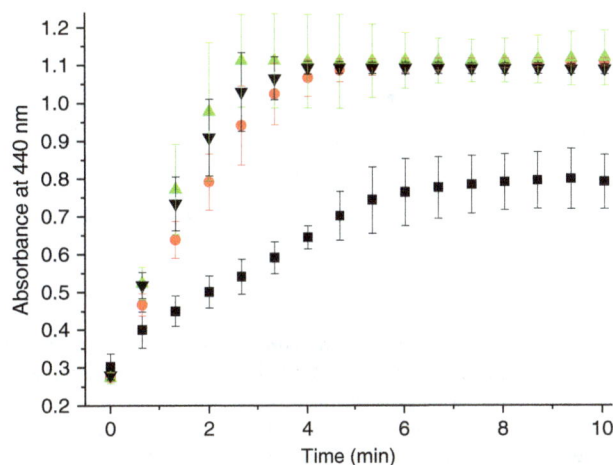

Figure 8 | Titration of NADH for the reduction of a water-soluble manganese porphyrin. Time courses for the reduction of Mn(III)TPyP (10 μM) in the presence FMN (10 μM) and modified PEI (250 μM in monomer) by following the absorbance of Mn(II)TPyP at 440 nm on the addition of 0.25 equiv. (black square), 0.5 equiv. (red circle), 1 equiv. (green up triangle) and 2 equiv. (blue down triangle) of NADH in deoxygenated water at 25 °C (all experiments were realized in triplicate).

This is in good agreement with the fact that the commercial polymer is positively charged in water, which allows its interaction with the FMN and NADH, favouring the hydride transfer. However, the lack of specificity for the interaction, combined with the absence of hydrophobic microenvironment, drastically slows down the reduction process. Therefore, the screening of polymers derivatized with various amount of octyl groups will be, in the future, a precious tool to better understand the effect of a locally hydrophobic microenvironment on the kinetics of electron transfer in water. Finally, it is worth to note that, on exposure to dioxygen, the oxidation of the Mn(II) porphyrin took place at the same rate, whether the experiment was performed with the modified or with the commercial polymer (Fig. 9b). This suggests that the oxidation kinetic mostly relies on the diffusion of dioxygen in water, and that the Mn porphyrin is not incorporated within the polymer but equally dispersed in solution in both cases.

Examples of artificial flavoreductase capable of reducing metal complexes are very scarce, and only two systems are provided

a

b

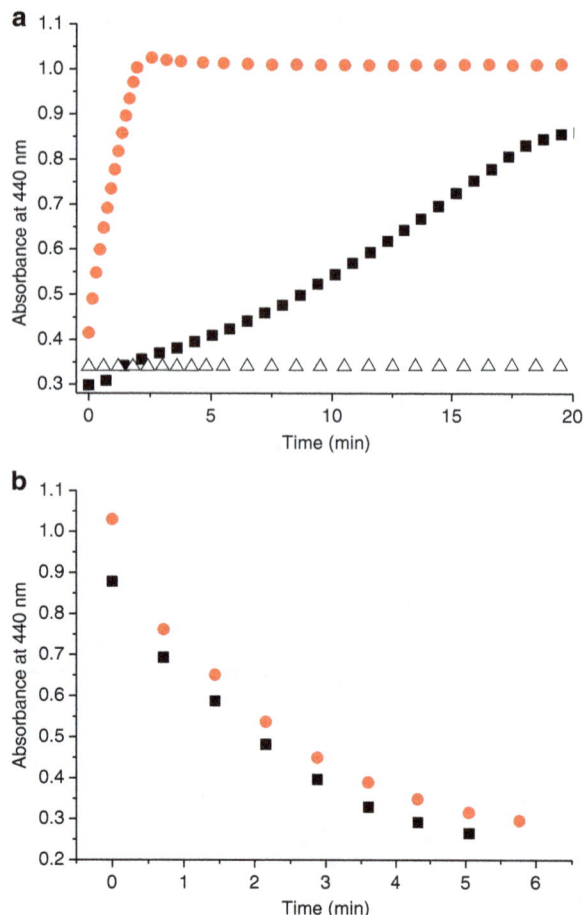

Figure 9 | Reduction and O₂ re-oxidation of a water-soluble manganese porphyrin. (**a**) Time course of the Mn(III)TPyP (10 μM) reduction on addition of NADH (4 equiv.) under inert conditions by following the formation of the reduced Mn(II)TPyP at 440 nm, in the presence of the modified PEI (250 μM) and FMN (10 μM; red circle), in the presence of the commercial PEI (250 μM) and FMN (10 μM; black square) and in the presence of FMN (10 μM) only (open triangle). (**b**) Time course of the Mn(II)TPyP (10 μM) oxidation on exposure to dioxygen in the presence of the modified PEI (250 μM; red circle) and in the presence of the commercial PEI (250 μM; black square) by following its absorbance at 440 nm.

with kinetic information about the reduction process. A synthetic multicomponent redox system, bearing a manganese porphyrin and an amphiphilic flavin, both inserted within a phospholipidic bilayer, was described in association with a pyruvate oxidase in charge of delivering electrons[12]. In this case, the appearance of the manganese(II) porphyrin was followed at 440 nm and was shown to take more than 3 h to be completed. Similarly, a flavocyclodextrin system was demonstrated to be able of reducing a manganese(III) porphyrin using synthetic nicotinamide with second-order rate constants sixfold better compared with the FMN in solution[38]. Since nicotinamide molecules used in those studies are not always the same, the best way to compare the activity of our polymer with the literature is to compare the values of the rate enhancement of the various systems with one of FMN in solution. In that case, the 'synzyme' appears to be three orders of magnitude better compared with the flavocyclodextrin[38].

Dioxygen activation in water. Finally, since the Mn(III)TPyP is efficiently reduced by the 'synzyme' in the presence of NADH and then (re)oxidized by dioxygen, we endeavoured to perform

Figure 10 | Schematic representation of the complete catalytic system. The artificial flavoenzyme collects electrons from NADH and provides a single-electron flow to the Mn(III) porphyrins to activate dioxygen and perform catalytic oxidation in water.

catalysis in water by simple use of dioxygen as the oxidant, using thioanisole as the substrate. In a typical experiment, 1,000 equiv. of substrate (100 mM final) were added to a water solution of the 'synzyme' (modified PEI + FMN) and Mn(III)TPyP (1 equiv. versus FMN; 0.1 mM final), before adding 10 equiv. of NADH (1 mM final). On addition of NADH, the reddish solution turned green after a few minutes, and the mixture was allowed to stir at room temperature for 5 h under oxygenated atmosphere (O₂ at 1 atm). The mixture was then filtered through a short silica gel column to remove the polymer, the FMN and the Mn porphyrin, before being analysed using gas chromatography. Under these conditions, the oxidation of thioanisole gave the sulfoxide as an unique product of the reaction, but the yield was not significantly higher than the control experiment without the polymer. Other control experiments, performed either without Mn porphyrin, or without NADH, did not generate oxidation product at all. Since the use of Mn(III)TPyP was not adapted to catalysis, we also tested a fluorinated Mn porphyrin (manganese(III)-*meso*-tetra (pentafluorophenyl)porphyrin (Mn(III)TF₅PP)), which is known as one of the best porphyrin for oxidation catalysis. In this case, the selectivity of the reaction did not change but the yield was improved to 28% (value obtained after subtracting the blank value obtained without the polymer). This rather hydrophobic porphyrin was initially solubilized in acetonitrile and added to the aqueous solution to obtain a 9:1 (water:MeCN) final mixture. It is worth noting that in the absence of the polymer, this Mn porphyrin precipitated a few minutes after its incorporation into the aqueous solution. Conversely, the solution was stable for hours in the presence of the polymer, demonstrating that the locally hydrophobic microenvironment of the polymer allows the solubilization of the Mn porphyrin in aqueous medium[39]. This modest yield of 28% for catalytic oxidation is not surprising since three major scientific locks have to be tackled simultaneously: (1) selective electron delivery at metal centres, (2) reductive activation of dioxygen at this metal centre and (3) catalysis in water. Nevertheless, these results clearly demonstrate the achievement, by a water-soluble handmade system, of a complete catalytic cycle reminiscent of the activities of iron monooxygenases such as cytochromes P450 (Fig. 10). This system is also simple enough to be easily adapted to various catalytic processes requiring a steady and accurate input of electrons.

Discussion

In summary, we have described the incorporation, by electrostatic interactions, of FMN cofactors into a water-soluble PEI bearing guanidinium and octyl groups. This macromolecular entity was demonstrated to be capable of efficiently collecting electron pairs from NADH and rapidly delivering single electrons towards redox cofactors such as manganese(III) porphyrin, with a rate enhancement of 4×10^3 compared with FMN in solution. This activity is fully reminiscent of cytochrome P450 reductase and opens new horizons in terms of green catalytic oxidation systems on the basis of the reductive activation of dioxygen at metal centres. As a proof of concept, this artificial reductase was associated with manganese(III) porphyrin in solution and was exposed to dioxygen in the presence of NADH. Under these conditions, the entire system was able to catalyse sulfoxidation reaction in water, at room temperature and using dioxygen as the sole source of oxygen. This system could easily be adapted to the use of other catalysts such as mononuclear or dinuclear iron complexes, but also copper complexes and other catalysts using electrons for their catalytic activity in water.

Methods

Instrumentation. All ultraviolet-visible measurements were performed with a Varian carry 300-bio ultraviolet-vis spectrophotometer in cuvettes equipped with septa for experiments under inert atmosphere. Fluorescence spectra were obtained with a Tecan infinite M200 pro-plate-reader. Gas chromatography analyses were performed using a SHIMADZU GC-2014A.

Synthesis of the artificial flavoenzyme. FMN, riboflavin and NADH were purchased from commercial suppliers and used without further purification. The TPyP and the TF$_5$PP were synthesized according to literature[40,41], and the metal insertion was carried out according to the method in ref. 42.

The modified polymer (guanidinium groups 0.2 equiv. per monomer/octyl groups 0.4 equiv. per monomer) was synthesized from multibranched PEI (25 kDa) that was dissolved in DMF to give a final PEI concentration of 8.6 mg ml^{-1} (200 mM in monomer residues). For derivatization with guanidinium groups, 4 equiv. of triethylamine per monomer was added to the PEI solution, and a freshly prepared DMF solution of 1-H-pyrazole-1-carboxamidin hydrochloride (1 ml at 1.2 mol l^{-1}) was added under vigorous stirring to 30 ml of the PEI solution. This reaction mixture was then left under stirring overnight at room temperature. For alkylation, 0.433 ml of iodooctane was added dropwise under vigorous stirring, which was continued for 4 days at room temperature. For purification, the crude reaction mixture was diluted (1:2) into hydrochloric acid (50 mM) and transferred to a dialysis tube (Spectra/Por membrane, Mw cutoff 14,000). The resulting solution was dialysed under slow stirring against each of the following buffers for at least 2 h: 20% EtOH in 50 mM HCl; 50 mM HCl; distilled water (twice); 50 mM NaOH (twice); and water (three times).

Reduction of the artificial flavoenzyme. For the reduction of FMN by NADH (with or without polymer), solutions of FMN (2.5 mM), modified polymer (20 mM), NADH (12.5 mM) and water were prepared separately and degassed for 1 h. In parallel, a ultraviolet-visible cuvette (2.5 ml), equipped with a septum, was degassed and a solution of FMN (50 μM) and modified PEI (1.2 mM) was prepared before injection of NADH (200 μM) through the septum. The same experiment was realized without polymer.

Electrochemistry. For electrochemistry experiments, a solution of FMN (0.1 mM) was prepared in HEPES buffer (10 mM). The square wave voltamogram was recorded under argon before and after the addition of modified polymer (2.5 mM) between 0 and -0.8 V. In absence of the polymer, the experiment was realized with KCl 10 mM as an electrolyte, which is not needed in the presence of the positively charged polymer. Measurements were realized with a carbon electrode, electrode of reference (Ag/AgCl) and counter electrode (Pt). (Step potential 0.005 V, amplitude 0.020 V, frequency 25 Hz.).

Reduction of a water-soluble manganese porphyrin. For the reduction of a water-soluble manganese porphyrin, an aqueous solution of FMN (10 μM), modified polymer (250 μM in monomer) and MnTPyP (10 μM) was prepared directly in a quartz cuvette (2.5 ml) equipped with a septum. In parallel, a solution of NADH (2.5 mM) was prepared in water, and both solutions were degassed for 30 min. Then, the kinetic spectrum for the reduction of MnTPyP was recorded at 20 °C under argon after the addition of NADH (40 μM). The same experiment was realized without polymer. After reduction, the cuvette was opened to the air to follow the oxidation process.

Dioxygen activation and catalysis. Catalysis experiments were realized in 1.5-ml vials in aqueous solutions of FMN (100 μM), MnTPyP (100 μM), modified polymer (2.5 mM in monomer) and thioanisole (100 mM) with a final volume of 1 ml. NADH (1 mM) was then added under vigorous stirring, and the mixture was left under stirring for 5 h at room temperature under O$_2$ atmosphere (1 atm). For GC analysis, acetophenone was added to the solution as internal standard and the mixtures were filtered through a short silica gel column (on Pasteur pipette) before injection in a Zebron ZB Semi Volatiles column (30 m × 0.25 mm × 0.25 μm). GC conditions were as follows: 100–130 °C, 5 °C min^{-1}, then 130–300 °C, 50 °C min^{-1}, then hold for 3 min. Injector and FID temperature was 300 °C. Retention times (min) are as follows: acetophenone, 4.03; thioanisole, 4.32; thioanisole sulfoxide, 7.56; and thioanisole sulfone, 8.04.

References

1. Merkx, M. *et al.* Dioxygen activation and methane hydroxylation by soluble methane monooxygenase: A tale of two irons and three proteins. *Angew. Chem. Int. Ed.* **40**, 2782–2807 (2001).
2. Meunier, B., de Visser, S. P. & Shaik, S. Mechanism of oxidation reactions catalyzed by cytochrome P450 enzymes. *Chem. Rev.* **104**, 3947–3980 (2004).
3. Wang, M. *et al.* Three-dimensional structure of NADPH–cytochrome P450 reductase: prototype for FMN- and FAD-containing enzymes. *Proc. Natl Acad. Sci. USA* **94**, 8411–8416 (1997).
4. Iyanagi, T., Xia, C. & Kim, J.-J. P. NADPH–cytochrome P450 oxidoreductase: prototypic member of the diflavin reductase family. *Arch. Biochem. Biophys.* **528**, 72–89 (2012).
5. Shen, A. L. & Kasper, C. B. Role of Ser457 of NADPH – cytochrome P450 oxidoreductase in catalysis and control of FAD oxidation – reduction potential. *Biochemistry (Mosc.)* **35**, 9451–9459 (1996).
6. Xia, C. *et al.* Conformational changes of NADPH-cytochrome P450 oxidoreductase are essential for catalysis and cofactor binding. *J. Biol. Chem.* **286**, 16246–16260 (2011).
7. Feiters, M. C., Rowan, A. E. & Nolte, R. J. M. From simple to supramolecular cytochrome P450 mimics. *Chem. Soc. Rev.* **29**, 375–384 (2000).
8. Van Esch, J., Roks, M. F. M. & Nolte, R. J. M. Membrane-bound cytochrome P-450 mimic. Polymerized vesicles as microreactors. *J. Am. Chem. Soc.* **108**, 6093–6094 (1986).
9. Schenning, A. P. H. J., Hubert, D. H. W., van Esch, J. H., Feiters, M. C. & Nolte, R. J. M. Novel bimetallic model system for cytochrome P450: effect of membrane environment on the catalytic oxidation. *Angew. Chem. Int. Ed. Engl.* **33**, 2468–2470 (1995).
10. Schenning, A. P. H. J., Lutje Spelberg, J. H., Hubert, D. H. W., Feiters, M. C. & Nolte, R. J. M. A supramolecular cytochrome P450 mimic. *Chem. Eur. J.* **4**, 871–880 (1998).
11. Tabushi, I. & Kodera, M. Flavin-catalyzed reductive dioxygen activation with N-methyldihydronicotinamide. *J. Am. Chem. Soc.* **108**, 1101–1103 (1986).
12. Groves, J. T. & Ungashe, S. B. Biocompatible catalysis. Enzymic reduction of metalloporphyrin catalysts in phospholipid bilayers. *J. Am. Chem. Soc.* **112**, 7796–7797 (1990).
13. Wilker, J. J., Dmochowski, I. J., Dawson, J. H., Winkler, J. R. & Gray, H. B. Substrates for rapid delivery of electrons and holes to buried active sites in proteins. *Angew. Chem. Int. Ed.* **38**, 89–92 (1999).
14. Dunn, A. R., Dmochowski, I. J., Winkler, J. R. & Gray, H. B. Nanosecond photoreduction of cytochrome P450cam by channel-specific Ru-diimine electron tunneling wires. *J. Am. Chem. Soc.* **125**, 12450–12456 (2003).
15. Tran, N.-H. *et al.* Light-initiated hydroxylation of lauric acid using hybrid P450 BM3 enzymes. *Chem. Commun.* **47**, 11936–11938 (2011).
16. Avenier, F. *et al.* Photoassisted generation of a dinuclear iron(III) peroxo species and oxygen-atom transfer. *Angew. Chem. Int. Ed.* **52**, 3634–3637 (2013).
17. Spetnagel, W. J. & Klotz, I. M. Catalysis of decarboxylation of oxalacetic acid by modified poly(ethylenimines). *J. Am. Chem. Soc.* **98**, 8199–8204 (1976).
18. Suh, J., Scarpa, I. S. & Klotz, I. M. Catalysis of decarboxylation of nitrobenzisoxazolecarboxylic acid and of cyanophenylacetic acid by modified polyethylenimines. *J. Am. Chem. Soc.* **98**, 7060–7064 (1976).
19. Suh, J. & Hong, S. H. Catalytic activity of Ni(II) – terpyridine complex in phosphodiester transesterification remarkably enhanced by self-assembly of terpyridines on poly(ethylenimine). *J. Am. Chem. Soc.* **120**, 12545–12552 (1998).
20. Liu, L. & Breslow, R. A potent polymer/pyridoxamine enzyme mimic. *J. Am. Chem. Soc.* **124**, 4978–4979 (2002).
21. Liu, L., Rozenman, M. & Breslow, R. Hydrophobic effects on rates and substrate selectivities in polymeric transaminase mimics. *J. Am. Chem. Soc.* **124**, 12660–12661 (2002).
22. Hollfelder, F., Kirby, A. J. & Tawfik, D. S. Efficient catalysis of proton transfer by synzymes. *J. Am. Chem. Soc.* **119**, 9578–9579 (1997).
23. Avenier, F., Domingos, J. B., Van Vliet, L. D. & Hollfelder, F. Polyethylene imine derivatives ('synzymes') accelerate phosphate transfer in the absence of metal. *J. Am. Chem. Soc.* **129**, 7611–7619 (2007).
24. Avenier, F. & Hollfelder, F. Combining medium effects and cofactor catalysis: metal-coordinated synzymes accelerate phosphate transfer by 108. *Chem. Eur. J* **15**, 12371–12380 (2009).
25. Haimov, A., Cohen, H. & Neumann, R. Alkylated polyethyleneimine/ polyoxometalate synzymes as catalysts for the oxidation of hydrophobic substrates in water with hydrogen peroxide. *J. Am. Chem. Soc.* **126**, 11762–11763 (2004).
26. Sevrioukova, I. F., Li, H., Zhang, H., Peterson, J. A. & Poulos, T. L. Structure of a cytochrome P450–redox partner electron-transfer complex. *Proc. Natl Acad. Sci. USA* **96**, 1863–1868 (1999).
27. Greaves, D., Deans, M., Galow T., R. H. & M. Rotello, V. Flavins as modular and amphiphilic probes of silica microenvironments. *Chem. Commun.* **9**, 785–786 (1999).

28. Jordan, B. J. *et al.* Polymeric model systems for flavoenzyme activity: towards synthetic flavoenzymes. *Chem. Commun.* **12**, 1248–1250 (2007).

29. Agasti, S. S. *et al.* Dendron-based model systems for flavoenzyme activity: towards a new class of synthetic flavoenzyme. *Chem. Commun.* **35**, 4123–4125 (2008).

30. Simtchouk, S., Eng, J. L., Meints, C. E., Makins, C. & Wolthers, K. R. Kinetic analysis of cytochrome P450 reductase from *Artemisia annua* reveals accelerated rates of NADPH-dependent flavin reduction. *FEBS J.* **280**, 6627–6642 (2013).

31. Jablonski, E. & DeLuca, M. Purification and properties of the NADH and NADPH specific FMN oxidoreductases from *Beneckea harveyi*. *Biochemistry (Mosc.)* **16**, 2932–2936 (1977).

32. Kuznetsov, V. Y. *et al.* The putidaredoxin reductase-putidaredoxin electron transfer complex: theoretical and experimental studies. *J. Biol. Chem.* **280**, 16135–16142 (2005).

33. Levine, H. L. & Kaiser, E. T. Oxidation of dihydronicotinamides by flavopapain. *J. Am. Chem. Soc.* **100**, 7670–7677 (1978).

34. Kaiser, E. T. & Lawrence, D. S. Chemical mutation of enzyme active sites. *Science* **226**, 505–511 (1984).

35. Saleh, F. S., Rahman, M. R., Okajima, T., Mao, L. & Ohsaka, T. Determination of formal potential of NADH/NAD + redox couple and catalytic oxidation of NADH using poly(phenosafranin)-modified carbon electrodes. *Bioelectrochemistry* **80**, 121–127 (2011).

36. Reichenbach-Klinke, R., Kruppa, M. & König, B. NADH model systems functionalized with Zn(II)-cyclen as flavin binding sitestructure dependence of the redox reaction within reversible aggregates. *J. Am. Chem. Soc.* **124**, 12999–13007 (2002).

37. Harriman, A. & Porter, G. Photochemistry of manganese porphyrins. Part 1.-Characterisation of some water soluble complexes. *J. Chem. Soc. Faraday Trans. 2* **75**, 1532–1542 (1979).

38. Tabushi, I. & Kodera, M. Flavocyclodextrin as a promising flavoprotein model. Efficient electron transfer catalysis by flavocyclodextrin. *J. Am. Chem. Soc.* **109**, 4734–4735 (1987).

39. Ren, Q.-Z., Yao, Y., Ding, X.-J., Hou, Z.-S. & Yan, D.-Y. Phase-transfer of porphyrins by polypeptide-containing hyperbranched polymers and a novel iron(iii) porphyrin biomimetic catalyst. *Chem. Commun.* **31**, 4732–4734 (2009).

40. Adler, A. D. *et al.* A simplified synthesis for meso-tetraphenylporphine. *J. Org. Chem.* **32**, 476–476 (1967).

41. Lindsey, J. S., Schreiman, I. C., Hsu, H. C., Kearney, P. C. & Marguerettaz, A. M. Rothemund and Adler-Longo reactions revisited: synthesis of tetraphenylporphyrins under equilibrium conditions. *J. Org. Chem.* **52**, 827–836 (1987).

42. Adler, A. D., Longo, F. R., Kampas, F. & Kim, J. On the preparation of metalloporphyrins. *J. Inorg. Nucl. Chem.* **32**, 2443–2445 (1970).

Acknowledgements

We thank the '*Agence Nationale de la Recherche*' (ANR BIOXICAT) and the '*Région Ile de France, réseau R2DS*' for financial support. We thank Hafsa Korri-Youssoufi for her help in the interpretation of electrochemical experiments.

Author contributions

F.A. and J.-P.M. designed the research and wrote the paper. Y.R. performed all the experiments apart from the synthesis of the manganese porphyrins, and R.R. synthesized the manganese porphyrins.

Additional information

6

In silico prediction and screening of modular crystal structures via a high-throughput genomic approach

Yi Li[1], Xu Li[1], Jiancong Liu[1], Fangzheng Duan[1] & Jihong Yu[1]

High-throughput computational methods capable of predicting, evaluating and identifying promising synthetic candidates with desired properties are highly appealing to today's scientists. Despite some successes, *in silico* design of crystalline materials with complex three-dimensionally extended structures remains challenging. Here we demonstrate the application of a new genomic approach to ABC-6 zeolites, a family of industrially important catalysts whose structures are built from the stacking of modular six-ring layers. The sequences of layer stacking, which we deem the genes of this family, determine the structures and the properties of ABC-6 zeolites. By enumerating these gene-like stacking sequences, we have identified 1,127 most realizable new ABC-6 structures out of 78 groups of 84,292 theoretical ones, and experimentally realized 2 of them. Our genomic approach can extract crucial structural information directly from these gene-like stacking sequences, enabling high-throughput identification of synthetic targets with desired properties among a large number of candidate structures.

[1] State Key Laboratory of Inorganic Synthesis and Preparative Chemistry, Jilin University, Qianjin Street 2699, Changchun 130012, China. Correspondence and requests for materials should be addressed to J.Y. (email: jihong@jlu.edu.cn).

Discovering new advanced materials, which is one of the most important tasks for materials scientists and chemists, still relies primarily on scientific intuition and trial-and-error experimentation[1]. In 2011, the US White House launched the Materials Genome Initiative aiming to develop high-throughput computer methods and data-sharing systems to complement and fully leverage existing experimental research on advanced materials. The incorporation of new computer and informatics tools has the potential to accelerate materials innovation in: (1) predicting a large number of unknown candidate compounds[2–15]; (2) evaluating the predicted compounds and removing the unrealizable ones[16–21]; and (3) screening the predicted compounds and identifying synthetic candidates with desired properties[22–31]. Despite all these successes, *in silico* materials innovation is still facing many challenges. Unlike the genes of organisms, encoding and decoding the structural information of many important crystalline materials remains very complicated. Meanwhile, the explicit structure-property relationships for many materials are not yet clear, so high-throughput identification of synthetic targets with desired properties among a large number of candidate structures is still challenging.

Fortunately, the structures of many crystalline materials can topologically be decomposed into a set of smaller and simpler building modules. In particular, many materials are built of well-defined parallel-stacked modular layers[14,32,33]. If each unique layer is assigned a predefined symbol, then the stacking of these layers can be expressed as a sequence of predefined symbols, just like the genes of organisms. Since each stacking sequence uniquely identifies a specific three-dimensional structure, we deem it the gene of the corresponding structure. Such gene-like one-dimensional stacking sequences can be easily processed by computers, so high-throughput enumeration, evaluation, and identification of theoretical structures with desired properties will be accessible. In this contribution, for the first time, we demonstrate the application of a new genomic approach to ABC-6 zeolites, a family of industrially important catalysts constructed from the stacking of modular 6-ring layers.

To date, over 150 types of ABC-6 zeolites with 28 distinct framework topologies have been discovered, among which cancrinite, sodalite and chabazite are the best-known representatives (Supplementary Table 1). The frameworks of all ABC-6 zeolites can be decomposed into parallel six-ring layers stacked along the *c*-direction in hexagonal unit cells, and the vertices of each 6-ring are corner-sharing TO_4 tetrahedra (T = Si, Al, or P and so on). An ABC-6 structure may consist of three types of six-ring layers, which are centred at the $(0,0,z)$, $(1/3,2/3,z)$, and $(2/3,1/3,z)$ axes, respectively. If we denote these three types of layers by letters A, B, and C, then the stacking sequences for cancrinite, sodalite and chabazite will be (AB), (ABC), and (AABBCC), respectively (Fig. 1). Meanwhile, the stacking of six-rings gives rise to various types of well-defined polyhedral cages in molecular dimensions, which are the most important structural features for ABC-6 zeolites (to avoid confusion, the stacking sequences for these polyhedral cages are given in lower case throughout this paper). These featured cages may hold various types of extraframework cations, anion groups and/or water molecules, which can be exchanged or removed, providing void space suitable for the adsorption, diffusion and reaction of many types of guest species[34–40]. For instance, chabazite and its synthetic counterparts are able to trap CO_2 in their featured cages, showing the highly desired capability for carbon capture from the atmosphere[36,37]; meanwhile, these zeolites are currently among the best industrial catalysts for methanol-to-olefin (MTO) reactions because of the confinement effect of their featured cages[41–43].

Due to these important applications, speculating how many unknown ABC-6 structures are realizable as new catalysts with desired properties is of great significance for the development of such materials. However, to answer this question is challenging. First, we need a highly efficient computational method to enumerate all possible ABC-6 structures. Second, we need to evaluate all enumerated structures and remove the unrealizable ones. More importantly yet more difficultly, we need a high-throughput structure screening method to identify candidate ABC-6 structures with desired properties according to functional needs. An early attempt was made towards answering this question, but failed in structure evaluation and structure identification[44].

Here we propose a new genomic approach towards the solution of these problems. In this work, we focus on the one-dimensional digital stacking sequences, that is, the genes of ABC-6 structures. By enumerating all possible stacking sequences, we are able to predict every ABC-6 topology that is chemically feasible. We have developed a ternary numeral coding system, in which each stacking sequence is expressed as a specific ternary numeral. To enumerate all possible stacking sequences, we went through all ternary numerals from the smallest one to the largest allowed and evaluated the chemical feasibility for each one of them. During this enumeration process, equivalent stacking sequences (for instance, (BCA), (CAB), (CBA), (ACB) and (ABCABC), and so on, are all equivalent to (ABC)) and chemically infeasible ones (for instance, (AAA) is chemically infeasible because each stacking layer in it is highly distorted from the ideal tetrahedral coordination) were removed. At the end of the enumeration, every one of our saved stacking sequences corresponded to a topologically unique and chemically feasible ABC-6 topology.

Besides structure enumeration, our genomic approach provides a high-throughput way to extract the most important structural information directly from the enumerated stacking sequences. For instance, our computer program can locate all constituent cages hidden in the stacking sequences, which are the most important structural features for ABC-6 zeolites. To do this, our computer program went through the corresponding stacking sequence back and forth to look for a string of any length that could be interpreted as a valid ABC-6 cage. Such a string should start and end with the same letter, and this letter should not appear in the middle of this string. By finding all such strings in a stacking sequence, we have located all constituent cages in every enumerated ABC-6 topology (Fig. 1). Besides constituent cages, some other structural features, such as the channels and the stacking compactness of six-ring layers are also important to ABC-6 zeolites. Channels link up ABC-6 cages to form a three-dimensional porous system, so their widths and orientations are crucial to the adsorption and diffusion of guest species. Some ABC-6 structures may possess narrow channels only, the openings of which are no wider than a six-ring; other structures may possess interconnecting 8-ring channels perpendicular to the *c*-axis or/and 12-ring channels running along the *c*-axis. Besides cages and channels, how compactly the six-ring layers are stacked is another important structural feature influencing the porosity and other related properties of ABC-6 zeolites. Highly compact stacking of six-ring layers leads to dense ABC-6 frameworks, whereas less compact stacking gives rise to frameworks with more accessible void spaces for guest species, which are highly desired for many applications. Because of the intrinsic nature of ABC-6 structures, compact stackings only occur between successive distinct layers, and those between successive identical layers are non-compact stackings. Here we define, for the first time, the stacking compactness of an ABC-6 structure as the difference in the numbers of compact and non-compact stackings divided by the total number of layer stackings. According to this definition,

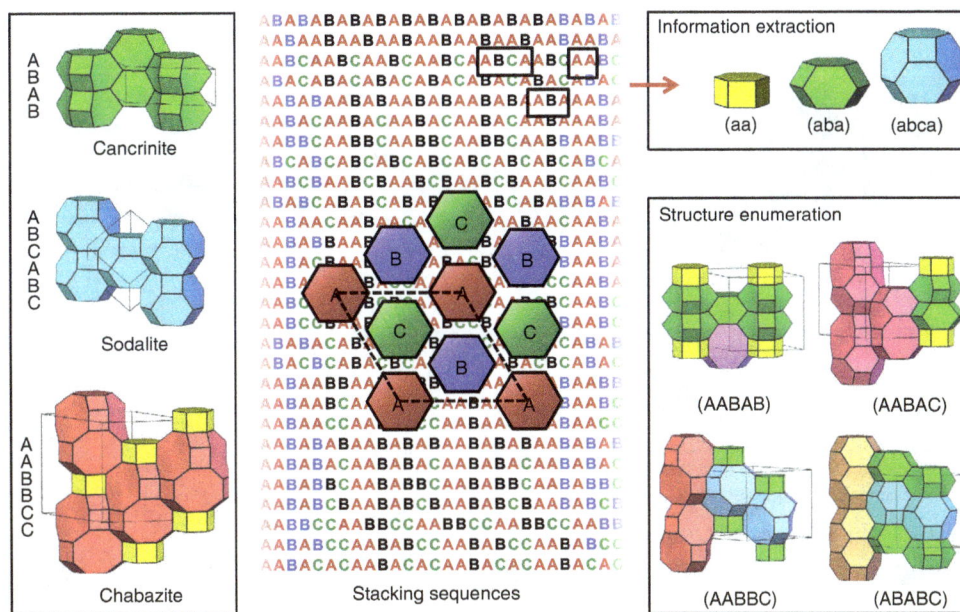

Figure 1 | Enumeration and interpretation of ABC-6 stacking sequences. All ABC-6 topologies are constructed from the stacking of three types of 6-ring layers (denoted by A, B, and C, respectively) along the c-direction in hexagonal unit cells. The sequences of the stacking of these modular layers determine the entire structures as well as the physical and chemical properties of ABC-6 zeolites, so we deem them the genes of this family. By enumerating every possible ternary stacking sequence, we are able to predict the structures of all topologically unique and chemically feasible ABC-6 structures. More importantly, lots of structural information, especially regarding the constituent cages that are crucial to the property and realizability of ABC-6 topologies, can be extracted directly from these stacking sequences. This figure shows the structures of cancrinite, sodalite and chabazite (left), the schematic drawing of the three types of six-ring layers with some of the enumerated stacking sequences (middle), some constituent cages determined from the stacking sequences (top right), and four enumerated ABC-6 structures comprised of five stacking layers (bottom right).

the highest stacking compactness of an ABC-6 structure is 1, corresponding to the densest framework where all layers are compactly stacked. The lowest stacking compactness is 0, corresponding to the most porous framework where only half of the layer stackings are compact. Via high-throughput interpretation of the stacking sequences, the information on channels and stacking compactness can be extracted by our computer program. Details regarding the enumeration and interpretation of ABC-6 stacking sequences can be found in the Methods section.

Results

Enumeration of ABC-6 structures. Considering the computational cost, we have enumerated 84,292 stacking sequences corresponding to all topologically unique and chemically feasible ABC-6 topologies comprised of N stacking layers ($N \leq 16$). The results are summarized in Supplementary Table 2. In all, 98.8% of the enumerated ABC-6 topologies possess 8-ring channels, far outnumbering the ones with 12-ring channels (0.2%) and the ones with 6-ring channels only (1.1%). The distribution of ABC-6 topologies among seven possible symmetries is also uneven. 95.7% of the ABC-6 topologies have the symmetry of $P3m1$, 2.3 and 1.7% belong to $P\text{-}3m1$ and $P\text{-}6m2$, respectively, and those belonging to other symmetries amount only to 0.3%. Most of the enumerated ABC-6 topologies consist of $5 \sim 9$ types of constituent cages.

From the stacking sequences we enumerated, we built the corresponding 84,292 atomic models. All of these models were fully optimized as silica polymorphs through a classic molecular mechanics method (see the Methods section and our online database[45] for more details). The framework energies relative to quartz for all of these models vary between 12.5 and

20.6 kJ (mol Si)$^{-1}$, and the framework densities vary between 15.6 and 18.7 Si nm^{-3} (Fig. 2a), agreeing well with those of existing ABC-6 zeolites. Moreover, statistics on Si–O, O–Si–O, and Si–O–Si distances in these models well obey the local interatomic distances (LIDs) criteria[19] recently discovered among all existing zeolites, indicating that all of our enumerated topologies are chemically feasible as tectosilicates (see the Methods section for more details). Figure 2b plots the framework density versus stacking compactness for 84,292 optimized ABC-6 models. The stacking compactness is proportional to framework density, just as it is defined. According to this plot, we are able to estimate the framework density of an unknown ABC-6 structure directly from its corresponding stacking sequence.

Grouping of ABC-6 structures. Figure 3a demonstrates the plot of lattice dimensions c versus a for all of the optimized atomic models. The a dimensions of the optimized models vary between 1.230 and 1.356 nm, and the c dimensions vary between $0.241 \times N$ and $0.259 \times N$ nm, where N is the number of stacking layers. Surprisingly, all of these models seem to cluster into several groups even for those with identical N. We believe that the grouping of ABC-6 models should arise from the discreteness of their stacking compactness values. For N-layered ABC-6 topologies, the stacking compactness may have $N/2 + 1$ or $(N-1)/2 + 1$ possible values, depending on whether N is an even or odd number. Figure 3b is the plot of c/a versus stacking compactness, showing the perfect grouping of 84,292 optimized ABC-6 models according to N and the stacking compactness. Thus, all ABC-6 topologies comprised of ≤ 16 stacking layers can be divided into 78 groups, 20 of which have at least one end member realized already (underlined with short bars in Fig. 3b). We can name each

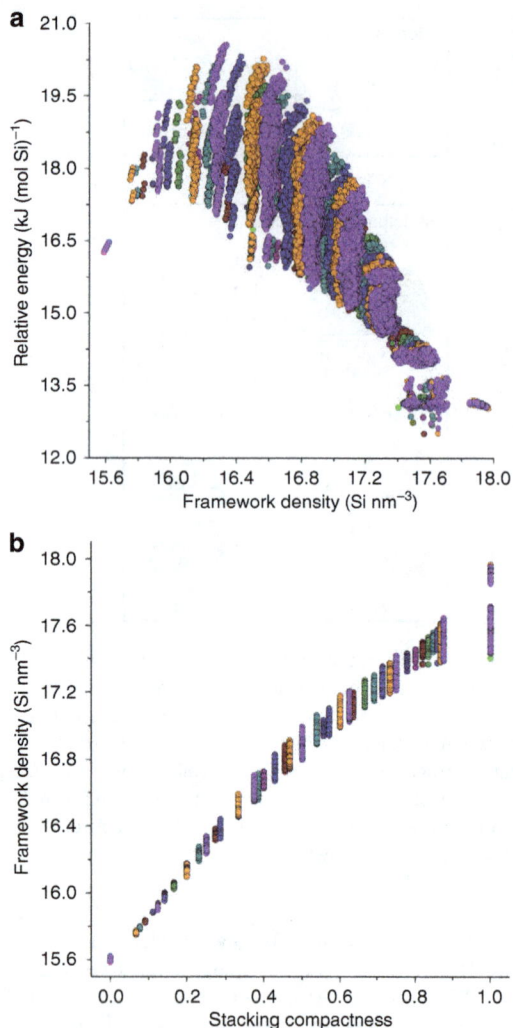

Figure 2 | Structural attributes of 84,292 optimized ABC-6 models.
(**a**) Framework energy versus framework density. The ranges of framework energies and framework densities are both consistent with those of existing ABC-6 zeolites. (**b**) Framework density versus stacking compactness. Stacking compactness reflects how compactly the six-ring layers are stacked in an ABC-6 structure, which is obviously proportional to framework density. ABC-6 models comprised of different numbers of stacking layers are shown in different colours in these two plots.

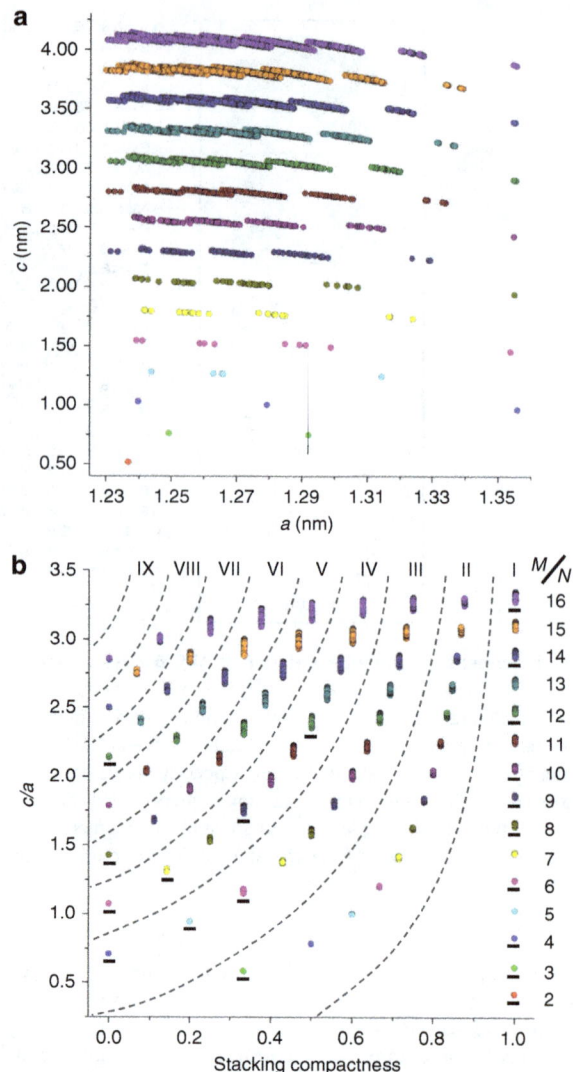

Figure 3 | Grouping of 84,292 optimized ABC-6 models. (**a**) Plot of lattice dimensions c versus a. (**b**) Plot of c/a versus stacking compactness. Each individual group can be named according to the number of stacking layers (N) and the rank of its corresponding stacking compactness (M). Groups having at least one end member realized already are underlined with short bars. ABC-6 models comprised of different numbers of stacking layers are shown in different colours in these two plots.

individual group in the form of N–M, where M (written as a Roman numeral) is the rank of its corresponding stacking compactness among all possible values for N-layered structures. For instance, six-layered ABC-6 structures may have four possible stacking compactness values, that is, 6/6, 4/6, 2/6, and 0/6. Thus, among all 6-layered structures, liottite ((ABABAC)) with the highest stacking compactness of 6/6 belongs to Group 6-I, erionite ((AABAAC)) and bellbergite ((AABCCB)) with a stacking compactness of 2/6 belong to Group 6-III, and chabazite ((AABBCC)) with the lowest stacking compactness of 0/6 belongs to Group 6-IV. Figure 3 can be used as a reference to determine the framework structures of new ABC-6 zeolites. When the lattice dimensions of a new ABC-6 zeolite are known, we may refer to these plots to determine which groups the new structure may belong to. Then, the most probable atomic models will be determined from these groups by examining whether their simulated X-ray diffraction patterns match the observed one.

Identification of the most realizable ABC-6 topologies. Although all of our enumerated ABC-6 topologies are chemically feasible as tectosilicates, only 23 of them have been realized as natural minerals or synthetic materials. Among these realized ABC-6 topologies, half possess six-ring channels only, contradicting the enumeration result that only 1.1% of the enumerated topologies do (Supplementary Table 2). We believe these contradictions arise from the fact that many of our enumerated topologies are not practically realizable. Thus far, we have only considered the chemical feasibility of the host frameworks, yet neglecting the contribution of extra-framework cations, anion groups, or water molecules inside the ABC-6 cages. As a matter of fact, all of the realized ABC-6 frameworks can only form when extra-framework species are present, implying that they are highly important to the formation of ABC-6 structures. Considering the strong host-guest interactions between ABC-6 cages and extra-framework species, we believe that these featured constituent cages may hold the key to improve our prediction.

After careful examination of the structural information we have extracted from the stacking sequences, we determine, for the first time, that all realized ABC-6 topologies are comprised of no more than four types of constituent cages, as is the case even for 36-layered kircherite, the most complex ABC-6 zeolite ever (Supplementary Table 1). This phenomenon is reasonable because every type of ABC-6 cage holds a specific collection of extra-framework species, which can form only under specific reaction conditions. Structures comprised of many types of cages can form only when the reaction conditions for all constituent cages are simultaneously fulfilled, which will be too difficult to occur in reality. Among the 23 already-realized ABC-6 topologies with ≤16 stacking layers, 2 are comprised of 1 type of cages, 6 comprised of 2, 14 comprised of 3, and the remaining 1 comprised of 4. In contrast, nearly 99% of the enumerated ABC-6 topologies are comprised of 5∼9 types of constituent cages (Supplementary Table 2). After removing all enumerated structures that are comprised of more than four types of cages, only 1,150 remained in the end (Table 1 and Supplementary Fig. 1; see our online database[45] for more details). Half of these 1,150 topologies possess six-ring channels only, which is consistent with the situation of realized ABC-6 zeolites. The cell dimensions, space groups, largest channel openings, framework energies, framework densities, stacking compactness and extracted constituent cages for these 1,150 ABC-6 structures are provided in Supplementary Data 1. In addition, we have calculated the theoretical solvent-accessible pore volumes and surface areas with respective to H_2O, H_2, CO_2, N_2, and CH_4 for these ABC-6 structures (Supplementary Data 1; see the Methods section for more details). The fractional pore volumes for these five important probe molecules are in the ranges of 5.24–11.93%, 4.05–9.91%, 2.48–7.40%, 1.55–5.75% and 1.21–5.09%, respectively, and the surface areas are in the ranges of 5.55–11.30 $Å^2 Si^{-1}$, 4.66–10.07 $Å^2 Si^{-1}$, 3.41–7.62 $Å^2 Si^{-1}$, 2.54–6.03 $Å^2 Si^{-1}$ and 2.17–5.39 $Å^2 Si^{-1}$, respectively. These data can be used to prescreen candidate structures for specific gas adsorption or separation applications. Among the 1,150 ABC-6 topologies constructed by no more than four types of constituent cages, 23 have already been realized. We deem the remaining 1,127 ABC-6 structures the most realizable synthetic candidates, because they are both chemically feasible and practically easy to form together with extraframework species. Recently, we have successfully realized two of these candidates, that is, magnesium aluminophosphate JU-60 and zinc aluminophosphate JU-61. These two new ABC-6 zeolites were both synthesized using 1,2-diaminocyclohexane as the structure-directing agent under hydrothermal conditions, and both of their structures were determined through single-crystal X-ray diffraction (see the Methods section for more details about their synthesis and structure determination). JU-60 belongs to Group 10-V, and its corresponding stacking sequence is (AABAACCBCC). JU-60 is comprised of four types of cages, including hexagonal prisms ((aa)), cancrinite cages ((aba)), chabazite cages ((abbcca)) and erionite cages ((abbcbba)), respectively (Fig. 4a). JU-61 belongs to Group 15-VII, and it is the first ABC-6 zeolite comprised of 15 stacking layers ((AABAABBCBBCCACC)). JU-61 consists of four types of cages, including the hexagonal prisms, cancrinite cages, gmelinite cages ((abba)), and a new type of ABC-6 cage ((abbcbbcca)), respectively (Fig. 4b). The synthesis of these new ABC-6 zeolites once again validates our prediction of the most realizable ABC-6 topologies.

Discussion

Focusing on the stacking sequences of ABC-6 zeolites, our genomic approach has provided a straightforward and reliable way to predict the most realizable synthetic candidates. More importantly, the key structural information, especially regarding the constituent cages, can be directly extracted from these stacking sequences, the genes of ABC-6 zeolites. Through a computer procedure similar to the enumeration of ABC-6 structures, we have enumerated 57 types of ABC-6 cages comprised of no more than 10 six-ring layers (Fig. 5 and Supplementary Fig. 3; see the Methods section for more details). As the physical and chemical properties of ABC-6 zeolites are mainly determined by their constituent cages, examining these ABC-6 cages enables the high-throughput screening of ABC-6 zeolites for specific applications. For instance, methanol-to-olefin (MTO) conversion over acidic zeolite catalysts has been an important non-petrochemical industrial process to produce highly demanded light olefins via natural gas, coal, or even

Table 1 | Numbers of topologically unique and practically realizable ABC-6 topologies*.

Number of constituent layers	Largest channel opening			Total
	6-ring	8-ring	12-ring	
2	0	0	1(1)	1(1)
3	1(1)	0	1(1)	2(2)
4	1(1)	1	1(1)	3(2)
5	1	2(1)	1	4(1)
6	2(1)	4(3)	1	7(4)
7	3	5(1)	2	10(1)
8	6(1)	11(1)	3	20(2)
9	7(1)	10(1)	3	20(2)
10	15(2)	20	6	41(2)
11	18	22	7	47
12	39(2)	35(2)	11	85(4)
13	46	41	14	101
14	100(1)	63	24	187(1)
15	130	57	28	215
16	254(1)	104	49	407(1)
Total	623(11)	375(9)	152(3)	1150(23)

*The numbers in brackets are the numbers of topologies that have already been realized.

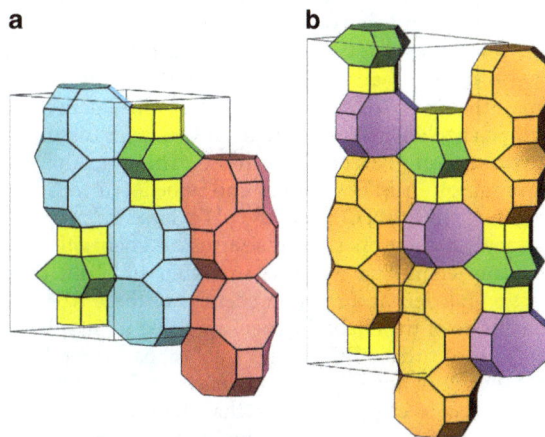

Figure 4 | Two new ABC-6 topologies recently realized by the authors. (a) Magnesium aluminophosphate JU-60 ((AABAACCBCC)) is constructed by four types of ABC-6 cages, being the first member in Group 10-V. (b) Zinc aluminophosphate JU-61((AABAABBCBBCCACC)) is the first member in Group 15-VII. The framework of JU-61 is also constructed from four types of ABC-6 cages, and one of them ((abbcbbcca)) has never been observed in any existing ABC-6 zeolites. Different types of ABC-6 cages are shown in different colours in this figure.

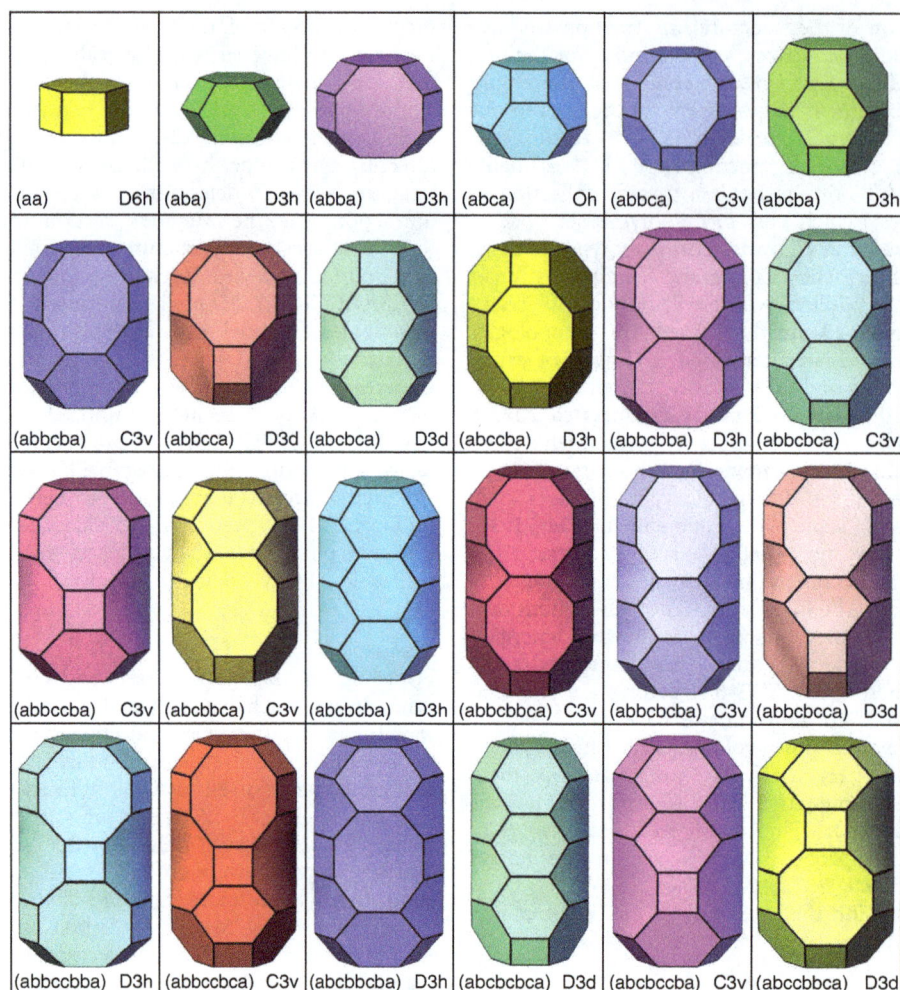

Figure 5 | Enumerated ABC-6 cages constructed by ≤8 six-ring layers. For each cage, the stacking sequence and the highest allowed symmetry are given. The stacking sequences representing the cages that have already been observed in existing ABC-6 materials are highlighted in red.

biomass[46,47]. Chabazite and its synthetic counterparts are currently among the best catalysts for MTO reactions, and the shape and size of their featured cage ((abbcca)) are believed to play the key role in this type of reactions by providing suitable confined void space[43]. To find new ABC-6 catalysts with better MTO performance than chabazite, we have performed density functional theory (DFT) calculations on the methylation of hexamethylbenzene within different ABC-6 cages assuming the same 'hydrocarbon pool' mechanism[48,49]. This reaction occurs at the beginning of an MTO process and is believed to be the key step to initiate the MTO process[50]. We have calculated seven ABC-6 cages that are in similar size to the chabazite cage and possess many eight-ring windows in favour of olefin diffusion. The reaction barriers and reaction energies of these ABC-6 cages, as well as those of the chabazite cage, are listed in Supplementary Table 9. Two of these ABC-6 cages ((abbccbba) and (abbccbca)) exhibit significantly lower reaction barriers and reaction energies than the chabazite cage, indicating that they may provide more suitable confinement effect on the hydrocarbon species than the chabazite cage (Supplementary Fig. 4). By checking the stacking sequences of the 1,127 most realizable synthetic candidates, we have found that only seven of them possess these 'superior' cages (Supplementary Data 1). In particular, two of these seven ABC-6 structures ((AABBAACCAABBCC) and (AABBAACCBBAABBCC)) possess large accessible pore volumes comparable to chabazite, making them the most promising synthetic candidates as new MTO catalysts.

Notably, we assume all enumerated ABC-6 topologies are silicate zeolites in this work. In fact, these topologies may also be realizable as other tetrahedrally coordinated materials, such as silicon sulfides, alkali halides, sp[3] carbon or silicon allotropes, and Zintl phases, which may have interesting mechanical, electronic, optical and chemical properties[51,52]. In particular, a series of zeolitic imidazolate frameworks with ABC-6 topologies have been reported recently, which exhibit the highly desired capability for the capture of fission product[53] and CO_2 (ref. 54). Moreover, our genomic approach is valid not only for ABC-6 structures but also for other crystalline materials that are constructed from the stacking of well-defined modular layers.

Methods

Enumeration of ABC-6 stacking sequences. Our computer programs for the enumeration and interpretation of ABC-6 stacking sequences were written in FORTRAN. To enumerate all possible stacking sequences of length N, our computer program went through every N-digit ternary numeral from the smallest one to the largest allowed. Because only unique stacking sequences were needed, we fixed the first digit of every ternary numeral to be '0' and enumerated the remaining $(N-1)$ digits. To guarantee that only the chemically feasible and topologically unique stacking sequences were retained, our computer program performed a two-step examination procedure for each numeral visited. First, our program checked if the current numeral consisted of three or more successive identical digits. If not, this numeral should represent a chemically feasible stacking sequence. Then, our program generated all equivalent numerals for the current one and examined whether any of these equivalent numerals was smaller than the current one. If not, then the current numeral should represent a new stacking sequence. Only the ternary numerals passing both examinations were saved by our computer program. This examination procedure was repeated until the largest allowed N-digit numeral was achieved. The enumeration of ABC-6 cages followed a similar procedure. The only difference was that the ternary numeral for a valid ABC-6 cage should start and end with the same digit, and this digit must be absent in the middle of this ternary numeral. To ensure that all our enumerated cages are topologically unique, we fixed the starting and ending digits to be '0' and enumerated the middle part with digits '1' and '2' only.

Structural information extraction. To extract the channel information from the stacking sequences, our computer program checked the following situations: (1) if a stacking sequence was comprised of only two types of letters, it represented an ABC-6 topology with 12-ring channels running along the c-direction; (2) if a stacking sequence consisted of successive identical letters, it indicated the existence of interconnecting 8-ring channels perpendicular to the c-direction; (3) other stacking sequences corresponded to ABC-6 topologies with six-ring channels only. To calculate the stacking compactness of an ABC-6 topology from its stacking sequence, our computer program counted the number of letters that were distinct from both of their neighbours and divided that value by N, the number of stacking layers.

Geometry optimization of ABC-6 models. The atomic models for 84,292 enumerated stacking sequences were built as silica polymorphs using Materials Studio (Accelrys Software Inc., 2005). The highest symmetries of these models were identified by the 'Find Symmetry' tool implemented in Materials Studio. These models were fully optimized without symmetry constraints by GULP[55] with the Sanders-Leslie-Catlow potentials[56]. All structural models were confirmed to have no imaginary phonon mode.

Evaluation of the optimized ABC-6 models. To evaluate the chemical feasibility of the optimized ABC-6 structures, we have also optimized 208 existing zeolites known to date[57] as silica polymorphs using the same empirical potentials. The framework densities and framework energies of our enumerated ABC-6 structures agreed well with those of 208 existing zeolite structures. Recently, we proposed a set of LIDs criteria[19], which have proved to be more effective and reliable for structure evaluation than other methods. According to these criteria, the means, standard deviations, and ranges of LIDs in a chemically feasible zeolite structure, including T–O, O–T–O, and T–O–T distances, should obey a set of relationships. In this work, the LIDs in 84,292 optimized ABC-6 structures were calculated using the program FraGen[58], and the results showed that all of these enumerated ABC-6 structures were chemically feasible. The LIDs for 1,150 most realizable ABC-6 structures are provided as Supplementary Data 2. The solvent-accessible pore volumes and surface areas for 1,150 most realizable ABC-6 structures were calculated using the 'Volume' tool implemented in Materials Studio. Rigid spheres with diameters of 2.65, 2.89, 3.30, 3.64 and 3.80 Å were used as the probes, corresponding to the kinetic diameters of H_2O, H_2, CO_2, N_2 and CH_4, respectively.

Synthesis of JU-60 and JU-61. Magnesium aluminophosphate JU-60 and zinc aluminophosphate JU-61 were both synthesized using 1,2-diaminocyclohexane (DACH) as the structure-directing agent under hydrothermal conditions. To synthesize JU-60, 0.1 g of pseudoboehmite (Al_2O_3, 74.3%) and 0.3 g of magnesium acetate were dispersed in 10 ml of H_2O with stirring for 2 h. A volume of 0.5 ml of DACH (99 wt %) was then added into the mixture with stirring, followed by the addition of 0.2 ml of H_3PO_4 (85 wt %). A homogeneous gel was formed with an overall molar composition of 1.0 MgO: 0.5 Al_2O_3: 2.1 H_3PO_4: 2.9 DACH: 404 H_2O. The gel was transferred into a 15-ml Teflon-lined stainless steel autoclave and heated at 180 °C for 3 days. The obtained crystals of JU-60 were separated by filtration, washed with distilled water and dried in air at room temperature. To synthesize JU-61, 0.25 g of pseudoboehmite (Al_2O_3, 62.5%) was dispersed in a mixture of 8 ml of H_2O and 0.4 ml of H_3PO_4 (85 wt%), followed by the addition of 0.41 g of $ZnCl_2$. After stirring for 2 h, 1.5 ml of DACH (99 wt%) was added. A homogeneous gel was formed after stirring for another 2 h, with an overall molar composition of 1.0 ZnO: 0.5 Al_2O_3: 2.0 H_3PO_4: 4.0 DACH: 152 H_2O. The gel was transferred into a 15-ml Teflon-lined stainless steel autoclave and heated at 180 °C for 5 days. The obtained crystals of JU-61 were separated by filtration, washed with distilled water and dried in air at room temperature.

X-ray structure determination for JU-60 and JU-61. Powder X-ray diffraction data were collected on a Rigaku D/max-2550 diffractometer with Cu Kα radiation ($\lambda = 1.5418$ Å). Single-crystal X-ray diffraction data were collected on a Bruker AXS SMART APEX II diffractometer using graphite-monochromated Mo Kα radiation ($\lambda = 0.71073$ Å) at the temperature of 23 ± 2 °C. Data processing was accomplished with the SAINT processing program. The framework structures of JU-60 and JU-61 were solved by direct methods and refined on F^2 by full matrix least-squares techniques with SHELXTL. Parts of the extra-framework species, such as DACH and water molecules, were located during least-squares refinement. JU-60 consists of three crystallographically distinct Al and three P sites. One of the Al site exhibited an average Al-O bond distance of 1.85 Å, indicating that it was half occupied by Mg. JU-61 consisted of five crystallographically distinct

tetrahedrally coordinated sites. Considering the restrictions of the odd number of layers and the Loewenstein's rule[59], we had to refine the structure of JU-61 assuming that all of the five tetrahedrally coordinated sites were co-occupied by disordered Al, P, and Zn. The occupancy ratio of Zn to Al was fixed to 2:3 according to the average bond distance in JU-61 (1.66 Å). To remove these disorders, we have also made several attempts to index JU-61 in a doubled unit cell, but the data collected in this way were not good enough for a feasible structure solution. The crystallographic tables, atomic coordinates, selected bond distances and angles, and powder X-ray diffraction patterns for JU-60 and JU-61 are provided in Supplementary Tables 3–8 and Supplementary Fig. 2.

Density functional theory calculations. All of the cage models were cut from the optimized ABC-6 structures. For each ABC-6 cage, one of the Si atoms in the eight-ring window was replaced by Al to produce the Brönsted acid site. The dangling bonds in all cages were saturated by H atoms. All atoms in ABC-6 cages and extra-framework species were fully optimized without any constraint at ONIOM(B3LYP/6–31G(d,p):AM1) level[60–62], where the acid site (SiO_3–O–AlO_2–OH–SiO_3 cluster) and extraframework species were in the high level (Supplementary Fig. 4). The achievement of energy minima or saddle points was checked by frequency calculations at the same level. The reaction barrier was calculated as the energy difference between the transition state and the reactant (hexamethylbenzene, methanol and the protonated ABC-6 cage). The reaction energies were calculated as the energy differences between the product (heptamethylbenzenium cation, water, and the deprotonated ABC-6 cage) and the reactant. To improve the precision of weak interaction energy calculations, we have performed single-point energy calculations at ωB97XD/6–31 + G(d,p) level[63] for all optimized models. All density functional theory calculations were carried out using the Gaussian 09 package[64].

References

1. Jain, A. et al. Commentary: The Materials Project: A materials genome approach to accelerating materials innovation. APL Mater. **1**, 011002 (2013).
2. Deem, M. W. & Newsam, J. M. Determination of 4-connected framework crystal structures by simulated annealing. Nature **342**, 260–262 (1989).
3. Delgado Friedrichs, O., Dress, A. W. M., Huson, D. H., Klinowski, J. & Mackayk, A. L. Systematic enumeration of crystalline networks. Nature **400**, 644–647 (1999).
4. Mellot Draznieks, C., Newsam, J. M., Gorman, A. M., Freeman, C. M. & Férey, G. De novo prediction of inorganic structures developed through automated assembly of secondary building units (AASBU method). Angew Chem. Int. Ed. **39**, 2270–2275 (2000).
5. Mellot-Draznieks, C., Dutour, J. & Férey, G. Hybrid organic-inorganic frameworks: Routes for computational design and structure prediction. Angew Chem. Int. Ed. **43**, 6290–6296 (2004).
6. Treacy, M. M. J., Rivin, I., Balkovsky, E., Randall, K. H. & Foster, M. D. Enumeration of periodic tetrahedral frameworks. II. Polynodal graphs. Microporous Mesoporous Mater. **74**, 121–132 (2004).
7. Férey, G., Mellot-Draznieks, C., Serre, C. & Millange, F. Crystallized frameworks with giant pores: are there limits to the possible? Acc. Chem. Res. **38**, 217–225 (2005).
8. Fischer, C. C., Tibbetts, K. J., Morgan, D. & Ceder, G. Predicting crystal structure by merging data mining with quantum mechanics. Nat. Mater. **5**, 641–646 (2006).
9. Woodley, S. M. & Catlow, R. Crystal structure prediction from first principles. Nat. Mater. **7**, 937–946 (2008).
10. O'Keeffe, M., Peskov, M. A., Ramsden, S. J. & Yaghi, O. M. The reticular chemistry structure resource (RCSR) database of, and symbols for, crystal nets. Acc. Chem. Res. **41**, 1782–1789 (2008).
11. Nørskov, J. K., Bligaard, T., Rossmeisl, J. & Christensen, C. H. Towards the computational design of solid catalysts. Nat. Chem. **1**, 37–46 (2009).
12. Oganov, A. R., Lyakhov, A. O. & Valle, M. How evolutionary crystal structure prediction works-and Why. Acc. Chem. Res. **44**, 227–237 (2011).
13. Pophale, R., Cheeseman, P. A. & Deem, M. W. A database of new zeolite-like materials. Phys. Chem. Chem. Phys. **13**, 12407–12412 (2011).
14. Dyer, M. S. et al. Computationally assisted identification of functional inorganic materials. Science **340**, 847–852 (2013).
15. Curtarolo, S. et al. The high-throughput highway to computational materials design. Nat. Mater. **12**, 191–201 (2013).
16. Foster, M. D. et al. Chemically feasible hypothetical crystalline networks. Nat. Mater. **3**, 234–238 (2004).
17. Walker, A. M., Slater, B., Gale, J. D. & Wright, K. Predicting the structure of screw dislocations in nanoporous materials. Nat. Mater. **3**, 715–720 (2004).
18. Sartbaeva, A., Wells, S. A., Treacy, M. M. J. & Thorpe, M. F. The flexibility window in zeolites. Nat. Mater. **5**, 962–965 (2006).
19. Li, Y., Yu, J. & Xu, R. Criteria for zeolite frameworks realizable for target synthesis. Angew Chem. Int. Ed. **52**, 1673–1677 (2013).

20. Combariza, A. F., Gomez, D. A. & Sastre, G. Simulating the properties of small pore silica zeolites using interatomic potentials. *Chem. Soc. Rev.* **42**, 114–127 (2013).

21. Li, Y. & Yu, J. New stories of zeolite structures: their descriptions, determinations, predictions, and evaluations. *Chem. Rev.* **114**, 7268–7316 (2014).

22. Greeley, J., Jaramillo, T. F., Bonde, J., Chorkendorff, I. & Nørskov, J. K. Computational high-throughput screening of electrocatalytic materials for hydrogen evolution. *Nat. Mater.* **5**, 909–913 (2006).

23. Yang, K., Setyawan, W., Wang, S., Buongiorno Nardelli, M. & Curtarolo, S. A search model for topological insulators with high-throughput robustness descriptors. *Nat. Mater.* **11**, 614–619 (2012).

24. Dubbeldam, D., Krishna, R., Calero, S. & Yazaydın, A. Ö. Computer-assisted screening of ordered crystalline nanoporous adsorbents for separation of alkane isomers. *Angew Chem. Int. Ed.* **51**, 11867–11871 (2012).

25. Lin, L.-C. *et al.* In silico screening of carbon-capture materials. *Nat. Mater.* **11**, 633–641 (2012).

26. Wilmer, C. E. *et al.* Large-scale screening of hypothetical metal-organic frameworks. *Nat. Chem.* **4**, 83–89 (2012).

27. Kim, J., Abouelnasr, M., Lin, L.-C. & Smit, B. Large-scale screening of zeolite structures for CO_2 membrane separations. *J. Am. Chem. Soc.* **135**, 7545–7552 (2013).

28. Kim, J. *et al.* New materials for methane capture from dilute and medium-concentration sources. *Nat. Commun.* **4**, 1694 (2013).

29. Colón, Y. J. & Snurr, R. Q. High-throughput computational screening of metal-organic frameworks. *Chem. Soc. Rev.* **43**, 5735–5749 (2014).

30. Bai, P. *et al.* Discovery of optimal zeolites for challenging separations and chemical transformations using predictive materials modeling. *Nat. Commun.* **6**, 5912 (2015).

31. Simon, C. M. *et al.* The materials genome in action: identifying the performance limits for methane storage. *Energy Environ. Sci.* **8**, 1190–1199 (2015).

32. Willhammar, T. *et al.* Structure and catalytic properties of the most complex intergrown zeolite ITQ-39 determined by electron crystallography. *Nat. Chem.* **4**, 188–194 (2012).

33. Esters, M. *et al.* Synthesis of inorganic structural isomers by diffusion-constrained self-assembly of designed precursors: a novel type of isomerism. *Angew Chem. Int. Ed.* **54**, 1130–1134 (2015).

34. Reinen, D. & Lindner, G.-G. The nature of the chalcogen colour centres in ultramarine-type solids. *Chem. Soc. Rev.* **28**, 75–84 (1999).

35. Lezhnina, M., Laeri, F., Benmouhadi, L. & Kynast, U. Efficient near-infrared emission from sodalite derivatives. *Adv. Mater.* **18**, 280–283 (2006).

36. Shang, J. *et al.* Discriminative separation of gases by a 'molecular trapdoor' mechanism in chabazite zeolites. *J. Am. Chem. Soc.* **134**, 19246–19253 (2012).

37. Hudson, M. R. *et al.* Unconventional, highly selective CO_2 adsorption in zeolite SSZ-13. *J. Am. Chem. Soc.* **134**, 1970–1973 (2012).

38. Xu, S. *et al.* Direct observation of cyclic carbenium ions and their role in the catalytic cycle of the methanol-to-olefin reaction over chabazite zeolites. *Angew. Chem. Int. Ed.* **52**, 11564–11568 (2013).

39. Xie, D. *et al.* SSZ-52, a zeolite with an 18-layer aluminosilicate framework structure related to that of the DeNOx catalyst Cu-SSZ-13. *J. Am. Chem. Soc.* **135**, 10519–10524 (2013).

40. Moliner, M., Martínez, C. & Corma, A. Synthesis strategies for preparing useful small pore zeolites and zeotypes for gas separations and catalysis. *Chem. Mater.* **26**, 246–258 (2014).

41. Olsbye, U. *et al.* Conversion of methanol to hydrocarbons: how zeolite cavity and pore size controls product selectivity. *Angew. Chem. Int. Ed.* **51**, 5810–5831 (2012).

42. Van Speybroeck, V. *et al.* Mechanistic studies on chabazite-type methanol-to-olefin catalysts: insights from time-resolved UV/Vis microspectroscopy combined with theoretical simulations. *Chem. Cat. Chem.* **5**, 173–184 (2013).

43. Li, X. *et al.* Confinement effect of zeolite cavities on methanol-to-olefin conversion: a density functional theory study. *J. Phys. Chem. C* **118**, 24935–24940 (2014).

44. Smith, J. V. & Bennett, J. M. Enumeration of 4-connected 3-dimensional nets and classification of framework silicates: the infinite set of ABC-6 nets: the Archimedean and σ-related nets. *Am. Mineral.* **66**, 777–788 (1981).

45. Li, Y. & Yu, J. Hypothetical Zeolite Frameworks. Available at <http://mezeopor.jlu.edu.cn/hypo/> (2015).

46. Haw, J. F., Song, W., Marcus, D. M. & Nicholas, J. B. The mechanism of methanol to hydrocarbon catalysis. *Acc. Chem. Res.* **36**, 317–326 (2003).

47. Van Speybroeck, V. *et al.* First principle chemical kinetics in zeolites: the methanol-to-olefin process as a case study. *Chem. Soc. Rev.* **43**, 7326–7357 (2014).

48. Dahl, I. M. & Solboe, S. On the reaction mechanism for hydrocarbon formation from methanol over SAPO-34: 1. Isotopic labeling studies of the co-reaction of ethene and methanol. *J. Catal.* **149**, 458–464 (1994).

49. Dahl, I. M. & Kolboe, S. On the reaction mechanism for hydrocarbon formation from methanol over SAPO-34: 2. Isotopic labeling studies of the co-reaction of propene and methanol. *J. Catal.* **161**, 304–309 (1996).

50. Lesthaeghe, D., De Sterck, B., Van Speybroeck, V., Marin, G. B. & Waroquier, M. Zeolite shape-selectivity in the gem-methylation of aromatic hydrocarbons. *Angew. Chem. Int. Ed.* **46**, 1311–1314 (2007).

51. Wang, H., Tse, J. S., Tanaka, K., Iitaka, T. & Ma, Y. Superconductive sodalite-like clathrate calcium hydride at high pressures. *Proc. Natl Acad. Sci. USA* **109**, 6463–6466 (2012).

52. Kim, D. Y., Stefanoski, S., Kurakevych, O. O. & Strobel, T. A. Synthesis of an open-framework allotrope of silicon. *Nat. Mater.* **14**, 169–173 (2015).

53. Sava, D. F. *et al.* Capture of volatile iodine, a gaseous fission product, by zeolitic imidazolate framework-8. *J. Am. Chem. Soc.* **133**, 12398–12401 (2011).

54. Nguyen, N. T. T. *et al.* Selective capture of carbon dioxide under humid conditions by hydrophobic chabazite-type zeolitic imidazolate frameworks. *Angew. Chem. Int. Ed.* **53**, 10645–10648 (2014).

55. Gale, J. D. GULP: Capabilities and prospects. *Z. Kristallogr.* **220**, 552–554 (2005).

56. Schröder, K.-P., Sauer, J., Leslie, M., Catlow, C. R. A. & Thomas, J. M. Bridging hydroxyl groups in zeolitic catalysts: a computer simulation of their structure, vibrational properties and acidity in protonated faujasites (H-Y zeolites). *Chem. Phys. Lett.* **188**, 320–325 (1992).

57. Baerlocher, C. & McCusker, L. B. Database of Zeolite Structures. Available at <http://www.iza-structure.org/databases/> (2015).

58. Li, Y., Yu, J. & Xu, R. FraGen: a computer program for real-space structure solution of extended inorganic frameworks. *J. Appl. Cryst.* **45**, 855–861 (2012).

59. Loewenstein, W. The distribution of aluminum in the tetrahedra of silicates and aluminates. *Am. Mineral.* **39**, 92–96 (1954).

60. Chung, L. W. *et al.* The ONIOM method and its applications. *Chem. Rev.* **115**, 5678–5796 (2015).

61. Tirado-Rives, J. & Jorgensen, W. L. Performance of B3LYP density functional methods for a large set of organic molecules. *J. Chem. Theory Comput.* **4**, 297–306 (2008).

62. Dewar, M. J. S., Zoebisch, E. G., Healy, E. F. & Stewart, J. J. P. AM1: a new general purpose quantum mechanical molecular model. *J. Am. Chem. Soc.* **107**, 3902–3909 (1985).

63. Chai, J.-D. & Head-Gordon, M. Long-range corrected hybrid density functionals with damped atom-atom dispersion corrections. *Phys. Chem. Chem. Phys.* **10**, 6615–6620 (2008).

64. Frisch, M. J. *et al.* Gaussian 09 (Gaussian, Inc., 2013).

Acknowledgements

This work was supported by the State Basic Research Project of China (Grant No. 2011CB808703) and the National Natural Science Foundation of China (Grant Nos. 91122029; 21273098; 21320102001). Y.L. acknowledges the support by Program for New Century Excellent Talents in University (NCET-13-0246).

Author contributions

J.Y. supervised and coordinated all aspects of the project. Y.L. wrote the computer programs and performed the enumeration, geometry optimization, evaluation, and high-throughput screening of all ABC-6 structures. X.L. performed density functional theory calculations for the selected ABC-6 cages. J.L. and F.D. synthesized JU-60 and JU-61.

Additional information

7

Covalency-reinforced oxygen evolution reaction catalyst

Shunsuke Yagi[1], Ikuya Yamada[1,2], Hirofumi Tsukasaki[3], Akihiro Seno[1], Makoto Murakami[3], Hiroshi Fujii[3], Hungru Chen[4], Naoto Umezawa[2,4], Hideki Abe[2,4], Norimasa Nishiyama[2,5] & Shigeo Mori[3]

The oxygen evolution reaction that occurs during water oxidation is of considerable importance as an essential energy conversion reaction for rechargeable metal–air batteries and direct solar water splitting. Cost-efficient ABO_3 perovskites have been studied extensively because of their high activity for the oxygen evolution reaction; however, they lack stability, and an effective solution to this problem has not yet been demonstrated. Here we report that the Fe^{4+}-based quadruple perovskite $CaCu_3Fe_4O_{12}$ has high activity, which is comparable to or exceeding those of state-of-the-art catalysts such as $Ba_{0.5}Sr_{0.5}Co_{0.8}Fe_{0.2}O_{3-\delta}$ and the gold standard RuO_2. The covalent bonding network incorporating multiple Cu^{2+} and Fe^{4+} transition metal ions significantly enhances the structural stability of $CaCu_3Fe_4O_{12}$, which is key to achieving highly active long-life catalysts.

[1] Nanoscience and Nanotechnology Research Centre, Osaka Prefecture University, Osaka 599-8570, Japan. [2] Precursory Research for Embryonic Science and Technology, Japan Science and Technology Agency, Tokyo 102-0075, Japan. [3] Department of Materials Science and Engineering, Osaka Prefecture University, Osaka 599-8531, Japan. [4] National Institute for Materials Science, Tsukuba 305-0044, Japan. [5] Deutsches Elektronen Synchrotron, Hamburg 22607, Germany. Correspondence and requests for materials should be addressed to S.Y. (email: s-yagi@21c.osakafu-u.ac.jp) or to I.Y. (email: i-yamada@21c.osakafu-u.ac.jp).

The oxygen evolution reaction (OER: $4OH^- \rightarrow O_2 + 2$ $H_2O + 4e^-$) is an energy conversion reaction that is essential for both the charging of rechargeable metal–air batteries and direct solar water splitting[1-5]. ABO_3 perovskite oxides are of particular interest because of their high catalytic OER activities, some of which are comparable to those of noble metal oxides such as RuO_2 and IrO_2 (refs 6–8). Along with reports on this high OER activity, many studies have been conducted to clarify the relationship between the electronic state and OER activity in perovskites[6,9-11]. Specifically, a simple descriptor of OER activity has been proposed by Suntivich et al.[9]; that is, the highest OER activity can be attained when the e_g occupancy of the B-site transition metal is close to unity. Transition metal ions with e_g^1 electron configurations enhance the covalency with oxygen ions, leading to effective charge transfer in the rate-determining steps. Cobalt-perovskites such as $Ba_{0.5}Sr_{0.5}Co_{0.8}Fe_{0.2}O_{3-\delta}$ (BSCF) have been widely investigated because of their intrinsically high OER activities, which are consistent with the above descriptor, but surface amorphization in OER cycles remains a serious issue[12]. Therefore, it is necessary to consider the intrinsic catalytic activity and stability separately. In this regard, perovskite oxides containing high-spin Fe^{4+} ions ($t_{2g}^3 e_g^1$ configuration) such as $CaFeO_3$ (CFO) and $SrFeO_3$ (SFO) are candidates for OER catalysts with high-catalytic activities. As it is proposed that the electronegativity, which tends to be enhanced in late 3d elements with high valences, serves to increase the metal–oxygen covalency[2,13], it is possible that the Fe^{4+} ions have higher OER activity than the nominally isoelectronic Mn^{3+} ions. The Co^{5+} and Ni^{6+} ions with nominal d^4 configuration are also expected to have higher OER activity, but the synthesis of perovskite-oxides-containing Co^{5+} and Ni^{6+} ions has not yet been reported. Further, to the best of our knowledge, Fe^{4+}-oxides have not been well investigated as OER catalysts to date. This is possibly because of their extreme synthesis conditions, as the majority of Fe^{4+}-oxides are synthesized under high pressures of above several GPa. As compounds synthesized under high pressure are metastable, they are likely to be excluded from the promising high-performance catalyst candidates. Thus, no reports on the testing of high-pressure synthesized compounds as electrochemical catalysts have been published[14]. Furthermore, the dissolution of metal ions seems unavoidable, because of the ionic characteristics of A-site alkaline-earth metal ions for $AFe^{4+}O_3$ perovskites, as in the case of $SrRuO_3$, for example, ref. 15.

Recent progress in high-pressure chemistry has enabled dramatic structural modifications, such as transitions from simple $A^{2+}Fe^{4+}O_3$ to quadruple $A^{2+}Cu_3^{2+}Fe_4^{4+}O_{12}$ perovskites (A = Ca, Sr; see crystal structures in Fig. 1b). $CaCu_3Fe_4O_{12}$ (CCFO) and its analogues exhibit unusual electronic properties, for example, charge disproportionation ($2Fe^{4+} \rightarrow Fe^{3+} + Fe^{5+}$)[16] in the case of CCFO, and the giant negative thermal expansion associated with second-order intersite charge transfer[17] in the case of $SrCu_3Fe_4O_{12}$. The electronic interactions between A'-Cu and B-Fe ions are predominant, where every oxide ion is connected to two B-site ions and one A'-site ion with strong covalency. This is because of the large overlapping that occurs between Cu (Fe) e_g and O 2p orbitals in square-planar (octahedral) coordination. In fact, the electron density distribution of CCFO obtained from our maximum entropy method analysis illustrates a substantial and widespread Fe–O–Cu network (Fig. 1b; details of this electron density analysis are given in the Supplementary Note 1). In contrast, the network is distributed only around Fe and O ions in a simple perovskite SFO. One can expect that the complex covalent bonding network in CCFO plays a significant role in determining its catalytic properties, as in the case of the photocatalytic activity

of Pt-loaded $CaCu_3Ti_4O_{12}$ (ref. 18). In this report, we show that Fe^{4+}-perovskite CCFO exhibits high OER catalytic activity, which is comparable to or exceeds that of state-of-the-art OER catalysts such as BSCF and the gold standard RuO_2. CCFO also possesses high stability under OER conditions over many cycles, owing to its enhanced covalent bonding network.

Results

Catalytic activity of Fe^{4+}-perovskites. The OER catalytic performance of the Fe^{4+}-perovskites CFO, SFO and CCFO is compared with that of BSCF and RuO_2, together with a nominally isoelectronic perovskite, $LaMnO_3$ (LMO), in Fig. 2. The tetra-valency of the Fe ions for CCFO was confirmed via Fe K-edge X-ray absorption spectra (Supplementary Fig. 2). To exclude geometrical effects, the current density per oxide surface area ($mA\, cm_{oxide}^{-2}$), in which the surface areas were determined using Brunauer–Emmett–Teller (BET) analysis, was adopted as the vertical axis in the voltammograms in this study (Supplementary Note 2, Supplementary Fig. 3, and Supplementary Table 1). As the OER activity of BSCF is strongly dependent on the synthesis conditions[9,12,19,20], two different BSCF samples calcined at 950 and 1,100 °C ($BSCF_{950}$ and $BSCF_{1100}$, respectively) were tested.

Figure 2a shows the obtained linear sweep voltammograms, and it can be seen that CCFO exhibits the highest OER activity of the catalysts tested here. The overpotential of CCFO for OER ($\eta = 0.31$ V), which was determined based on the onset potentials at $0.5\, mA\, cm_{oxide}^{-2}$, is the lowest of the examined substances, while its specific activity (current density at 1.6 V versus RHE) is the highest (Fig. 2a,b). The CCFO Tafel slope ($51\, mV\, dec^{-1}$) is as low as those of the SFO and CFO (63 and $47\, mV\, dec^{-1}$, respectively; Fig. 2c). The excellent properties of CCFO, which are attributed to the presence of the Fe^{4+} ions, exceed those of $BSCF_{1100}$ and RuO_2. On the other hand, the BSCF performance reported by Suntivich et al.[9] is superior to that of the CCFO examined in this study. This is attributed to the difference in the synthesis conditions of these particular samples, because BSCF has exhibited different OER performance in a number of reports[9,12,19,20] (see also, the XRD profiles of the tested BSCF samples in Supplementary Fig. 4). Thus, we are unable to definitively conclude that CCFO exhibits superior OER performance to BSCF in this study, and further investigations are required in order to compare the intrinsic OER activities of these substances. However, the intrinsic superiority of the Fe^{4+} ions can be confirmed by comparison between $AFe^{4+}O_3$ (A = Ca, Sr) and non-Fe^{4+} oxides. The overpotentials of SFO and CFO are 0.41 and 0.39 V, respectively, which are comparable to those of BSCF ($\eta = 0.38$ and 0.36 V for $BSCF_{950}$ and $BSCF_{1100}$, respectively) and are lower than that of RuO_2 ($\eta = 0.49$ V). By comparing the specific activities (current densities at 1.6 V, Fig. 2b), it can be seen that SFO and CFO have high activities comparable to those of $BSCF_{1100}$ and RuO_2. On the other hand, LMO exhibits poor OER catalytic activity; its overpotential cannot be defined because of the overly small current density, while the specific activity is only a centesimal fraction of that exhibited by $AFeO_3$. This is because the secondary descriptor, electronegativity[2,13], predominates the OER activity in these cases.

As the B-site ion (Fe^{4+}) is identical among CFO, SFO and CCFO, the excellent OER activity of CCFO can be attributed to its particular structure, that is, it is a quadruple perovskite incorporating ordered A'-site Cu ions. It should be noted that a reference Cu^{2+}–Fe^{3+} complex oxide examined in this study, $CuFe_2O_4$ spinel, did not exhibit high OER activity (Supplementary Fig. 5), thus, the combination of a $Cu^{2+}O_4$ square and $Fe^{4+}O_6$ octahedron is a key factor enhancing the OER activity, as will be discussed below.

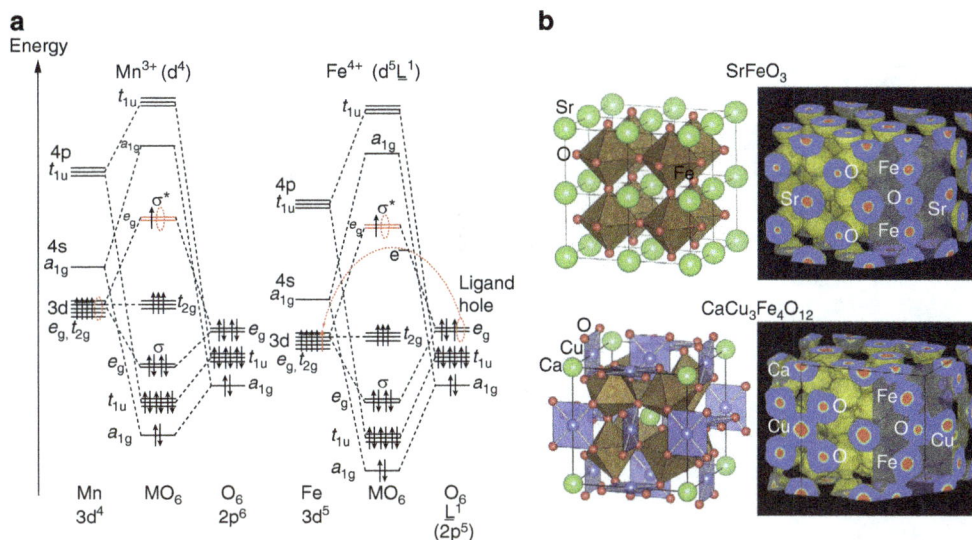

Figure 1 | Electronic and crystal structures of SFO and CCFO perovskites. (**a**) Schematic illustration of molecular orbitals for regular $Mn^{3+}O_6$ and $Fe^{4+}O_6$ octahedra. The Mn^{3+}- and Fe^{4+}- ion 3d-orbital energy levels are higher and lower than those of the O 2p orbitals, respectively. Therefore, the highest occupied molecular orbitals σ^* generated from the e_g and 2p orbitals have 3d and 2p characteristics for the Mn^{3+} and Fe^{4+} ions. The holes at the σ^* orbitals are due to the e_g and O 2p orbitals in the former and latter, respectively, resulting in different representations of d^4 and $d^5\underline{L}^1$ for Mn^{3+} and Fe^{4+}, respectively, where $\underline{L}$ denotes a ligand hole at the O 2p orbital[25]. The π-bonds between the t_{2g} and 2p orbitals are neglected for simplicity. (**b**) Crystal structures and 3D electron density maps of SFO and CCFO. SFO is crystallized in a cubic ABO_3-type perovskite structure, and CCFO is crystallized in a cubic quadruple $AA'_3B_4O_{12}$-type structure with a $2a_0 \times 2a_0 \times 2a_0$ unit cell (a_0: a-axis length of a simple ABO_3 perovskite). In these types of perovskites, the A-sites are occupied by alkaline, alkaline-earth or rare-earth metal ions, the A'-sites by Jahn–Teller active ions such as Cu^{2+} and Mn^{3+}, and the B-sites by d-block transition metal ions. 3D electron density maps of SFO (equi-density level: $0.4\,Å^{-1}$) and CCFO (equi-density level: $0.5\,Å^{-1}$) were obtained from maximum entropy method analysis of synchrotron X-ray powder diffraction data. The shaded cross-sections indicate the (110) and $\left(1\ \tfrac{4}{3}\ 0\right)$ planes of SFO and CCFO, respectively. The widespread covalent network incorporating the Cu, Fe and O ions is exemplified by CCFO. These illustrations were drawn using the VESTA3 program[26]. The synchrotron X-ray powder diffraction patterns and Rietveld refinement results are shown in Supplementary Fig. 1 and Supplementary Note 1.

Stability of Fe^{4+}-perovskites. The Fe^{4+}-perovskite stability under OER conditions was tested. Fig. 2d–f show the cyclic voltammograms (CV) of SFO, CFO and CCFO for continuous 100 cycles. In the SFO case, the OER current density is suppressed even in the anodic sweep of the first cycle, because of the degradation of the SFO; this implies the dissolution of metal ions[15]. On the other hand, the increase in current density for 100th cycle is possibly attributed to the increase in electrochemical surface area in amorphization[12]. For the CFO, the OER current density increases slightly in the first $\sim$10 cycles, and then gradually decreases over the 100 cycles. As can be seen in the Tafel plots of the 3rd and 100th cycles for the SFO and CFO (Fig. 2g,h), both the Tafel slopes are increased after 100 cycles; this clearly suggests the degradation of the SFO and CFO under OER conditions (also see the increases in overpotentials for 100 cycles in Supplementary Note 3 and Supplementary Fig. 6). Considering the fact that the highest occupied molecular orbitals are dominated by the O 2p orbitals in SFO and CFO, as shown in Fig. 1a, the above results are explained by the trend that the highly elevated O 2p band centre (or the deep Fe 3d orbital) increases the activity but decreases the stability, as suggested for cobalt perovskites[6]. However, CCFO is remarkably stable up to 100 cycles, in spite of the fact that it has the same electronic configuration as SFO and CFO. For CCFO, the current density increases in the first $\sim$10 cycles and remains almost unchanged. The CCFO Tafel slope does not vary significantly, rather it improves slightly over the 100 cycles (Fig. 2i). This corresponds to a slight improvement in the catalytic activity. Thus, we conclude that CCFO is an excellent OER catalyst that satisfies both the activity and stability requirements.

To determine the difference in stability for these Fe^{4+}-perovskites, the surface structures of SFO, CFO and CCFO were investigated using high-resolution transmission electron microscopy (HRTEM) both before and after the 100-cycle OER measurements. Figure 3 shows surface HRTEM images of SFO, CFO and CCFO samples, as-synthesized, as-cast, and after the 100-cycle OER measurements. Well-crystalline surface structures can be observed for all the as-synthesized powders and the bulk crystallinity of all the samples was retained after the 100-cycle OER measurements (see also, the electron diffraction patterns in Supplementary Fig. 7). However, thin amorphous layers ($\sim$5 nm) formed on the surfaces of all the as-cast catalysts. Further amorphization gradually occurred during the OER cycles in the case of the SFO and CFO samples, resulting in thick amorphous layers about 20 nm after 100 cycles. These amorphous layers caused suppression of the OER reaction of these two perovskites with decreased current density as shown in Fig. 2d,e. In contrast, CCFO retained the thin amorphous layer ($\sim$5 nm) even after 100 cycles, and no erosion was observed. If the amorphous layers isolated only the catalyst surfaces from the electrolyte, the catalytic activities would converge on the lower levels equally. However, the experimental results suggest that the amorphous layers reflect the bulk properties to some extent. The significant improvement in the CCFO stability in comparison with that of $AFe^{4+}O_3$ can be attributed to the transformation of the covalent bonding networks. In a simple $AFe^{4+}O_3$ perovskite, the B-site Fe^{4+} ions are covalently bound to the oxide ions, whereas the A-site alkaline-earth metal ions have ionic characteristics[15]. Thus, the A-site ions are easily dissolved in the electrolyte during OER[12]. In contrast, the covalent Cu–O bonds in square-planar units of CCFO aid formation of the covalent bonding network and prevent progressive amorphization during the OER measurements.

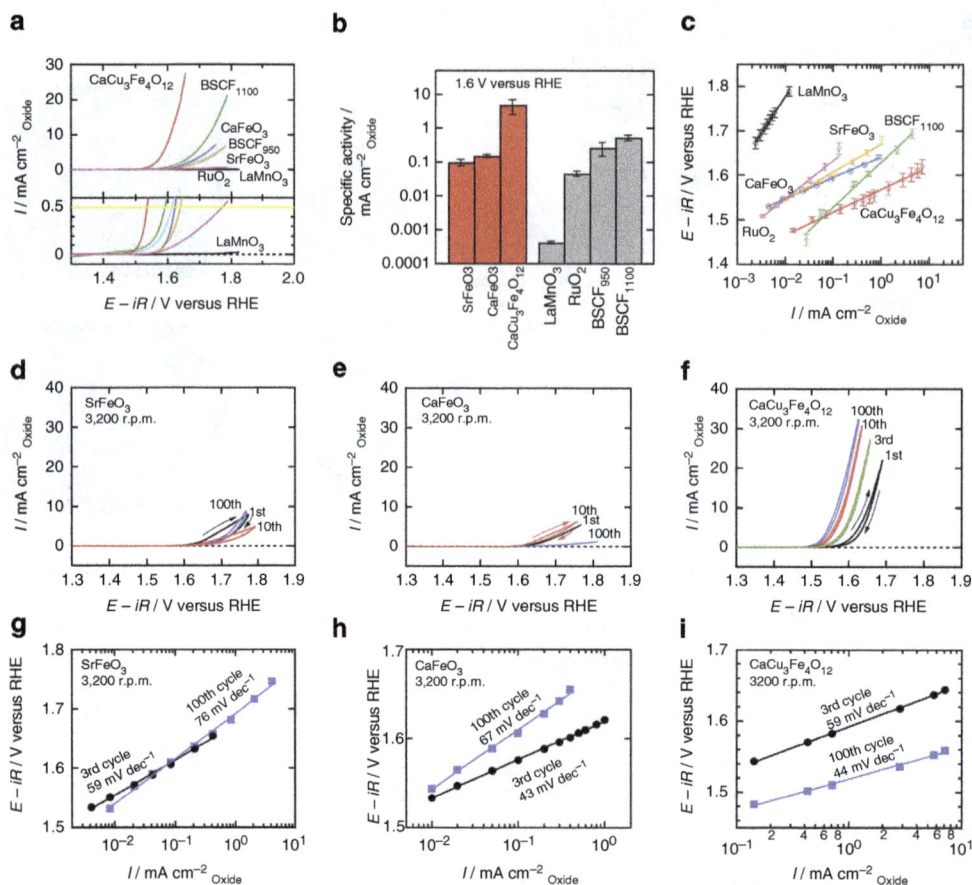

Figure 2 | OER catalytic performance of Fe^{4+}-perovskites and references. (**a**) Linear sweep voltammograms for OER for SFO, CFO, CCFO, LMO, BSCF and RuO$_2$. The overpotential (η) of each catalyst was determined from the onset potential, E_{onset} (V versus RHE); E_{onset} is the potential at 0.5 mA cm$^{-2}_{oxide}$ and $\eta = E_{onset} - 1.23$ (V). (**b**) Specific activities (current density at 1.6 V versus RHE) for SFO, CFO, CCFO, LMO, BSCF and RuO$_2$. (**c**) Tafel plots for SFO, CFO, CCFO, LMO and BSCF. The error bars show the s.d. of three independent measurements. All data in (**a-c**) were obtained from the third cycle. Cyclic voltammograms of (**d**) SFO, (**e**) CFO and (**f**) CCFO for 100 cycles. Cycle dependence of Tafel slopes for (**g**) SFO, (**h**) CFO and (**i**) CCFO. Hundred continuous cycle measurements were performed with a higher disk rotation rate of 3,200 r.p.m. to prevent adhesion of the O$_2$ bubbles to the electrode. In (**b,c**) the error bars correspond to the s.d. obtained from three independent measurements.

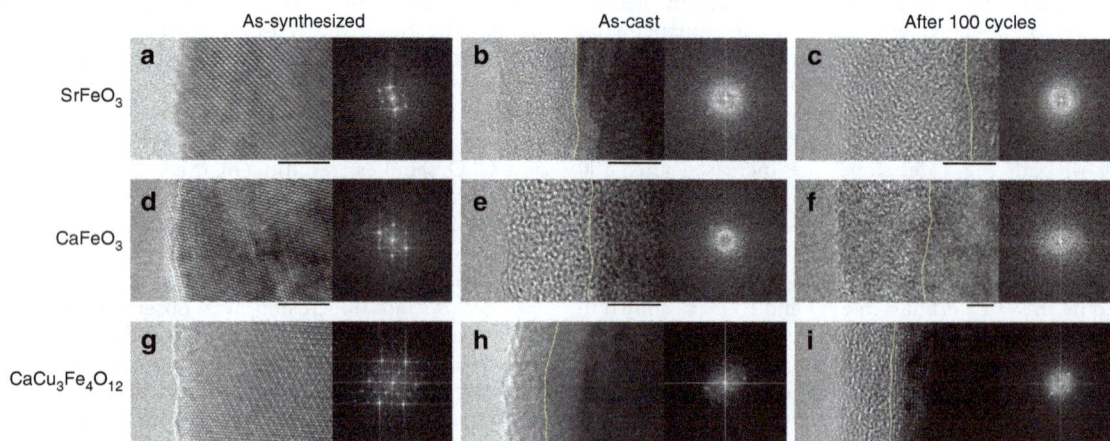

Figure 3 | HRTEM and fast fourier transform (FFT) images of perovskite oxides before and after OER measurements. The boundaries between the crystalline and amorphous regions are divided by orange dotted lines. All the FFT images were obtained from surface regions of ~10 × 10 nm^2. Scale bar: 5 nm.

Discussion

Here, we demonstrate the structural features of CCFO that are associated with OER catalytic activity. When we assume that the local crystal structures of CCFO are reflected on the surface at a certain level, several factors that increase the catalytic activity are considered. Fig. 4 proposes three possible OER routes for SFO

and CCFO. The left route is the conventional Eley-Rideal (ER)-type mechanism for SFO and CCFO. In the ER-type mechanism, OH$^-$ adsorbates are bound to B-site Fe ions on the surface (Fig. 4a), in which the rate-determining step is considered to be the formation of the O–O bond (reaction 2) or the subsequent deprotonation (reaction 3)[9] along with the redox

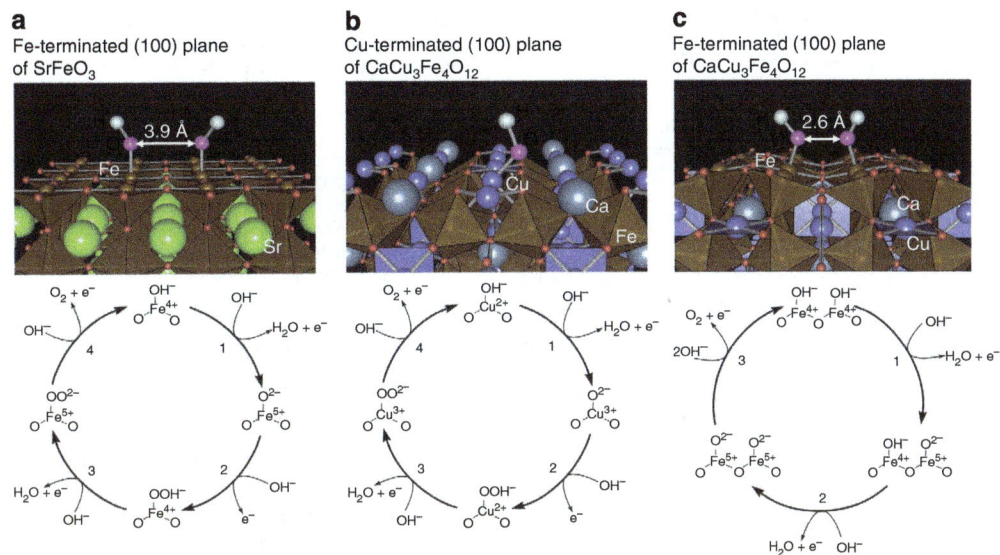

a
Fe-terminated (100) plane
of $SrFeO_3$

b
Cu-terminated (100) plane
of $CaCu_3Fe_4O_{12}$

c
Fe-terminated (100) plane
of $CaCu_3Fe_4O_{12}$

Figure 4 | OH⁻ adsorbed surfaces for SFO and CCFO and corresponding OER mechanism. (a) OH^- adsorbates on Fe-terminated (100) plane of SFO for Fe-mediated route (ER type). The interatomic distance between the nearest neighbouring OH^- adsorbates is ~ 3.9 Å. **(b)** OH^- adsorbate on (Ca,Cu)O-terminated (100) planes of CCFO for Cu-mediated route (ER type). **(c)** OH^- adsorbates on FeO_2-terminated (100) planes of CCFO for Fe-mediated route (LH type). The interatomic distance between the nearest neighbouring OH adsorbates is ~ 2.6 A. In all cases, the Cu^{2+}/Cu^{3+} or Fe^{4+}/Fe^{5+} redox couple acts as the reaction mediator under the assumption that the adsorbed OH^- ions occupy the original oxygen sites in the crystal structure.

reaction of the B-site ions. In both cases, the electron charges are transferred to the Fe ions efficiently through strong Fe–O covalent bonds. An almost identical mechanism is most likely valid on the Cu-terminated surface of CCFO (Fig. 4b). The Cu and Fe ions can tolerate the redox reactions in the Cu^{2+}/Cu^{3+} and Fe^{4+}/Fe^{5+} states, leading to the stable and high OER activity exhibited by CCFO. It should be noted that the Langmuir-Hinshelwood (LH)-type reaction can occur through the direct formation of the O–O bond between the neighbouring oxygen atoms connected to the nearest neighbouring Fe ions, because of the short distance (Fig. 4c). The oxygen–oxygen distance is shortened to ~ 2.6 Å by heavily bent Fe–O–Fe bonds ($\sim 140°$) for CCFO. This oxygen–oxygen distance is comparable to that of α-Mn_2O_3, in which the LH-type mechanism is thought to dominate[21]. In contrast, the oxygen–oxygen distances for simple cubic perovksite SFO is ~ 3.9 Å because of the linear Fe–O–Fe bonds ($=180°$). This oxygen–oxygen distance is too large to permit the oxygen atoms to interact with each other and form oxygen molecules. In the LH-type reaction, one of the two possible rate-determining steps in the ER-type reaction (that is, the deprotonation of the oxyhydroxide group to form peroxide ions) is skipped, resulting in the acceleration of the reaction. Thus, the realization of the LH-type reaction is another major specificity of CCFO.

In summary, the Fe^{4+}-perovskite CCFO exhibits promising OER activity. Further, CCFO has a widespread covalent bonding network that enhances its stability. The cationic arrangements of this substance provide a further increase in the OER activity. These findings indicate that the covalent network consisting of multiple transition metal ions in CCFO plays a crucial role in the activity and stability of the OER catalysis. In addition, the various unexplored A'-B ion couplings in quadruple perovskites may provide further high-performance, high-stability and cost-effective OER catalysts.

Methods

Sample preparation. SFO, CFO and CCFO were synthesized via a high-pressure synthesis method. LMO and BSCF were obtained using a polymerized method,

while $CuFe_2O_4$ was synthesized via the inverse coprecipitation method. RuO_2 (99.9%) was used as purchased from RARE METALLIC, Co, Ltd. The sample preparation details are given in the Supplementary Methods.

Characterization. X-ray diffraction patterns of reference oxides were obtained using a laboratory X-ray diffractometer (Rigaku Ultima IV) with Cu Kα radiation. Synchrotron X-ray powder diffraction patterns of the Fe^{4+}-perovskites were obtained at the SPring-8 BL02B2 beamline. Fe K-edge X-ray absorption spectra of the CCFO and Fe^{3+} reference oxides were collected at room temperature and in absorption mode at the SPring-8 BL01B1 beamline. Crystal structure refinements of CFO, SFO and CCFO were conducted based on the obtained synchrotron X-ray powder diffraction data using a Rietveld refinement program RIETAN-FP[22]. Electron density analysis of SFO and CCFO was performed using the Dysnomia maximum entropy method program[23]. HRTEM images were collected using a JEOL JEM-2100F.

Preparation of catalyst inks. The catalyst inks were prepared by reference to the methods reported by Suntivich et al.[9,24] and Jung et al.[7] K^+ ion-exchanged Nafioni was used as a immobilizing binder, which did not prevent the transport of dissolved O_2 to the catalyst surface. A ~ 3.33 wt.% K^+ ion-exchanged Nafion suspension was prepared by mixing a 5 wt.% proton-type Nafion suspension (Sigma-Aldrich) and 0.1 M KOH aqueous solution at 2:1 by volume. The pH of the 5 wt.% proton-type Nafion suspension was initially ~ 1 and 2 and was changed to ~ 11 after mixing. The catalyst inks of the perovskites and reference oxides (RuO_2 and IrO_2, Sigma-Aldrich) were prepared by mixing 50 mg of oxide, 10 mg of acetylene black (AB), and 0.3 mL of ~ 3.33 wt.% K^+ ion-exchanged Nafion suspension. The volumes of the inks were adjusted to 10 mL by the addition of tetrahydrofuran (Sigma-Aldrich). Thus, the final concentration of the catalyst inks was 5 $mg_{oxide} mL_{ink}^{-1}$, 1 $mg_{AB} mL_{ink}^{-1}$ and ~ 1 $mg_{Nafion} mL_{ink}^{-1}$. A rotating ring-disk electrode (BAS Inc, Japan) consisting of a glassy carbon (GC) disk of 0.4 cm in diameter and a Pt ring part of 0.7 and 0.5 cm outer and inner diameter, respectively, was used as a working electrode after mirror polishing with 0.05 µg alumina slurry (BAS Inc). Then, 6.4 µL of catalyst ink was drop-cast onto the GC disk part ($0.2 \times 0.2 \times \pi cm^2$). The catalyst layer on the GC disk part was dried overnight in vacuum at room temperature, and was composed of 0.25 $mg_{oxide} cm_{disk}^{-2}$, 0.05 $mg_{AB} cm_{disk}^{-2}$ and ~ 0.05 $mg_{Nafion} cm_{disk}^{-2}$.

Electrochemical characterization. Electrochemical characterization was conducted with a rotating-disk electrode rotator (RRDE-3 A, BAS Inc) at an electrode rotation rate of 1,600 or 3,200 r.p.m. in combination with a bipotentiostat (ALS Co, Ltd., Japan). For all experiments, a Pt wire electrode and Hg/HgO electrode (International Chemistry Co, Ltd., Japan) filled with a 0.10 M KOH aqueous solution (Nacalai Tesque, Inc, Japan) were used as the counter and reference electrodes, respectively. All measurements were conducted under O_2 saturation at room temperature (~ 25 °C), which fixed the equilibrium potential of the O_2/H_2O redox couple to 0.304 V versus Hg/HgO (or 1.23 V versus RHE). For the catalysis evaluation of the perovskites for OER, the potential of the catalyst-modified GC part was controlled from 0.3–0.9 V versus Hg/HgO (1.226–1.826 V versus RHE) at 10 mV s^{-1}. For all measurements, the current density was iR-corrected ($R = \sim 43\,\Omega$) using the measured solution resistance, and capacitance-corrected by taking the average between the anodic and cathodic scans[9]. All the OER currents are shown relative to the surface area of the oxide catalysts estimated using BET analysis (BELSORP-max, BEL Japan, Inc, Japan).

References

1. Fabbri, E., Habereder, A., Walter, K., Kötz, R. & Schmidt, T. J. Developments and perspectives of oxide-based catalysts for the oxygen evolution reaction. *Catal. Sci. Technol.* **4**, 3800–3821 (2014).
2. Hong, W. T. *et al.* Toward the rational design of non-precious transition metal oxides for oxygen electrocatalysis. *Energy Environ. Sci.* **8**, 1404–1427 (2015).
3. Katsounaros, I., Cherevko, S., Zeradjanin, A. R. & Mayrhofer, K. J. J. Oxygen electrochemistry as a cornerstone for sustainable energy conversion. *Angew. Chem. Int. Ed.* **53**, 102–121 (2014).
4. Wang, Z. -L., Xu, D., Xu, J. -J. & Zhang, X. -B. Oxygen electrocatalysts in metal-air batteries: from aqueous to nonaqueous electrolytes. *Chem. Soc. Rev.* **43**, 7746–7786 (2014).
5. Subbaraman, R. *et al.* Trends in activity for the water electrolyser reactions on 3d M(Ni,Co,Fe,Mn) hydr(oxy)oxide catalysts. *Nat. Mater.* **11**, 550–557 (2012).
6. Grimaud, A. *et al.* Double perovskites as a family of highly active catalysts for oxygen evolution in alkaline solution. *Nat. Commun.* **4**, 2439 (2013).
7. Jung, J. -I., Jeong, H. Y., Lee, J. -S., Kim, M. G. & Cho, J. A bifunctional perovskite catalyst for oxygen reduction and evolution. *Angew. Chem. Int. Ed.* **53**, 4582–4586 (2014).
8. Lee, Y., Suntivich, J., May, K. J., Perry, E. E. & Shao-Horn, Y. Synthesis and activities of rutile IrO_2 and RuO_2 nanoparticles for oxygen evolution in acid and alkaline solutions. *J. Phys. Chem. Lett.* **3**, 399–404 (2012).
9. Suntivich, J., May, K. J., Gasteiger, H. A., Goodenough, J. B. & Shao-Horn, Y. A perovskite oxide optimized for oxygen evolution catalysis from molecular orbital principles. *Science* **334**, 1383–1385 (2011).
10. Calle-Vallejo, F. *et al.* Number of outer electrons as descriptor for adsorption processes on transition metals and their oxides. *Chem. Sci* **4**, 1245–1249 (2013).
11. Bockris, J. O'M. & Otagawa, T. The electrocatalysis of oxygen evolution on perovskites. *J. Electrochem. Soc.* **131**, 290–302 (1984).
12. May, K. J. *et al.* Influence of oxygen evolution during water oxidation on the surface of perovskite oxide catalysts. *J. Phys. Chem. Lett.* **3**, 3264–3270 (2012).
13. Suntivich, J. *et al.* Estimating hybridization of transition metal and oxygen states in perovskites from O K-edge X-ray absorption spectroscopy. *J. Phys. Chem. C* **118**, 1856–1863 (2014).
14. van Eldik, R. & Klärner, F. -G. (eds) *High Pressure Chemistry: Synthetic, Mechanistic, and Supercritical Applications* (Wiley-VCH, 2008).
15. Chang, S. H. *et al.* Functional links between stability and reactivity of strontium ruthenate single crystals during oxygen evolution. *Nat. Commun.* **5**, 4191 (2014).
16. Yamada, I. *et al.* A perovskite containing quadrivalent iron as a charge-disproportionated ferrimagnet. *Angew. Chem. Int. Ed.* **47**, 7032–7035 (2008).
17. Yamada, I. *et al.* Giant negative thermal expansion in the iron perovskite $SrCu_3Fe_4O_{12}$. *Angew. Chem. Int. Ed.* **50**, 6579–6582 (2011).
18. Clark, J. H. *et al.* Visible light photo-oxidation of model pollutants using $CaCu_3Ti_4O_{12}$: an experimental and theoretical study of optical properties, electronic structure, and selectivity. *J. Am. Chem. Soc.* **133**, 1016–1032 (2011).
19. Jung, J.-I. *et al.* Fabrication of $_{0.5}Sr_{0.5}Co_{0.8}Fe_{0.2}O_{3-\delta}$ catalysts with enhanced electrochemical performance by removing an inherent heterogeneous surface film layer. *Adv. Mater.* **27**, 266–271 (2015).
20. Mohamed, R. *et al.* Electrocatalysis of perovskites: The influence of carbon on the oxygen evolution activity. *J. Electrochem. Soc.* **162**, F579–F586 (2015).
21. Ramírez, A. *et al.* Evaluation of MnO_x, Mn_2O_3, and Mn_3O_4 electrodeposited films for the oxygen evolution reaction of water. *J. Phys. Chem. C* **118**, 14073–14081 (2014).
22. Izumi, F. & Momma, K. Three-dimensional visualization in powder diffraction. *Solid State Phenom.* **130**, 15–20 (2007).
23. Momma, K., Ikeda, T., Belik, A. A. & Izumi, F. Dysnomia, a computer program for maximum-entropy method (MEM) analysis and its performance in the MEM-based pattern fitting. *Powder Diffr.* **28**, 184–193 (2013).
24. Suntivich, J., Gasteiger, H. A., Yabuuchi, N. & Shao-Horn, Y. Electrocatalytic measurement methodology of oxide catalysts using a thin-film rotating disk electrode. *J. Electrochem. Soc.* **157**, B1263–B1268 (2010).
25. Bocquet, A. E. *et al.* Electronic structure of $SrFe^{4+}O_3$ and related Fe perovskite oxides. *Phys. Rev. B Condens. Matter* **45**, 1561–1570 (1992).
26. Momma, K. & Izumi, F. VESTA 3 for three-dimensional visualization of crystal, volumetric and morphology data. *J. Appl. Crystallogr.* **44**, 1272–1276 (2011).

Acknowledgements

The authors would like to thank Mr Masaaki Fukuda for his advice on chemical reactions, Dr Masaichiro Mizumaki for his assistance with the X-ray absorption measurements, Dr Ya Xu and Mr Jun-ya Sakurai for their assistance with BET analyses and Drs. Mikio Takano and Shu Yamaguchi for fruitful discussion. The synchrotron radiation experiments were performed at SPring-8 under the approval of the Japan Synchrotron Radiation Research Institute (Proposal Nos. 2013A1188, 2014B1128 and 2014B1129). This work was supported by a Grant-in-Aid for Scientific Research (B 15H04169) from the Japan Society for the Promotion of Science, the Ministry of Education, Culture, Sports, Science and Technology of Japan.

Author contributions

S.Y. and I.Y. conceived, designed and co-wrote the paper and contributed equally to this work. I.Y., M.M., A.S. and H.F. synthesized the catalysts. S.Y., I.Y., M.M., A.S. and H.F. conducted electrochemical measurements. H.T. and S.M. conducted HRTEM observation. All authors discussed the results and commented on the manuscript.

Additional information

Competing financial interests: The authors declare no competing financial interests.

8

Platinum–nickel frame within metal-organic framework fabricated *in situ* for hydrogen enrichment and molecular sieving

Zhi Li[1], Rong Yu[2], Jinglu Huang[2], Yusheng Shi[1], Diyang Zhang[1], Xiaoyan Zhong[2], Dingsheng Wang[1], Yuen Wu[1,3] & Yadong Li[1,3]

Developing catalysts that provide the effective activation of hydrogen and selective absorption of substrate on metal surface is crucial to simultaneously improve activity and selectivity of hydrogenation reaction. Here we present an unique *in situ* etching and coordination synthetic strategy for exploiting a functionalized metal-organic framework to incorporate the bimetallic platinum–nickel frames, thereby forming a frame within frame nanostructure. The as-grown metal-organic framework serves as a 'breath shell' to enhance hydrogen enrichment and activation on platinum–nickel surface. More importantly, this framework structure with defined pores can provide the selective accessibility of molecules through its one-dimensional channels. In a mixture containing four olefins, the composite can selectively transport the substrates smaller than its pores to the platinum–nickel surface and catalyse their hydrogenation. This molecular sieve effect can be also applied to selectively produce imines, which are important intermediates in the reductive imination of nitroarene, by restraining further hydrogenation via cascade processes.

[1] Department of Chemistry and Collaborative Innovation Center for Nanomaterial Science and Engineering, Tsinghua University, Beijing 100084, China. [2] Beijing National Center for Electron Microscopy, School of Materials Science and Engineering, Tsinghua University, Beijing 100084, China. [3] Center of Advanced Nanocatalysis (CAN-USTC), University of Science and Technology of China, Hefei, Anhui 230026, China. Correspondence and requests for materials should be addressed to Y.W. (email: yuenwu@ustc.edu.cn) or to Y.L. (email: ydli@mail.tsinghua.edu.cn).

Hydrogenation reaction is a fundamental component in metal catalysis. How to improve the absorption and dissociation of H_2 on metal surface is strictly related to the activity of hydrogenation reactions. Expanding the exposed metal sites as much as possible is an effective strategy to optimize the usage of precious metal, which is also a benefit for the activation of H_2. For that sake, multitudinous structures such as hollow[1,2], porous[3], concave[4,5] metallic structure with high surface area-to-volume ratio have been developed. Among them, frame-structured metal material has been demonstrated as one promising catalyst not only because all the reactive corners and edges can be maintained, but also the three-dimensional (3D) molecular accessibility which can facilitate the contact between H_2 and metal[6-8]. Apart from the activity, promoting the selectivity of hydrogenation is another key concern for the design of nanocatalysts, which is mainly dependent on the absorption of substrate on metal surface. Owing to intensive research efforts focusing on surface science and catalysis, substantial factors such as crystal facet[9], exposed defects[10], interfaces[11] and surface ligands[12] have been discovered to influence the selective absorption of substrate on metal surface. Learning from the nature that carries out enzymatic transformations with excellent shape- and size selectivity, we believe metal-organic frameworks (MOFs) possessing tunable porosity and 3D nanoframe structure may impart molecular sieving to metal catalyst by controlling the diffusion of substrate, thereby tuning the selectivity of hydrogenation. Hence, the frame motif can be extended by coating a shell of MOF on the surface of metal frame to achieve unique frame within frame (frame @ frame) structure, which may endow new chances to achieve H_2 enrichment and molecular sieving in metal catalysis simultaneously.

Dealloying process is a top-down strategy to carve the bimetallic structure at nanoscale. Driven by the different chemical reactivity of two metallic species, this versatile method has gained great success hitherto in constructing the bimetallic nanoframe[6,8] structures. In contrast, the fabrication of MOFs is based on the cooperative assembly of organic linker and metal ions, which can be termed as bottom-up strategy. There is a severe drawback (often neglected) if dealloying strategy is adopted to construct bimetallic nanoframe. The carving process for generating interior vacancies and surface defects is strictly related to the dissolution of active metals. By whatever means necessary including oxidative etching which usually utilize oxidant[13-15] or galvanic reaction involving the replacement between two different metals[16,17], the active metals are consistently converted to ionic counterpart and abandoned in most cases[18].

Herein, we take the advantages of both top-down and bottom-up strategies, using organic linkers to capture the abandoned Ni^{2+} ion during the dealloying process, to build a shell of MOFs on the surface of Pt–Ni alloy *in situ*. This unique frame @ frame nanostructure is expected to inherit the desirable properties of both Pt–Ni frames and MOFs.

Results

Synthesis and characterization. To produce this unique frame @ frame structure, we firstly prepared Ni-rich Pt–Ni alloy according to our previously reported method[19]. The starting Pt–Ni polyhedrons exhibit excellent monodispersity with average size of ~20 nm and uniform truncated octahedral morphology (Fig. 1a). The polyvinyl pyrrolidone (PVP)-capped Ni-rich Pt–Ni nanoparticles (NPs) were submersed in dimethylformamide (DMF) to form a turbid solution, followed by adding a solution of 2,5-dioxidoterephthalate. During 12-h solvothermal process, three representative samples at 0.5, 4 and 12 h were collected and

observed by transmission electron microscopy (TEM). In the first 0.5 h, a shell with lower contrast had emerged on the surface of Pt–Ni NPs (Fig. 1b). Successively, the initial truncated octahedral Pt–Ni alloy would evolve into hybrid structure that opens nanoframe located within a readily formed overlayer, while maintaining its original symmetry (Fig. 1c,d). That is, a well-defined MOF of Ni_2dobdc (dobdc^{4-} (2,5-dioxidoterephthalate)), commonly known as Ni-MOF-74 (refs 20), was expected to be weaved on the surface of Pt–Ni nanoframes. The TEM image in low magnification demonstrated that the etched Pt–Ni frames could be incorporated fully within the matrices of *in situ*-formed MOF in a well-dispersed manner (Supplementary Fig. 1). The inherent process may relate to the following equations:

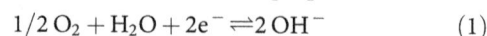

$$1/2\,O_2 + H_2O + 2e^- \rightleftharpoons 2\,OH^- \qquad (1)$$

$$Ni\,(0) - 2e^- \rightleftharpoons Ni^{2+} \qquad (2)$$

$$2Ni^{2+} + dobdc^{4-} \rightleftharpoons Ni_2dobdc \qquad (3)$$

In detail, the oxidative etching of Pt–Ni alloy and the *in situ* nucleation of MOF-74 are two major interactive processes in this chemical etching, to some content maintaining synchronization. The two oxidation–reduction reactions shown in equations (1) and (2) are assigned to the electron transfer from Ni(0) to oxygen. Pt is a relatively inert element to oxygen compared with Ni, which determines the different diffusion rate during the etching process. The intrinsic formation mechanism of nanoframe may follow the Kirkendall effect[16,21]. In other words, the outward diffusion of Ni^{2+} and inward spread of voids dominate the whole etching process. Considering the unique Pt-segregated surface/Ni-segregated core structure reported[19], the surface Pt shell cannot retain integrity as the etching of Ni proceeds, thereby generating the cavities inside the shell and driving the segregation of Pt at the frames. Apart from the electron transfer in oxidative etching of Ni to Ni^{2+}, the species transfer of Ni from Pt–Ni nanoframe to Ni-MOF-74 is crucial to this frame @ frame structures. As schematically illustrated in Fig. 1, the Ni^{2+} on the surface of Pt–Ni nanoframe can be captured by the near-neighbour organic linker to form MOFs *in situ*. Once the precipitation of Ni-MOF-74 in equation (3) occurred, the equilibrium of oxidative etching reactions would be broken and shifted towards the generation of Ni^{2+} (equation (2)), thus largely accelerating the etching rate. To be accompanied by the Pt–Ni polyhedrons being etched to framework, another MOF would readily emerge and encapsulate the metallic framework within their matrices.

The composition evolution from Ni-rich Pt–Ni polyhedra to Pt-rich Pt–Ni frame and the nucleation of MOFs can be verified by X-ray diffraction (XRD) patterns, the inductively coupled plasma atomic emission spectroscopy (ICP-AES) and energy-dispersive X-ray spectra (Supplementary Table 1, Supplementary Figs 2 and 3). Compared with the initial Pt–Ni polyhedra, a set of peaks belonging to Ni-MOF-74 emerged after the structural evolution, indicating the successful coating of MOF on the etched Pt–Ni frame. The corresponding peaks derived from the organic groups of MOF overlayer, evidenced by the Fourier transform infrared spectra, perfectly match the signals of pure Ni-MOF-74 synthesized by previously reported method[22] (Supplementary Fig. 4). After treating this composite with dilute acetic acid, the coating MOF can be removed by cutting off the coordination between phenolate and carboxylic oxygen atoms and Ni^{2+}, thus leaving a bare Pt-Ni-framed structure (Supplementary Fig. 5). From the XRD patterns belonging to Pt–Ni frame, the face-centred cubic peaks for (111), (200) and (220) facets were found to shift towards lower 2θ values due to the increasing d spacing, which demonstrated the dissolution of Ni from parent Ni-rich Pt–Ni alloy. X-ray photoelectron spectroscopy (XPS) was used to

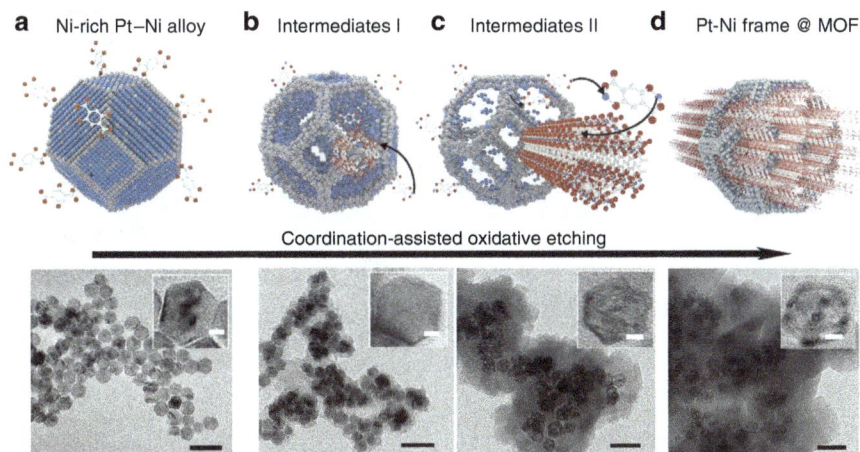

Figure 1 | Scheme and corresponding TEM images of the coordination-assisted oxidative etching process. (**a**) Initial solid Pt–Ni polyhedra. (**b**) Pt–Ni frame @ MOF intermediates I. (**c**) Pt–Ni frame @ MOF intermediates II. (**d**) Final Pt–Ni frame @ MOF. The scale bars, 50 nm. (Insets are the magnified TEM images. The scale bars, 5 nm).

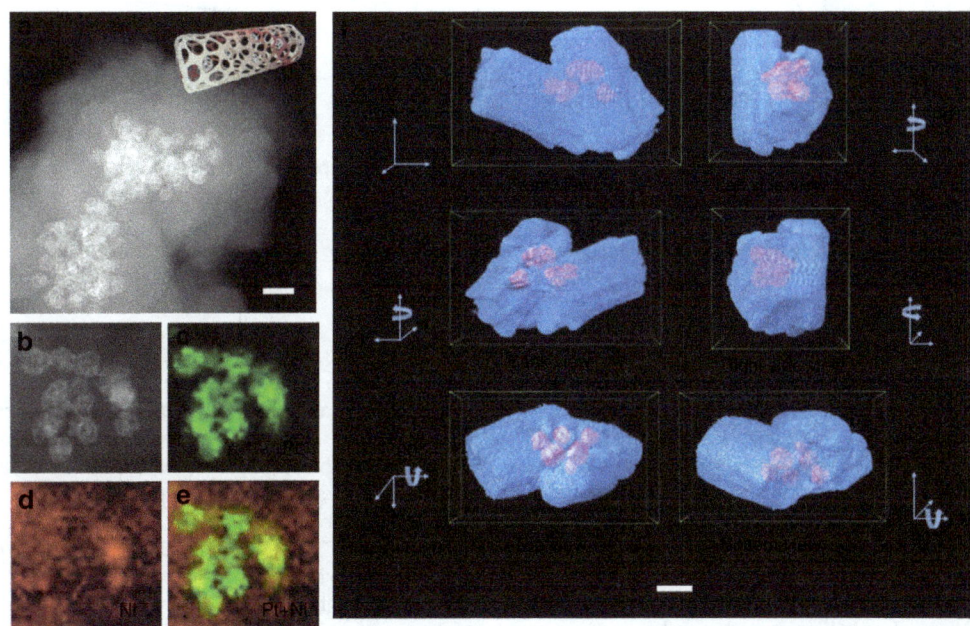

Figure 2 | Characterization of Pt–Ni frame @ MOF. (**a**) HAADF-STEM image and ideal model of Pt–Ni frame @ MOF. The Pt–Ni frames were easily distinguished from the shrouding MOF matrix based on their different contrast. (**b-e**) Energy-dispersive X-ray elemental mapping results of Pt–Ni frame @ MOF, suggesting that Ni is homogeneously distributed throughout the entire nanostructure and Pt is concentrated where initial Pt–Ni alloy is situated. (**f**) Six projected images of three-dimensional visualization of tomographic reconstruction images of Pt–Ni frame @ MOF, demonstrating that the Pt–Ni frames were fully enshrouded by MOF-74. The scale bars, 50 nm.

trace the valence evolution in this coordination-assisted chemical etching (Supplementary Fig. 6). The Ni 2p and Pt 4f spectra of Pt–Ni polyhedra and Pt–Ni frame revealed that the surface Ni was partly oxidized to Ni^{2+} and most of the surface Pt retained metallic state. In the case of Pt–Ni frame @ MOF, the spectra for Ni showed the oxidation state was predominant, which was attributed to depletion of Ni(0) by coordination process. Meanwhile, the peaks assigned to metallic Pt were largely depressed, suggesting the seamless coating of MOFs will block the activated photoelectrons from coated Pt–Ni frame during the XPS measurements.

The intrinsic morphology and spatial distribution of the frame @ frame structure were examined by the high-angle annular dark-field scanning transmission electron microscope

(HAADF-STEM) for which they were well-suited because of the mitigatory electron radiation damage relative to bright field (Fig. 2)[23,24]. The unique features of encapsulated Pt–Ni frame including hollow interior and the interconnecting edges in space can be clearly seen in Fig. 2a. The frame structure was robust enough to resist the chemical etching process, lattice strain from the interface of Pt–Ni and MOF, and the post treatment such as centrifugation and drying. The corresponding elemental maps tell the Pt concentrated at where initial Ni-rich Pt–Ni alloy situated, whereas the Ni distribute homogeneously throughout the entire architecture (Fig. 2b–e). These results coincided to the line scan profiles (Supplementary Fig. 7) and reinforced that Ni could be preferentially etched from the Pt–Ni alloy versus Pt because of its more tendency towards oxidative etching in the presence of

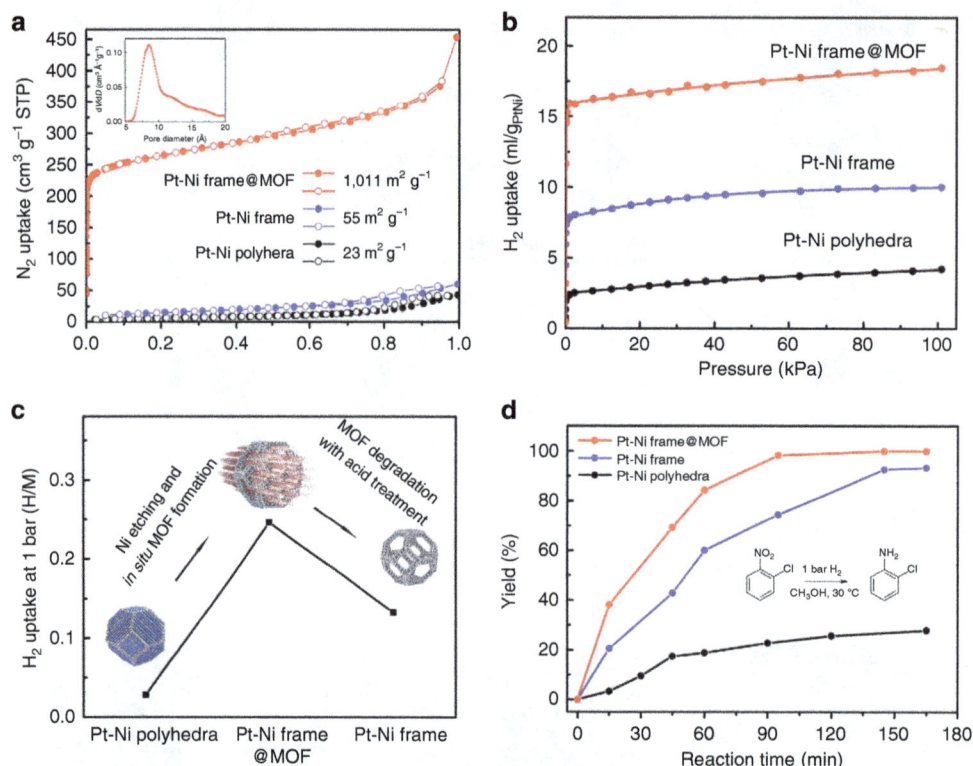

Figure 3 | Gas-sorption properties and catalytic hydrogenation efficiencies. (**a**) Nitrogen-sorption isotherms at 77 K up to 1 bar. The filled and open symbols represent adsorption and desorption curves, respectively (**b**) H_2 adsorption isotherms at 273 K up to 1 bar normalized by the mass of metal. Detailed interpretation of the calculation process is available in the ESI. (**c**) Comparison of H_2 uptake at 273 K and 1 bar among three catalysts. (**d**) Yield (%) of 2-chloroaniline as a function of time in the selective hydrogenation of 1-chloro-2-nitrobenzene with Pt–Ni polyhedra, Pt–Ni frame and Pt–Ni frame @ MOF.

organic linkers. To provide more detailed information that describes the 3D imaging of real structure, tomographic data were reconstructed based on a series of 2D HAADF-STEM images, which were taken at consecutive tilt angles from $-72°$ to $72°$ with each $4°$ tilt increment (Supplementary Fig. 8)[25]. The 3D tomography of frame @ frame structure (Supplementary Fig. 9) was achieved based on the differences in Z-contrast between Pt–Ni frame and MOF. Six projected images of 3D visualization of tomographic reconstruction of Pt–Ni frame @ MOF showed that the Pt–Ni frame were fully enshrouded by well-defined MOF-74 (Fig. 2f). More detailed animated voxel recording the rotation of tomogram is available in Supplementary Movie 1. Time-tracking scanning electron microscope images along with the structural evolution also support the seamless encapsulation of Pt–Ni NPs by grown MOFs (Supplementary Fig. 10). It is notable that this coordination-assisted chemical etching presented here can be readily generalized to octahedral Ni-rich Pt–Ni polyhedra (Supplementary Fig. 11) to construct another type of 'concave Pt–Ni alloy @ frame' structure (Supplementary Fig. 12).

Gas-sorption and catalytic properties. As is well-known, MOF-74, which owns characteristic honeycomb structure composed of 1D channel and open metal sites, is a widely used framework material due to its excellent chemical robustness and thermal stability[20]. Based on the profiles of thermogravimetric analysis (Supplementary Fig. 13), the linker decomposition starts at about $350°C$ for Pt–Ni frame @ MOF composite, which is in good agreement with the as-prepared pure Ni-MOF-74. This grown MOF-74 not only enables evident confinement to

strengthen the rigidity of Pt–Ni frames, but also behaves as a 'breath shell' to largely enhance the uptake and enrichment of gas molecules. From the N_2 adsorption and desorption isotherms on these catalysts, the frame @ frame structure exhibits 1 order of magnitude enhancement in Brunauer–Emmett–Teller surface area than the starting Pt–Ni polyhedra and bare Pt–Ni frame (Fig. 3a). According to the IUPAC definition, the sharp uptake at P/P_0 from 10^{-5} to 10^{-1} indicates a standard type I isotherm with characteristic of 8.6-Å micropores and the additional uptake at high relative pressure of $P/P_0 = 0.9$ implies the existence of macro pores generated by packing of frame @ frame nanostructures.

The H_2 enrichment properties of frame @ frame structure were studied by H_2 adsorption isotherms at 273 K, which were shown in Fig. 3b. By eliminating the amount of hydrogen absorption on pure Ni-MOF-74, the number of hydrogen atoms absorbed on each metallic atom ($Pt(0) + Ni(0)$) were normalized for Pt–Ni polyhedra, Pt–Ni frame and Pt–Ni frame @ MOFs separately (Supplementary Figs 14 and 15). An obvious increment from 0.03 H per metal atom absorbed in bare Pt–Ni polyhedra to 0.25 H per metal atom in Pt–Ni frame @ MOFs was observed after the encapsulation of MOF-74. In contrast, the removal of grown MOF-74 by acid treatment would result in a 46% degradation of H_2 storage for the Pt–Ni frame. We found that this H_2 enrichment can be applied to facilitate the catalytic efficiency of hydrogenation reaction. In our case of study, the selective hydrogenation of 1-chloro-2-nitrobenzene, which represents an important industrial conversion[26,27], was selected to probe the structure-activity relationship of as-prepared frame @ frame catalysts. Since the diffusion problem for 1-chloro-2-nitrobenzene is expected to be negligible through the large pore apertures of

MOF-74, the enhanced H_2 storage does confer this frame @ frame catalyst even higher catalytic efficiency compared with the bare Pt–Ni frame. Actually according to the TEM and XRD measurements, this frame @ frame catalyst can retain its structural stability and catalytic activity after 10 runs of recycle measurement without significant decrease in 2-chloroaniline selectivity (Supplementary Figs 16–18).

Discussion

Accordingly, this well-defined textural property may endow frame @ frame catalyst with molecular sieving to achieve efficient and selective hydrogenation for target products, if the size of substrate or product were ingeniously designed. Hydrogenation of a mixture containing styrene, 2,4,6-trimethylstyrene, trans-stilbene and 4,4'-dimethyl-trans-stilbene with different sizes (Fig. 4b) has been conducted to investigate the substrate-size selectivity in MOFs. As a comparison, catalysts composed of bare Pt–Ni nanoframe and Pt–Ni frame directly loaded on Ni-MOF-74 (Supplementary Fig. 19) were synthesized to further elucidate the effect of molecular sieving. Styrene molecules (8.4 Å) are small enough to diffuse through the pore apertures of MOF shells onto the Pt–Ni frame surface without hindrance. Therefore, it is reasonable that the frame @ frame structure can catalyse the hydrogenation of styrene with higher activity than Pt–Ni frame and Pt–Ni frame on MOF (Fig. 4b), resulting from the H_2 enrichment. If the size of molecules increased, the diffusion of substrates would be strictly limited by the uniform pores of Ni-MOF-74. This confinement will significantly result in efficiency decay for Pt–Ni frame @ MOF catalyst by retarding the

contact between olefins and Pt–Ni nanoframe inside. Apart from the deceased diffusion rates, the diffusion posture in channel could be limited, which was also detrimental to the hydrogenation of C=C bond in the middle if these olefins had to lay down in the channel[28]. In contrast, the hydrogenation of olefins catalysed by both of the Pt–Ni frame and Pt–Ni frame on MOF were barely affected because of the sufficient exposed Pt and Ni atoms outside.

Imines are important chemical intermediates due to their good electrophilicity for many important condensation, reduction and addition reactions[29–31]. The one-pot cascade reductive imination of nitroarenes with carbonyl compounds is highly attractive. Three main chemical transformations should be taken into consideration in this process: reduction of nitroarene to generate aniline, condensation of aniline and aldehyde and hydrogenation of the imine (Fig. 4c). Some substituted reductants such as carbon monoxide (CO) or CH_3OH instead of H_2 were usually introduced to prevent the over reductions of imines in previous reports[29,32]. Comparably, using hydrogen as a reductant to achieve one-pot reductive imination of nitroarene, which involves obvious size evolution from reactants to products, is more desirable and suitable for evaluating the performance of H_2 enrichment and molecular sieving. As such, the catalytic behaviours of three catalysts with different topological structures (Pt–Ni frame, Pt–Ni frame on MOF and Pt–Ni frame @ MOF) were studied towards this cascade reaction. In the first step, the diffusion of nitroarene is unaffected for all of these catalysts because this small molecule can quickly diffuse onto the metal surface. Once aniline is formed through a reductive process, it will condense with aldehydes to provide target imines. In line with our conjecture, the frame @

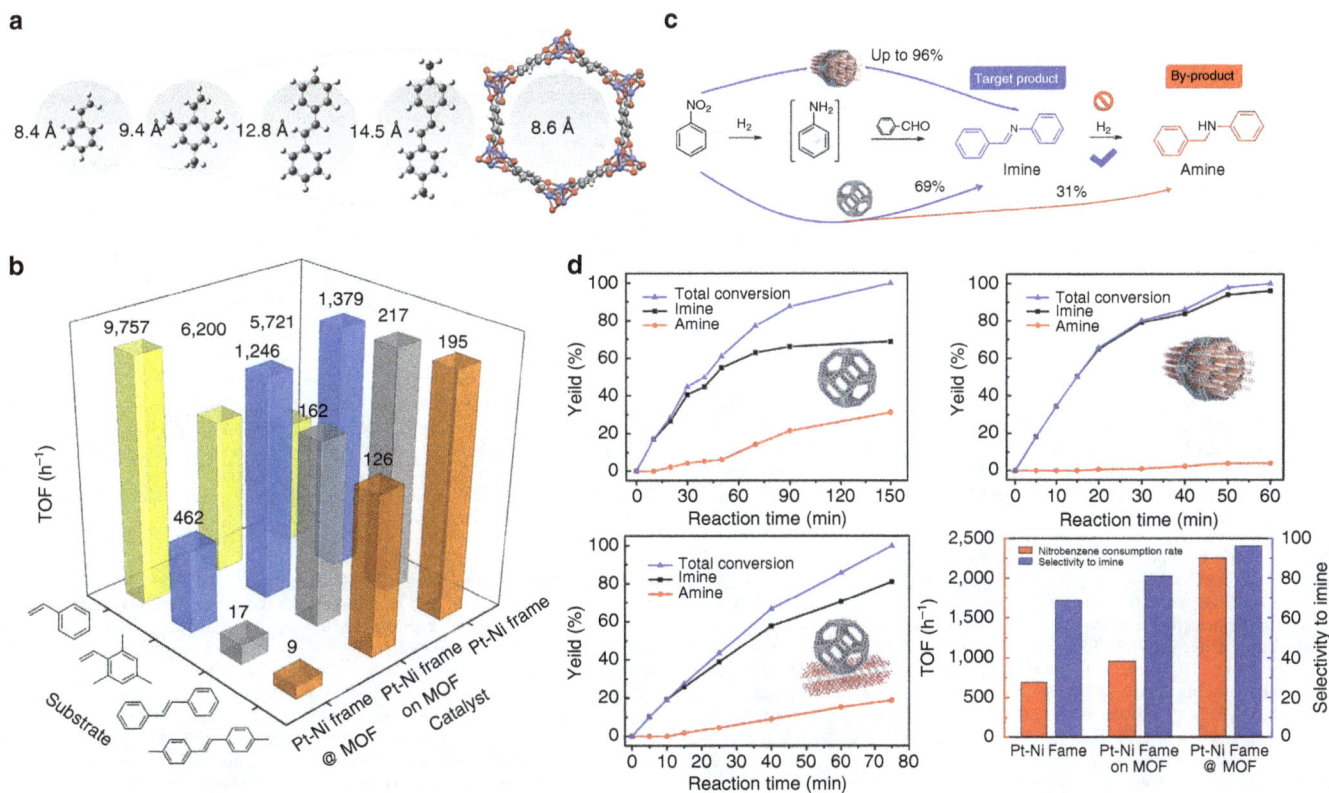

Figure 4 | Size-selective catalytic behaviors of Pt–Ni frame @ MOF (a) Scheme showing the comparison of the maximum diameters of four representative substrates with the pore diameter of Ni-MOF-74. (b) Hydrogenation of styrene, 2,4,6-trimethylstyrene, trans-stilbene and 4,4'-dimethyl-trans-stilbene catalysed by three catalysts. (c) Scheme showing the size-selective catalysis in reductive imination of nitrobenzene. (d) Kinetic curves, TOFs and selectivity to imine in the cascade reductive imination of nitrobenzene catalysed by Pt–Ni frame, Pt–Ni frame on MOF and Pt–Ni frame @ MOF. The TOF values were calculated on the basis of the active sites measured from the CO chemsorption experiments.

frame structure is the most efficient catalyst, whose catalytic activity (based on surface metal atoms measured by CO titration, Supplementary Table 3) reaches 3.3 and 2.4 times higher than that of bare Pt–Ni nanoframe and Pt–Ni frame on MOF catalysts, respectively, to produce imines due to the superior H_2 enrichment (Fig. 4d). The most important finding for this unique catalyst is that the over-reduction of imine, which produces N-phenylbenzylamine, can be effectively avoided. It is inferred that the uniform micropores derived from the grown MOFs can realize the size selectivity and suppress the diffusion of imine towards interior Pt–Ni frame once the condensation process is finished. Combining the outstanding H_2 enrichment and molecular sieving derived from frame @ frame catalyst, imines could be efficiently and selectively produced in a cascade reaction by artfully modulating the reaction process. In addition, Ni nanocrystal and Ni-MOF-74 were tested to be catalytically inactive in the hydrogenation of 1-chloro-2-nitrobenzene, hydrogenation of styrene and reductive imination of nitrobenzene (Supplementary Table 2). The commercial Pt/C showed relatively higher hydrogenation efficiency (with same Pt loading) in the selective hydrogenation of 1-chloro-2-nitrobenzene. However, measuring the surface atoms by CO titration, the turnover frequencies (TOFs) towards hydrogenation of mixed olefins and reductive imination of nitroarenes are relatively lower for Pt/C. Further, the selectivity of commercial Pt/C to target imine product is much lower than that of Pt–Ni frame @ MOF, resulting from its non-restricted contact with substrate (Supplementary Fig. 20).

In conclusion, a novel frame @ frame structure can be sophisticatedly directed by combining the oxidative etching of Pt–Ni alloy and *in situ* precipitation of MOF together. The present design criteria enable this open structure efficient and multifunctional catalyst, namely, maximized the use of precious Pt at the active corners and edges of Pt–Ni bimetallic nanoframes, increased H_2 enrichment which allows for more facile reactivity towards hydrogenation reaction, excellent molecular-size selectivity that originate from the grown microporous metal-organic frameworks. These findings based on the structural evolution of bimetallic nanostructure may also be applicable to many other bimetallic NPs @ MOFs catalysts, possibly offering a hint to simultaneously tune the activity, selectivity and durability.

Methods

Reagents. Analytical grade benzyl alcohol was obtained from Beijing Chemical Reagents, China. Pt(acac)$_2$ (99%), Ni(acac)$_2$ (99%), PVP (molecular weight (MW) = 8,000), nitrobenzene, benzaldehyde, benzoic acid, styrene, 2,4,6-trimethylstyrene, *trans*-stilbene, biphenyl, anisole and Platinum (5% on carbon) were purchased from Alfa Aesar. 2,5-dihydroxyterephthalic acid, 4,4′-dimethyl-*trans*-stilbene and 1-chloro-2-nitrobenzene were acquired from TCI. Aniline was purchased from J.K Scientific. All of the chemicals used in this experiment were analytical grade and used without further purification.

Characterizations. The crystalline structure and phase purity were determined using a Rigaku RU-200b X-ray powder diffractometer with CuKa radiation (*l* = 1.5418 Å). The composition of the product was measured by the ICP-AES and energy-dispersive X-ray spectra. The catalysts' sizes and morphologies were analysed on a Hitachi H-800 TEM and a FEI Tecnai G2 F20 S-Twin high-resolution TEM. XPS experiments were performed on a ULVAC PHI Quantera microprobe. Binding energies (BE) were calibrated by setting the measured BE of C 1s to 284.8 eV. H_2 adsorption isotherms were measured using a Quantachrome Autosorb-1 volumetric instrument at 273 K. The temperature was maintained at 273 K during measurements by putting excess ice with water in Dewar flask. All sample was degassed over 8 h at 423 K under vacuum to remove adsorbed gas or moisture. Requisite amount of hydrogen was injected into the volumetric set-up at volumes required to achieve a targeted set of pressures[33]. N_2 sorption isotherms were performed in a Quantachrome Autosorb-1 at 77 K up to 1 bar. Before measurement, all samples were degassed over 8 h at 423 K under vacuum. Brunauer–Emmett–Teller surface area were obtained by analysing nitrogen adsorption isotherm. Pore size distributions were determined from the adsorption data based on the Horvath–Kwazoe model for cylinder pore geometry. Fourier transform infrared spectra were recorded on a Bruker-VERTEX 70 spectrometer. Thermogravimetry analyses were performed on Netzsch STA 449F3 thermogravimetric analyser over a temperature range of 40–850 °C at a heating rate of 10 °C min^{-1} in nitrogen atmosphere. Scanning electron microscopy was performed with a Hitachi SU-8010 instrument. A FEI Titan 80–300 TEM equipped with a spherical aberration (Cs) corrector for the objective lens working at 300 kV was used for collecting the HAADF-STEM tomography tilt series, which consisted of 37 HAADF-STEM images at the tilt range from −72° to 72° at a tilt increment of 4°. Simultaneous iterative reconstruction technique in FEI Inspect3D software was used for 3D reconstruction. Chimera software was employed to generate the 3D volume rendering of the reconstructions and analysis of the volumes.

Preparation of truncated octahedral Pt-Ni alloy and octahedral Pt-Ni alloy. In a typical synthesis of Pt-Ni truncated octahedral nanocrystals, Pt(acac)$_2$ (40 mg), PVP (MW = 8,000) (400 mg), Ni(acac)$_2$ (250 mg) and aniline (0.5 ml) were dissolved in 25 ml of benzyl alcohol, followed by 10 min of vigorous stirring. The resulting homogeneous green solution was transferred into a 50-ml Teflon-lined stainless-steel autoclave. The sealed vessel was then heated at 180 °C for 12 h before it was cooled down to room temperature. The products were separated via centrifugation and further purified by an ethanol–acetone mixture. In a typical synthesis of Pt-Ni octahedral nanocrystals, Pt(acac)$_2$, (40 mg), PVP (MW = 8,000, 400 mg), Ni(acac)$_2$ (250 mg) and benzoic acid (250 mg) were dissolved in 25 ml of benzyl alcohol, followed by 10 min of vigorous stirring. The resulting homogeneous green solution was transferred into a 50-ml Teflon-lined stainless-steel autoclave. The sealed vessel was then heated at 180 °C for 12 h before it was cooled down to room temperature. The products were separated via centrifugation and further purified by an ethanol–acetone mixture.

Preparation of *in situ*-grown Pt-Ni frame @ Ni-MOF-74. In a typical synthesis of Pt-Ni frame @ Ni-MOF-74, as-prepared truncated octahedron-shaped Pt-Ni nanoalloys (containing 1 mg Pt; based on inductively coupled plasma mass spectrometry (ICP-MS) measurement)) were dispersed in 1 ml DMF, and an appropriate amount of dihydroxyterephthalic acid (35 mg in 7.5 ml DMF) was added. The resulting solution was transferred into a 10-ml Telfon-lined stainless-steel autoclave. After stirring for 10 min, the sealed vessel was heated at 110 °C for 12 h before it was cooled down to room temperature. The *in situ*-grown Pt-Ni frame @ Ni-MOF-74 (Pt-Ni frame @ MOF) was obtained after washing and centrifugation by deionized water and methanol for several times. The obtained Pt-Ni frame @ Ni-MOF-74 was kept immersed in methanol for 5 days; the solvent was changed for fresh methanol once a day. Finally, the Pt-Ni frame @ Ni-MOF-74 was heated under vacuum at 150 °C and stored in a dry box for further use. In a typical synthesis of concave Pt-Ni @ Ni-MOF-74, the as-prepared octahedron-shaped Pt-Ni nanoalloys (containing 1 mg Pt; based on ICP-MS measurement) were dispersed in 0.5 ml DMF, and 7 ml DMF containing DOT (55 mg) and PVP (MW = 30,000, 80 mg) was added. The resulting solution was transferred into a 10-ml Telfon-lined stainless-steel autoclave. After stirring for 10 min, the sealed vessel was then heated at 100 °C for 12 h before it was cooled down to room temperature. The *in situ*-grown concave Pt-Ni @ Ni-MOF-74 was obtained after washing and centrifugation by deionized water and methanol for several times. The obtained concave concave Pt-Ni @ Ni-MOF-74 was kept immersed in methanol for 5 days; the solvent was changed for fresh methanol once a day. Finally, the concave Pt-Ni @ Ni-MOF-74 was heated under vacuum at 150 °C and stored in a dry box for further use.

Preparation of bare Pt-Ni frame. In a typical procedure, Pt-Ni frame @ MOF (containing 1 mg Pt; based on ICP-MS measurement)) was dispersed in 10 ml H_2O. Into this solution, 10 ml dilute acetic acid (50%) was added. The resulting solution was stirred vigorously for 8 h in 30 °C to achieve complete decomposition of Ni-MOF-74. The Pt-Ni frames were obtained after washing and centrifugation by deionized water and methanol.

Preparation of Ni-MOF-74 and Pt-Ni frame on Ni-MOF-74. Pure Ni-MOF-74 was synthesized by a modified condition from the literature procedures[34]. To a solution of 2,5-dihydroxyterephthalic acid (148 mg, 0.75 mmol) in THF (2.5 ml), a solution of nickel(II) acetate tetrahydrate (375 mg, 1.5 mmol) in water (25 ml) was added. The suspension was stirred and ultrasonicated until homogenous. The resulting solution was transferred into a 10-ml Telfon-lined stainless-steel autoclave. The sealed vessel was then heated at 110 °C in a preheated oven for 72 h before it was cooled down to room temperature. Ni-MOF-74 was obtained after washing and centrifugation with deionized water and methanol. In a typical preparation of Pt-Ni frame on Ni-MOF-74 (Pt-Ni frame on MOF), bare Pt-Ni frames were dispersed in methanol to form a suspension with a concentration of 5 mg Pt per 10 ml methanol. Into this solution, 20 mg Ni-MOF-74 was added and stirred at room temperature for 8 h. The composites were then separated via centrifugation and washed with methanol for several times. The obtained Pt-Ni frame on MOF was kept immersed in methanol for 5 days; the solvent was changed for fresh methanol once a day. Finally, the Pt-Ni frame @ Ni-MOF-74 was heated under vacuum at 150 °C and stored in a dry box for further use.

Typical procedure for the catalytic hydrogenation of 1-chloro-2-nitrobenzene. About 35 µl 1-chloro-2-nitrobenzene (0.3 mmol) in 1.5 ml methanol and the catalyst (contain 0.005 mmol Pt, 1.6 mol %) were added in a 10-ml round flask. The round flask was purged with H_2 to completely remove air from the reactor. Then, the reaction mixture was stirred at 30 °C under 1 bar H_2. The progress of the reaction was monitored by gas chromatography (GC)-MS and the extent of conversion was determined on the basis of the ratio of area of substrate and product by an external standard method.

Typical procedure for the catalytic hydrogenation of olefins. Hydrogenation of olefins was carried out in THF solution under 1 bar H_2. In a typical procedure, the catalysts containing 0.0025 mmol Pt was loaded into a 10-ml round flask and THF (1.5 ml) was added to the reactor. The mixture was sonicated homogenously before mixed olefins (styrene, 2,4,6-trimethylstyrene, *trans*-stilbene and 4,4′-dimethyl-*trans*-stilbene) (0.1 mmol for each component) were introduced. Afterward, 0.1 mmol of biphenyl was also added as internal standard. The round flask was purged with H_2 to completely remove air from the reactor and the reaction was allowed to proceed at 30 °C under 1 bar H_2. The progress of the reaction was monitored by GC-MS. Hydrogenation rates for olefins were calculated on the basis of the consumption rates for the substrates. The TOFs were calculated using the following equation:

$$\text{TOF} = \frac{\text{Substrate hydrogenated(mol)}}{N \text{ active sites(mol)} \cdot \text{time(h)}} \text{ in[molecules per active site per h]} \quad (4)$$

The active sites of the catalysts were measured by CO titration experiments.

Typical procedure for the cascade reductive imination of nitroarenes. First, 22 µl nitrobenzene (0.2 mmol) and 35 µl benzaldehyde (0.3 mmol) in 2 ml ethanol and the catalyst (contain 0.005 mmol Pt, 2.5 mol %) were added in a 10-ml round flask. The round flask was purged with H_2 to completely remove air from the reactor. Then, the reaction mixture was stirred at 30 °C under 1 bar H_2. The progress of the reaction was monitored by GC-MS[29,35] using anisole as an internal standard. Reaction rates for cascade reductive imination of nitroarenes were calculated on the basis of the consumption rates for the nitrobenzene. Conversion was defined as the mole ratio of converted nitrobenzene to starting nitrobenzene. The selectivity of imines based on nitrobenzene was calculated according to equation (5).

$$S = \frac{\text{Imines in products}}{\text{Imines in products} + \text{by-products derived from nitrobenzene in products}} \quad (5)$$
$$\times 100\%$$

Calculated maximum diameters of selected molecules. For olefins considered in this research, theoretical computations were performed with a Gaussian 03W programme package using density functional theory. Geometry optimizations and harmonic vibrational frequencies are computed with the B3LYP functional. Molecular lengths were measured as the distance between the two farthest apart atoms plus an estimate of the van der Waals radii of hydrogen (1.2 Å)[36,37].

CO titration experiments. The numbers of active sites on the surface of catalysts were determined from CO titration using a catalyst analyser (BEL-A, Japan) with a mass spectrometer (Inprocess Instruments, GAM200) as detector at 323 K. Prior to CO titration, the catalysts (containing *ca.* 2.5–3 mg metal) were treated at 423 K for 60 min and then cooled to 323 K under an argon flow (40 ml min^{-1}). The CO uptake was measured by the decrease in the peak areas induced by chemsorption compared with the area of a calibrated volume. The metal dispersion was calculated assuming a stoichiometry of one CO molecule per surface metal atom (metal atom = Pt, Ni).

References

1. Xia, Y. N. *et al.* Gold nanocages: from synthesis to theranostic applications. *Acc. Chem. Res.* **44**, 914–924 (2011).
2. Yavuz, M. S. *et al.* Gold nanocages covered by smart polymers for controlled release with near-infrared light. *Nat. Mater.* **8**, 935–939 (2009).
3. Wittstock, A., Zielasek, V., Biener, J., Friend, C. & Bäumer, M. Nanoporous gold catalysts for selective gas-phase oxidative coupling of methanol at low temperature. *Science* **327**, 319–322 (2010).
4. Wu, Y. *et al.* A strategy for designing a concave Pt-Ni alloy through controllable chemical etching. *Angew. Chem. Int. Ed.* **51**, 12524–12528 (2012).
5. Cui, C., Gan, L., Heggen, M., Rudi, S. & Strasser, P. Compositional segregation in shaped Pt alloy nanoparticles and their structural behaviour during electrocatalysis. *Nat. Mater.* **12**, 765–771 (2013).
6. Chen, C. *et al.* Highly crystalline multimetallic nanoframes with three-dimensional electrocatalytic surfaces. *Science* **343**, 1339–1343 (2014).
7. Xia, B. Y., Wu, H. B., Wang, X. & Lou, X. W. One-Pot Synthesis of cubic PtCu$_3$ nanocages with enhanced electrocatalytic activity for the methanol oxidation reaction. *J. Am. Chem. Soc.* **134**, 13934–13937 (2012).
8. Wu, Y. *et al.* Sophisticated construction of Au islands on Pt-Ni: an ideal trimetallic nanoframe catalyst. *J. Am. Chem. Soc.* **136**, 11594–11597 (2014).
9. Xiao, B. *et al.* Copper nanocrystal plane effect on stereoselectivity of catalytic deoxygenation of aromatic epoxides. *J. Am. Chem. Soc.* **137**, 3791–3794 (2015).
10. Bourikas, K., Kordulis, C. & Lycourghiotis, A. Titanium dioxide (anatase and rutile): surface chemistry, liquid–solid interface chemistry, and scientific synthesis of supported catalysts. *Chem. Rev.* **114**, 9754–9823 (2014).
11. Fu, Q. *et al.* Interface-confined ferrous centers for catalytic oxidation. *Science* **328**, 1141–1144 (2010).
12. Wu, B., Huang, H., Yang, J., Zheng, N. & Fu, G. Selective hydrogenation of alpha,beta-unsaturated aldehydes catalysed by amine-capped platinum-cobalt nanocrystals. *Angew. Chem. Int. Ed.* **51**, 3440–3443 (2012).
13. Matanovic, I., Garzon, F. H. & Henson, N. J. Theoretical study of electrochemical processes on Pt-Ni alloys. *J. Phys. Chem. C* **115**, 10640–10650 (2011).
14. Shui, J. l., Chen, C. & Li, J. C. M. Evolution of nanoporous Pt-Fe alloy nanowires by dealloying and their catalytic property for oxygen reduction reaction. *Adv. Funct. Mater.* **21**, 3357–3362 (2011).
15. Xiong, Y. *et al.* Understanding the role of oxidative etching in the polyol synthesis of Pd nanoparticles with uniform shape and size. *J. Am. Chem. Soc.* **127**, 7332–7333 (2005).
16. González, E., Arbiol, J. & Puntes, V. F. Carving at the nanoscale: sequential galvanic exchange and kirkendall growth at room temperature. *Science* **334**, 1377–1380 (2011).
17. Sun, Y. G. & Xia, Y. N. Mechanistic study on the replacement reaction between silver nanostructures and chloroauric acid in aqueous medium. *J. Am. Chem. Soc.* **126**, 3892–3901 (2004).
18. Blonder, G. Simple model for etching. *Phys. Rev. B* **33**, 6157–6168 (1986).
19. Wu, Y., Cai, S., Wang, D., He, W. & Li, Y. Syntheses of water-soluble octahedral, truncated octahedral, and cubic Pt-Ni nanocrystals and their structure-activity study in model hydrogenation reactions. *J. Am. Chem. Soc.* **134**, 8975–8981 (2012).
20. Chen, B. *et al.* Rod packings and metal-organic frameworks constructed from rod-shaped secondary building units. *J. Am. Chem. Soc.* **127**, 1504–1518 (2005).
21. Yin, Y. *et al.* Formation of hollow nanocrystals through the nanoscale Kirkendall effect. *Science* **304**, 711–714 (2004).
22. Liu, J., Tian, J., Thallapally, P. K. & McGrail, B. P. Selective CO$_2$ capture from flue gas using metal-organic frameworks—a fixed bed study. *J. Phys. Chem. C* **116**, 9575–9581 (2012).
23. Buseck, P. R., Cowley, J. M. C. & Eyring, L. (ed.) *High-resolution transmission electron microscopy and associated techniques* (Oxford University Press, New York, 1988).
24. de Jonge, N. & Ross, F. M. Electron microscopy of specimens in liquid. *Nat. Nanotechnol.* **6**, 695–704 (2011).
25. Zhong, X. Y. *et al.* Three-dimensional quantitative chemical roughness of buried ZrO$_2$/In$_2$O$_3$ interfaces via energy-filtered electron tomography. *Appl. Phys. Lett.* **100**, 101604 (2012).
26. Wei, H. *et al.* FeOx-supported platinum single-atom and pseudo-single-atom catalysts for chemoselective hydrogenation of functionalized nitroarenes. *Nat. Commun.* **5**, 5634 (2014).
27. Wang, Y. *et al.* Phase-transfer interface promoted corrosion from PtNi$_{10}$ nanoctahedra to Pt$_4$Ni nanoframes. *Nano Res.* **8**, 140–155 (2015).
28. Guo, Z. *et al.* Pt nanoclusters confined within metal–organic framework cavities for chemoselective cinnamaldehyde hydrogenation. *ACS Catal.* **4**, 1340–1348 (2014).
29. Huang, J. *et al.* Direct one-pot reductive imination of nitroarenes using aldehydes and carbon monoxide by titania supported gold nanoparticles at room temperature. *Green Chem.* **13**, 2672–2677 (2011).
30. Shi, M. & Xu, Y.-M. Catalytic, Asymmetric Baylis-Hillman reaction of imines with methyl vinyl ketone and methyl acrylate. *Angew. Chem. Int. Ed.* **41**, 4507–4510 (2002).
31. Uematsu, N., Fujii, A., Hashiguchi, S., Ikariya, T. & Noyori, R. Asymmetric transfer hydrogenation of imines. *J. Am. Chem. Soc.* **118**, 4916–4917 (1996).
32. Xiang, Y., Meng, Q., Li, X. & Wang, J. In situ hydrogen from aqueous-methanol for nitroarene reduction and imine formation over an Au-Pd/Al$_2$O$_3$ catalyst. *Chem. Commun.* **46**, 5918–5920 (2010).
33. Li, G. *et al.* Shape-dependent hydrogen-storage properties in Pd nanocrystals: which does hydrogen prefer, octahedron (111) or cube (100)? *J. Am. Chem. Soc.* **136**, 10222–10225 (2014).
34. Liu, J., Tian, J., Thallapally, P. K. & McGrail, B. P. Selective CO$_2$ capture from flue gas using metal–organic frameworks—a fixed bed study. *J. Phys. Chem. C* **116**, 9575–9581 (2012).
35. Pintado-Sierra, M., Rasero-Almansa, A. M., Corma, A., Iglesias, M. & Sánchez, F. Bifunctional iridium-(2-aminoterephthalate)–Zr-MOF chemoselective catalyst for the synthesis of secondary amines by one-pot three-step cascade reaction. *J. Catal.* **299**, 137–145 (2013).

36. Jae, J. *et al.* Investigation into the shape selectivity of zeolite catalysts for biomass conversion. *J. Catal.* **279**, 257–268 (2011).
37. Li, J. R., Kuppler, R. J. & Zhou, H. C. Selective gas adsorption and separation in metal-organic frameworks. *Chem. Soc. Rev.* **38**, 1477–1504 (2009).

Acknowledgements

This work was supported by the State Key Project of Fundamental Research for Nanoscience and Nanotechnology (2011CB932401 and 2011CBA00500), the National key Basic Research Program of China (2012CB224802) and the National Natural Science Foundation of China (Grant No. 21221062, 21171105, 21322107 and 21131004). This work made use of the resources of the Beijing National Center for Electron Microscopy. We thank B.Q. Xu and K.Q. Sun for helpful discussions.

Author contributions

Z.L. performed the experiments, collected and analysed the data, and wrote the paper. R.Y., J.H. and X.Z. helped with HRTEM and electron tomography analyses. Y.S. helped with the CO titration experiments and analyses. Y.L. and Y.W. conceived the experiments, planned synthesis, analysed results and wrote the paper.

Additional information

Competing financial interests: The authors declare no competing financial interests.

9

Cobalt-centred boron molecular drums with the highest coordination number in the CoB_{16}^- cluster

Ivan A. Popov[1,*], Tian Jian[2,*], Gary V. Lopez[2], Alexander I. Boldyrev[1] & Lai-Sheng Wang[2]

The electron deficiency and strong bonding capacity of boron have led to a vast variety of molecular structures in chemistry and materials science. Here we report the observation of highly symmetric cobalt-centered boron drum-like structures of CoB_{16}^-, characterized by photoelectron spectroscopy and *ab initio* calculations. The photoelectron spectra display a relatively simple spectral pattern, suggesting a high symmetry structure. Two nearly degenerate isomers with D_{8d} (**I**) and C_{4v} (**II**) symmetries are found computationally to compete for the global minimum. These drum-like structures consist of two B_8 rings sandwiching a cobalt atom, which has the highest coordination number known heretofore in chemistry. We show that doping of boron clusters with a transition metal atom induces an earlier two-dimensional to three-dimensional structural transition. The CoB_{16}^- cluster is tested as a building block in a triple-decker sandwich, suggesting a promising route for its realization in the solid state.

[1] Department of Chemistry and Biochemistry, Utah State University, Logan, Utah 84322, USA. [2] Department of Chemistry, Brown University, Providence, Rhode Island 02912, USA. * These authors contributed equally to this work. Correspondence and requests for materials should be addressed to A.I.B. (email: a.i.boldyrev@usu.edu) or to L.-S.W. (email: lai-sheng_wang@brown.edu).

Boron, the fifth element in the periodic table, possesses such diverse chemical structures and bonding that are second only to carbon. Bulk boron consists of connected three-dimensional (3D) cages in many of its allotropes[1,2] and boron-rich borides[3,4]. However, for isolated clusters it was computationally shown[5,6] that icosahedral cage structures of B_{12} and B_{13} were unstable, even though they were initially proposed as possible candidates for these two clusters[7]. Over the past decade, small anionic boron clusters have been systematically characterized both experimentally and theoretically to exhibit planar or quasi-planar structures in their ground states up to B_{27}^- (refs 8–10). Recent works show that anionic boron clusters continue to be two-dimensional (2D) at B_{30}^- (ref. 11), B_{35}^- (ref. 12) and B_{36}^- (ref. 13). The 2D-to-3D transition was suggested to occur at B_{20} for neutral[14], and at B_{16}^+ for cationic clusters[15]. Very recently it is shown that the transition from 2D to fullerene-like 3D structures occurs in negatively charged boron clusters at about 40 boron atoms in B_{39}^- (ref. 16) and B_{40}^- (ref. 17). Due to the nearly spherical shapes of these clusters, they have been named borospherenes. Doping boron clusters with a single metal atom opens a new avenue to create clusters with novel structures and chemical bonding. It has been experimentally observed that various transition metal atoms can be placed inside of monocyclic boron rings to form beautiful molecular wheel-type structures $(M\textcircled{C}B_n^-)$[18], following an electronic design principle inspired by the doubly σ and π aromatic B_9^- cluster[19]. It was shown that the $Nb\textcircled{C}B_{10}^-$ and $Ta\textcircled{C}B_{10}^-$ clusters possess the record coordination number of 10 in the planar environment for the central metal atom[20]. These clusters have pushed the limits of structural chemistry.

Here we report the observation of a large metal-doped boron cluster of CoB_{16}^-, which is produced using a laser vaporization cluster source and characterized by photoelectron spectroscopy (PES). Extensive computational searches reveal that there are two nearly degenerate structures for CoB_{16}^-, which are indistinguishable at the highest level of theory employed. They both possess tubular double-ring framework and give similar photoelectron spectral patterns. The structures can be viewed as two B_8 rings sandwiching a Co atom, reminiscent of a drum and giving rise to the highest coordination number known in chemistry thus far.

Results

Experimental results. The photoelectron spectra of CoB_{16}^- at two photon energies are displayed in Fig. 1. The lowest binding energy band (X) represents the electron detachment transition from the anionic ground state to that of neutral CoB_{16}. The higher binding energy bands, A, B, ..., denote detachment transitions to the excited states of neutral CoB_{16}. The vertical detachment energies (VDEs) for all observed bands are given in Table 1, where they are compared with the calculated VDEs.

The 266 nm spectrum (Fig. 1a) reveals three well-resolved PES bands for CoB_{16}^-. The band X gives rise to a VDE of 2.71 eV. The adiabatic detachment energy (ADE) for band X was evaluated from its onset to be 2.48 eV, which also represents the electron affinity of neutral CoB_{16}. The width of band X suggests an appreciable geometry change between the ground electronic state of CoB_{16}^- and the ground electronic state of CoB_{16}. Following a relatively large energy gap, an intense and broad band A is observed at a VDE of 3.45 eV and a close-lying band B at a VDE of 3.78 eV. The 193 nm spectrum (Fig. 1b) shows nearly continuous signals beyond 4 eV. The sharp spikes above 5 eV in the high binding energy side of the 193 nm spectrum are due to statistical noises because of low electron counts. An intense and broad band C is clearly observed at a VDE of 4.86 eV. Two more bands can be tentatively identified at higher binding energies, D (VDE: ∼5.3 eV) and E (VDE: ∼5.6 eV). Overall, the PES spectral

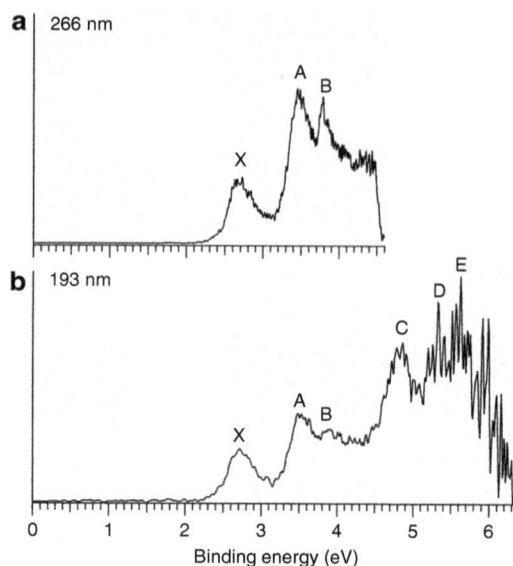

Figure 1 | Photoelectron spectra. Photoelectron spectra (**a**) at 266 nm (4.661 eV) and (**b**) at 193 nm (6.424 eV) of CoB_{16}^-.

pattern is relatively simple, suggesting that the framework of the CoB_{16}^- cluster is likely to have high symmetry.

Theoretical results and comparison with experiment. Extensive structural searches were initially done at the PBE0/3-21G level of theory with the follow-up calculations ($\Delta = 25$ kcal mol^{-1}) at the PBE0/Def2-TZVP level of theory, which led to two similar drum-like structures: isomer **I** (D_{8d}, $^3A_2$) and isomer **II** (C_{4v}, $^1A_1$) identified as the global minima for CoB_{16}^- (Fig. 2). These two highly symmetric structures, consisting of a central Co atom sandwiched by two B_8 monocyclic rings, are found to be almost degenerate at various levels of theory (Supplementary Fig. 1 and Supplementary Table 1). Clearly, the method dependency of predicting relative energies of the low-lying structures for CoB_{16}^- suggests the importance of comparison with experiment in determining the global minimum. We previously studied how optimized geometries of small boron clusters differed at density functional theory (DFT) and CCSD(T) levels of theory[21,22]. We found that B3LYP/6-311 + G* geometries are quite close (within 0.03 Å between nearest boron atoms) to those at the CCSD(T)/6-311 + G* level of theory. We also compared the geometries of boron clusters at PBE0/6-311 + G* and B3LYP/6-311 + G*, and found that they are also very close[23]. Therefore, PBE0/3-21G level of theory was used for the preliminary search and PBE0/Def2-TZVP for the final optimized geometries of CoB_{16}^-. The highest level of theory employed (ROCCSD(T)/6-311 + G(2df)//PBE0/Def2-TZVP (this abbreviation means that single-point energy calculations were performed at ROCCSD(T)/6-311 + G(2df) using optimized UPBE0/Def2-TZVP geometries here and elsewhere) indicates 1.4 kcal mol^{-1} energy difference including zero-point energy corrections (Supplementary Fig. 1). This small value is in the range of the theoretical errors for such a complex transition-metal-doped boron cluster. Therefore, isomers **I** and **II** should be considered to be degenerate based on our calculations. Figure 2 shows the small differences in bond distances between isomers **I** and **II**; the latter is not significantly distorted from the D_{8d} symmetry. The B–B bond lengths of the B_8 rings for both isomers are in the range of 1.55–1.63 Å, similar to the corresponding values (1.56 Å) in the $Co\textcircled{C}B_8^-$ molecular wheel[18]. The nearest isomer **III** (C_2, ^1A) is 8.7 kcal mol^{-1} higher in energy at the ROCCSD(T) method and represents a

Table 1 | Experimental and theoretical vertical electron detachment energies (VDEs) in eV of CoB_{16}^-.

VDE (exp.)*	Isomer I $(1a_1^2 1e_1^4 1e_2^4 1b_2^2 1e_3^4 2e_1^4 2a_1^2 1e_4^4 2e_2^4$ $3a_1^2 3e_2^4 2e_3^4 3e_1^4 4e_1^4 5e_1^4 4a_1^2 2b_2^2 4e_2^2)$						Isomer II $(1a_1^2 1e^4 1b_2^2 1b_1^2 2a_1^2 2e^4 3e^4 3a_1^2 2b_2^2 1a_2^2 4a_1^2$ $2b_1^2 5a_1^2 4e^4 3b_2^2 3b_1^2 5e^4 6e^4 7e^4 6a_1^2 7a_1^2 4b_2^2)$					
	UPBE0†		UB3LYP‡		ROCCSD (T)§		UPBE0†		UB3LYP‡		ROCCSD (T)§	
	MO	VDE (theo.)	MO	VDE (theo.)	MO	VDE (theo.)	MO	VDE (theo.)	MO	VDE (theo.)	MO	VDE (theo.)
X	2.71 (5)	$4e_2$	2.58	$4e_2$	2.49	$4e_2$	2.59	$4b_2$	2.53	$4b_2$	2.47	$4b_2$	2.61
A	3.45 (3)	$2b_2$	2.97	$2b_2$	2.91	$2b_2$	3.28	$7a_1$	3.09	$7a_1$	3.02	$7a_1$	—‖
B	3.78 (3)	—		—			—‖	—		—			—‖
C	4.86 (5)	—		—			—‖	—		—			—‖
D	5.3 (1)	—		—			—‖	—		—			—‖
E	5.6 (1)	—		—			—‖	—		—			—‖

*Numbers in parentheses indicate the uncertainties of the last digit. The ADE of the X band is measured to be 2.48(5) eV.
†The VDEs were calculated at the UPBE0/6-311 + G(2df)//UPBE0/Def2-TZVP level of theory. Spin contamination was found to be very small.
‡The VDEs were calculated at the UB3LYP/6-311 + G(2df)//UPBE0/Def2-TZVP level of theory. Spin contamination was found to be very small.
§The VDEs were calculated at the ROCCSD(T)/6-311 + G(2df)//PBE0/Def2-TZVP level of theory, because the UHF wave function has a very high spin-contamination.
‖VDE could not be calculated at this level of theory.

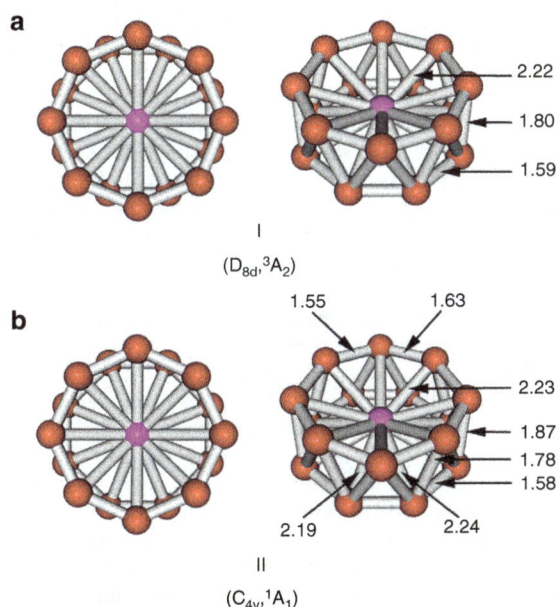

Figure 2 | Two views of isomer I and isomer II of the CoB_{16}^- cluster. The point group symmetries and spectroscopic states of isomer I (**a**) and isomer II (**b**) are shown in parentheses. Sticks drawn between atoms help visualization and do not necessarily represent classical 2c–2e B–B or Co–B bonds here and elsewhere. All distances are in Å.

distorted drum-like structure composed of two B_7 rings with two B atoms outside the drum (Supplementary Fig. 1). In fact, the majority of the low-lying isomers within 20 kcal mol^{-1} (Supplementary Fig. 1) represent various derivatives (drum-like or possessing principal geometrical features of the drum-like structure) of isomers **I** and **II**, showing the stability of the drum-like structures. It should be noted that there are significant bonding interactions between the two B_8 rings and between the Co atom and all 16 B atoms in both isomers **I** and **II** (*vide infra*). Interestingly, the drum structure in a quintet state (isomer **XIV** in Supplementary Fig. 1) appears to be the most stable one out of all other quintet isomers. It should be mentioned that there were two previous DFT calculations on similar drum-like structures of neutral boron clusters doped with transition metal atoms[24,25].

To facilitate comparisons between the experimental and theoretical results, we calculated low-lying VDEs of isomers **I** and **II** of CoB_{16}^- using three methods (Table 1). We found that

the VDEs computed using the two DFT methods are not very impressive; but we observed good agreement between the theoretical VDEs at ROCCSD(T)/6-311 + G(2df) and the experimental data for the first two detachment channels (Table 1). Since isomer **I** is open shell, the electron detachment energy from the doubly degenerate $4e_2$-HOMO should lead to a doublet final state for the neutral. The computed VDE at ROCCSD(T) is 2.59 eV, compared with the experimental VDE of 2.71 eV. The next electron detachment from the non-degenerate $2b_2$-HOMO-1 should lead to both a quartet and a doublet final state, with the quartet being lower in energy. The calculated VDE for the quartet final state at ROCCSD(T) is 3.28 eV, compared with the VDE of the A band at 3.45 eV. Unfortunately, we were not able to calculate any higher VDEs because of the limitation of the ROCCSD(T) method. However, we believe that the good agreement between experiment and theory for the first two VDEs provides sufficient credence for the identified drum-like isomer **I** for the CoB_{16}^- cluster.

Isomer **II** gives very similar theoretical VDEs as isomer **I** at all three levels of theory, consistent with the similarities in their geometries. Since isomer **II** is a closed shell species, we were able to calculate only the first VDE value at the ROCCSD(T) method as 2.61 eV, also in good agreement with the experimental data. Furthermore, the calculated ADEs of isomer **I** (2.45 eV) and isomer **II** (2.43 eV) (PBE0/Def2-TZVP) are in excellent agreement with the experimentally measured ADE value of 2.48 eV. We should point out that there is a Jahn–Teller distortion for the neutral CoB_{16} drum-like structure of isomer **I**, consistent with the broad X band observed in the PES spectra (Fig. 1). Indeed, the calculated relaxed neutral CoB_{16} structure $\mathbf{I^0}$ (Supplementary Fig. 2 and Supplementary Table 1) has lower symmetry (C_{2v}), as one would expect for the Jahn–Teller distorted structure due to the occupation of the doubly degenerate HOMO ($4e_2$) of isomer **I** by a single electron. In fact, the HOMO ($4b_2$) of isomer **II** originates from the HOMO ($4e_2$) of isomer **I** when one of the doubly degenerate orbitals is doubly occupied. Therefore, the detachment of one electron from the doubly occupied HOMO ($4b_2$) of isomer **II** leads to the same neutral structure $\mathbf{I^0}$. The high relative energy of isomer **III**, as well as its appreciably higher theoretical first VDE of 3.65 eV (Supplementary Table 2), makes this cluster unlikely to be populated in the molecular beam in any appreciable amount.

Discussion

Tubular (or drum-like) boron clusters have been of interest for many years, because they can be considered as the embryos for boron nanotubes[14]. However, such drum-like structures have never been observed experimentally for bare boron clusters, even though they have been shown to be stable computationally[14,26–29]. For instance, the B_{20} cluster was first suggested as the global minimum on the basis of theoretical calculations[14], but it has not been observed or confirmed experimentally[29]. Tubular structures were also studied for the bare B_{16}^+, B_{16}, B_{16}^- and B_{16}^{2-} species[15,30]. For the B_{16}^+ cationic cluster, the tubular structure was suggested to be the global minimum[15], whereas the tubular structures of both B_{16} and B_{16}^- were found to be high-energy isomers[30]. Clearly, the strong coordination interactions with the Co atom significantly stabilize the tubular B_{16} to give the drum-like global minima (structures **I** and **II**) for CoB_{16}^-. Bare anionic boron clusters are found to be 2D up to B_{36}^- (ref. 13), while some transition-metal-doped anionic boron clusters are found to preserve the planar boron framework on metal doping[31,32]. The largest experimentally observed metal-doped boron cluster (CoB_{12}^-) maintains a similar planar geometry for the B_{12} moiety[32]. Hence, the doping of the Co atom induces an earlier 2D-to-3D transition for boron clusters, as shown by the 3D isomers **I** and **II** of CoB_{16}^-. In fact, the CoB_{16}^- drum structure represents the highest coordination number known in chemistry today. The previous highest coordination number known experimentally was 15 for $[Th(H_3BNMe_2BH_3)_4]$ (ref. 33), though theoretical studies have suggested the highest coordination numbers of 15 in $PbHe_{15}^{2+}$ (ref. 34) and 16 in the Friauf–Laves phases in $MgZn_2$ or $MgNi_2$ (ref. 35). Endohedral fullerenes $(M@C_{60})$ have been observed[36,37], but the metal atom in those cases interacts with the C_{60} shell primarily ionically and it does not stay in the centre of C_{60}.

It is interesting to point out that the B–B distances in the B_8 rings of both isomers **I** and **II** of CoB_{16}^- and the bare tubular B_{16} are very similar (Supplementary Table 3). To gain insight into the chemical bonding of the CoB_{16}^- drums, we performed chemical bonding analyses for isomers **I** and **II** using the Adaptive Natural Density Partitioning (AdNDP) method[38], which is an extension of the popular Natural Bond Orbital method[39]. It should be noted that the bonding in some double-ring tubular boron clusters has been discussed previously[9,40,41].

Since isomer **I** has two unpaired electrons, we used the unrestricted AdNDP (UAdNDP) analysis, which enables treatments of the α and β electrons separately. To obtain an averaged result for a bond (Fig. 3), we added the UAdNDP results for the α and β electrons of the same type of bonds. According to the UAdNDP analysis results, the 58 valence electrons in CoB_{16}^- can be divided into four sets. The first set (Fig. 3a,b) consists of localized bonding elements, while the other three sets (Fig. 3c–g, h–j, k–o) are composed of delocalized bonding elements. In the first set, the UAdNDP analysis for isomer **I** revealed the following localized bonding elements: one lone pair (1c–2e bond) (Fig. 3b) of $3d_{z^2}$-type on Co with an occupation number (ON) of 1.98 $|e|$ and sixteen 2c–2e B–B σ-bonds (Fig. 3a) with ON = 1.84 $|e|$ within each B_8 ring (all superimposed onto the B_{16} fragment in Fig. 3), which can also be viewed as 3c–2e bonds with the ON = 1.96 $|e|$ responsible for the bonding between the boron rings. In the last case, a boron atom from the neighbouring ring contributes somewhat (0.12 $|e|$) to the formation of the 3c–2e σ-bond. The 2c–2e B–B σ-bonds are very similar to the peripheral B–B bonds found in all 2D boron clusters[8–10]. The second set includes five delocalized σ bonds (denoted as σ + σ), which are formed from delocalized σ bonds between the two B_8 rings. Since the σ orbitals between the two boron rings overlap positively, we designate them as σ + σ in the *second* set, which constitutes σ-aromaticity according to the $4n + 2$ $(n = 2)$ Hückel rule. The three delocalized 16c–2e σ + σ bonds (Fig. 3c–e) with ON = 1.82–1.86 $|e|$ involve only σ-bonding within the boron rings, whereas the two delocalized 17c–2e σ + σ bonds (Fig. 3f,g) come primarily from the $3d_{xy}$ and $3d_{x^2-y^2}$ AOs of Co interacting

with the boron rings. It should be noted that the direct covalent interactions between Co and the B_{16} unit via the $3d_{xy}$ and $3d_{x^2-y^2}$ AOs of Co are found to be around 0.6 $|e|$ according to the AdNDP analysis. The third set (Fig. 3h–j) shows three delocalized σ–σ bonds, which represent bonding interactions within each ring, but anti-bonding interactions between the two boron rings. This set of delocalized bonds also constitutes σ-aromaticity according to the $4n + 2$ $(n = 1)$ Hückel rule. In the *third* set, the 16c–2e σ–σ bond (Fig. 3h) involves mainly the two boron rings, whereas the two 17c–2e σ–σ bonds (Fig. 3i,j) involve interactions between the $3d_{xz}$ and $3d_{yz}$ AOs of Co with the boron rings. The direct covalent interaction of the $3d_{xz}$ and $3d_{yz}$ AOs of Co with the boron kernel is assessed to be around 0.5 $|e|$. The *last* set includes five delocalized bonds, which represent π–π interactions between the boron rings: three 16c–2e π–π bonds (Fig. 3k–m) with ON = 1.98–2.00 $|e|$ and two 16c–1e π–π bonds (Fig. 3n,o) with ON = 1.00 $|e|$ (one unpaired electron on each bond). The eight π electrons in the *last* set suggest π-aromaticity according to the $4n$ rule $(n = 2)$ for triplet states. Therefore, the stability of isomer **I** of CoB_{16}^- can be considered to be due to the double σ- and π-aromaticity and bonding interactions of the 3d AOs of Co with the B_8 rings.

As expected, isomer **II** of CoB_{16}^-, which is close in energy and geometry to isomer **I**, has almost the same bonding pattern as that of isomer **I** (Supplementary Fig. 3). All the bonding elements found in isomer **I** are also found in isomer **II** except for the *last* set (Supplementary Fig. 3k–n). Since isomer **II** is closed shell, eight electrons in the *last* set are observed to form four 16c–e π–π bonds with ON = 1.98–2.00 $|e|$, rendering this isomer π-antiaromatic. Hence, isomer **II** exhibits conflicting aromaticity (σ-aromatic and π-antiaromatic), which leads to some distortion to C_{4v} symmetry compared to the D_{8d} symmetry of the doubly aromatic isomer **I**. As was mentioned earlier, the HOMO ($4b_2$) of isomer **II** originates from the HOMO ($4e_2$) of isomer **I** when one of the doubly degenerate orbitals is doubly occupied. Indeed, occupation of only one degenerate MO by two electrons causes the electronic instability, which causes the geometric rearrangement of isomer **II** lowering the D_{8d} symmetry to C_{4v}.

To understand the interactions between Co and the tubular B_{16} host, we have performed AdNDP analyses for the neutral B_{16} tubular isomer (Supplementary Fig. 4). Similar to isomers **I** and **II** of CoB_{16}^-, the AdNDP analyses give 16 2c–2e B–B σ-bonds with ON values of 1.70 $|e|$ within the two B_8 rings. The encapsulation of Co strengthens the B–B σ-bonds within each B_8 ring in CoB_{16}^-, but weakens the inter-ring interactions, compared with the bare B_{16}, as reflected by their ON values (Fig. 3 and Supplementary Fig. 4) and the B–B bond lengths (Supplementary Table 3). The remaining 16 electrons in B_{16} participate in delocalized bonding: five 16c–2e σ + σ bonds and three 16c–2e π–π bonds, rendering the tubular B_{16} doubly σ- and π-aromatic. The major difference in chemical bonding between the drum-like B_{16} and CoB_{16}^- comes from two factors: (1) the formation of an additional set (Fig. 3h–j) of the delocalized σ–σ bonds in CoB_{16}^-; and (2) participation of Co 3d AOs in the two 17c–2e σ + σ bonds (Fig. 3f,g). Both factors are consistent with structural changes between CoB_{16}^- and B_{16}. There are strong bonding interactions between Co and the B_{16} host in CoB_{16}^- to stabilize the tubular B_{16} structure, because the global minimum of B_{16} is planar[30].

Isomer **I** of CoB_{16}^- is open shell with two unpaired electrons, whereas isomer **II** can be viewed as a result of Jahn–Teller distortion from isomer **I**. Addition of two electrons to isomers **I** or **II** would create a closed shell and doubly aromatic CoB_{16}^{3-} species with D_{8d} symmetry. Our calculations indeed confirmed this hypothesis: CoB_{16}^{3-} was found to be a minimum on the potential energy surface with very similar bond distances as in isomer **I** (Supplementary Table 3). The triply charged CoB_{16}^{3-} species can be electronically stabilized by external akali metal

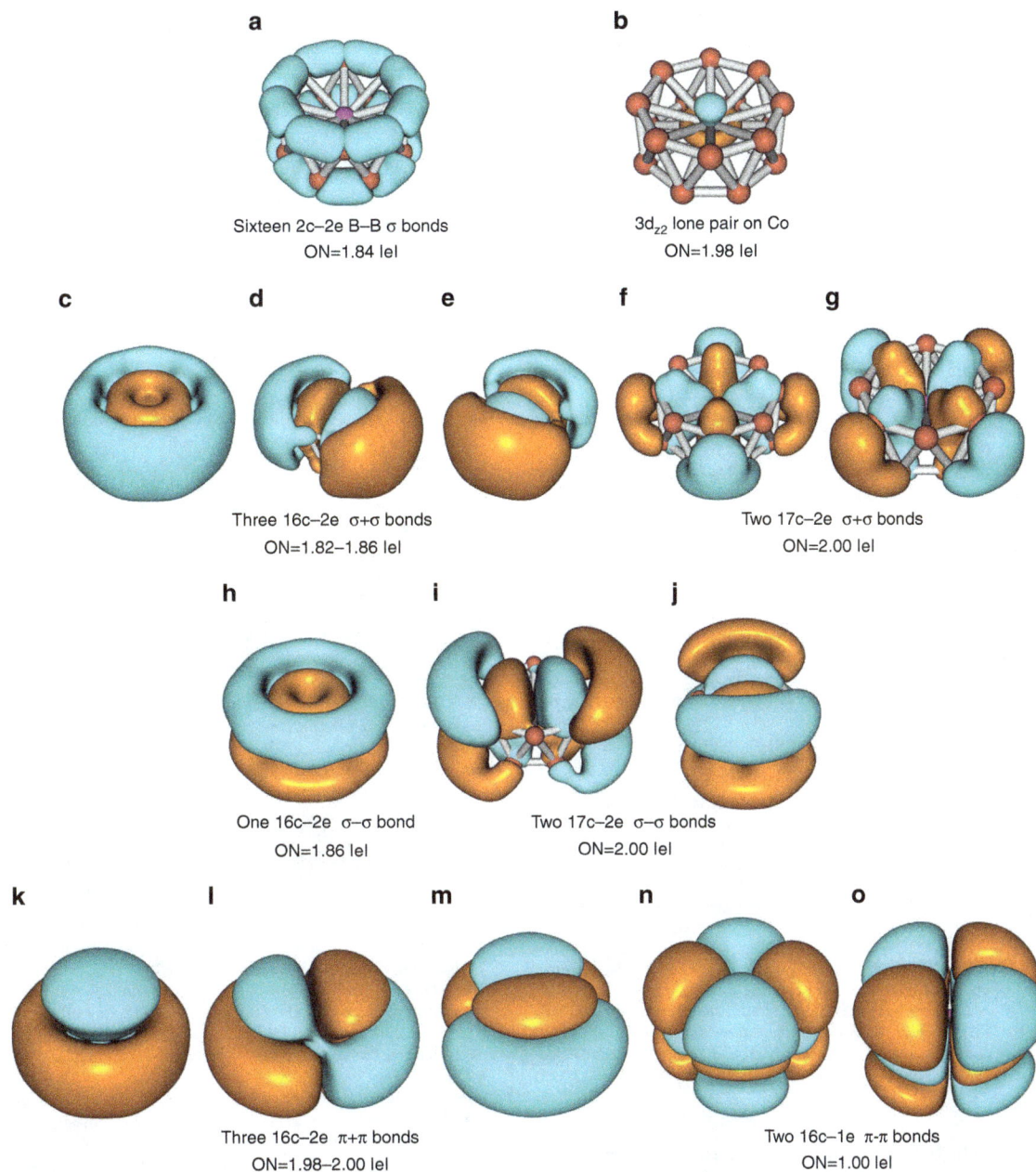

Figure 3 | Chemical bonding picture. (a–o) The overall chemical bonding picture (**a–o**) obtained for the isomer I of the CoB_{16}^- molecular drum via the UAdNDP analysis. ON denotes occupation number here and elsewhere.

cations, such as in $Na_2CoB_{16}^-$. Since ligation would be needed to ultimately synthesize CoB_{16}^-, we considered a triple-decked $[CoB_{16}(CaCp)_2]^-$ sandwich complex (Supplementary Fig. 5), using the divalent Ca atoms and the aromatic $C_5H_5^-$ (Cp^-) ligands. It should be mentioned that similar $[CpLiB_6LiCp]^{2-}$ triple-decked complex[42] with the double anti-aromatic B_6^{2-} unit was previously suggested to be stable and viable experimentally. We found that the $[CoB_{16}(CaCp)_2]^-$ triple-decked complex was a minimum on the potential energy surface with high electronic stability. All the B–B and Co–B bond lengths were found to be almost the same as in isomers I and II of CoB_{16}^- (Supplementary Table 3). We have further performed AdNDP analyses and found that the triple-decked sandwich complex exhibits exactly the same chemical bonding pattern as the parent CoB_{16}^- (Supplementary Figs 6–8). The Natural Population Analysis (NPA) charge on Ca was found to be $+1.54$, consistent with the initial hypothesis and the charge-transfer

nature of the triple-decked $[CoB_{16}(CaCp)_2]^-$ sandwich complex. Thus, the CoB_{16}^- molecular drum can serve as a building block for the design of novel cluster-assembled nanomaterials. The high stability of the CoB_{16}^- drum structures may also help the search for new metal-boride phases containing various boron ring units[43].

We have produced and characterized a large Co-doped boron cluster, CoB_{16}^-, using photoelectron spectroscopy and quantum-chemical calculations. Extensive computational searches established two high symmetry (D_{8d} and C_{4v}) drum-like structures with Co sandwiched by two B_8 rings as nearly degenerate global minima. The CoB_{16}^- molecular drums represent the highest coordination for a metal atom known in chemistry and opens new possibilities for designing novel boron-based nanomaterials. First, the CoB_{16}^- drums may be considered as the embryo to make filled boron nanotubes due to the significant B–B bonding between the two B_8 rings. Second, there are possibilities to observe larger doped-boron clusters with even higher coordination number to further push the limit of coordination number in chemistry. Third, we have demonstrated one possibility to use CoB_{16}^- as a building block of new cluster-assembled nanomaterials in a triple-decked complex.

Methods

Experimental methods. The experiment was carried out using a magnetic-bottle PES apparatus equipped with a laser vaporization cluster source[44]. Briefly, the CoB_{16}^- anion clusters were produced by laser vaporization of a cold-pressed target composed of Co and isotopically enriched ^{11}B. Bismuth was added as a binder and it also provided a convenient calibrant (Bi^-) for the PES experiment. Clusters formed in the nozzle were entrained in a He carrier gas and underwent a supersonic expansion to form a collimated cluster beam. The He carrier gas was seeded with 5% Ar for better cooling of the entrained clusters[22]. The anionic clusters were extracted from the collimated cluster beam and analysed by a time-of-flight mass spectrometer. The CoB_{16}^- anion clusters were mass selected and decelerated before being photodetached by a laser beam at 193 nm (6.424 eV) from an ArF excimer laser or 266 nm (4.661 eV) from a Nd:YAG laser. Photoelectrons were collected at nearly 100% efficiency by a magnetic bottle and analysed in a 3.5 m long flight tube. The resolution of the apparatus, $\Delta Ek/Ek$, was about than 2.5%, that is, ~25 meV for 1 eV electrons.

Theoretical methods. Search for the global minimum of CoB_{16}^- was performed using the Coalescence Kick program[45] at the PBE0/3-21G level of theory[46,47]. The Coalescence Kick algorithm generated ~10,000 trial structures for each spin multiplicity (singlet, triplet and quintet), followed by geometry optimization. Low-lying isomers within 25 kcal mol^{-1} were further refined at a more expansive basis set, Def2-TZVP[48]. For each structure, vibrational frequencies were calculated and imaginary frequencies were followed to ensure that the isomer corresponded to a true minimum on the potential energy surface. Spin contamination was found to be <10% in all DFT calculations. For selected isomers, we performed additional geometry optimization at various DFT levels, as well as more accurate single-point coupled-cluster calculations [ROCCSD(T)/6-311 + G(2df)], to reliably establish the relative energy ordering. Vertical detachment energies of the three lowest energy structures were calculated at three different methods (UPBE0, UB3LYP and ROCCSD(T)) to compare with the experimental data. The VDEs were obtained as the difference in energy between the ground state of the anion and selected low-lying electronic states of the neutral molecule at the geometry of the anion. All calculations were done using GAUSSIAN-09 (ref. 49).

To understand the chemical bonding, we carried out electron localization analyses using the AdNDP method[38] at the PBE0/6-31G(d) level of theory. Previously, AdNDP results have been shown to be insensitive to the level of theory or basis set used[50]. The AdNDP analysis is based on the concept of electron pairs as the main elements of chemical bonds. It represents the molecular electronic structure in terms of n-centre two-electron (nc-2e) bonds, recovering the familiar lone pairs (1c-2e) and localized 2c-2e bonds or delocalized nc-2e bonds ($3 \leq n \leq$ total number of atoms in the system). The MOLEKEL 5.4.0.8 program[51] is used for molecular structure and AdNDP bond visualizations.

References

1. Albert, B. & Hillebrecht, H. Boron: elementary challenge for experimenters and theoreticians. *Angew. Chem. Int. Ed. Engl.* **48**, 8640–8668 (2009).
2. Oganov, A. R. *et al.* Ionic high-pressure form of elemental boron. *Nature* **457**, 863–867 (2009).
3. Kuhlmann, U. & Werheit, H. Improved Raman effect studies on boron carbide ($B_{4.3}C$). *Phys. Status Solidi (b)* **175**, 85–92 (1993).
4. Nelmes, R. J. *et al.* Observation of inverted-molecular compression in boron carbide. *Phys. Rev. Lett.* **74**, 2268–2271 (1995).
5. Kawai, R. & Weare, J. H. Instability of the B_{12} icosahedral cluster: rearrangement to a lower energy structure. *J. Chem. Phys.* **95**, 1151–1159 (1991).
6. Boustani, I. A comparative study of *ab initio* SCF-CI and DFT. Example of small boron clusters. *Chem. Phys. Lett.* **233**, 273 (1995).
7. Hanley, L., Whitten, J. L. & Anderson, S. L. Collision-induced dissociation and ab initio studies of boron cluster ions: determination of structures and stabilities. *J. Phys. Chem.* **92**, 5803–5812 (1988).
8. Alexandrova, A. N., Boldyrev, A. I., Zhai, H. J. & Wang, L. S. All-boron aromatic clusters as potential new inorganic ligands and building blocks in chemistry. *Coord. Chem. Rev.* **250**, 2811–2866 (2006).
9. Sergeeva, A. P. *et al.* Understanding boron through size-selected clusters: structure, chemical bonding, and fluxionality. *Acc. Chem. Res.* **47**, 1349–1358 (2014).
10. Li, W. L., Pal, R., Piazza, Z. A., Zeng, X. C. & Wang, L. S. B_{27}^-: appearance of the smallest boron cluster containing a hexagonal vacancy. *J. Chem. Phys.* **142**, 204305 (2015).
11. Li, W. L., Zhao, Y. F., Hu, H. S., Li, J. & Wang, L. S. B_{30}^-: A quasiplanar chiral boron cluster. *Angew. Chem. Int. Ed. Engl.* **53**, 5540–5545 (2014).
12. Li, W. L. *et al.* The B_{35} cluster with a double-hexagonal vacancy: a new and more flexible structural motif for borophene. *J. Am. Chem. Soc.* **136**, 12257–12260 (2014).
13. Piazza, Z. A. *et al.* Planar hexagonal B_{36} as a potential basis for extended single-atom layer boron sheets. *Nat. Commun.* **5**, 3113 (2014).
14. Kiran, B. *et al.* Planar-to-tubular structural transition in boron clusters: B_{20} as the embryo of single-walled boron nanotubes. *Proc. Natl Acad. Sci. USA* **102**, 961–964 (2005).
15. Oger, E. *et al.* Boron cluster cations: transition from planar to cylindrical structures. *Angew. Chem. Int. Ed. Engl.* **46**, 8503–8506 (2007).
16. Chen, Q. *et al.* Experimental and theoretical evidence of an axially chiral borospherene. *ACS Nano.* **9**, 754–760 (2015).
17. Zhai, H. J. *et al.* Observation of an all-boron fullerene. *Nat. Chem* **6**, 727–731 (2014).
18. Romanescu, C., Galeev, T. R., Li, W. L., Boldyrev, A. I. & Wang, L. S. Transition-metal-centered monocyclic boron wheel clusters ($M©B_n$): a new class of aromatic borometallic compounds. *Acc. Chem. Res.* **46**, 350–358 (2013).
19. Zhai, H. J., Alexandrova, A. N., Birch, K. A., Boldyrev, A. I. & Wang, L. S. Hepta- and octa-coordinated boron in molecular wheels of 8- and 9-atom boron clusters: observation and confirmation. *Angew. Chem. Int. Ed. Engl.* **42**, 6004–6008 (2003).
20. Galeev, T. R., Romanescu, C., Li, W. L., Wang, L. S. & Boldyrev, A. I. Observation of the highest coordination number in planar species: decacoordinated $Ta©B_{10}$ and $Nb©B_{10}$ anions. *Angew. Chem. Int. Ed. Engl.* **51**, 2101–2105 (2012).
21. Zhai, H. J., Wang, L. S., Alexandrova, A. N. & Boldyrev, A. I. Electronic structure and chemical bonding of B_5^- and B_5 by photoelectron spectroscopy and ab initio calculations. *J. Chem. Phys.* **117**, 7917–7924 (2002).
22. Alexandrova, A. N., Boldyrev, A. I., Zhai, H. J. & Wang, L. S. Electronic structure, isomerism, and chemical bonding in B_7^- and B_7. *J. Phys. Chem. A* **108**, 3509–3517 (2004).
23. Piazza, Z. A. *et al.* A photoelectron spectroscopy and ab initio study of B_{21}^-: negatively charged boron clusters continue to be planar at 21. *J. Chem. Phys.* **136**, 104310 (2012).
24. Xu, C., Cheng, L. J. & Yang, J. L. Double aromaticity in transition metal centered double-ring boron clusters $M@B_{2n}$ (M = Ti, Cr, Fe, Ni, Zn; n = 6, 7, 8). *J. Chem. Phys.* **141**, 124301 (2014).
25. Tam, N. M., Pham, H. T., Duong, L. V., Pham-Ho, M. P. & Nguyen, M. T. Fullerene-like boron clusters stabilized by an endohedrally doped iron atom: B(n)Fe with n = 14, 16, 18 and 20. *Phys. Chem. Chem. Phys.* **17**, 3000–3003 (2015).
26. An, W., Bulusu, S., Gao, Y. & Zeng, X. C. Relative stability of planar versus double-ring tubular isomers of neutral and anionic boron cluster B_{20} and B_{20}^-. *J. Chem. Phys.* **124**, 154310 (2006).
27. Sergeeva, A. P. *et al.* B_{22}^- and B_{23}^-: all-boron analogues of anthracene and phenanthrene. *J. Am. Chem. Soc.* **134**, 18065–18073 (2012).
28. Popov, I. A., Piazza, Z. A., Li, W. L., Wang, L. S. & Boldyrev, A. I. A combined photoelectron spectroscopy and *ab initio* study of the quasi-planar B_{24}^- cluster. *J. Chem. Phys.* **139**, 144307 (2013).
29. Romanescu, C., Harding, D. J., Fielicke, A. & Wang, L. S. Probing the structures of neutral boron clusters using IR/VUV two color ionization: B_{11}, B_{16}, and B_{17}. *J. Chem. Phys.* **137**, 014317 (2012).
30. Sergeeva, A. P., Zubarev, D. Y., Zhai, H.-J., Boldyrev, A. I. & Wang, L. S. A photoelectron spectroscopic and theoretical study of B_{16}^- and B_{16}^{2-}: an all-boron naphthalene. *J. Am. Chem. Soc.* **130**, 7244–7246 (2008).
31. Li, W. L., Romanescu, C., Piazza, Z. A. & Wang, L. S. Geometrical requirements for transition-metal-centered aromatic boron wheels: the case of VB_{10}^-. *Phys. Chem. Chem. Phys.* **14**, 13663–13669 (2012).
32. Popov, I. A., Li, W. L., Piazza, Z. A., Boldyrev, A. I. & Wang, L. S. Complexes between planar boron clusters and transition metals: a photoelectron spectroscopy and *ab initio* study of CoB_{12}^- and RhB_{12}^-. *J. Phys. Chem. A* **118**, 8098–8105 (2014).
33. Daly, S. R. *et al.* Synthesis and properties of a fifteen-coordinate complex: the thorium aminodiboranate [$Th(H_3BNMe_2BH_3)_4$]. *Angew. Chem. Int. Ed. Engl* **49**, 3379–3381 (2010).
34. Hermann, A., Lein, M. & Schwerdtfeger, P. The Gregory-Newton problem of kissing sphere applied to chemistry: the search for the species with the highest coordination number. *Angew. Chem. Int. Ed. Engl.* **46**, 2444–2447 (2007).
35. Komura, Y. & Tokunaga, K. Structural studies of stacking variants in Mg-base Friauf-Laves phases. *Acta Crystallogr. Sect. B* **36**, 1548–1554 (1980).
36. Wang, L. S. *et al.* The electronic structure of $Ca@C_{60}$. *Chem. Phys. Lett.* **207**, 354–359 (1993).
37. Popov, A. A., Yang, S. & Dunsch, L. Endohedral fullerenes. *Chem. Rev.* **113**, 5989–6113 (2013).
38. Zubarev, D. Y. & Boldyrev, A. I. Developing paradigms of chemical bonding: adaptive natural density partitioning. *Phys. Chem. Chem. Phys.* **10**, 5207–5217 (2008).

39. Foster, J. P. & Weinhold, F. Natural bond orbitals. *J. Am. Chem. Soc.* **102**, 7211–7218 (1980).

40. Yuan, Y. & Cheng, L. J. B_{14}^{2+}: a magic number double-ring cluster. *J. Chem. Phys.* **137**, 044308 (2012).

41. Johansson, M. P. On the strong ring currents in B_{20} and neighboring boron toroids. *J. Phys. Chem. C.* **113**, 524–530 (2009).

42. Yang, L.-M., Wang, J., Ding, Y.-H. & Sun, C.-C. Sandwich-like compounds based on bare all-boron cluster B_6^{2-}. *Phys. Chem. Chem. Phys.* **10**, 2316–2320 (2008).

43. Fokwa, B. P. T. & Hermus, M. All-boron planar B_6 ring in the solid-state phase $Ti_7Rh_4Ir_2B_8$. *Angew. Chem. Int. Ed. Engl.* **51**, 1702–1705 (2012).

44. Wang, L. S., Cheng, H. S. & Fan, J. W. Photoelectron spectroscopy of size-selected transition metal clusters: Fe_n^-, $n = 3 - 24$. *J. Chem. Phys* **102**, 9480–9493 (1995).

45. Sergeeva, A. P., Averkiev, B. B., Zhai, H. J., Boldyrev, A. I. & Wang, L. S. All-boron analogues of aromatic hydrocarbons: B_{17}^- and B_{18}^-. *J. Chem. Phys.* **134**, 224304 (2011).

46. Adamo, C. & Barone, V. Toward reliable density functional methods without adjustable parameters: The PBE0 model. *J. Chem. Phys.* **110**, 6158–6170 (1999).

47. Binkley, J. S., Pople, J. A. & Hehre, W. J. Self-consistent molecular orbital methods. 21. Small split-valence basis sets for first-row elements. *J. Am. Chem. Soc.* **102**, 939–947 (1980).

48. Weigend, F. & Ahlrichs, R. Balanced basis sets of split valence, triple zeta valence and quadruple zeta valence quality for H to Rn: design and assessment of accuracy. *Phys. Chem. Chem. Phys.* **7**, 3297–3305 (2005).

49. Frisch, M. J. *et al. GAUSSIAN09, Revision B.01* (Gaussian, Inc., 2009).

50. Sergeeva, A. P. & Boldyrev, A. I. The chemical bonding of Re_3Cl_9 and $Re_3Cl_9^{2-}$ revealed by the adaptive natural density partitioning analyses. *Comm. Inorg. Chem.* **31**, 2–12 (2010).

51. Varetto, U. *MOLEKEL 5.4.0.8* (Swiss National Supercomputing Centre, 2009).

Acknowledgements

This work was supported by the National Science Foundation (CHE-1263745 to L.S.W. and CHE-1361413 to A.I.B.). Computer, storage and other resources from the Division of Research Computing in the Office of Research and Graduate Studies at Utah State University are gratefully acknowledged.

Author contributions

L.S.W. and A.I.B. designed the research. I.A.P. and A.I.B performed and analyzed the calculations. L.S.W., T.J. and G.V.L. designed experiments and analysed the experimental data. All authors contributed to the interpretation and discussion of the data. I.A.P. and T.J. wrote the manuscript.

Additional information

Competing financial interests: The authors declare no competing financial interests.

10

Mesoporous MnCeO$_x$ solid solutions for low temperature and selective oxidation of hydrocarbons

Pengfei Zhang[1], Hanfeng Lu[2], Ying Zhou[2], Li Zhang[1], Zili Wu[1], Shize Yang[3], Hongliang Shi[3], Qiulian Zhu[2], Yinfei Chen[2] & Sheng Dai[1,4]

The development of noble-metal-free heterogeneous catalysts that can realize the aerobic oxidation of C–H bonds at low temperature is a profound challenge in the catalysis community. Here we report the synthesis of a mesoporous Mn$_{0.5}$Ce$_{0.5}$O$_x$ solid solution that is highly active for the selective oxidation of hydrocarbons under mild conditions (100–120 °C). Notably, the catalytic performance achieved in the oxidation of cyclohexane to cyclohexanone/cyclohexanol (100 °C, conversion: 17.7%) is superior to those by the state-of-art commercial catalysts (140–160 °C, conversion: 3-5%). The high activity can be attributed to the formation of a Mn$_{0.5}$Ce$_{0.5}$O$_x$ solid solution with an ultrahigh manganese doping concentration in the CeO$_2$ cubic fluorite lattice, leading to maximum active surface oxygens for the activation of C–H bonds and highly reducible Mn^{4+} ions for the rapid migration of oxygen vacancies from the bulk to the surface.

[1] Chemical Sciences Division, Oak Ridge National Laboratory, Oak Ridge, Tennessee 37831, USA. [2] Institute of Catalytic Reaction Engineering, College of Chemical Engineering, Zhejiang University of Technology, Hangzhou 310014, China. [3] Materials Science and Technology Division, Oak Ridge National Laboratory, Oak Ridge, Tennessee 37831, USA. [4] Department of Chemistry, University of Tennessee, Knoxville, Tennessee 37996, USA. Correspondence and requests for materials should be addressed to H.L. (email: luhf@zjut.edu.cn) or to S.D. (email: dais@ornl.gov).

Aerobic oxidation has been considered as one of the most fundamental processes throughout organic synthesis and industrial chemistry[1-9]. Nowadays, realizing the selective oxidation of sp^3 C–H bonds at low temperatures represents a critical challenge in the petroleum industry, because the current methods for the activation of C–H bonds generally require high temperature (for example, $\sim 600\,°C$ for propane dehydrogenation) and excessive energy input, often resulting in uncontrolled product selectivity and undesirable cokes[10-20]. Among all C–H activation processes, the liquid-phase oxidation of cyclohexane to KA oil (K: cyclohexanone, A: cyclohexanol, production $> 2 \times 10^6$ ton per year) is widely deployed in Nylon-6 and Nylon-6,6 production[21]. The industrial process proceeds with homogeneous Co/Mn carboxylate salts at $140–160\,°C$ using 0.9– 1.0 MPa air as an oxidant[3]. To minimize the overoxidation of KA oil to by-products, cyclohexane conversion is preferentially limited to $< 5\%$. Figure 1a summaries representative pathways to caprolactam (monomer for Nylon-6); the low cyclohexane conversion is definitely a bottleneck of the state-of-art technologies. This situation prompted catalysis scientists to explore the possibility of developing new catalysts, for example, N-hydroxyphthalimide[22], metalloporphyrins[23], transition metal ions-substituted molecular sieve catalysts[24], supported gold catalysts[25-27] and carbon-based catalysts[28,29]. However, several important issues are still unresolved, such as catalyst recycling and separation, the use of H_2O_2 or tert-butylhydroperoxide oxidants (the desired oxidant is air or O_2) or dependence on noble metal elements. From the standpoint of chemical kinetics, the development of a heterogeneous catalyst that functions at lower temperature, may prevent deep radical oxidation to a large degree, ideally achieving a higher cyclohexane conversion.

Recently, MnO_x–CeO_2 hybrid catalysts with multiple redox states and high oxygen storage capacity have exhibited superior performance in several types of catalytic oxidation, such as ammonia oxidation, combustion of volatile organic compounds and CO oxidation[30-33]. Compared with either MnO_x or CeO_2 (ceria), the significant decrease in reaction temperature enabled by the MnO_x–CeO_2 composite is very appealing, which directly evidences the synergistic interaction of MnO_x and CeO_2 with more active oxygen species. These 'reactive' oxygen species (for example, O_2^-, O_2^{2-} and O^-) are generated exactly at the interface between the MnO_x and ceria lattice, the so-called $Mn_yCe_{1-y}O_x$ solid solution. Since the formation of a –Mn–O–Ce– bond would reduce the Coulomb interaction of $Mn^{\delta+}$–$O^{\gamma-}$ or $Ce^{\delta+}$–$O^{\gamma-}$, the formation energy of oxygen vacancies can be greatly lowered[34]. Several approaches to MnO_x–CeO_2 catalysts— such as the co-precipitation[31,32], sol-gel[35], combustion[33], surfactant-assisted precipitation[36] and hydrothermal methods[37] —have been developed. Unfortunately, traditional methods of preparing a MnO_x–CeO_2 catalyst often lead to the formation of multiphases with limited $Mn_yCe_{1-y}O_x$ solid solution, which is only observed at the interfaces between MnO_x and ceria nano-crystals. Recently, Yang and co-workers reported a general route to phase-pure transition-metal-substituted ceria nanocrystals via solution-based pyrolysis of bimetallic Schiff base complexes, but the ratio of transition metal substitution (10 mol %) is somewhat low[38]. Given that the solid solution phase of a MnO_x–CeO_2 catalyst is responsible for the low-temperature redox activity, a $Mn_{0.5}Ce_{0.5}O_x$ solid solution with 50% manganese atoms doping into a ceria lattice may be an ideal candidate for catalytic oxidation, because maximum active oxygen species are expected in such a structure. In the view of synthetic chemistry, the biggest challenge for constructing a $Mn_{0.5}Ce_{0.5}O_x$ solid solution with as high as 50% cerium atoms substituted by manganese atoms but retaining the cubic fluorite structure lies in controlling the homogenization with –Mn–O–Ce–bonds throughout the backbone (Fig. 1b).

In this contribution, we report an efficient, sustainable approach to a homogeneous $Mn_{0.5}Ce_{0.5}O_x$ solid solution, whose ideal structure with Mn^{4+} ions in the ceria matrix is suggested by X-ray diffraction (XRD), X-ray photoelectron spectroscopy (XPS), scanning electron transmission microscopy–X-ray energy dispersive spectroscopy (STEM–XEDS) mapping analysis and H_2 temperature-programmed reduction (H_2-TPR). To the best of our knowledge, it is the first time for the ultrahigh concentration of Mn^{4+} ion to be stabilized in a ceria lattice[30-38]. The essence of the current strategy for fabricating a uniform $Mn_{0.5}Ce_{0.5}O_x$ solid solution is the slow hydrolysis of Mn/Ce precursors at the surfaces of ionic liquid 'supermolecular' networks. Surprisingly, a mesoporous structure with a high surface area is observed for the $Mn_{0.5}Ce_{0.5}O_x$ solid solution after ionic liquids removal. This structure is highly advantageous in heterogeneous catalysis, since it can expose more surface oxygen species, and faster mass diffusion/transfer can be expected[39]. This versatile soft-templating method for well-defined mesopores can cover various oxide solid solutions even transition metal perovskites such as $Co_{0.5}Ce_{0.5}O_x$, $Cu_{0.2}Mn_{0.3}Ce_{0.5}O_x$, and $YMnO_3$. We show the outstanding activity of a $Mn_{0.5}Ce_{0.5}O_x$ solid solution catalyst in the low temperature, heterogeneous oxidation of cyclohexane ($100\,°C$, conversion: 17.7%, selectivity for KA oils: 81%) with molecular oxygen as the oxidant. It is significantly superior to the results of current technology ($140–160\,°C$, conversion: 3–5%); this process could be extended to the selective oxidation of various allylic or benzyl C–H bonds with the corresponding alcohols/ketones as products. This study provides a simple general strategy to obtain a mesoporous $Mn_{0.5}Ce_{0.5}O_x$ solid solution catalyst that can make selective, O_2-based oxidation of sp^3 C–H bonds at mild temperatures possible.

Results

Fabrication of $Mn_{0.5}Ce_{0.5}O_x$ solid solutions. The detailed route to the $Mn_{0.5}Ce_{0.5}O_x$ catalyst is shown in Fig. 1c. In the present model system, manganese (II) acetate, cerium (IV) methoxyeth-oxide, 1-butyl-3-methylimidazolium bis(trifluoromethanesul-fonyl)imide ($BmimTf_2N$) and ethanol were mixed at a ratio of (1.0:1.8:1.6:8.0 w/w/w/w), and stirred at room temperature for 2 h. The dark red homogeneous solution was poured into a petri dish to evaporate solvents at $50\,°C$ for 24 h, followed by solidifi-cation of the sample at $200\,°C$ for 2 h with the formation of a primary metal oxo matrix around hydrophobic $BmimTf_2N$ via electrostatic interaction; and a frizzy solid film formed (Supplementary Fig. 1). It should be emphasized that the initial treatment temperature ($200\,°C$) was higher than values used during surfactants or block copolymers-induced processes ($\sim 95–120\,°C$)[40]. The good thermal stability of $BmimTf_2N$ (decomposition temperature: $> 350\,°C$, Supplementary Fig. 2) results in its high tolerable temperature, which allows a higher condensation degree of Mn/Ce precursors for strong backbones, and therefore affords the possibility of recycling $BmimTf_2N$. In previous methods for forming mesoporous metal oxides, the surfactants or block copolymers used as soft templates usually cannot be removed and recycled before calcinations; otherwise, the porosity would collapse[39,40]. However, the organic templates cannot survive during high-temperature treatment (for example, $500\,°C$) and this sacrificial behaviour obstructs their industrial application. In contrast, the structure-directing $BmimTf_2N$ template can be easily extracted and recovered by refluxing in ethanol (Supplementary Figs 3-4), resulting in $Mn_{0.5}Ce_{0.5}O_x@200$. The as-made sample was thermally treated at $500\,°C$ for 2 h in air ($Mn_{0.5}Ce_{0.5}O_x@500$).

Characterization of mesoporous metal oxides. Figure 1b illustrates the evolution of crystal structures upon doping of

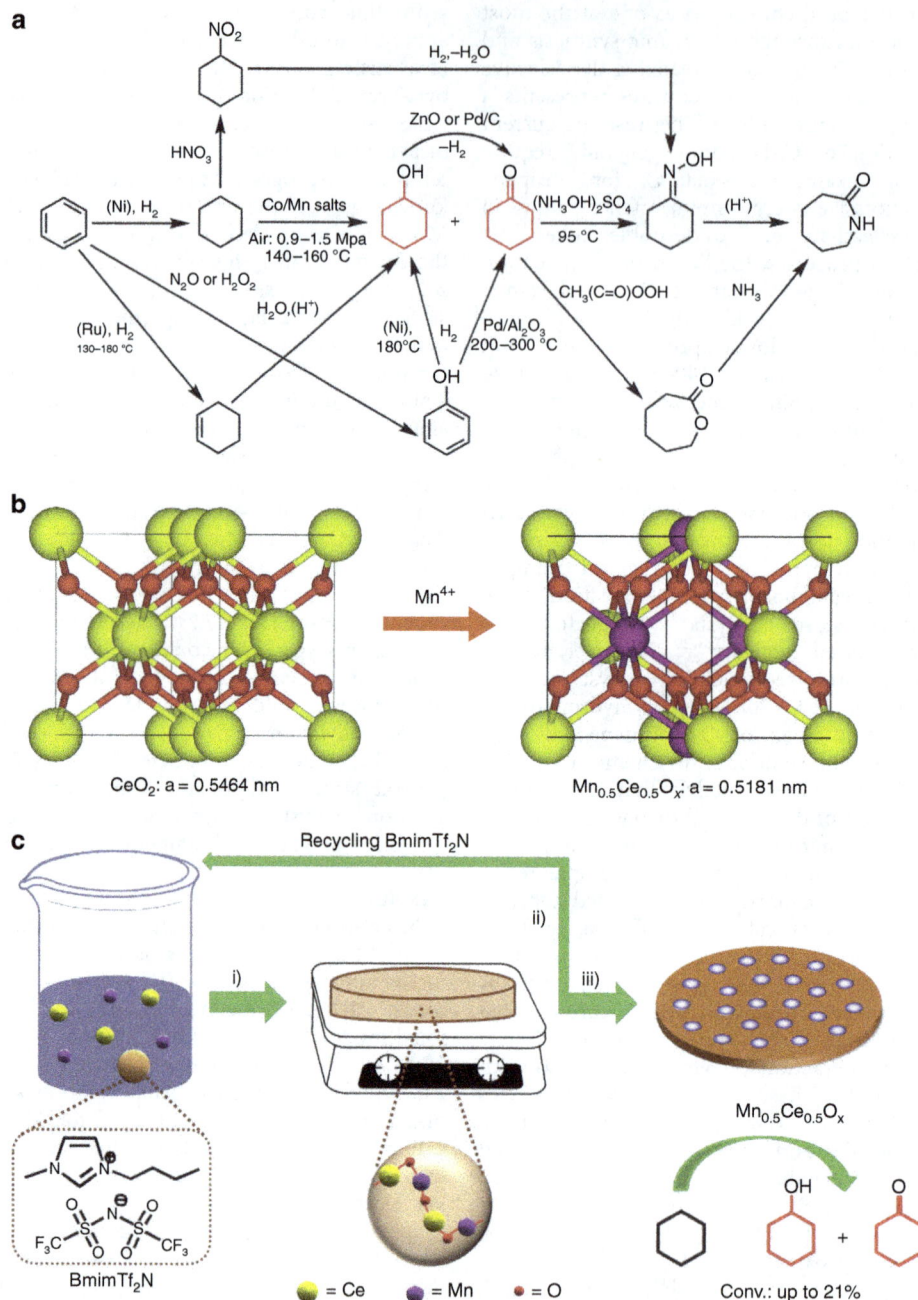

Figure 1 | Synthetic and catalytic strategies. (**a**) A summary of state-of-art processes for caprolactam production (monomer for Nylon-6); (**b**) the evolution of doping 50% Mn^{4+} ions into a CeO_2 lattice; (**c**) a solvent evaporation-induced self-assembly between metal salts and hydrophobic ionic liquid, reaction conditions: manganese (II) acetate, cerium (IV) methoxyethoxide and 1-butyl-3-methylimidazolium bis(trifluoromethanesulfonyl)imide (BmimTf$_2$N) in ethanol: (i) stirring at room temperature for 2 h, and pouring into a petri dish at 50 °C for 24 h and 200 °C for 2 h, (ii) removing and recycling the BmimTf$_2$N by Soxhlet extraction in ethanol (24 h), (iii) thermal treatment in air oven at 500 °C for 2 h.

50% Mn^{4+} ions into a ceria lattice, and a density functional theory calculation of structural models showed the change in the optimized lattice parameter a. Compared with ceria ($a = 0.5464$ nm), an ideal $Mn_{0.5}Ce_{0.5}O_x$ with a symmetrical Mn^{4+} substitution undergoes shrinkage along the a axis ($a = 0.5181$ nm). This is reasonable because the ionic radii of manganese ions (Mn^{4+}: 0.053 nm; Mn^{3+}: 0.065 nm; Mn^{2+}: 0.083 nm) are smaller in size than those of cerium ions (Ce^{4+}: 0.097 nm; Ce^{3+}: 0.114 nm). Indeed, the XRD pattern of the $Mn_{0.5}Ce_{0.5}O_x$@500 sample showed a clear shift towards a higher Bragg angle compared with pure ceria and its corresponding lattice parameter a calculated by a (111) peak at 29.795° ($a = 0.5194$ nm)

was very close to the above theoretical result ($a = 0.5181$ nm), revealing the possible replacement of Ce^{4+} by Mn^{4+} in the cubic fluorite structure (Fig. 2a, Supplementary Fig. 5, Supplementary Table 1). Extremely broad diffraction peaks for (111), (220) and (311) reflections of the $Mn_{0.5}Ce_{0.5}O_x$@500 sample were observed. The average crystalline size was 1.4 nm, calculated by the Scherrer equation. The small crystalline size can be attributed to the confined hydrolysis and condensation of Mn/Ce precursors templated by the heterogeneous BmimTf$_2$N structure[41]. In addition, a partially crystalline structure has already formed in the $Mn_{0.5}Ce_{0.5}O_x$@200 sample.

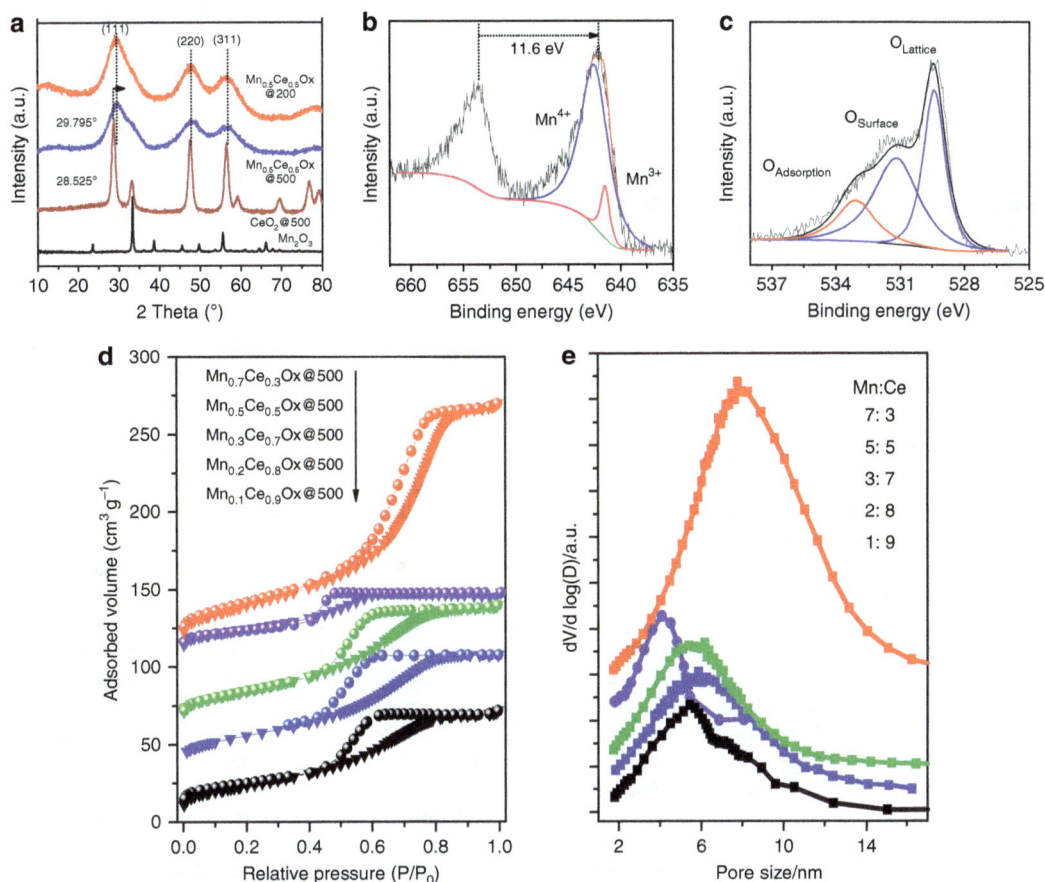

Figure 2 | Structural characterizations of catalysts. (**a**) XRD patterns of $Mn_{0.5}Ce_{0.5}O_x@200$, $Mn_{0.5}Ce_{0.5}O_x@500$, $CeO_2@500$ and Mn_2O_3. (**b**) XPS spectra of Mn 2p and (**c**) XPS spectra of O 1s of $Mn_{0.5}Ce_{0.5}O_x@500$. (**d**) N_2 sorption isotherm curves of $Mn_yCe_{1-y}O_x@500$ samples at 77 K; For clarity, the isotherm curves for $Mn_{0.2}Ce_{0.8}O_x@500$, $Mn_{0.3}Ce_{0.7}O_x@500$, $Mn_{0.5}Ce_{0.5}O_x@500$ and $Mn_{0.7}Ce_{0.3}O_x@500$ were offset by 30, 60, 105 and 105 $cm^3 g^{-1}$, respectively. (**e**) pore size distributions of $Mn_yCe_{1-y}O_x@500$ samples.

To study the oxidation state of surface species, XPS spectra for the Mn 2p and O 1s core levels of the $Mn_{0.5}Ce_{0.5}O_x@500$ sample were recorded and are shown in Fig. 2b,c. The XPS curve of Mn 2p exhibited two peaks at 653.7 and 642.1 eV, which can be attributed to the Mn $2p_{1/2}$ and Mn $2p_{3/2}$ states, respectively. The spin orbit splitting is $\Delta E = 11.6$ eV, close to the value of MnO_2 (11.7 eV)[42]. In addition, the Mn $2p_{3/2}$ peak is fitted with a Shirley background and Gaussian-Lorenz model functions, and two peaks at 641.5 and 642.6 eV can be obtained, based on standard binding energy and previous literatures[32,42]. The observed binding energies suggest the co-existence of Mn^{3+} and Mn^{4+} ions, but Mn^{4+} species with 87% content dominate the surface, in accordance with the structural model discussed above. Meanwhile, the O 1s spectrum with a shoulder peak is very broad, possibly owing to the overlapping contributions of various oxygen species. The curve was then resolved with the model discussed above and fitted into three peaks. The peaks at 529.4, 531.2 and 533.1 eV are ascribed to lattice oxygen atoms (O^{2-}, denoted as O_α), surface oxygen species (for example, O_2^-, O_2^{2-}, O^-, denoted as O_β) and chemisorbed water and/or carbonates (denoted as O_γ), respectively. It is well recognized in the literatures that the O_β species from defective sites with an unsaturated structure are of great importance in the catalytic oxidation process[32,37]. The surface atomic concentration was then calculated by integrating the peak areas of different oxygen species. The atomic ratio of "reactive" oxygen species (O_β) can reach 44.1%, arguing for the great potential of this solid solution in catalytic oxidations.

The porous nature of $Mn_{0.5}Ce_{0.5}O_x$ samples was evaluated by nitrogen sorption measurements at 77 K. The $Mn_{0.5}Ce_{0.5}O_x@200$ sample was dominated by micropores with remarkable N_2 uptake at low relative pressure and its specific surface area calculated by the Brunauer–Emmett–Teller (BET) method was 467 $m^2 g^{-1}$ (Supplementary Fig. 6). The rich porosity should be directed during the removal of $BmimTf_2N$. It also can be concluded that the backbone of the $Mn_{0.5}Ce_{0.5}O_x$ sample formed at 200 °C is strong enough to withstand the high pressure of molecular packing. Both XRD patterns and Fourier-transform infrared spectra of the $Mn_{0.5}Ce_{0.5}O_x@200$ sample suggest that it is an oxide precursor with acetate anions incorporated in the matrix (Fig. 2a, Supplementary Fig. 7). A weak coordination-induced network containing $Mn(OAc)_2$ and partially dehydrated cerium hydroxide were proposed for the $Mn_{0.5}Ce_{0.5}O_x@200$ sample, wherein the close connection between manganese and cerium ions is the key to restructuring into a $Mn_{0.5}Ce_{0.5}O_x$ solid solution during calcination. This confined restructuring can prevent the formation of separate bulk manganese or cerium oxide phases (Supplementary Fig. 8)[38].

Thermal treatment of the $Mn_{0.5}Ce_{0.5}O_x@200$ sample led to pore expansion, as shown by the pore size distributions of samples at different temperatures (200, 400, 500 and 600 °C); the pore expansion is possibly the result of the progressive growth of nanocrystals (Supplementary Fig. 9)[43]. It is Interesting that the $Mn_{0.5}Ce_{0.5}O_x@500$ material possessed a characteristic type IV sorption isotherm with a H_1 hysteresis loop, including a sharp capillary condensation step at $p/p_0 = 0.4$–0.5. The pore diameter located in 3–6 nm with a narrow distribution, derived from the sorption branch of the isotherm by using Barrett–Joyner–Halenda model (Fig. 2d,e). The $Mn_{0.5}Ce_{0.5}O_x$@500 was a typically mesoporous material with a BET surface area of $89\,m^2\,g^{-1}$. A series of mixed oxide solutions with a different Mn/Ce atomic ratio (1:9, 2:8, 3:7, 7:3) were also prepared, and mesoporous structures with high surface areas were observed for those samples (Supplementary Table 2). The $Mn_{0.7}Ce_{0.3}O_x$@500 sample possessed a specific surface area of $125\,m^2\,g^{-1}$ with large mesopores around 10 nm. Moreover, the current solvent evaporation-induced assembly of binary $Mn_{0.5}Ce_{0.5}O_x$ around the BmimTf$_2$N template can easily be extended to more metal-oxide combinations with similar mesoporous structures, such as: $Co_{0.5}Ce_{0.5}O_x$@500 (using another Period 4 transition metal: $S_{BET} = 52\,m^2\,g^{-1}$, pore size: ~ 3 nm; Supplementary Figs 10–11), YMnO$_3$@700 (transition metal perovskite: $S_{BET} = 56\,m^2\,g^{-1}$, pore size: ~ 7.5 nm; Supplementary Fig. 12), and $Cu_{0.2}Mn_{0.3}Ce_{0.5}O_x$@500 (ternary metal oxide: $S_{BET} = 78\,m^2\,g^{-1}$, pore size: ~ 4 nm; Supplementary Figs 13–14). In some cases, the pore size of the target material (for example, SiO_2) can be precisely tailored on a mesoporous scale (for example, 3–40 nm), via adjusting the mass ratio between precursor molecules and BmimTf$_2$N (Supplementary Fig. 15). Given that a large BmimTf$_2$N aggregation is responsible for generating distances/pores between the primary oxide particles, polymerized BmimTf$_2$N was then synthesized, which could lead to wider mesopores (Supplementary Figs 16–17).

The transmission electron microscopy (TEM) and STEM in high-angular dark field mode (STEM-HAADF) images directly witness the evolution of $Mn_{0.5}Ce_{0.5}O_x$ samples at different treatment temperatures. The $Mn_{0.5}Ce_{0.5}O_x$@200 sample was rich in porosity with apparent pores of around 1–3 nm, in agreement with the value by nitrogen sorption measurement (Fig. 3a,b,e,f). Actually, the ionothermal synthesis of carbon materials (200 °C) in BmimTf$_2$N solvent also resulted in a porosity within microporous domains. The clusters/aggregations of BmimTf$_2$N, formed during interaction with the precursors, are more or less within 1–3 nm (ref. 44). This is understandable, since the density functional theory studies of ionic liquids suggest that imidazolium cations can form extended hydrogen bond interactions with up to three anions, leading to highly structured ionic liquid clusters of the minimal free energy[44–46]. High-resolution TEM (HRTEM) image of $Mn_{0.5}Ce_{0.5}O_x$@200 showed some lattice fringes, and diffuse rings in the selected area electron diffraction patterns were observed (Fig. 3c). Therefore, the initial crystalline structure with a poor crystallinity has formed even at 200 °C. To verify the compositional details of $Mn_{0.5}Ce_{0.5}O_x$@200, STEM–XEDS mapping analysis was carried out (Fig. 3d). In a 50×50 nm region, the Mn and Ce X-ray signals were evenly distributed and the atomic ratio of Mn:Ce was around 1:1 by energy-dispersive X-ray spectroscopy (EDS) with drift-corrected spectral imaging.

Thermal treatment at 500 °C can enlarge the pore size into the mesoporous range, as indicated by the TEM image of $Mn_{0.5}Ce_{0.5}O_x$@500; the sample contains a high degree of the interstitial porosity between interconnected nanocrystals (Fig. 3g). It is worthy to note that no phases for separate MnO_x particles can be observed by HRTEM, further revealing the formation of a homogeneous solid solution. The HRTEM image of $Mn_{0.5}Ce_{0.5}O_x$@500 exhibits clear lattice fringes and well-defined ring (in the electron diffraction pattern) structures of the (111), (220) and (311) planes for cubic ceria, implying extremely poor crystallinity and small crystal size, in accordance with the broad peaks in the XRD pattern (Fig. 3h). The homogeneous distribution of Mn and Ce atoms was also indicated by the STEM–XEDS mapping analysis (Fig. 3i–m).

The H$_2$-TPR profiles of the $Mn_{0.5}Ce_{0.5}O_x$@500 sample are displayed in Fig. 4. Only one reduction peak was observed and located at ~ 250 °C, a much lower temperature than the values from pure ceria (> 500 °C) or MnO_x (350–600 °C; Fig. 4a)[31–33]. The reduction temperature of the $Mn_{0.5}Ce_{0.5}O_x$@500 sample was also lower than those of the hybrid oxides with different Mn/Ce contents ($Mn_{0.1}Ce_{0.9}O_x$@500, $Mn_{0.7}Ce_{0.3}O_x$@500; Supplementary Fig. 18). This decreased reduction temperature can be attributed to the formation of a $Mn_{0.5}Ce_{0.5}O_x$ solid solution with maximum –Mn–O–Ce– bonds, which can greatly lower the oxygen vacancy formation energy and enhance the mobility of oxygen species from the bulk to the surface to a large degree[34]. With CuO as the standard material, the H$_2$ consumption of a $Mn_{0.5}Ce_{0.5}O_x$@500 sample can reach $4.22\,mmol\,g^{-1}$ and such a high value clearly suggests the large amount of the 'active' oxygen species. If the chemical composition of our catalyst is assumed to be $Mn^{4+}Ce^{4+}O_{4-x}$, the X value, based on the consumed H$_2$, is calculated to be 1.1, in turn evidencing the doping of Mn^{4+} into the ceria lattice. Thus the H$_2$-TPR peak can be assigned to the highly reducible manganese species with direct reduction from Mn^{4+} to Mn^{2+}, along with partial surface Ce^{4+} reduction[47]. It is interesting that the reduction peak starts at 75 °C, in other words, that the oxygen vacancy is forming at such a low temperature, allowing the possibility of low-temperature catalysis. To probe the reversibility of active oxygen at low temperature, multiple redoxes of the $Mn_{0.5}Ce_{0.5}O_x$@500 sample from 60 to 160 °C were carried out via a H$_2$ reduction–aerobic oxidation cycle (Fig. 4b). During three cycles, the H$_2$-TPR curves kept to the same trend and a similar amount of H$_2$ consumption. By combining the unique properties of the current solid solution (for example, abundant active oxygen species, redox activity at low temperature and good stability) and the characteristic features of mesoporous materials (large pore size and high surface area), $Mn_{0.5}Ce_{0.5}O_x$@500 contains most of the prerequisites for a noble metal-free heterogeneous catalyst to realize low-temperature selective oxidation of hydrocarbons by O$_2$.

Aerobic oxidation of cyclohexane by $Mn_yCe_{1-y}O_x$ Catalysts. Initial attempts to optimize the aerobic oxidation of cyclohexane were performed at 100 °C in the presence of different catalysts. A blank run without catalysts did not give any products in 4 h, suggesting that the auto-oxidation of cyclohexane by molecular oxygen cannot proceed under such a condition (Table 1, Entry 1). When catalysed by $Mn_{0.5}Ce_{0.5}O_x$@500, the oxidation of cyclohexane occurred at 100 °C with a moderate conversion (6.5%) and a remarkable selectivity (95%) for KA oil (Table 1, Entry 2). As a noble metal-free solid catalyst, $Mn_{0.5}Ce_{0.5}O_x$@500 indeed drives the aerobic oxidation of cyclohexane at a relatively low temperature. It should be emphasized that controlled oxidations of cyclohexane with CeO_2@500, MnO_x@500 or the physical mixture of CeO_2@500 and MnO_x@500 cannot proceed, confirming the synergistic action of manganese and cerium species in a $Mn_{0.5}Ce_{0.5}O_x$@500 solid solution (Table 1, Entries 3–5). The $Mn_{0.5}Ce_{0.5}O_x$@200 sample was also active for this process, which is reasonable since an initial crystalline structure has already formed at 200 °C (Table 1, Entry 6). The mixed oxides with various Mn/Ce atomic ratios (1:9, 2:8, 3:7 and 7:3) were also tested in the cyclohexane oxidation (Table 1, Entries 7–10). The optimal ratio was $\sim 1:1$, in accordance with the H$_2$-TPR results. The maximum –Mn–O–Ce– bonds throughout the matrix of the $Mn_{0.5}Ce_{0.5}O_x$@500 solid solution may be responsible for its high activity, because more oxygen vacancies can be expected at low temperature.

The reaction temperature had a strong effect on the oxidation of cyclohexane. The cyclohexane conversion increased as the

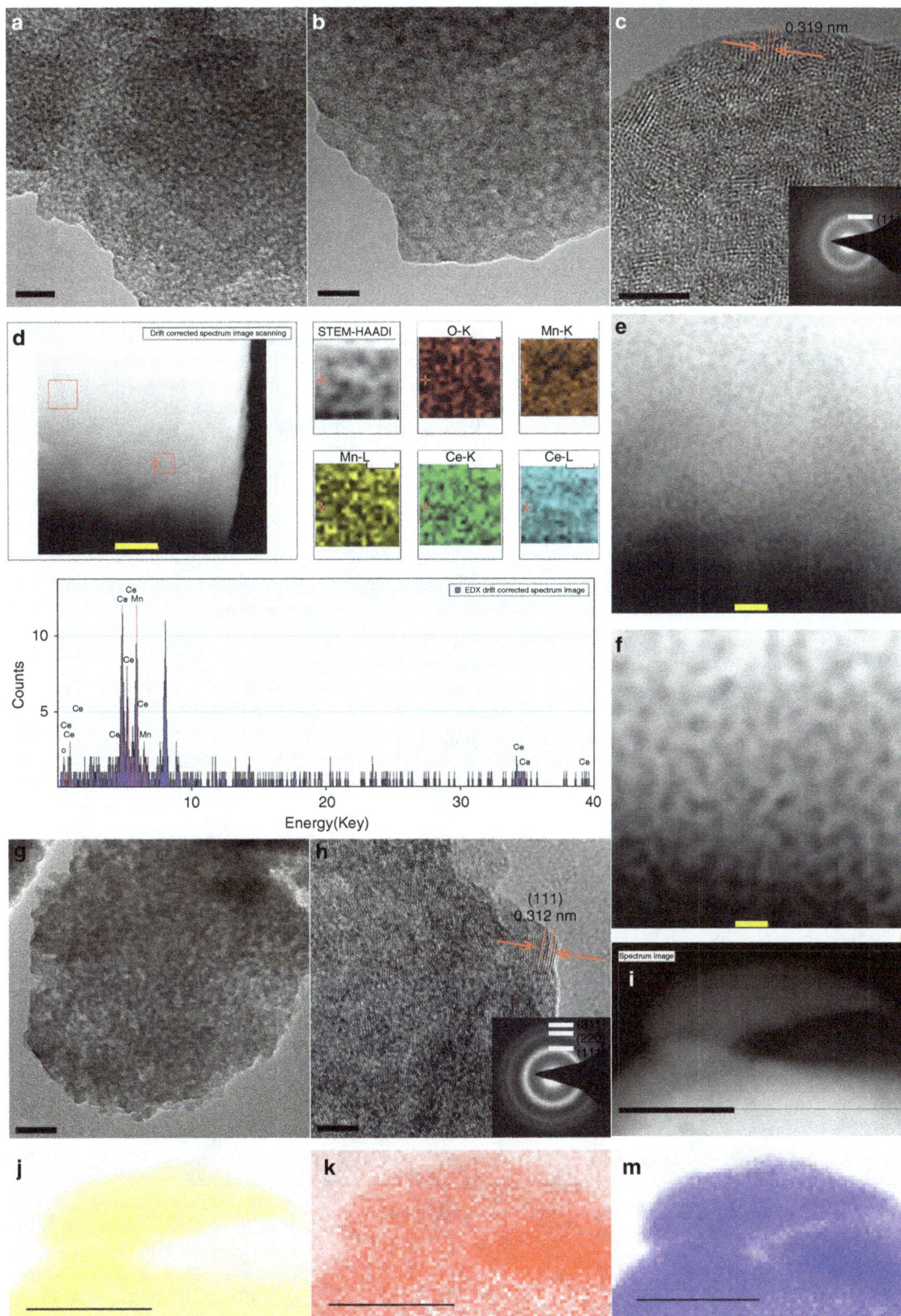

Figure 3 | Studies of the catalyst by electron microscopy. (**a–c**) TEM/HRTEM images of $Mn_{0.5}Ce_{0.5}O_x$@200 sample, the scale bar are 20, 10 and 5 nm, respectively. the inset in **c** is an electron microscopy pattern. (**d**) STEM-HAADF image of $Mn_{0.5}Ce_{0.5}O_x$@200, scale bar, 100 nm; the corresponding XEDS of the O-K, Mn-K, Mn-L, Ce-K, Ce-L signals and XEDS. (**e,f**) STEM-HAADF image of $Mn_{0.5}Ce_{0.5}O_x$@200 sample, scale bar, 20 nm and 10 nm, respectively. (**g,h**) TEM/HRTEM images of $Mn_{0.5}Ce_{0.5}O_x$@500 sample, scale bar, 20 and 5 nm, the inset in **h** is an electron microscopy pattern. (**i**) STEM-HAADF image of $Mn_{0.5}Ce_{0.5}O_x$@500 sample and the corresponding elemental mapping for Ce (**j**), Mn (**k**), O (**m**). Scale bar, 50 nm.

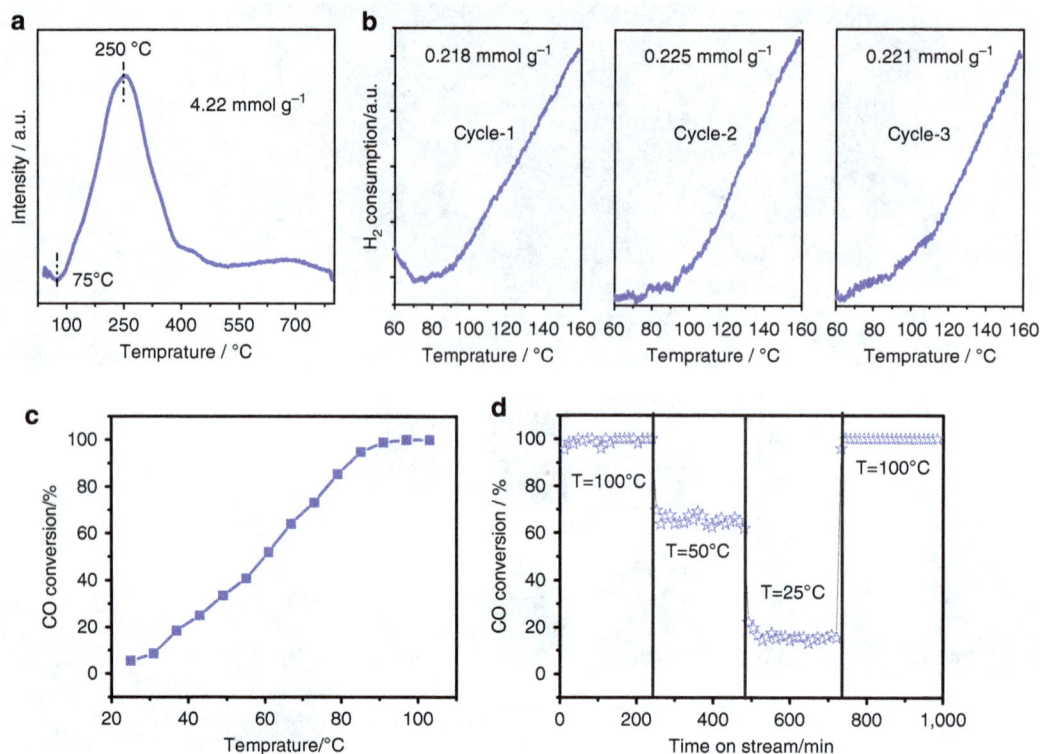

Figure 4 | Redox property of the catalyst. (**a**) H_2-TPR curve of $Mn_{0.5}Ce_{0.5}O_x$@500 catalyst; (**b**) H_2-TPR curve of $Mn_{0.5}Ce_{0.5}O_x$@500 catalyst during H_2 reduction-aerobic oxidation cycles; (**c**) Catalytic CO oxidation at different temperature over $Mn_{0.5}Ce_{0.5}O_x$@500 catalyst; (**d**) Stability of $Mn_{0.5}Ce_{0.5}O_x$@500 catalyst under CO oxidation and its catalytic performance during varied temperatures.

Table 1 | Selective oxidation of cyclohexane under different conditions*.

Entry	Catalyst	T (°C)	t (h)	Conv. (%)	Sel. (%)	K/A[†]
1	Blank	100	4 h	<0.1%	—	—
2	$Ce_{0.5}Mn_{0.5}O_x$@500	100	4 h	6.5%	95%	4.8
3	CeO_2@500	100	4 h	<0.1%	—	—
4	MnO_x@500	100	4 h	<0.1%	—	—
5	CeO_2@500 (50 wt%) MnO_x@500 (50 wt%)	100	4 h	<0.1%	—	—
6	$Ce_{0.5}Mn_{0.5}O_x$@200	100	4 h	5.1%	92%	4.2
7	$Ce_{0.1}Mn_{0.9}O_x$@500	100	4 h	0.4%	98%	7.5
8	$Ce_{0.2}Mn_{0.8}O_x$@500	100	4 h	2.3%	96%	5.8
9	$Ce_{0.3}Mn_{0.7}O_x$@500	100	4 h	2.6%	97%	6.1
10	$Ce_{0.7}Mn_{0.3}O_x$@500	100	4 h	4.8%	98%	5.2
11	$Ce_{0.5}Mn_{0.5}O_x$@500	80	4 h	1.0%	>99%	>99.0
12	$Ce_{0.5}Mn_{0.5}O_x$@500	120	4 h	10.5%	84%	3.5
13	$Ce_{0.5}Mn_{0.5}O_x$@500	150	4 h	18.8%	52%	5.4
14	$Ce_{0.5}Mn_{0.5}O_x$@500	100	8 h	13.5%	90%	3.3
15	$Ce_{0.5}Mn_{0.5}O_x$@500	100	12 h	17.7%	81%	3.6
16	$Ce_{0.5}Mn_{0.5}O_x$@500	100	16 h	21.8%	63%	6.5
17[‡]	$Ce_{0.5}Mn_{0.5}O_x$@500; in argon	100	4 h	<0.1%	—	—
18[§]	$Ce_{0.5}Mn_{0.5}O_x$@500; hydroquinone 50 mg	100	4 h	<0.1%	—	—

Conv., conversion; Sel., selectivity.
*Reaction conditions: cyclohexane 10 mmol (842 mg), catalyst 30 mg, CH_3CN 3 ml O_2 10 bar. Selectivity = [cyclohexanol + cyclohexanone]/[consumed cyclohexane] × 100; conversion = [consumed cyclohexane]/[initial cyclohexane] × 100, respectively.
†K/A = the molar ratio between cyclohexanone and cyclohexanol.
‡In Argon.
§With hydroquinone 50 mg as additive.

temperature increased from 80 to 150 °C; at the same time, a decreased selectivity for KA oil was observed (Table 1, Entries 11–13). It is interesting that oxidation can proceed at a temperature as low as 80 °C, which is in good agreement with the observation in H_2-TPR that the active oxygen species is available above 75 °C. With the development of processes for low-temperature cyclohexane oxidation in mind, we focused on

catalytic oxidation at 100 °C. The optimization of reaction time suggested that the reaction time of 12 h seemed to be a suitable time, and a 17.7% cyclohexane conversion with 81% selectivity for KA oil was obtained (Table 1, Entries 14–16). To probe the reaction pathway, two controlled runs were then performed. When the catalytic oxidation was carried out in argon, no detectable products were observed, giving evidence that molecular

Table 2 | Selective oxidation of different hydrocarbons by a Mn/Ce catalyst[*].

Entry	Substrate	T (°C)	t (h)	Conv. (%)	Product (sel. %)	
1		110	4	53.2	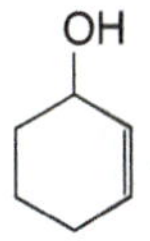 26 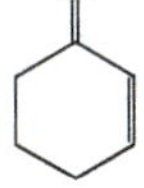	52
2	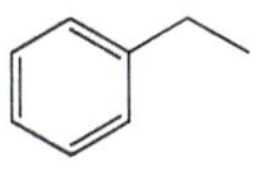	120	6	20.3	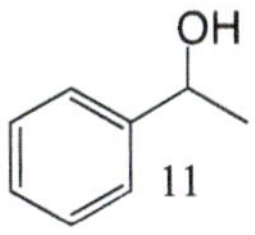 11 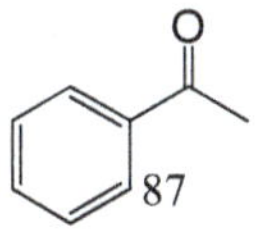	87
3	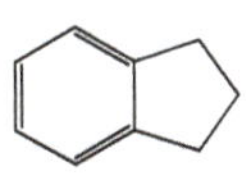	110	4	75.4	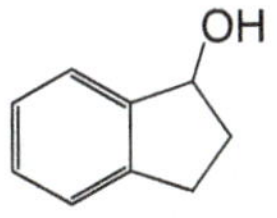 17 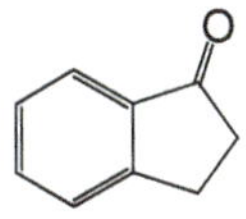	65
4	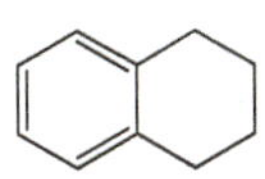	120	4	44.8	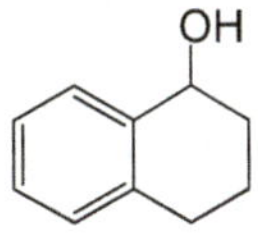 26 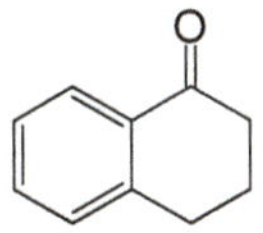	58
5	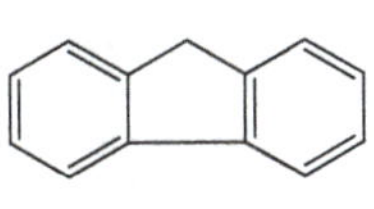	110	4	86.3	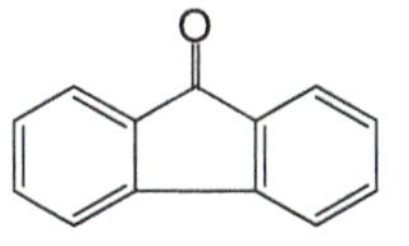 95	
6	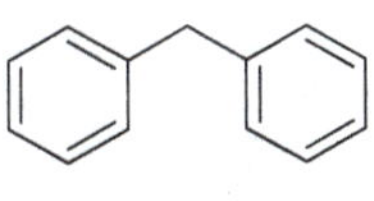	120	6	36.2	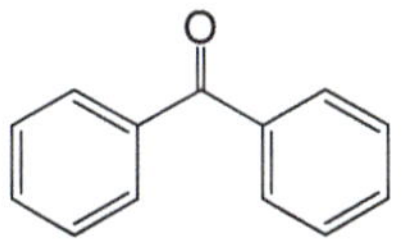 92	

Conv., conversion; Sel., selectivity.
*Reaction conditions: substrate 1 mmol, anisole 1 mmol (internal standard), $Ce_{0.5}Mn_{0.5}O_x$@500 catalyst 30 mg, O_2 10 bar, CH_3CN 5 ml.

oxygen is the principal oxygen donor in the system (Table 1, Entry 17). In addition, the catalytic oxidation would be quenched, if hydroquinone, a free-radical scavenger, was added into the reaction system, which implied that the oxidation of cyclohexane may proceed through a radical chain mechanism (Table 1, Entry 18). The stability of the $Mn_{0.5}Ce_{0.5}O_x$@500 catalyst was then investigated by cyclohexane oxidation for 4 h. After each run, the catalyst was recovered by centrifugation, and then carefully transferred into a reactor by the reaction solvent. The $Mn_{0.5}Ce_{0.5}O_x$@500 worked well in at least 20 runs without significant activity loss, suggesting that the oxidation should run in a heterogeneous manner and it is a prerequisite for practical applications (Supplementary Figs 19–20). A possible reaction mechanism was then purposed, based on the results above, *in situ* diffuse reflectance infrared spectroscopy (DRIFTS) and *in situ* Raman spectra (Supplementary Fig. 21, Supplementary Note 1).

Aerobic oxidation of hydrocarbons and CO. To probe the potential of this $Mn_{0.5}Ce_{0.5}O_x$@500 solid solution, various hydrocarbons with sp^3 C–H bonds were oxygenated at 110–120 °C (Table 2). Cyclohexene was oxidized with a moderate conversion to the mixture of 2-cyclohexen-1-one and 2-cyclohexen-1-ol (Table 2, Entry 1). The oxidation of ethylbenzene proceeded with high selectivity to acetophenone, although the ethylbenzene conversion was somewhat low (Table 2, Entry 2). Catalysed by $Mn_{0.5}Ce_{0.5}O_x$@500 catalyst, the indane oxidation afforded a

conversion of 75.4%, with 1-indanol and 1-indanone as the main products (Table 2, Entry 3). The catalyst also worked well in the oxidation of tetralin, a key step in the commercial production of α-naphthol (Table 2, Entry 4)[48]. Fluorene and diphenylmethane with a large molecular size could be transformed into fluorenone and diphenylmethanone, with high selectivity (Table 2, Entries 5–6). Therefore, it is probably fair to say that the $Mn_{0.5}Ce_{0.5}O_x$@500 solid solution is a general catalyst for aerobic oxidation of allylic- or benzyl sp^3 C–H bonds at relatively low temperature.

Actually, the same target for low temperature oxidation is also pursued in catalytic combustion, such as CO oxidation[32,36,38]. Encouraged by the interesting activity of $Mn_{0.5}Ce_{0.5}O_x$@500 in O_2 activation, we undertook a study of CO oxidation reaction over a $Mn_{0.5}Ce_{0.5}O_x$@500 catalyst. The profile for CO conversion as a function of reaction temperature is presented in Fig. 4c. The $Mn_{0.5}Ce_{0.5}O_x$@500 catalyst enables the 100% CO conversion at around 90 °C. It should be highlighted that the T_{50} (Temperature at which the 50% CO conversion is achieved) by $Mn_{0.5}Ce_{0.5}O_x$@500 catalyst (60 °C) is lower than MnO_x–CeO_2 catalysts by other methods (co-precipitation method: 127 °C, surfactant-assisted method: 95 °C, hydrothermal method: 105 °C, citrate sol-gel method: 160 °C, Supplementary Table 3). The high activity of the $Mn_{0.5}Ce_{0.5}O_x$@500 catalyst is attributed to the abundant superoxide species formed on the surface of the solid solution. The stability of $Mn_{0.5}Ce_{0.5}O_x$@500 catalyst was also investigated, and it was found that 100% CO conversion at 100 °C can be preserved for 240 min (Fig. 4d). Moreover, the catalytic activity of the $Mn_{0.5}Ce_{0.5}O_x$@500 sample was stable at different temperatures, and showed a rapid response to the temperature change. Notably, the $Mn_{0.5}Ce_{0.5}O_x$@500 catalyst is active for CO oxidation even at room temperature ($\sim$19% CO conversion).

Discussion

In summary, we have shown the successful construction of mesoporous $MnCeO_x$ solid solutions via a simple, effective and sustainable self-assembly strategy, which has at the same time been recognized in the fabrication of other hybrid metal oxides with well-defined mesopores. Experimental results reported herein, illustrate that the aerobic oxidation of cyclohexane to KA oil by $Mn_{0.5}Ce_{0.5}O_x$@500 catalyst can proceed above 80 °C without any noble metal catalysts or sacrificial additives, and under optimized reaction conditions (100 °C), 17.7% cyclohexane conversion with 81% selectivity for KA oil was obtained. This finding could reinvigorate research into such a process for commercial exploitation, and thus make cyclohexane oxidation by a heterogeneous catalyst viable. In addition, selective oxidation of allylic or benzyl C-H bonds in various hydrocarbons were realized by the $Mn_{0.5}Ce_{0.5}O_x$@500 catalyst using molecular oxygen as an oxidant. The versatility of $Mn_{0.5}Ce_{0.5}O_x$@500 catalyst was also witnessed in CO oxidation with outstanding activity at a relatively low temperature (100% conversion at 90 °C).

Actually, the exceptional activity of the as-made catalyst can be the result of forming a $Mn_{0.5}Ce_{0.5}O_x$ solid solution—which has been confirmed by a structural model, an XRD pattern, XPS analysis, TEM images, STEM–XEDX mapping analysis and an H_2-TPR study—with several unique characteristics: (1) A high proportion (44.1%) of active oxygen species on the surface to promote O–O/C–H bond activation; (2) the introduction of 50 mol% Mn^{4+} ions into ceria matrix for the formation of maximum solid solution phases that can lower the energy for oxygen vacancy formation and benefit the rapid migration of oxygen vacancies from the bulk to the surface, thus continuing the activation of gas oxygen molecules; (3) a mesoporous structure for fast mass transfer/diffusion, and rich porosity to expose any more active sites ready for interaction with cyclohexane/O_2. We expect that the $Mn_{0.5}Ce_{0.5}O_x$ solid solution will provide a mild strategy for cyclohexane oxidation, and the manner of self-assembly with ionic liquids will inspire more designs of mesoporous oxide solid solutions for specific tasks in the near future.

Methods

Synthesis of $Mn_{0.5}Ce_{0.5}O_x$ solid solution. In a typical synthesis of mesoporous $Mn_{0.5}Ce_{0.5}O_x$ solid solution oxides, 6.16 g of cerium (IV) methoxyethoxide (18–20% in methoxyethoxide, Gesta), 0.63 g $Mn(OOCCH_3)_2 \cdot 6H_2O$ (99%, Aldrich) and 1.0 g of ionic liquid (BmimTf$_2$N) were dissolved in 5.0 ml of ethanol. The solution was stirred at room temperature for 2 h until $Mn(OOCCH_3)_2 \cdot 6H_2O$ was completely dissolved. Subsequently, ethanol (5.0 ml) was added slowly with stirring. The mixed solution was gelled in an open petridish at 50 °C for 24 h and aged at 200 °C for 2 h, and a solid film was obtained. The ionic liquid was extracted by refluxing the sample with ethanol in a Soxhlet extractor for 24 h. The as-made sample ($Mn_{0.5}Ce_{0.5}O_x$@200) was thermally treated at 500 °C for 2 h with the heating rate of 1 K min^{-1} in air, and the final sample denoted as $Mn_{0.5}Ce_{0.5}O_x$@500. Other metal oxides were prepared by the same process except with different metal precursors. The materials were characterized by N_2 adsorption (TriStar, Micromeritics) at 77 K, powder XRD (Panalytical Empyrean diffractometer with Cu Kα radiation $k = 1.5418$ A° operating at 45 kV and 40 mA), thermogravimetric analysis (TGA 2950, TA Instruments), Fourier-transform infrared spectrum (PerkinElmer Frontier FTIR spectrometer) and H_2-TPR (Auto chem II, Micromeritics).

Typical procedure for the catalytic oxidation of CO. Catalytic CO oxidation was carried out in a fixed-bed reactor (U-type quartz tube) with inner diameter of 4 mm at atmospheric pressure. A 30 mg catalyst supported by quartz wool was loaded in the reactor. The feed gas of 1% CO balanced with dry air passed though the catalyst

bed at a flow rate of 10 ml min^{-1}, corresponding to a gas hourly space velocity of 20,000 ml (h gcat)$^{-1}$.

Typical procedure for the catalytic oxidation of cyclohexane. Catalytic oxidations of cyclohexane under pressured O_2 were carried out in a Teflon-lined stainless steel batch reactor (PARR Instrument, USA). Typically, cyclohexane (10 mmol; calculated by weight), CH_3CN (3 ml) and catalysts used as described in the manuscript were loaded into the reactor (total volume: 100 ml). The reactor was sealed, and then purged with O_2 to replace the air for three times. The O_2 pressure was increased to 1 MPa, and then the reactor was heated to the desired temperature in 15 min. Then, the reaction was carried out for the desired time with stirring (stirring rate: 1,500 r.p.m.). After reaction, the reactor was placed in ice water to quench the reaction, and the products were analysed by gas chromatography (GC) with internal standard (2-butanone). The structure of products and by-products was identified using Perkin Elmer GC–MS (Clarus 680-Clarus SQ 8C) spectrometer by comparing retention times and fragmentation patterns with authentic samples.

Typical procedure for the catalytic oxidation of other hydrocarbons. In a typical oxidation, 1 mmol substrate, 1 mmol anisole (internal standard), 5 ml CH_3CN and 30 mg $Mn_{0.5}Ce_{0.5}O_x$@500 catalyst were added into a Teflon-lined stainless steel batch reactor. The reactor was sealed and purged with O_2 to replace the air for three times. After increasing the O_2 pressure to 1 MPa, the reactor was heated to the desired temperature in 20 min. Then, the reaction was carried out for the desired time with magnetic stirring (stirring rate: 1,500 r.p.m.). After the reaction, the reactor was placed in ice water to quench the reaction, and the products were analysed by GC and GC–MS.

Method for *in situ* DRIFTS. *In situ* DRIFTS measurement was performed on a Nicolet Nexus 670 spectrometer equipped with a MCT detector cooled by liquid nitrogen and an *in situ* chamber (HC-900, Pike Technologies) which allows the sample heated up to 900 °C. The exiting stream was analysed by an online quadrupole mass spectrometer (OmniStar GSD-301 O_2, Pfeffer Vacuum). Before measurement, the $Mn_{0.5}Ce_{0.5}O_x$@500 powder (100 mg) was treated *in situ* at 500 °C in 20% O_2/He (30 min) with a flow rate of 25 ml min^{-1} to eliminate water traces. After cooling to room temperature in a He flow (20 ml min^{-1}), the background spectrum was collected for spectral correction, and background peaks were also collected at 100 and 150 °C, respectively. Then, cyclohexane stream (by bubbling with He 20 ml min^{-1}) was introduced to the *in situ* chamber for adsorption and reaction.

Method for raman spectroscopy. The procedure for Raman spectra collection: Raman spectra were excited with a 532 nm laser (LAS-NY532/50) and collected with Horiba JobinYvon HR800 (800mm optical length), with a diffraction grating of 600 grooves per mm, the scattered light was detected with a charge-coupled device, cooled to 203 K for thermal-noise reduction. The Raman spectra of samples were collected from 25 to 150 °C in the range of 100–4,000 cm^{-1} with two accumulations for each spectrum.

References

1. ten Brink, G.-J., Arends, I. W. C. E. & Sheldon, R. A. Green, catalytic oxidation of alcohols in water. *Science* **287**, 1636–1639 (2000).
2. Enache, D. *et al.* Solvent-free oxidation of primary alcohols to aldehydes using Au-Pd TiO2 catalysts. *Science* **311**, 362–365 (2006).
3. Recupero, F. & Punta, C. Free radical functionalization of organic compounds catalyzed by N-hydroxyphthalimide. *Chem. Rev.* **107**, 3800–3842 (2007).
4. Corma, A. *et al.* Exceptional oxidation activity with size-controlled supported gold clusters of low atomicity. *Nat. Chem.* **5**, 775–781 (2013).
5. Ma, C. *et al.* Mesoporous Co3O4 and Au/Co3O4 catalysts for low-temperature oxidation of trace ethylene. *J. Am. Chem. Soc.* **132**, 2608–2613 (2010).
6. Kamata, K., Yonehara, K., Nakagawa, Y., Uehara, K. & Mizuno, N. Efficient stereo- and regioselective hydroxylation of alkanes catalysed by a bulky polyoxometalate. *Nat. Chem.* **2**, 478–483 (2010).
7. Punniyamurthy, T., Velusamy, S. & Iqbal, J. Recent advances in transition metal catalyzed oxidation of organic substrates with molecular oxygen. *Chem. Rev.* **105**, 2329–2363 (2005).
8. Shannon, S. S. Palladium-catalyzed oxidation of organic chemicals with O_2. *Science* **309**, 1824–1826 (2005).
9. Milo, A., Neel, A. J., Toste, F. D. & Sigman, M. S. A data-driven approach to mechanistic elucidation in chiral anion catalysis. *Science* **347**, 737–743 (2015).
10. Kesavan, L. *et al.* Solvent-free oxidation of primary carbon-hydrogen bonds in toluene using Au-Pd alloy nanoparticles. *Science* **331**, 195–199 (2011).
11. Liu, Y.-J. *et al.* Overcoming the limitations of directed C-H functionalizations of heterocycles". *Nature* **515**, 389–393 (2014).
12. Chen, G. *et al.* Interfacial effects in iron-nickel hydroxide–platinum nanoparticles enhance catalytic oxidation. *Science* **344**, 495–499 (2014).

13. Marimuthu, A., Zhang, J. & Linic, S. Tuning selectivity in propylene epoxidation by plasmon mediated photo-switching of cu oxidation state. *Science* **339**, 1590–1593 (2013).

14. White, M. C. Adding aliphatic C–H bond oxidations to synthesis. *Science* **335**, 807–809 (2012).

15. Yuan, C. et al. Metal-free oxidation of aromatic carbon–hydrogen bonds through a reverse-rebound mechanism. *Nature* **499**, 192–196 (2013).

16. Ghavtadze, N., Melkonyan, F. S., Gulevich, A. V., Huang, C. & Gevorgyan, V. Conversion of 1-alkenes into 1,4-diols through an auxiliary-mediated formal homoallylic C–H oxidation. *Nat. Chem.* **6**, 122–125 (2014).

17. Fu, Q. et al. Interface-confined ferrous centers for catalytic oxidation. *Science* **328**, 807–809 (2012).

18. Zope, B. N., Hibbitts, D. D., Neurock, M. & Davis, R. J. Reactivity of the gold/water interface during selective oxidation catalysis. *Science* **330**, 74–78 (2010).

19. Frei, H. Selective hydrocarbon oxidation in zeolites. *Science* **313**, 309–310 (2006).

20. Das, S., Incarvito, C. D., Crabtree, R. H. & Brudvig, G. W. Molecular recognition in the selective oxygenation of saturated C–H bonds by a dimanganese catalyst. *Science* **312**, 1941–1943 (2006).

21. Zhou, W. et al. Highly selective liquid-phase oxidation of cyclohexane to KA oil over Ti-MWW catalyst: evidence of formation of oxyl radicals. *ACS Catal.* **4**, 53–62 (2014).

22. Ishii, Y., Iwahama, T., Sakaguchi, S., Nakayama, K. & Nishiyama, Y. Alkane oxidation with molecular oxygen using a new efficient catalytic system: *N*-hydroxyphthalimide (NHPI) combined with Co(acac)$_n$ (n = 2 or 3). *J. Org. Chem.* **61**, 4520–4526 (1996).

23. Guo, C. C. et al. Effective catalysis of simple metalloporphyrins for cyclohexane oxidation with air in the absence of additives and solvents. *Appl. Catal. A Gen.* **246**, 303–309 (2003).

24. Dugal, M., Sankar, G., Raja, R. & Thomas, J. M. Designing a heterogeneous catalyst for the production of adipic acid by aerial oxidation of cyclohexane. *Angew. Chem. Int. Ed.* **39**, 2310–2313 (2000).

25. Liu, Y., Hironori Tsunoyama, H., Akita, T., Xie, S. & Tsukuda, T. Aerobic Oxidation of cyclohexane catalyzed by size-controlled Au clusters on hydroxyapatite: size effect in the sub-2 nm regime. *ACS Catal.* **1**, 2–6 (2011).

26. Hughes, M. D. et al. Tunable gold catalysts for selective hydrocarbon oxidation under mild conditions. *Nature* **437**, 1132–1135 (2005).

27. Turner, M. et al. Selective oxidation with dioxygen by gold nanoparticle catalysts derived from 55-atom clusters. *Nature* **454**, 981–983 (2008).

28. Li, X.-H., Chen, J.-S., Wang, X., Sun, J. & Antonietti, M. Metal-free activation of dioxygen by graphene/g-C3N4 nanocomposites: functional dyads for selective oxidation of saturated hydrocarbons. *J. Am. Chem. Soc.* **133**, 8074–8077 (2011).

29. Yu, H. et al. Selective catalysis of the aerobic oxidation of cyclohexane in the liquid phase by carbon nanotubes. *Angew. Chem. Int. Ed.* **50**, 3978–3982 (2011).

30. Chen, Z. et al. Recent advances in manganese oxide nanocrystals: fabrication, characterization, and microstructure. *Chem. Rev.* **112**, 3833–3855 (2012).

31. Qi, G., Yang, R. & Chang, R. MnOx-CeO$_2$ mixed oxides prepared by co-precipitation for selective catalytic reduction of NO with NH$_3$ at low temperatures. *Appl. Catal. B* **51**, 93–106 (2004).

32. Venkataswamy, P., Rao, K. N., Jampaiah, D. & Reddy, B. M. Nanostructured manganese doped ceria solid solutions for CO oxidation at lower temperatures. *Appl. Catal. B* **162**, 122–132 (2015).

33. Delimaris, D. & Ioannides, T. VOC oxidation over MnOx-CeO$_2$ catalysts prepared by a combustion method. *Appl. Catal. B* **84**, 303–312 (2008).

34. Cen, W., Liu, Y., Wu, Z., Wang, H. & Weng, X. A theoretic insight into the catalytic activity promotion of CeO2 surfaces by Mn doping. *Phys. Chem. Chem. Phys.* **14**, 5769–5777 (2012).

35. Wang, X., Kang, Q. & Li, D. Catalytic combustion of chlorobenzene over MnOx-CeO2 mixed oxide catalysts. *Appl. Catal. B* **86**, 166–175 (2009).

36. Zou, Z., Meng, M. & Zha, Y. Surfactant-assisted synthesis, characterization, and catalytic oxidation mechanisms of the mesoporous MnOx-CeO$_2$ and Pd/MnOx-CeO$_2$ catalysts used for CO and C$_3$H$_8$ oxidation. *J. Phys. Chem. C* **114**, 468–477 (2010).

37. Wang, Z. et al. Catalytic removal of benzene over CeO$_2$-MnOx composite oxides prepared by hydrothermal method. *Appl. Catal. B* **138-139**, 253–259 (2013).

38. Elias, J. S., Risch, M., Giordano, L., Mansour, A. N. & Yang, S.-H. Structure, bonding, and catalytic activity of monodisperse, transition-metal-substituted CeO2 nanoparticles. *J. Am. Chem. Soc.* **136**, 17193–17200 (2014).

39. Wan, Y., Yang, H. & Zhao, D. Y. "Host-Guest" chemistry in the synthesis of ordered nonsiliceous mesoporous materials. *Acc. Chem. Rev.* **39**, 423–432 (2006).

40. Yang, P. D. et al. Generalized syntheses of large-pore mesoporous metal oxides with semicrystalline frameworks. *Nature* **396**, 152–155 (1998).

41. Wang, Y. T. & Voth, G. A. Unique spatial heterogeneity in ionic liquids. *J. Am. Chem. Soc.* **127**, 12192–12193 (2005).

42. Wagner, C. D., Riggs, W. M., Davis, L. E., Moulder, J. F. & Muilenberg, G. E. *Handbook of X-Ray Photoelectron Spectroscopy* (Perkin-Elmer Corp., 1979).

43. Poyraz, A. S., Kuo, C.-H., Biswas, S., King'ondu, C. & Suib, S. L. A general approach to crystalline and monomodal pore size mesoporous materials. *Nat. Commun.* **4**, 2952 (2013).

44. Zhang, P. F. et al. Updating biomass into funtional carbon materials in ionothermal manner. *ACS Appl. Mater. Interfaces* **6**, 12515–12522 (2014).

45. Wang, Y., Li, H. R. & Han, S. J. The chemical nature of the + C-H•••X- (X = Cl or Br) interaction in imidazolium halide ionic liquids. *J. Chem. Phys.* **124**, 044504 (2006).

46. Mele, A., Tran, C. D. & Lacerda, S. H. D. The structure of a room-temperature ionic liquid with and without trace amounts of water: the role of C-H•••O and C-H•••F interactions in 1-n-butyl-3-methylimidazolium tetrafluoroborate. *Angew. Chem. Int. Ed.* **42**, 4364–4366 (2003).

47. Lu, H.-F., Zhou, Y., Han, W.-F., Huang, H.-F. & Chen, Y.-F. High thermal stability of ceria-based mixed oxide catalysts supported on ZrO$_2$ for toluene combustion. *Catal. Sci. Technol.* **3**, 1480–1484 (2013).

48. Llabrés i Xamena, F. X., Casanova, O., Galiasso Tailleur, R., Garcia, H. & Corma, A. Metal organic frameworks (MOFs) as catalysts: a combination of Cu2 + and Co2 + MOFs as an efficient catalyst for tetralin oxidation. *J. Catal.* **255**, 220–227 (2008).

Acknowledgements

P.F.Z., L.Z., Z.L.W. and S.D. were supported by the U.S. Department of Energy, Office of Science, Basic Energy Sciences, Chemical Sciences, Geosciences, and Biosciences Division. The DRIFTS study was conducted at the Center for Nanophase Materials Sciences, which is a DOE Office of Science User Facility. H.L.S. was supported by the Department of Energy, Office of Science, Basic Energy Sciences, Materials Sciences and Engineering Division. H.F.L., Y.Z., Q.L.Z. and Y.F.C. were supported by the Natural Science Foundation of China (NO. 21107096, 21506194), the Natural Science Foundation of Zhejiang province (No. LY14E080008) and the commission of Science and Technology of Zhejiang province (No. 2013C03021).

Author contributions

P.Z., H.L. and S.D. conceived and designed the experiments. P.Z. and H.L. performed all the experiments and analysed all the data. L.Z. and Z.W. carried out the *in situ* diffuse reflectance infrared spectroscopy. Y.Z. and Q.Z. took part in the XPS, HRTEM and Raman tests. S.Y. performed the STEM mapping for the Mn$_{0.5}$Ce$_{0.5}$O$_x$@500 sample. H.S. completed the DFT calculation. Y.C. and S.D. discussed the results and commented on the manuscript. P.Z., H.L. and S.D. co-wrote the paper.

Additional information

11

Bipyramid-templated synthesis of monodisperse anisotropic gold nanocrystals

Jung-Hoon Lee[1,*], Kyle J. Gibson[1,*], Gang Chen[1] & Yossi Weizmann[1]

Much of the interest in noble metal nanoparticles is due to their plasmonic resonance responses and local field enhancement, both of which can be tuned through the size and shape of the particles. However, both properties suffer from the loss of monodispersity that is frequently associated with various morphologies of nanoparticles. Here we show a method to generate diverse and monodisperse anisotropic gold nanoparticle shapes with various tip geometries as well as highly tunable size augmentations through either oxidative etching or seed-mediated growth of purified, monodisperse gold bipyramids. The conditions employed in the etching and growth processes also offer valuable insights into the growth mechanism difficult to realize with other gold nanostructures. The high-index facets and more complicated structure of the bipyramid lead to a wider variety of intriguing regrowth structures than in previously studied nanoparticles. Our results introduce a class of gold bipyramid-based nanoparticles with interesting and potentially useful features to the toolbox of gold nanoparticles.

[1] Department of Chemistry, The University of Chicago, 929 East 57th Street, Chicago, Illinois 60637, USA. * These authors contributed equally to this work. Correspondence and requests for materials should be addressed to Y.W. (email: yweizmann@uchicago.edu).

Noble metal nanoparticles have become an integral part of the emerging field of nanotechnology, with a wide variety of potential applications, including surface-enhanced Raman spectroscopy (SERS), drug delivery and therapeutics, catalysis and non-linear optics[1-5]. Due to the structure- and size-dependent character of localized surface plasmon resonance and local field enhancement[6-8], precise control over the synthesis of nanostructures allows for a variety of programmable designs and highly tunable optical properties of the metallic metals. In addition, the ability to attain monodisperse colloidal nanoparticles in high yield is a critical step in the widespread use of these material. Gold bipyramids have shown remarkable size and shape monodispersity[9]. By theoretical calculation, stronger local field enhancement is expected in bipyramids than in nanorods or other shapes owing to the sharp tips[10]. However, the direct synthesis only yields roughly 30% bipyramids, with the other shape impurities being nanorod ($\sim$10%) and pseudo-spherical particles ($\sim$60%)[9]. The yield has been improved slightly using surfactant (cetyltributylammonium bromide) with larger headgroups than cetyltrimethylammonium bromide (CTAB)[11], but pure gold bipyramids are yet unrealized using synthetic approaches alone. Recently, purification through depletion-induced flocculation has been shown to successfully separate gold nanorods from spherical nanoparticles[12]. Herein, we demonstrate purification of a range of gold bipyramid sizes through depletion flocculation using benzyldimethyl-hexadecylammonium chloride (BDAC) as the surfactant, and the purified product can further be used as a seed to craft other monodisperse nanoparticles (Fig. 1).

Results

Purifying and enlarging the bipyramids. Gold bipyramids were synthesized according to the method by Liu and Guyot-Sionnest[9] using seed-mediated growth, and subsequently purified by depletion flocculation. BDAC was chosen for the purification due to the significantly higher micelle concentration than CTAB at the same concentrations (roughly 2.6 times more, see Supplementary Note 1). According to the theoretical model proposed by Park, the high micelle concentration induces flocculation at much lower surfactant concentrations, helping to avoid certain issues that can generally arise at high surfactant

concentrations such as high solution viscosity, solubility issues and the unpredictable transitions from spherical micelles to rod-like or worm-like micelles[12]. Figure 2a shows representative ultraviolet–visible (ultraviolet–vis) spectra and transmission electron microscopy (TEM) image for highly purified gold bipyramids over 90% (see also Fig. 2c for spectra 1–5 and Supplementary Fig. 1a for other sizes). The purification utilizes depletion attraction forces to selectively flocculate nanopaticles with high facial surface area, in this case the bipyramids. The strength of the attractive force is proportional to the volume of pure water generated during the approach of the two nanoparticles, which is likewise proportional to the possible contact area of the nanoparticles. Therefore, particles with large possible contact areas, such as the bipyramids, will selectively flocculate, while those with low possible contact areas, such as the spherical impurities, will remain in the supernatant (1: as-synthesized bipyramids, 2: supernatant, 3: purified bipyramids, see also Supplementary Fig. 1 and Methods for detailed procedure and TEM images). Bipyramids over 100 nm become increasingly difficult to separate from pseudo-spherical impurities as the facial surface area of both bipyramids and pseudo-spherical impurities are increased, resulting in undesirable co-flocculation (Supplementary Fig. 1c).

On obtaining pure, monodisperse gold bipyramids, a variety of other gold nanostructures with novel shapes or changes in size were synthesized from the bipyramids in a process similar to seed-mediated growth. Because the 'seed' used in this case is monodisperse, the product obtained through these transformations is also monodisperse in both shape and size, with the polydispersities ranging from only 2 to 5%. Using the purified bipyramid as a seed, the limitations to size of both the synthesis and separation can be overcome, allowing an increase in particle size by over 100 nm with high purity. TEM images in Fig. 2b and ultraviolet–Vis spectra 6–10 in Fig. 2c show the range of sizes of the regrown bipyramids obtained by adding either a different amount of bipyramid seeds or adjusting the concentration of reactants in growth solution (see Supplementary Tables 1 and 2 for synthetic conditions and detailed size measurements). The full width at half maximum of the longitudinal surface plasmon resonance (LSPR) peak is between 58 and 153 nm and these values, which compare favourably to that of nanorods ($\sim$100–200 nm or 6.2–12.4 eV)[13,14], also confirm the monodispersity of the bipyramids. It has also been observed that the aspect ratio can be altered slightly during the regrowth by either changing the pH or the amount of AgNO$_3$ in the growth solution (Supplementary Fig. 2b).

Bipyramid dumbbells. In addition to enlarging the bipyramids, regrowth of bipyramids by changing the synthetic conditions introduces several new structures to be added to the toolbox of gold nanoparticles, but also offers powerful insights into the growth mechanism yet unrealized with nanostructures of simpler shape. Regrowth of gold nanospheres and gold nanorods is well studied, and the transformation of nanorods to dumbbells has revealed much about the intricacies of nanoparticle growth conditions. However, both the nanosphere and nanorod fail to offer the complex crystal facets and twinned shape that the bipyramid provides. We first compared the regrowth of bipyramids using different surfactants (CTAB, cetyltrimethylammonium chloride (CTAC) and BDAC) as well as their mixtures. For standard growth solution with individual surfactants, size augmentation was dominant over any structural changes. Size augmentation of bipyramids with CTAB provided the best shape uniformity and surface smoothness, compared with both BDAC and CTAC (Fig. 3a, condition 1, and Supplementary Fig. 3). These results can

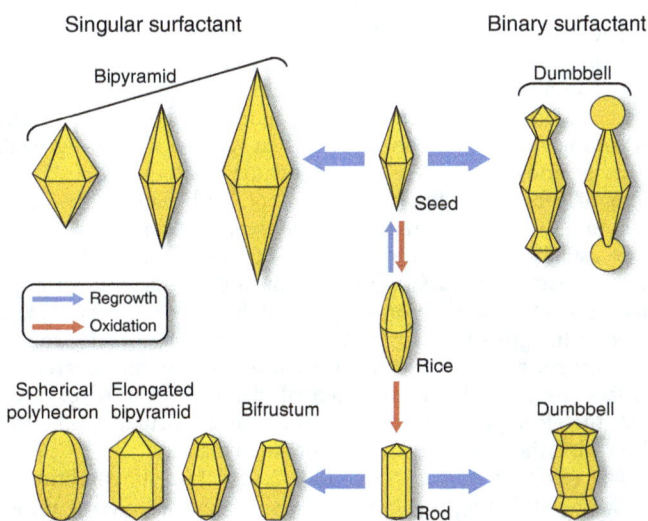

Figure 1 | Schematic representation of the shape control of the bipyramids. The various pathways to generate unique structures originating from the bipyramid are shown.

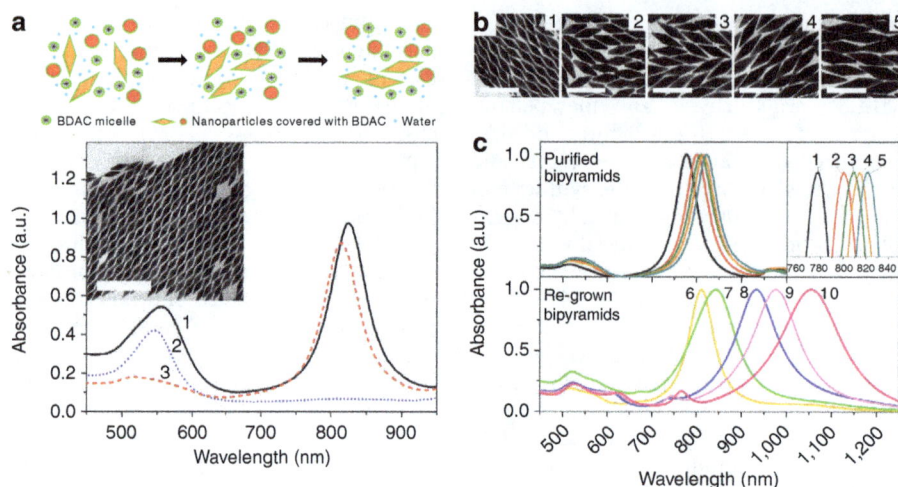

Figure 2 | Various sizes of bipyramids generated from the purification and regrowth. (**a**) Schematic illustration, ultraviolet–vis–near-infrared spectrum and TEM image of bipyramids resultant from the purification by depletion flocculation with BDAC. (**b**) The various sizes of bipyramids are regrown from the original bipyramid seed in **a**. Bipyramids shown in **b** (1–5) correspond to spectra 6–10 in **c**. (**c**) The normalized extinction spectra of purified (1–5) and regrown bipyramids (6–10) are shown. The inset shows an enlarged spectrum of the LSPR for the purified bipyramids. Full width at half maximum of LSPR peaks were measured as 58, 60, 63, 60, 59, 73, 113, 121, 126 and 154 nm or 21.4, 20.7, 19.7, 20.7, 21.0, 17.0, 11.0, 10.3, 9.8 and 8.1 eV for 1–10, respectively.

be explained by considering the different binding affinities of CTAB and BDAC, that is, $Br^- > Cl^-$, and also the degree of underpotential deposition in the presence of Ag^+ ions onto the gold surface[9,15–17]. The regrowth with CTAC resulted in random shape growth and aggregation because centrifugation with CTAC to remove remaining CTAB reduces particle stability, which is consistent with reported result for nanorod growth using CTAC (Supplementary Fig. 3)[18,19]. According to reports of nanorod growth using cationic headgroup with bromide anion as a counterpart, larger headgroups lead to nanorods of higher aspect ratio and slowed growth rate, implying a higher binding affinity and resulting in a more stable bilayer on the particle surface[11,15,20,21]. The results from our observation and literature suggest that the affinity of surfactants in the presence of Ag^+ ions could be extended to CTAB > BDAC > CTAC.

Systematic studies were conducted with the three possible binary surfactant combinations, that is, CTAB/CTAC, CTAB/BDAC and BDAC/CTAC (Supplementary Fig. 4). Despite changing the molar ratios, overgrowth at the tips was not observed and only somewhat observed for BDAC/CTAB or CTAC/BDAC, respectively (Supplementary Fig. 4a). Intriguingly, the molar ratio of surfactants in binary systems showed a huge influence on the regrowth of bipyramids, especially at the tip region. Previously, mixed surfactants as capping agents have been used for adjusting the aspect ratio of gold nanorods[15,21], Ag-tipped overgrowth of gold nanorods[22] and tetrahedral-like gold nanotripods[23]. However, the role of each component for the growth is still ambiguous due to the complexity of interactions. As mentioned previously, the binding affinity of CTAB is greater than that of CTAC. However, as the ratio of CTAC:CTAB is increased to between 90:1 and 900:1, an equilibrium exists where CTAC occupies some of the surface area. Evidently, the CTAC is localized to the tip region where the binding preference of CTAB over CTAC is minimal. In this case, growth at the less protected ends capped by CTAC results in the observed tip overgrowth utilizing the shape-directing properties of the halide (Fig. 3b, condition 6)[15–17,22]. When the ratio is either too high or too low or when pure CTAB is used, the tip growth is not observed as size augmentation becomes the dominant growth mode as the surface is covered uniformly in a single surfactant (Fig. 3a, condition 1, and Supplementary Fig. 4b,c). This experimental observation is

qualitatively consistent with reported results for adjusting aspect ratio of gold nanorod and Ag-tipped overgrowth of gold nanorods using a BDAC/CTAB mixed surfactant system[15,22].

By carefully modifying the reaction conditions, we were also able to control the tip shape of dumbbell-like bipyramids to further confirm the role of components in the growth solution. Figure 3 shows TEM images and ultraviolet–vis–near-infrared spectra of a variety of monodisperse regrown structures from bipyramids with singular or binary surfactants. We observed marked changes at the bipyramid tips including spherical-, rod- and bipyramid-like tips with binary surfactants. The standard growth solution of bipyramids in these experiments consisted of 91.9 µM of $HAuCl_4$, 18.4 µM of $AgNO_3$, 0.0184 N of HCl and 147 µM of L-ascorbic acid. The molar ratio of [ascorbic acid]/[$HAuCl_4$] was 1.6 in Fig. 3 to ensure complete reduction of Au^{3+} to Au^0 (ref. 22). If less ascorbic acid is used, remaining Au^{3+} acts as an oxidant, and an etching process becomes dominant, resulting in shortened bipyramids (Supplementary Fig. 5)[24]. The major factors for the shape changes in these reactions were HCl and $AgNO_3$. It is reported that the reduction rate of Au^{3+} can be increased by increasing the pH of the solution, which can cause fast deposition along the existing crystal face, leading to growth of the nanorod in all directions having shorter aspect ratio[25,26]. The pH of the growth solution is 1.74 for the case of standard growth solution and 3.62 for the growth solution without HCl. Under both of these conditions, reduction of Ag^+ can be neglected due to the increased redox potential of ascorbic acid[27]. It is now established that a monolayer of silver can be deposited onto a particular surface of gold during the growth through underpotential deposition, thus stabilizing the surface and slowing the growth rate[9,16,17,28]. The growth rate of all facets in the structure can be altered depending on the degree of surface coverage by underpotential deposition of silver ions, and therefore differing growth rates of particular facets will determine the final structure of the nanoparticle.

The results in Fig. 3a, condition 2, and Fig. 3b, condition 7, show predominant growth along the short axis with sharp apices. As compared with growth in standard growth solution, a faster growth rate is expected in all directions for bipyramids without HCl at pH 3.62. However, the growth rate at the tip is greatly inhibited, because of the presence of Ag^+ ions, which can more

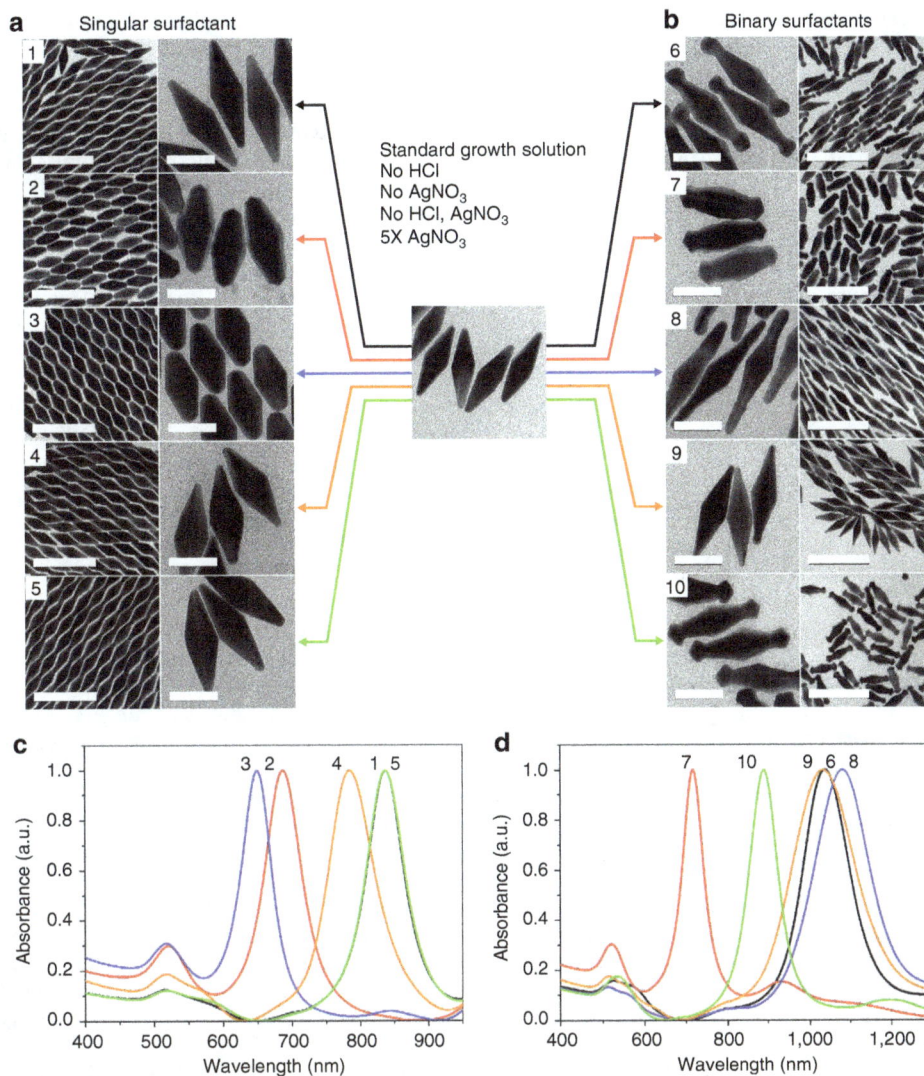

Figure 3 | Shape-controlled bipyramids regrown with either singular and binary surfactants are shown. (**a,b**) TEM images of the monodisperse, regrown structures from purified bipyramid seeds with singular surfactant (conditions 1–5) and binary surfactants (conditions 6–10). The coloured arrows indicate the specific conditions for regrowth (see also Supplementary Table 1 for detailed conditions). Scale bars, 200 nm (low magnification); 50 nm (high magnification). (**c,d**) Normalized ultraviolet–vis–near-infrared spectra of regrown structures with singular surfactant (1–5 in **c** correspond to 1–5 in **a**) and binary surfactants (6–10 in **d** correspond to 6–10 in **b**), respectively. See Supplementary Table 3 for detailed size measurements.

easily access the tip region, resulting in underpotential deposition. The high curvature of the tip region results in lower surfactant packing density than on the bipyramid faces, allowing additional room for Ag^+ to deposit in the intersurfactant regions[22]. This is resulted in a stepped facet formed along the long axis of the bipyramids, thereby sharpening the bipyramid tip for both[9]. The growth for binary surfactants showed increased overgrowth at the tip than for singular case because it has less coverage of CTAB layer accelerating the growth along the tip region. For binary surfactants, adding more or less $AgNO_3$ without HCl resulted in little enlogation and unrestricted growth of the width, especially at the tips, indicating that increased inhibition of growth along the long axis can greatly alter the shape of particle (Supplementary Fig. 6a)

Likewise, in the experiments without $AgNO_3$ at pH 1.76, fast growth is expected due to the absence of Ag^+ ions, however, slower than the case without both HCl and $AgNO_3$ due to the effect of the pH on the reducing reduction power of ascorbic acid. The reaction proceeds with a rate that allows for controlled growth due to the low pH, but the absence of Ag^+ ions, normally

yielding a stepped crystal facet at this pH for bipyramids, results in facilitated growth of a nanorod-like structure[9,16,17]. Without Ag^+ ions to block certain growth facets, bipyramids grown in a singular CTAB surfactant system were synthesized with low aspect ratio (~ 2.3) and poorly defined tips; due to the even coverage of the singular surfactant and lack of Ag^+ underpotential deposition, growth occurred evenly in all directions (Fig. 3a, condition 3). Interestingly, rod-like tips were formed in the binary surfactant system (Fig. 3b, condition 8). Similar to the above case, it is believed that overgrowth at the tip region was induced from the reduced surface coverage of CTAB in the binary system. In the absence of Ag^+, the overgrowth was accelerated and bypassed the stepped structure to form the rod-like structures at the tips.

Figure 3a, condition 4, and Fig. 3b, condition 9, show the regrowth of bipyramids without both HCl and $AgNO_3$. In this case, a much faster growth rate is expected than the former two cases, either without HCl or without $AgNO_3$. Due to a greatly faster deposition and the absence of Ag^+ inhibition, rapid addition of Au^0 atoms occurs on the stepped surface of bipyramids evenly in all directions. This results in particle maintaining its original bipyramid shape for both the singular and binary surfactant systems. The aspect ratio for binary surfactants was slightly higher than the singular case because of the lessened CTAB coverage as mentioned above, resulting in a small preference for growth at the tip and a slightly elongated particle.

Meanwhile, additional $AgNO_3$ ($5\times$ more than standard growth conditions) shows significant morphological changes at the tip region in binary surfactants, given less coverage of CTAB at the tip than the singular surfactant. Distinct crystalline structures identical to bipyramids were formed at the tips at both ends (Fig. 3b, condition 10). However, negligible changes were observed from singular CTAB surfactant (Fig. 3a, condition 5). These results indicate that insufficient protection from CTAC surfactant can be more sensitive for inhibition of growth from underpotential deposition of Ag^+, allowing for easier access to the tip end, resulting in significant shape changes. The addition of more acid in this case results in negligible changes for both singular and binary surfactants (Supplementary Figs 2 and 6b). The ultraviolet–vis–near-infrared spectra in Fig. 3c,d reflects the various structural changes of regrown bipyramids dependent on the growth conditions. Because most of regrown structures show changes in length, significant peak shift of longitudinal plasmon peaks with narrow widths were observed.

The crystalline structure of individual regrown bipyramids with singular and binary surfactants was determined by high-resolution TEM (HR-TEM) and fast Fourier transform (FFT) patterns (Fig. 4). Figure 4b shows HR-TEM images and the corresponding FFT patterns of enlarged bipyramids resulting from the regrowth with a singular surfactant and standard growth solution. The magnified images in Fig. 4b show clear lattice fringes at the tips and middle of the particles. It shows that both fringes are parallel to the growth axis and twinned along the long axis. The spacing between lattice fringes is confirmed as 0.234 nm by direct measurement, corresponding to the (111) facets of the bipyramids[9]. The FFT pattern in Fig. 4b shows a diffraction pattern of the bipyramid corresponding to orientation 1 in Fig. 4a. The indexed reflections in the FFT pattern correspond to the lattice parameters: $d_{111} = 0.234$ nm, $d_{220} = 0.143$ nm, $d_{222} = 0.118$ nm, d_{200} and $d_{020} = 0.204$ nm, $d_{311} = 0.122$ nm, $d_{400} = 0.102$ nm, and $d_{420} = 0.091$ nm. Reflections indexed as $(1\bar{1}1)$, (311) and (220) are scattered from T3 and T4. Reflections indexed as $(\bar{2}00)$, (020) and (220) are from T1. The remaining reflections that are not indexed are induced by multiple scattering effect. All measured values are within an error range of $\pm 2\%$ compared with bulk data and agree well with reported results, indicating that regrown bipyramids are penta-twinned with face-centred cubic structure[29–32]. All of the regrown bipyramids with either singular or binary surfactants are confirmed to have the same FFT pattern as a bipyramid in Fig. 4b (Supplementary Fig. 7). The tip angles (θ) of bipyramids in Fig. 4b were measured to be $27.02 \pm 1.35°$, which correspond to (117) high-index facets having an average step length (s) of ~ 3.5 atoms (Supplementary Fig. 7a)[9]. However, the identity of this high-index facet varies as the tip angles differ from that of bipyramids[9,33–35]. If the tip angle is smaller than bipyramids in Fig. 4b, the average step length (s) can be >3.5, which can be indexed as $\{11l\}$, where $l > 7$. Likewise, when the tip angle is larger than that of bipyramids, the average step length (s) can be <3.5, which can be indexed as $\{11l\}$, where $l < 7$. Interestingly, HR-TEM images and FFT of the dumbbell-like bipyramid in Fig. 4h show multiply twinned spherical structures at both tips. (The black arrows in Fig. 4h indicate the twin planes of the spherical tips.) The circular reflections in Fig. 4h correspond to the lattice parameters: $d_{111} = 0.229$ nm, $d_{220} = 0.143$ nm and $d_{200} = 0.203$ nm, respectively. This circular

type of FFT pattern resembles spherical particles having pentagonal symmetry, also known as decahedra[29,30,32,36].

Oxidative etching of bipyramids. Gold bipyramids can also undergo oxidative etching utilizing molecular oxygen dissolved in solution. Cycling oxidations and reductions of gold atoms can slowly shape the nanoparticle through a time-dependent aging process[37]. Additive oxidants such as hydrogen peroxide have been shown to accelerate the etching process and to reshape nanoparticles through atomic substraction and addition, but result in nanoparticles with poor shape and size dispersity[11,38]. Taking advantage of the purity of the gold bipyramids and their inherent monodispersity, this etching process has been adapted to create other monodisperse structures. Because of the high-energy nature of the bipyramid facets and sharp tips, these highly reactive particles are susceptible to oxygen as an etchant despite it being a weak oxidant. As seen in Fig. 5a, bipyramids begin to etch at the tips and are continuously sculpted to rice shaped and eventually rod shaped as the reaction progresses at 120 °C and in the presence of 0.1 M BDAC. Ultraviolet–vis spectra in Fig. 5b show continuous changes that are consistent with the structural changes as seen in the TEM images. In the presence of BDAC as a protective agent and etchant, the purified gold bipyramids facilitate a four-electron oxidation of four Au^0 atoms to Au^+, forming the gold chloride salt and simultaneously reducing molecular oxygen to water. The high temperature of the system then allows for disproportionation at the surface of the nanoparticle, depositing Au^0 atoms back onto the surface in low-energy and low-defect areas[37]. This process cycles, continuously removing atoms at defect areas and replacing them in defect-free areas, until the particle reaches an energy minimum, in this case a gold nanorod. At room temperature, a similar kinetic process still occurs, but resulting in a blue-shift of the LSPR peak of less than a couple nanometres every day (Supplementary Fig. 8a). This indicates the thermodynamic nature of the reaction, and that the etching of the bipyramid to the nanorod could be further accelerated by increasing the temperature. When CTAB is used as a protective agent and etchant, the reaction slows, requiring nearly an order of magnitude longer to obtain the analogous structures to the ones obtained from heating for 30 mins in Fig. 5a (Supplementary Fig. 8b). These reaction rates are qualitatively consistent with generally accepted reactivates of halides[37]. It is noteworthy that the etching reaction rate can be increased by decreasing the concentration to the critical micelle concentration of the surfactant (1 mM), which results in lessened protection of the nanoparticles from the oxidative species (Supplementary Fig. 8c).

To extend our regrowth strategy to other structures, we further synthesize regrown structures using the monodisperse etched structures as seeds with both singular and binary surfactants. Figure 6 shows TEM images and ultraviolet–vis spectra of regrown structures using the rod shape particles from etching as seeds. The growth behaviours from etched particles using unitary and binary surfactants were confirmed to be very similar, likely to due to the absence of the sharp tips that affect the tip of gold bipyramids. Interestingly, the rice shape particles from the etching can be reversed to the original structure of bipyramids when the standard growth solution is used for the regrowth (Supplementary Fig. 9). On the other hand, the regrowth of rod shape particles with standard growth solution cannot be fully reversed to the bipyramids, instead forming a bifrustum, ultimately lacking the sharp tips of a bipyramid (Fig. 6a condition 1). Applying the same growth conditions as bipyramids, new types of nanostructures from bifrustum to short dumbbell were synthesized using the

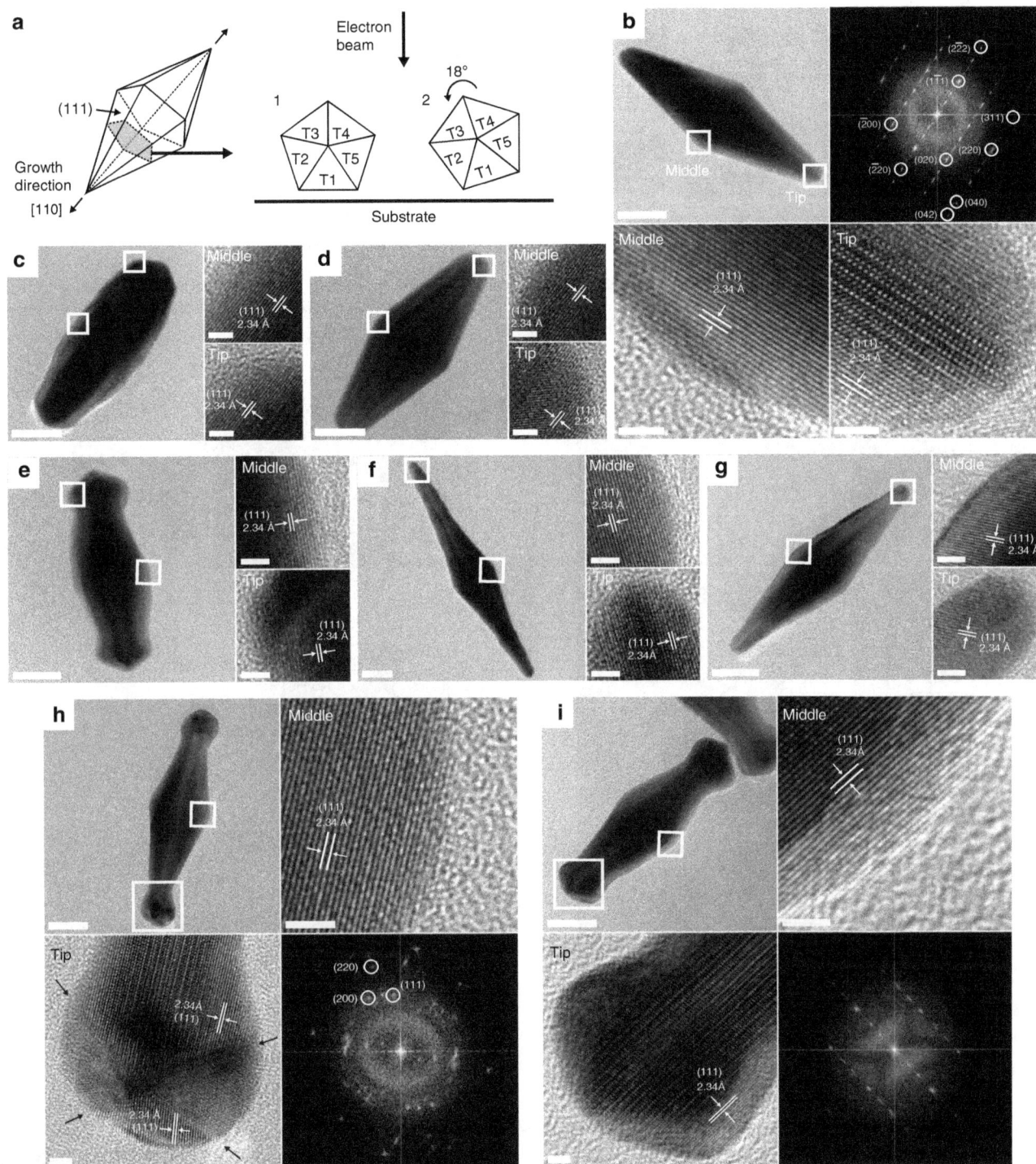

Figure 4 | HR-TEM images and FFT patterns of regrown bipyramids with singular and binary surfactants. (a) Schematic illustrations for cyclic penta-tetrahedral twinning of bipyramids. The grey area shows the cross-section of the bipyramid perpendicular to the growth direction. Each twinning plane is labelled from T1 to T5. Schematic on the right of the panel shows most of the possible orientations of bipyramids on the substrate with respect to the beam direction. Regrown bipyramids in Fig. 3a—conditions 1–3 for **b–d**, respectively; Fig. 3b—conditions 7–9 for **e–g**, respectively; Fig. 3b—conditions 6 and 10 for **h** and **i**, respectively. Magnified images showing the lattice fringes represent the areas marked with white boxes at the tip and middle of particles. The black arrows in **h** indicate the twin planes of the spherical tips in the dumbbell-shaped structures. Scale bars, 20 nm (low-magnification images); 2 nm (high-magnification images).

etched particles, with plasmonic resonances covering the short wavelength region between 500 and 800 nm. All of the regrown structures from the etched rod-shaped particles using singular or binary surfactants are also confirmed to have similar FFT patterns as the bipyramids in Fig. 4b (Supplementary Fig. 7).

Discussion

The regrowth strategy we have described has successfully synthesized various, novel gold nanoparticle geometries based on the bipyramid shape. The purity and monodispersity of the bipyramid results in a likewise pure and uniform product ranging

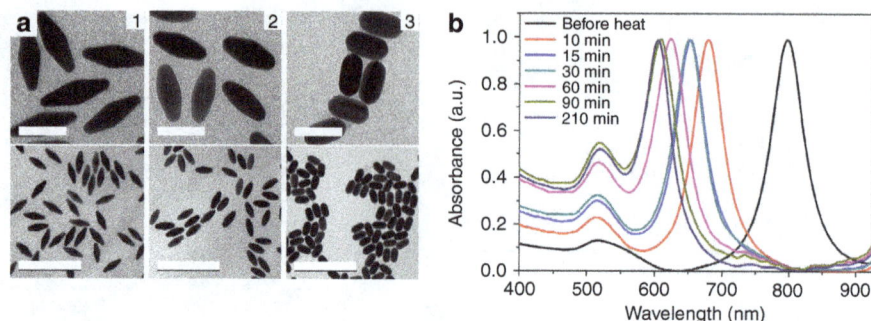

Figure 5 | Etched structures from oxidative etching with BDAC surfactant and heating at 120 °C. (**a**) TEM images of the etched structures resulting from heating for 10, 30 and 90 min, shown from 1 to 3. Scale bars, 50 nm (high magnification); 200 nm (low magnification). (**b**) Normalized ultraviolet–vis–near-infrared spectra of etched structures resultant of increasing the heating time.

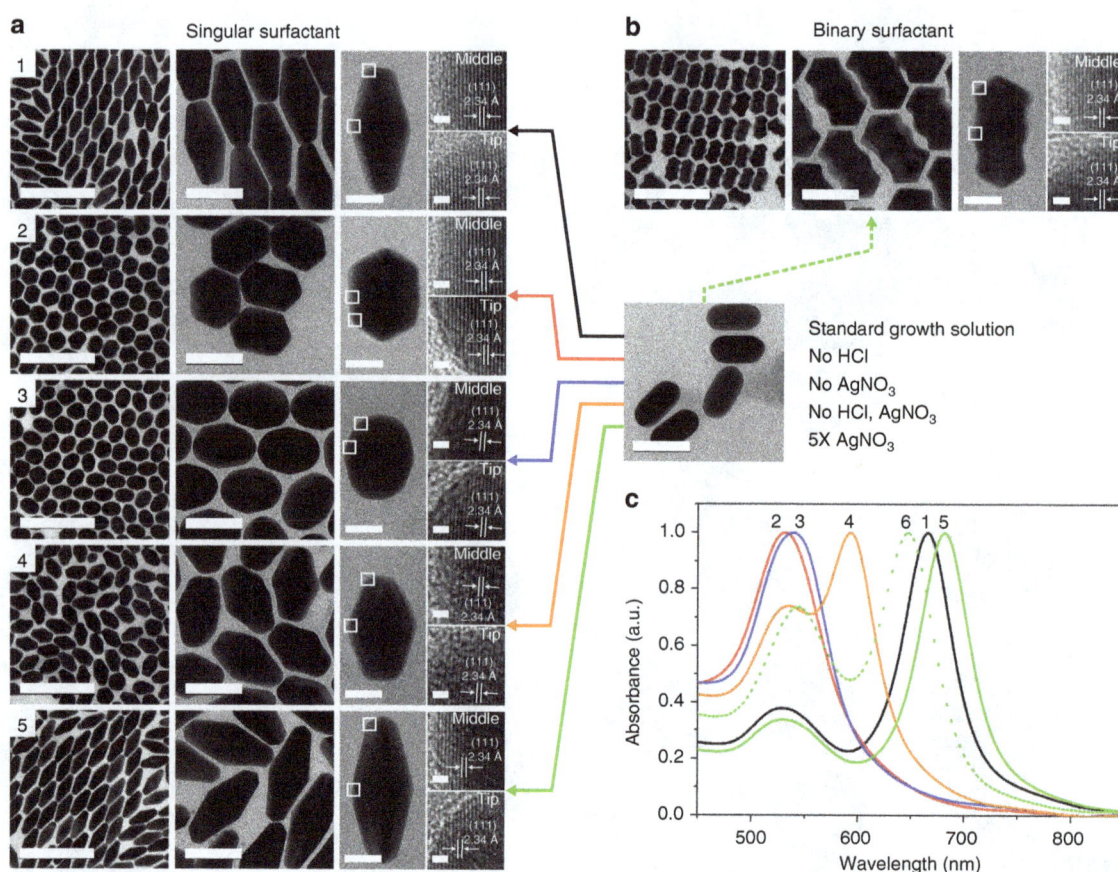

Figure 6 | Controlling the shape of oxidatively etched nanorods through the regrowth with singular and binary surfactants. (**a,b**) TEM and HR-TEM images of monodisperse, regrown structures from the oxidatively etched nanorods in Fig. 5a (90 min heating) with singular surfactant (**a**) and binary surfactants (**b**). The coloured arrows indicate the specific conditions for regrowth (See also Supplementary Table 1 for detailed conditions). Magnified images showing the lattice fringes represent the areas marked with white boxes at the tip and middle of particles. (**c**) Normalized ultraviolet–vis–near-infrared spectra of the regrown structures with singular surfactant (1–5 correspond to 1–5 in panel a) and binary surfactants (6 corresponds to **b**) are shown. See Supplementary Table 3 for detailed size measurements. Scale bars, 200, 50, 20 and 1 nm (from low to high magnification respectively).

from augmented bipyramids, dumbbells with spherical, pointed, and rod-like tips, bifrustums and spherical polyhedra. Regrowth of bipyramids in a single surfactant can be controllably tuned to give a length of well over 200 nm and a longitudinal plasmonic resonance peak well into the near-infrared range, as well as allowing for control of the aspect ratio. Regrowth with binary surfactant systems offer insight into the localization of particular surfactants on the particle surface. When combined with varying

concentrations of acid and silver offer the opportunity to yield unique geometries. In addition, the high-energy nature of the bipyramid structure allows for controllable oxidative etching using only molecular oxygen as the etchant to create highly monodisperse nanorice and nanorod structures, which have additionally been regrown using the same procedure. Finally, the various growth conditions starting from the highly monodisperse but twinned structure of the bipyramids allows to create noble metal nanoparticles of a wide range of size and shape with unprecedented narrow distribution, and this will be key in future efforts to assemble these structures for shaping the optical response for potential applications in self-assembly, non-linear optics and surface-enhanced Raman spectroscopy (SERS).

Methods

Materials and instruments. All chemicals were purchased from commercial suppliers and used without further purification. CTAB (Bioxtra, ≥99.0%), BDAC (cationic detergent), cetyltrimethylammonium chloride, citric acid trisodium salt dihydrate (≥99.5%, BioUltra, for molecular biology) (CTAC, ≥98.0%), hydrogen tetrachloroaurate trihydrate (HAuCl$_4$•3H$_2$O) and L-ascorbic Acid (Bioxtra, ≥98.0%) were purchased from Sigma Aldrich. Silver nitrate (AgNO$_3$, ≥99.8%) and sodium borohydride (NaBH$_4$, ≥99%) were purchased from Fluka. Hydrochloric acid (HCl, 1 N) was purchased from Fisher scientific. Nanopure water (18.2 MΩ, Barnstead Nanopure, Thermo Scientific, MA, USA) was used in all experiments. The glass vials were purchased from Kimble chase (4 and 20 ml, NJ, USA). All glasswares were cleaned using freshly prepared aqua regia (HCl:HNO$_3$ in a 3:1 ratio by volume) followed by rinsing with copious amounts of water. RCT Basic (IKA, NC, US) was used for magnetic stirring. Ultraviolet–vis–near-infrared spectra were measured with Synergy H4 (Biotek, VT, USA) and Cary 5000 ultraviolet–vis–near-infrared (Agilent, CA, USA). The formvar/carbon-coated copper grid (Ted Pella, Inc. Redding, CA, USA) and TEM (Tecnai G2 F30 Super Twin microscope, 300 kV and Tecnai G2 Spirit, 200 kV, FEI, OR, USA) were used for the TEM analysis.

Synthesis and purification of gold bipyramids. The gold bipyramids were synthesized according to the literature procedure[9]. Briefly, gold seeds were prepared with 18.95 ml of ultrapure water, 0.25 ml of 10 mM HAuCl$_4$ and 0.5 ml of freshly prepared 10 mM sodium citrate, followed by adding 0.3 ml of fresh and ice-cold 10 mM NaBH$_4$ under 500 rpm with magnetic stirrer at room temperature. The reaction mixture was stirred for 2 h, and aged for a week before use (After aging, seed solutions are stable for a month and give reproducible results for the synthesis of bipyramids). The bipyramids were grown in a solution containing 10 ml of 0.1 M CTAB, 0.5 ml of 0.01 M HAuCl$_4$, 0.1 ml of 0.01 M AgNO$_3$, 0.2 ml of 1 N HCl, 0.08 ml of 0.1 M ascorbic acid, and varying amounts of seed solution; shown were seed volumes of 110, 100, 95, 80 and 60 µl corresponding to 1–5 in Supplementary Fig. 1a and Supplementary Table 2. The solution was gently stirred at 400 r.p.m., and was kept in an oil bath at 30 °C for 2 h. The colloid was centrifuged at 13,000 g at 30 °C for 15 min, and washed with 10 ml of 1 mM CTAB, performed twice. After removing the supernatant, the precipitate was redispersed in 3 ml of 1 mM CTAB solution for further purification. Volumes of 0.5 M BDAC solution and ultrapure water were added to the 3 ml of crude bipyramid solution to obtain 10 ml of solution with the desired BDAC concentration. The concentration of BDAC desired is dependent on the size of the gold bipyramids and was determined experimentally (Supplementary Fig. 1). Concentrations of BDAC used were 230, 260, 310, 320 and 350 mM corresponding to the bipyramids prepared with 60, 70, 95, 100 and 110 µl of seed solutions, respectively. The solution was mixed and left undistributed in an incubator at 30 °C for 11 h. The resulting pink supernatant was carefully removed, and 3 ml of 1 mM CTAB was added to the vial to redisperse the precipitate. The vial was then sonicated for 1 min. The resulting purified solution (brown in colour) was centrifuged at 8,000 g for 8 min and washed with 1 ml of 1 mM CTAB, repeated twice, to remove the excess BDAC. Finally, the purified bipyramids were redispersed in 1.5 ml of 1 mM CTAB solution to be used for all regrowth reactions. The purified bipyramids in Fig. 2a prepared with 80 µl of seed solution were used as seeds for all regrowth reactions.

Controlling the size of bipyramids. The size of gold bipyramids can be controlled using either a different concentration of growth solution or different amount of purified seed solution. For the typical preparation of regrowth solution, 0.9 ml of 0.1 M CTAB solution was kept for 5 min in an oil bath at 30 °C with magnetic stirring at 400 r.p.m. HAuCl$_4$, AgNO$_3$, HCl and ascorbic acid were then added sequentially as detailed in Supplementary Table 1 and kept for 5 min. Finally, an amount of purified bipyramid seed solutions (Supplementary Table 1) was added and kept for 2 h. To maintain the reaction volume constant, the varied amount of purified bipyramid seed solutions were adjusted to 0.1 ml with 1 mM CTAB. The resulting solution was centrifuged at 7,000 g for 8 min and washed with 1 mM CTAB, repeated twice, then redispersed in 1 mM CTAB for further characterization.

Controlling the shape of nanoparticles. Various tips and unique shape of structures can be controlled with single or binary surfactants along with modified growth solutions as detailed in Supplementary Table 1. The procedure with a single surfactant is similar to the enlarging of the bipyramid, except the condition adding chemicals. For the synthesis without HCl, AgNO$_3$ or both, the same volume of nanopure water is added to the regrowth solution in its stead. For excess amount of AgNO$_3$, a higher concentration of solution is used with the same volume as the standard regrowth solution. Hundred microlitre of purified bipyramid seed

solution in 1 mM CTAB is added to the prepared regrowth solution and kept for 2 h. For the typical preparation of regrowth solution with binary surfactants, 0.1 M CTAC was kept for 5 min in an oil bath at 30 °C with magnetic stirring at 400 r.p.m. Hundred microlitre of purified bipyramid seed solution is centrifuged at 7,000 g and redispersed in 0.1 ml of desired concentration of CTAB solution to adjust the ratio between the surfactants (CTAC and CTAB). The reactants with purified bipyramid seed were added same as above and kept at 30 °C for 2 h to complete the reaction. The resulting solution was centrifuged at 7,000 g for 8 min and washed with 1 mM CTAB, repeated twice, then redispersed in 1 mM CTAB for further characterization.

Oxidative etching of bipyramids. For oxidative etching, 100 µl of purified bipyramids in 1 mM CTAB was added to 900 µl of 100 mM BDAC solution with no other reagents. A glass vial with screw cap was sealed with teflon tape to prevent the leakage of vapour during the heating. The sealed vial was placed in oil bath pre-heated at 120 °C and kept under stirring at 300 r.p.m. with magnetic stirrer. (Use caution when heating the sealed container as pressure will build inside the flask. Properly sealing the vial is also crucial to controlling the reaction speed, as any leaks will accelerate the reaction.) After 10, 15, 30, 60, 90 and 210 min, the vials were cooled to room temperature with a water bath to halt the oxidative process. The resulting solution was centrifuged at 8,000 g for 8 min and washed with 1 mM CTAB, repeated twice and then redispersed in 100 µl of 1 mM CTAB for further regrowth and characterization. See Supplementary Table 1 for the detailed conditions for regrowth.

References

1. Lee, J.-H. *et al.* Tuning and maximizing the single-molecule surface-enhanced Raman scattering from DNA-tethered nanodumbbells. *ACS Nano* **6**, 9574–9584 (2012).
2. Liu, G. L. *et al.* A nanoplasmonic molecular ruler for measuring nuclease activity and DNA footprinting. *Nat. Nanotechnol.* **1**, 47–52 (2006).
3. Lee, J.-H., You, M.-H., Kim, G.-H. & Nam, J.-M. Plasmonic nanosnowmen with a conductive junction as highly tunable nanoantenna structures and sensitive, quantitative and multiplexable surface-enhanced Raman scattering probes. *Nano Lett.* **14**, 6217–6225 (2014).
4. Ghosh, P., Han, G., De, M., Kim, C. K. & Rotello, V. M. Gold nanoparticles in delivery applications. *Adv. Drug. Deliv. Rev.* **60**, 1307–1315 (2008).
5. Darbha, G. K. *et al.* Selective detection of mercury (ii) ion using nonlinear optical properties of gold nanoparticles. *J. Am. Chem. Soc.* **130**, 8038–8043 (2008).
6. Eustis, S. & El-Sayed, M. A. Why gold nanoparticles are more precious than pretty gold: Noble metal surface plasmon resonance and its enhancement of the radiative and nonradiative properties of nanocrystals of different shapes. *Chem. Soc. Rev.* **35**, 209–217 (2006).
7. Prodan, E., Radloff, C., Halas, N. J. & Nordlander, P. A hybridization model for the plasmon response of complex nanostructures. *Science* **302**, 419–422 (2003).
8. Savage, K. J. *et al.* Revealing the quantum regime in tunnelling plasmonics. *Nature* **491**, 574–577 (2012).
9. Liu, M. & Guyot-Sionnest, P. Mechanism of silver(i)-assisted growth of gold nanorods and bipyramids. *J. Phys. Chem. B* **109**, 22192–22200 (2005).
10. Liu, M., Guyot-Sionnest, P., Lee, T.-W. & Gray, S. K. Optical properties of rodlike and bipyramidal gold nanoparticles from three-dimensional computations. *Phys. Rev. B* **76**, 235428 (2007).
11. Kou, X. *et al.* Growth of gold bipyramids with improved yield and their curvature-directed oxidation. *Small* **3**, 2103–2113 (2007).
12. Park, K., Koerner, H. & Vaia, R. A. Depletion-induced shape and size selection of gold nanoparticles. *Nano Lett.* **10**, 1433–1439 (2010).
13. Ye, X., Zheng, C., Chen, J., Gao, Y. & Murray, C. B. Using binary surfactant mixtures to simultaneously improve the dimensional tunability and monodispersity in the seeded growth of gold nanorods. *Nano Lett.* **13**, 765–771 (2013).
14. Chen, H., Shao, L., Li, Q. & Wang, J. Gold nanorods and their plasmonic properties. *Chem. Soc. Rev.* **42**, 2679–2724 (2013).
15. Nikoobakht, B. & El-Sayed, M. A. Preparation and growth mechanism of gold nanorods (nrs) using seed-mediated growth method. *Chem. Mater.* **15**, 1957–1962 (2003).
16. Langille, M. R., Personick, M. L., Zhang, J. & Mirkin, C. A. Defining rules for the shape evolution of gold nanoparticles. *J. Am. Chem. Soc.* **134**, 14542–14554 (2012).
17. Lohse, S. E., Burrows, N. D., Scarabelli, L., Liz-Marzán, L. M. & Murphy, C. J. Anisotropic noble metal nanocrystal growth: The role of halides. *Chem. Mater.* **26**, 34–43 (2013).
18. Garg, N., Scholl, C., Mohanty, A. & Jin, R. The role of bromide ions in seeding growth of Au nanorods. *Langmuir* **26**, 10271–10276 (2010).
19. Vigderman, L., Khanal, B. P. & Zubarev, E. R. Functional gold nanorods: Synthesis, self-assembly, and sensing applications. *Adv. Mater.* **24**, 4811–4841 (2012).

20. Kou, X. *et al.* Growth of gold nanorods and bipyramids using cteab surfactant. *J. Phys. Chem. B* **110**, 16377–16383 (2006).

21. Kou, X. *et al.* One-step synthesis of large-aspect-ratio single-crystalline gold nanorods by using CTPAB and CTBAB surfactants. *Chem. Eur. J* **13**, 2929–2936 (2007).

22. Park, K. & Vaia, R. A. Synthesis of complex Au/Ag nanorods by controlled overgrowth. *Adv. Mater.* **20**, 3882–3886 (2008).

23. Ali Umar, A. & Oyama, M. High-yield synthesis of tetrahedral-like gold nanotripods using an aqueous binary mixture of cetyltrimethylammonium bromide and hexamethylenetetramine. *Cryst. Growth Des.* **9**, 1146–1152 (2008).

24. Rodríguez-Fernández, J., Pérez-Juste, J., Mulvaney, P. & Liz-Marzán, L. M. Spatially-directed oxidation of gold nanoparticles by Au(iii) − CTAB complexes. *J. Phys. Chem. B* **109**, 14257–14261 (2005).

25. Kim, F., Sohn, K., Wu, J. & Huang, J. Chemical synthesis of gold nanowires in acidic solutions. *J. Am. Chem. Soc.* **130**, 14442–14443 (2008).

26. Sohn, K. *et al.* Construction of evolutionary tree for morphological engineering of nanoparticles. *ACS Nano* **3**, 2191–2198 (2009).

27. Pal, T. *et al.* Organized media as redox catalysts. *Langmuir* **14**, 4724–4730 (1998).

28. Personick, M. L., Langille, M. R., Zhang, J. & Mirkin, C. A. Shape control of gold nanoparticles by silver underpotential deposition. *Nano Lett.* **11**, 3394–3398 (2011).

29. Lisiecki, I. *et al.* Structural investigations of copper nanorods by high-resolution tem. *Phys. Rev. B* **61**, 4968–4974 (2000).

30. Lisiecki, I. Size shape, and structural control of metallic nanocrystals. *J. Phys. Chem. B* **109**, 12231–12244 (2005).

31. Johnson, C. J., Dujardin, E., Davis, S. A., Murphy, C. J. & Mann, S. Growth and form of gold nanorods prepared by seed-mediated, surfactant-directed synthesis. *J. Mater. Chem.* **12**, 1765–1770 (2002).

32. Zhou, G. *et al.* Growth of nanobipyramid by using large sized Au decahedra as seeds. *ACS Appl. Mater. Interfaces* **5**, 13340–13352 (2013).

33. Xiao, J. *et al.* Synthesis of convex hexoctahedral pt micro/nanocrystals with high-index facets and electrochemistry-mediated shape evolution. *J. Am. Chem. Soc.* **135**, 18754–18757 (2013).

34. Tian, N., Zhou, Z.-Y. & Sun, S.-G. Platinum metal catalysts of high-index surfaces: From single-crystal planes to electrochemically shape-controlled nanoparticles. *J. Phys. Chem. C* **112**, 19801–19817 (2008).

35. Mettela, G., Boya, R., Singh, D., Kumar, G. V. P. & Kulkarni, G. U. Highly tapered pentagonal bipyramidal Au microcrystals with high index faceted corrugation: Synthesis and optical properties. *Sci. Rep.* **3**, 1793 (2013).

36. Elechiguerra, J. L., Reyes-Gasga, J. & Yacaman, M. J. The role of twinning in shape evolution of anisotropic noble metal nanostructures. *J. Mater. Chem.* **16**, 3906–3919 (2006).

37. Long, R., Zhou, S., Wiley, B. J. & Xiong, Y. Oxidative etching for controlled synthesis of metal nanocrystals: atomic addition and subtraction. *Chem. Soc. Rev.* **43**, 6288–6310 (2014).

38. Tsung, C.-K. *et al.* Selective shortening of single-crystalline gold nanorods by mild oxidation. *J. Am. Chem. Soc.* **128**, 5352–5353 (2006).

Acknowledgements

We acknowledge funding support from the University of Chicago. K.J.G. is also supported by the CBI Training Grant (NIH 5T32GM008720). We thank Jeremy Gendler for his work on the purification and statistics. We also thank Philippe Guyot-Sionnest, Dmitri Talapin and E.W. Malachosky for the many helpful discussions.

Author contributions

J.-H.L., K.J.G. and Y.W. conceived the idea. J.-H.L. designed and initiated the experiment. J.-H.L. developed the synthetic method. K.J.G. and G.C. developed the purification design and K.J.G. performed the purification steps. G.C. measured HR-TEM and EDS. J.-H.L., K.J.G. and Y.W. analysed data and wrote the manuscript. Y.W. supervised the project. All authors discussed and approved the manuscript.

Additional information

12

An *in situ* self-assembly template strategy for the preparation of hierarchical-pore metal-organic frameworks

Hongliang Huang[1,2], Jian-Rong Li[2], Keke Wang[1], Tongtong Han[1], Minman Tong[1], Liangsha Li[1], Yabo Xie[2], Qingyuan Yang[1], Dahuan Liu[1] & Chongli Zhong[1]

Metal-organic frameworks (MOFs) have recently emerged as a new type of nanoporous materials with tailorable structures and functions. Usually, MOFs have uniform pores smaller than 2 nm in size, limiting their practical applications in some cases. Although a few approaches have been adopted to prepare MOFs with larger pores, it is still challenging to synthesize hierarchical-pore MOFs (H-MOFs) with high structural controllability and good stability. Here we demonstrate a facile and versatile method, an *in situ* self-assembly template strategy for fabricating stable H-MOFs, in which multi-scale soluble and/or acid-sensitive metal-organic assembly (MOA) fragments form during the reactions between metal ions and organic ligands (to construct MOFs), and act as removable dynamic chemical templates. This general strategy was successfully used to prepare various H-MOFs that show rich porous properties and potential applications, such as in large molecule adsorption. Notably, the mesopore sizes of the H-MOFs can be tuned by varying the amount of templates.

[1] State Key Laboratory of Organic-Inorganic Composites, Beijing University of Chemical Technology, Beijing 100029, China. [2] Beijing Key Laboratory for Green Catalysis and Separation, Department of Chemistry and Chemical Engineering, College of Environmental and Energy Engineering, Beijing University of Technology, Pingleyuan 100, Chaoyang, Beijing 100124, China. Correspondence and requests for materials should be addressed to J.-R.L. (email: jrli@bjut.edu.cn) or to C.Z. (email: zhongcl@mail.buct.edu.cn).

Porous materials have been attracting intense research interest due to their broad range of possible applications[1]. Tailoring structures and pore properties with rational design and controllability of porous materials are crucial for their specific applications, but challenging in practical preparation. Traditional porous solids such as zeolites, activated carbon and mesoporous silica are relatively difficult in modifying and tailoring their structures and functions, in particular at the molecular level, whereas newly developed metal-organic frameworks (MOFs), composed of organic linkers and inorganic nodes, are recognized to be easy in this respect[2,3]. This type of materials has shown great potential in various applications including adsorption, separation, catalysis, sensing and so on[4–6]. So far, the related research mainly focuses on the microporous MOFs. In particular, some stable microporous MOFs were reported in recent years[7–9], which greatly promoted the practical applications of these new materials. The small pore size in microporous MOFs benefits the adsorption and separation of small molecules, but restricts their diffusion and also prevents larger molecules from accessing the MOF channels, thus greatly limiting their applications in some cases[10]. Thereby, the design and preparation of MOF materials with larger pore sizes are imperative, yet challenging to date.

Two approaches have mainly been developed to 'enlarge' pores of MOFs: (i) construct MOFs by using large building units (metal clusters and/or organic ligands)[11–14] and (ii) fabricate MOF materials with large pores as crystal 'defect'[10,15]. With respect to the former, the ligand-extension strategy was widely adopted[16–19]. However, the pore size in the resulting periodic nanostructures of MOFs is still limited to be smaller than 10 nm (ref. 17). In particular, with the increase of pore size, the frameworks usually become unstable in most cases. On the other hand, the ligand extension normally also results in the interpenetration of the structures, which would dramatically decrease the pore size[16,17]. Therefore, increasing the pore size of MOFs and keeping their framework stable remain a great challenge in design and synthesis. In addition, from the viewpoint of cost, synthesis of large linkers is too expensive for practical applications of resulting MOF materials. For the latter approach, the ligands can be cheap but the fabrication methods are pivotal and difficult to follow in most cases.

Alternatively, template methods have been explored to prepare stable hierarchical-pore MOFs (H-MOFs) containing both micropores and mesopores/macropores[20]. As we know that the hard template and the soft template methods have been widely adopted in preparing mesoporous and other hierarchical-pore materials. For the hard template method, as calcinations or acid etchants are often required to remove the templates[21], MOFs can hardly be kept stable during this process, thus limiting its application in fabricating H-MOFs. For the soft template method, surfactants or block copolymers are used as templates, being easily manipulated and feasible for some MOFs, and thus has been used in preparing some H-MOFs[20,22–26]. For example, cetyltrimethylammonium bromide has been used as the template to prepare hierarchical-pore $Cu_3(BTC)_2$ ($H_3BTC = 1,3,5$-benzenetricarboxylic acid) in water–ethanol or ionic liquid systems[20,23–25]. Similarly, triblock copolymers such as P123 and F127, which are widely used to prepare meso-porous silicon, have also been used to prepare H-MOFs[27–31]. However, as most MOFs were synthesized in polar solvents, such as N,N-dimethylformamide (DMF), N,N-dimethylacetamide and N,N-diethylformamide, traditional surfactants are not able to play the role of a template due to their amphipathy. Therefore, the soft template method also has a limitation in preparing H-MOFs. Other methods for preparing H-MOFs by crystal 'defect' include gelation[32], CO_2-expanded liquids[33], pseudomorphic replication

by transformation of oxide[34] and so on. However, these innovative approaches are complicated to operate, being comparatively difficult to extend to other MOFs.

For the template method, it is crucial to select appropriate template in the preparation of porous materials. The template should not only have a good interaction/affinity with reaction precursors but also should be able to remove easily, keeping the structure of targeted porous materials intact. In addition, for some practical applications, bad stability of a material often is viewed as extremely negative. However, the weakness of the instability in a material can also be a positive factor in some cases[35]. Motivated by the context described above, herein we propose to use metal-organic assemblies (MOAs) including MOFs as the templates to prepare H-MOFs through an in situ self-assembly approach, where both the targeted porous materials and the templates belong to coordination complexes, being compatible in structural nature and reaction activity. Simultaneously, we can take advantage of their relatively different stabilities to get desired H-MOFs, that is, unstable MOAs as templates and stable MOFs as parents for targeting H-MOFs.

It has been well-documented that although MOF-5 is stable in several solvents[16], the sensitivity towards moisture and acid can lead to its structural collapse and decomposition[36,37]. In contrast, some MOFs are chemically, thermally and mechanically stable, such as UiO-66(Zr)[38], which can maintain its crystal structure even in an acid solution. These differences in the stability stimulate us to try to use water- or acid-sensitive MOFs, such as MOF-5 as the potential template to synthesize stable H-UiO-66(Zr).

Here we suppose that if this hypothesis can be accomplished, a lot of labile MOAs could be used as template sources for the preparation of various stable H-MOFs. Theoretically, in a self-assembly reaction process, the reversibility of coordination bonds in these coordination complexes could keep the MOAs template forming and disappearing during the stable MOFs forming and growing. Thus, the MOAs could indeed act as a chemical dynamic template to direct the construction of H-MOFs. In this work, we demonstrate this idea, a new template-based strategy, to prepare stable H-MOFs by adopting an in situ self-assembly synthesis method (Fig. 1).

Results

Preparation of H-MOFs through two-step reactions. A series of proof-of-concept experiments were performed. First, we used MOF-5 (Fig. 2a) as a template precursor to prepare H-UiO-66(Zr) (see Fig. 2d for the structure of UiO-66(Zr)) through a two-step reaction process. Nano-sized MOF-5 particles were first synthesized and mixed with $ZrCl_4$ and terephthalic acid in DMF. The mixture was then heated under solvothermal reaction condition, similar to that for synthesizing UiO-66(Zr). After the reaction, the resulting product template@H-UiO-66(Zr) was washed with acid aqueous solution to get targeted material H-UiO-66(Zr). Powder X-ray diffraction (PXRD) measurements show that the resulting H-UiO-66(Zr) has the same diffraction patterns as the parent UiO-66(Zr) (Fig. 2e and Supplementary Fig. 18). The N_2 adsorption at 77 K demonstrates the formation of the hierarchical-pore material with both micropores and mesopores (Supplementary Fig. 5). The detail of the preparation process can be seen in the Supplementary Methods section.

Considering such a fact that the mesopore size (about 11 nm) of the H-UiO-66(Zr) is much smaller than that of originally added MOF-5 particles with the size in the range of 370 ~ 520 nm (Supplementary Figs 1 and 2, the evaluated polydispersity index is 0.1), we suspect that there exists a 'decomposition and/or rearrangement' of MOF-5 particles in the reaction system.

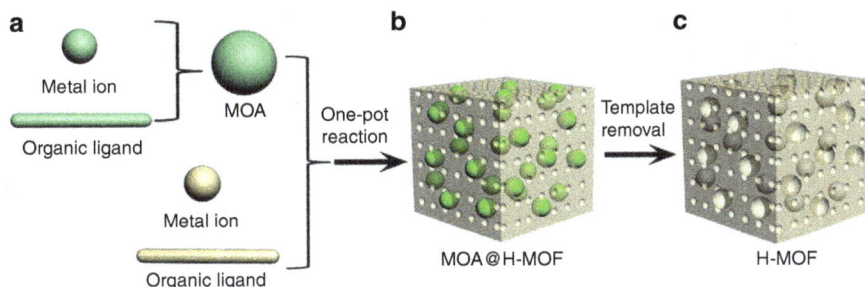

Figure 1 | Schematic representation for the preparation of H-MOF. (a) *In situ* self-assembly of MOA through the reaction between metal ion and organic ligand. (b) MOA@H-MOF composite formed by one-pot self-assembly reaction. (c) H-MOF formed through removing MOA template.

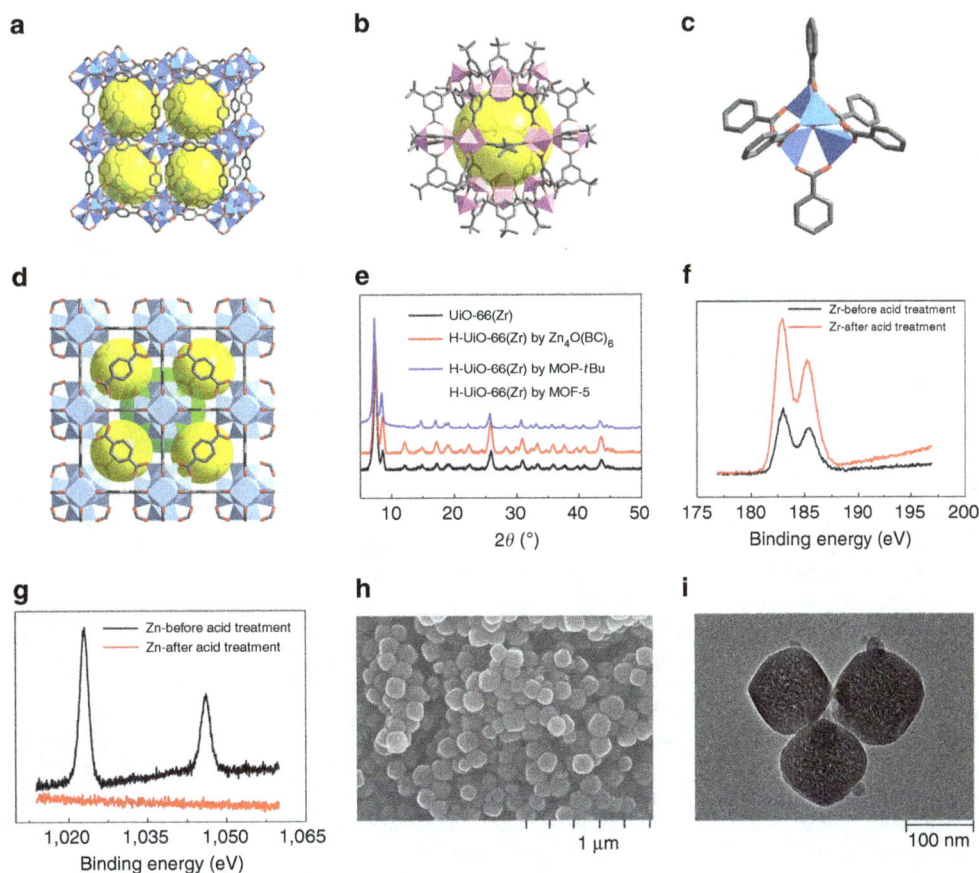

Figure 2 | Template-based preparation and characterization of H-UiO-66(Zr). (a) The structure of MOF-5, (b) the structure of metal-organic polyhedron (MOP-*t*Bu), (c) the structure of $Zn_4O(BC)_6$ and (d) the structure of UiO-66(Zr). Colour scheme: Zn atom, light blue polyhedron; Cu atom, pink polyhedron; Zr atom, cinerous polyhedron; C atom, grey; and O atom, red. All H atoms have been omitted for clarity. The yellow and green spheres represent the void inside of MOF and MOP. (e) PXRD patterns of H-UiO-66(Zr) prepared using different template precursors. (f,g) The XPS spectra of template@H-UiO-66(Zr) and H-UiO-66(Zr) prepared with MOF-5 as the template precursor over the Zr 3*d* and Zn 2*p* spectral regions, respectively. (h) SEM image and (i) transmission electron microscope image of H-UiO-66(Zr) prepared with MOF-5 as the template precursor.

This is indeed coincident with the characteristic of MOF-5 as a typical coordination complex, where MOF-5 particles were in a self-assembly stage of dynamic equilibrium due to the reversibility of the involved coordination bonds[39,40]. The general elemental analysis (EA) and inductive-coupled plasma (ICP) emission spectra analysis for the template@H-UiO-66(Zr) gave the contents of Zr 27.14, Zn 11.05, C 35.65% and H 1.57%; however, for H-UiO-66(Zr) they are Zr 31.24, Zn 0.24, C 33.49% and H 2.27%, respectively. The mole ratio of C/Zn in the 'template' was estimated (by assuming that all the Zr comes from H-UiO-66(Zr) and H-UiO-66(Zr) has the same element contents

as that parent UiO-66(Zr)) to be 3.5, being smaller than that in MOF-5 (theoretical value is 6) (Supplementary Table 2). This result indicates that in the formation of H-UiO-66(Zr), the actual template is not MOF-5 particles but is some newly *in situ*-regenerated MOAs. This judgment can also be supported by the fact that no diffraction peaks of the MOF-5 were observed in template@H-UiO-66(Zr) (Supplementary Fig. 19). Besides, it was also found that the MOF-5 particles decompose within 1 h in the ZrCl₄ DMF solution even at room temperature, which further reveals the 'decomposition and/or rearrangement' of MOF-5 in the reaction system. On the basis of these observations, the fact of a complicated self-assembly process could be thus justified. These generated MOAs act thus as the 'templates' to direct the formation of template@UiO-66(Zr). During acid treatment, the 'template' was removed. Therefore, the acid treatment indeed acts as an activation process to clear the pores of the H-MOF.

As shown in Fig. 2f,g, X-ray photoelectron spectroscopy (XPS) spectra demonstrate that the template@H-UiO-66(Zr) sample contains both Zn and Zr, but after acid treatment almost no Zn was detected (on the sample surface), while Zr remained. Based on the ICP analysis, a little bit Zn was identified in the H-UiO-66(Zr), which might be attributed to the Zn ions anchoring (through coordinating with some groups, such as carboxylate) in mesopore surfaces of H-UiO-66(Zr). These coordination groups were generated from the 'defect' of the UiO-66(Zr) framework. In addition, as given above, the C and H contents in H-UiO-66(Zr) were close to the theoretical values (C 34.6% and H 1.68%) of parent UiO-66(Zr). Moreover, the Fourier transform infrared (FT-IR) spectrum indicates no free terephthalic acid remained in the pores of H-UiO-66(Zr) (Supplementary Fig. 31).

Furthermore, high-angle annular dark-field scanning transmission electron microscopy (HAADF-STEM) analysis combined with energy dispersive X-ray spectroscopy (EDX) mapping reveals that Zr, O and C elements were homogeneously distributed in the whole template@H-UiO-66(Zr) sample but Zn element was heterogeneous (Supplementary Fig. 40). In addition, scanning electron microscope (SEM) and transmission electron microscope images show that the H-UiO-66(Zr) sample is of small nanoparticles ($\sim$100 nm) and has a sponge-like morphology, which implies the existence of crystal defect (Fig. 2h,i). These results suggest the existence of irregularly distributed mesopores in H-UiO-66(Zr). All these results confirm the existence of the Zn-based templates in template@H-UiO-66(Zr) and these templates can be almost completely removed by the acid treatment to give the H-MOF material.

In addition, there exist hysteresis loops in the N_2 adsorption isotherms of template@H-UiO-66(Zr) and H-UiO-66(Zr), being indicative of mesoporous structures of both materials (Supplementary Fig. 6). Before the acid treatment, the mesoporous structure may be created by removal of part of soluble template fragments in the sample treatment process through the DMF washing. However, the surface area and pore volume are low in template@H-UiO-66(Zr). After acid treatment, the material became clearly more porous and the mesopore was much larger than that before treatment. This result also implies that the acid treatment indeed is just an activation process to clear the pores as discussed above. It should also be pointed out that although UiO-66(Zr) is not stable in phosphoric acid solution[41], it is stable in diluted HCl solution[38]. As shown in Supplementary Fig. 8, N_2 adsorption isotherms indicate no mesopore in UiO-66(Zr) after soaking in diluted HCl solution (pH 1) for 12 h.

To further confirm the associated self-assembly process in the preparation of H-MOFs using this new method and its universal accessibility, another two types of MOAs were also chosen as the template precursors to carry out the synthesis experiments, including simple complex molecules and metal-organic polyhedra (MOPs) molecules. These molecule-based MOAs can be used as the supramolecular building units in constructing MOFs[16,42,43]. For the former, we explored $Zn_4O(BC)_6$ (BC, benzenecarboxylate; Fig. 2c) to again prepare H-UiO-66(Zr). $Zn_4O(BC)_6$ is a primary coordination complex about 2 nm in size, which is the secondary building unit of MOF-5 structure. Specifically, the $Zn_4O(BC)_6$ was first prepared by the reaction of $Zn(NO_3)_2 \cdot 6H_2O$ and benzoic acid in DMF. After the reaction, the resulting solution was cooled down to room temperature. Then, $ZrCl_4$ and terephthalic acid were added and the mixture was heated under

reaction conditions similar to those for the synthesis of parent UiO-66(Zr). Figure 2e and Supplementary Fig. 16 show that the resulting material also has the same PXRD pattern as that of the parent UiO-66(Zr). After washing with acid, the resulting H-UiO-66(Zr) was harvested as confirmed by the PXRD, XPS spectra, ICP, EA, thermal gravimetric analysis (TGA) and N_2 adsorption/desorption isotherms (Supplementary Figs 3,16,24 and 36). XPS spectrum indicates that almost no Zn exists in the H-UiO-66(Zr) sample surface after acid treatment. The EA and ICP analyses gave the contents of Zr 27.95, Zn 12.75, C 45.32 and H 3.15% in the template@H-UiO-66(Zr); however, for H-UiO-66(Zr) they are Zr 31.08, Zn 0.25, C 33.12 and H 2.05%. The evaluated mole ratio of C/Zn of the 'template' in template@H-UiO-66(Zr) is 6.8, which is again much smaller than that in $Zn_4O(BC)_6$ molecule (theoretically, the mole ratio of C/Zn is 10.5). This result indicates that there is a 'decomposition and/or rearrangement' of the $Zn_4O(BC)_6$ molecule in the in situ self-assembly reaction, the $Zn_4O(BC)_6$ is not the true template. In addition, as given above the C and H contents in H-UiO-66(Zr) close to the theoretical value (C 34.6 and H 1.68%) of parent UiO-66(Zr), suggesting the removal of most templates during the acid treatment. Moreover, again there is a little Zn in generated H-UiO-66(Zr), probably due to the same reason as proposed above. On the other hand, HAADF-STEM image and EDX mapping also show that Zr, C and O elements were homogeneously distributed in the whole template@H-UiO-66(Zr) but Zn element was heterogeneous (Supplementary Fig. 41).

To further explore the components of the molecule-based MOA templates in fabricating H-UiO-66(Zr) and the template removability, we prepared the H-UiO-66(Zr) by using Zn_4O $(BC-CH_3)_6$ ($BC-CH_3$, 4-methylbenzenecarboxylate) template precursor. As shown in Supplementary Fig. 38, based on the chemical shift at 20 p.p.m. the ^{13}C-NMR spectrum indicates the existence of the $BC-CH_3$ in template@H-UiO-66(Zr). After the acid treatment, this peak disappeared, indicating the removal of the template (Supplementary Fig. 39). Similarly, we also prepared the H-UiO-66(Zr) by using $Zn_4O(BC-NO_2)_6$ ($BC-NO_2$, 4-nitrobenzoic acid). As shown in Supplementary Fig. 32, based on the wavenumber at 1,522 and 1,437 cm^{-1}, the FT-IR spectrum indicates the existence of the $BC-NO_2$ in template@H-UiO-66(Zr). However, after the acid treatment these peaks disappeared, indicating the removal of the template. In terms of the above analysis, it was demonstrated that both metal ions and organic portions in template can be efficiently removed by the acid treatment, making sure the purity of resulting H-UiO-66(Zr).

In the case of MOP-tBu (Fig. 2b)[44], experimentally, as-synthesized MOP-tBu, $ZrCl_4$ and terephthalic acid were mixed and dispersed in DMF. The mixture solution was heated to address the reaction under the conditions similar to those for preparing parent UiO-66(Zr). Again, the resulting material has the same PXRD patterns as the UiO-66(Zr) and acid-treated sample exhibits the characteristic of the H-UiO-66(Zr) (Fig. 2e, Supplementary Figs 4,17,25 and 42). These results indicate that the MOPs can also act as the template precursors to construct H-MOFs.

Above proof-of-concept experiments reveal that MOAs with different original structural complexity and size, including primary complex molecule, supramolecular MOPs and structurally extended MOFs all can act as the template precursors for fabricating H-MOFs. Interestingly, the pore size of resulting H-MOFs is independent on the size of these initially used MOAs or their particles, suggesting that there exists a 'decomposition and/or rearrangement' of them to form new MOAs, which act as the in situ self-assembly templates in the construction of these H-MOFs in given reaction systems.

Preparation of H-MOFs through one-pot reaction. On the base of above interesting findings, we also tried to prepare H-MOFs by a one-pot reaction approach, so as to further confirm the *in situ* self-assembly template mechanism, again on H-UiO-66(Zr). One-pot reaction of $Zn(NO_3)_2 \cdot 6H_2O$, excess terephthalic acid and $ZrCl_4$ in DMF was conducted under solvothermal conditions similar to those for the synthesis of parent UiO-66(Zr). After acid washing, PXRD and N_2 adsorption confirm that the resulting material is H-UiO-66(Zr) (Supplementary Figs 12a and 22a). With the same route, we also checked the applicability of this strategy in using $Zn_4O(BC)_6$ and MOP-*t*Bu precursors as template sources in one-pot reaction. The product based on $Zn_4O(BC)_6$ precursors was obtained from the reaction of $Zn(NO_3)_2 \cdot 6H_2O$, benzoic acid, $ZrCl_4$ and terephthalic acid in DMF. After washing with acid, H-UiO-66(Zr) was obtained as confirmed by the PXRD and N_2 adsorption (Supplementary Figs 10a and 20a). It should be pointed out that the ligand exchange between formed MOFs and free ligands could exist in the solution of this system such as observed in UiO-66, Materials of Institute Lavoisier (MILs) and zeolitic imidazolate frameworks (ZIFs) systems[45,46]. Indeed, the UiO-66(Zr) with ligand missing-linker defects can also be prepared by adding given amount of modulators such as acetic acid, benzoic acid and trifluoroacetic acid in the reaction system[47–49]. However, the resulting new pores in these 'defect' UiO-66(Zr) was small and no obvious hysteresis loop was observed in their N_2 adsorption–desorption isotherms. In addition, we also tried to prepare H-UiO-66(Zr) with only adding HBC in the reaction system through the one-pot reaction between terephthalic acid and $ZrCl_4$. N_2 adsorption indicates that no mesopore was generated in resulting material (Supplementary Fig. 7). Similarly, H-UiO-66(Zr) can also be prepared through the reaction of $Cu(NO_3)_2 \cdot 3H_2O$, 5-*t*-butyl-1,3-benzenedicarboxylic acid, $ZrCl_4$ and terephthalic acid in DMF, followed by acid washing (Supplementary Figs 11a and 21a). All these results demonstrate that a one-pot reaction of MOA template precursors and targeted H-MOF precursors is also feasible in the preparation of H-MOFs. As we expected, the

'formation and degradation/rearrangement' of the MOAs in the *in situ* one-pot reaction enable them to play a template role in the formation of mesoporous structures of resulting H-MOFs.

To examine the universality of this *in situ* self-assembly method, we performed additional synthesis experiments on a couple of other typical MOFs with good physicochemical stabilities, including ZIF-8, MIL-101(Cr), DUT-5 and several functionalized UiO-66(Zr) by one-pot reaction. As expected, 19 H-MOFs were successfully prepared and characterized by EA, ICP, PXRD, FT-IR, TGA and N_2 adsorption (see Supplementary Information). Some structural features of eight representative H-MOFs are listed in Table 1 and the complete data are provided in Supplementary Table 1.

As a whole, it was found that in the preparation, actual templates can not be identified in all cases. Combining element analysis and infrared characterizations, we found that almost no organic residues from the template or free ligands were left in pores of resulting H-MOFs after the acid treatment. It was also demonstrated that the crystallinity of all H-MOFs become to be bad compared with their parent MOFs, probably due to generating numerous defects in the structures of H-MOFs. The lost of crystallinity of H-MOFs in some cases has also been confirmed in literatures[28,32]. Furthermore, the crystallinity and the porosity are directly related to the template amounts as discussed in detail below. These H-MOFs have also a better purity, even if incomparable with their parent MOFs in some cases. For TGA results, some as-synthesized samples present several weight-loss steps, which could be basically ascribed to (1) the loss of a small quantity of free guest solvent molecules in pores of H-MOFs at lower temperature range, (2) the lose of high boiling point solvent molecules and coordinated water molecules in the H-MOFs structures at higher temperature range, and (3) the decomposition of the H-MOFs frameworks. The second step usually was combined by the third, representing a sequential weight loss in a broad temperature range. It was also found that TGA curves of some functional H-MOFs, for example, H-UiO-66-NH$_2$(Zr), exhibited a continuous weight loss similar to its

Table 1 | Pore features of eight representative H-MOFs (prepared by one-pot reaction) and their parent MOFs.

H-MOF and MOF	MOA template precursor type	S_{BET}* (m²g⁻¹)	S_{micro}† (m²g⁻¹)	S_{micro}/S_{meso}‡	V_t§ (m²g⁻¹)	V_{micro}‖ (m²g⁻¹)	V_{micro}/V_{meso}¶	Micro pore size (Å)#	Meso pore size** (Å)
H-ZIF-8	MOF-5	1,611	1,117	2.26	0.82	0.29	0.55	12.7	38.5
ZIF-8		1,737	1,737		0.7			11	
H-MIL-101(Cr)	ZIF-8	441	157	0.55	0.58	0.07	0.13	51	123
MIL-101(Cr)		2,927	2,357		1.49			29/34	
H-DUT-5	In-BPDC	1,183	737	1.65	0.91	0.68	2.96	11.0	39.4
DUT-5		1,652	1,424		0.82			11.1	
H-UiO-66(Zr)	Zn$_4$O(BC)$_6$	917	300	0.49	0.87	0.13	0.17	11.8	38
UiO-66(Zr)		1,204	1,024		0.59			8/11	
H-UiO-66-NH$_2$(Zr)	IRMOF-3	600	327	1.19	0.49	0.14	0.38	11.8/14.7	56.2
UiO-66-NH$_2$(Zr)		1,070	1,052		0.42			7.4/9.5	
H-UiO-66-Cl(Zr)	Zn$_4$O(BC)$_6$	758	334	0.79	0.81	0.15	0.23	12	56.2
UiO-66-Cl(Zr)		794	750		0.33			5.8/7.5	
H-UiO-66-Br(Zr)	MOP-*t*Bu	558	221	0.66	0.50	0.11	0.29	11	55.5
UiO-66-Br(Zr)		806	615		0.43			5.6/7.3	
H-UiO-66(Hf)	MOP-*t*Bu	505	171	0.51	0.77	0.09	0.12	11.2	52.5
UiO-66(Hf)		890	739		0.43			8/11	

H-MOF, hierarchical-pore MOF; MOA, metal-organic assembly; MOF, metal-organic framework.
*S_{BET} is the Brunauer-Emmett-Teller (BET) specific surface area.
†S_{micro} is the *t*-plot-specific micropore surface area calculated from the N_2 adsorption–desorption isotherm.
‡S_{meso} is the specific mesopore surface area estimated by subtracting S_{micro} from S_{BET}.
§V_t is the total specific pore volume determined by using the adsorption branch of the N_2 isotherm at $P/P_0 = 0.99$.
‖V_{meso} is the specific mesopore volume obtained from the Barrett-Joyner-Halenda (BJH) cumulative specific adsorption volume of pores of 1.70–300.00 nm in diameter.
¶V_{micro} is the specific micropore volume calculated by subtracting V_{meso} from V_t.
#The micropore diameter is determined by the density functional theory (DFT) method.
**The mesopore diameter is determined from the local maximum of the BJH distribution of pore diameters obtained in the adsorption branch of the N_2 isotherm at 77 K.

parent UiO-66-NH$_2$(Zr) at the whole temperature range, which might be ascribed to the introduction of polar group in these MOFs (Supplementary Fig. 30)[38].

In addition, we also noticed that the template not only can create mesopore for H-MOFs but also can affect the microporosity of the MOFs (Table 1). For example, the micropore surface area of H-UiO-66(Zr) (prepared by using Zn$_4$O(BC)$_6$ precursors as the template source) is 300 m^2 g^{-1}, which is much smaller than that of parent UiO-66(Zr) (1,024 m^2 g^{-1}), whereas the micropore size is 11.8 and 14.1 Å, being larger than those in UiO-66(Zr) (8 Å and 11 Å, corresponding to the tetrahedral and octahedral cages, respectively). For other H-MOFs, the similar results are observed. That is, the introduction of templates can decrease the micropore surface area and increase the micropore diameter at different degrees of final H-MOFs. In particular, for the H-MIL-101(Cr), there should have been various mesopores with the sizes across the two mesoporous cages (29 and 34 Å) in parent MIL-101(Cr). As a result, no step adsorption similar to that in MIL-101(Cr) was observed in H-MIL-101(Cr) (Supplementary Fig. 12j).

It should be pointed out that the development of reliable methods to rationally prepare H-MOFs with controllable/tailorable structures and properties is much more challenged. In this work, we have definitely confirmed that the mesopore sizes of the resulting H-MOFs are tunable by varying the amount of the MOAs templates. Taking H-UiO-66(Zr) as an example, Fig. 3 shows that the mesopore diameters of prepared materials can vary from 40 to 300 Å, depending on the amount of Zn$_4$O(BC)$_6$ or MOF-5 template precursors used in synthesis. It must be pointed out here that based on our experimental results, only when the amounts of Zn$_4$O(BC)$_6$ template precursors were controlled in the range of 0.25 ~ 0.75 equiv., the decrease of template precursor amounts can increase the mesopore sizes in the resulting H-UiO-66(Zr). As shown in Supplementary Figs 9 and 23, the H-UiO-66(Zr) became of poor crystallinity and even amorphous if the template precursors were more than 0.75 equiv.,

while no mesopore structure was created when less than 0.25 equiv. This observed relationship between mesopore size and the amount of template precursor may be related to the size of the formed template fragments under different precursor concentration conditions. In general, in crystallization process, a lower concentration of the precursor will lead to a lower nucleation rate, which results in a lower concentration of nuclei in the reaction system. A limited number of nuclei grow slowly in the reaction system, consequently resulting in bigger crystal particles. Otherwise, a higher concentration of the precursor can lead to a higher nucleation rate and much more nuclei, thus smaller crystal particles can be expected at a higher concentration of the precursor. As a consequence, an increase of the concentration (amount) of precursor in the reaction mixture usually resulted in a decrease of the generated template particle size in a given concentration range[50]. To further verify the tunability of mesoporosity of H-MOFs prepared by this approach, the mesopore sizes of H-ZIF-8, H-MIL-101(Cr) and H-DUT-5 synthesized with different amounts of template precursor were also tested (Supplementary Figs 13–15). It can be seen clearly from these results that the mesopore sizes of these H-MOFs are dependent on the amount of template precursors used in each case.

As stated above, the ligand-extension strategy for expanding pore size of MOFs often induces the interpenetration and instability of the MOF frameworks. Nevertheless, using our *in situ* self-assembly template method that takes the advantage of relative stability of different MOAs, stable mesoporous structures of H-MOFs can be easily created. Furthermore, the resulting H-MOFs with tunable mesopore sizes are stable just similar to their parents, as confirmed in the case of H-UiO-66(Zr) and UiO-66(Zr) (Supplementary Fig. 43).

Large-molecule adsorption in H-MOFs. We know that adsorbents with large pores could be used in the capture of large

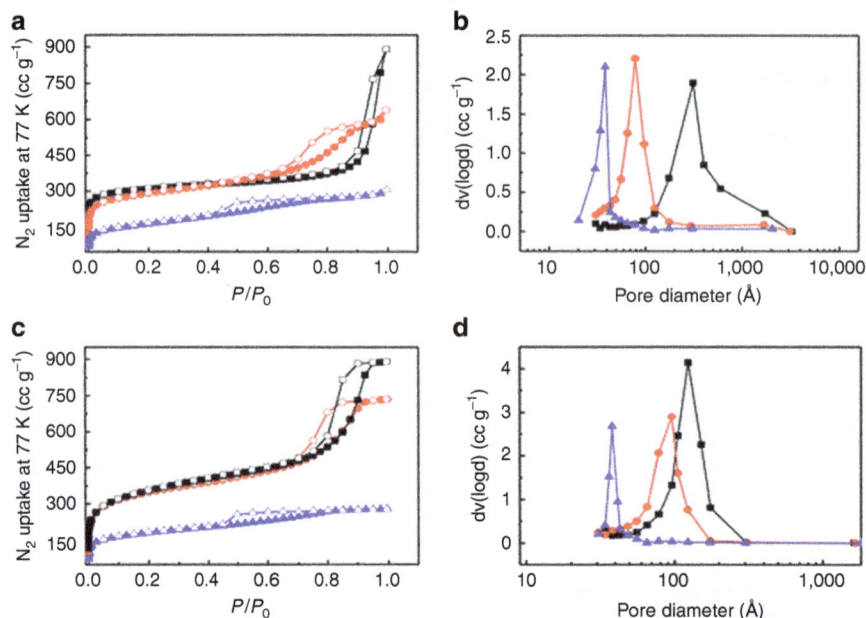

Figure 3 | Tuning the mesopore size through changing the amount of template. (**a,c**) N$_2$ adsorption–desorption isotherms at 77 K and (**b,d**) pore size distributions of H-UiO-66(Zr) prepared with different amounts of template (preparation conditions: (**a,b**) 1 ml (black curve), 2 ml (red curve) and 3 ml (blue curve) of nanosized MOF-5 suspension solution in DMF as the template precursor (the MOF-5 concentration in the suspension solution is about 12 mg ml^{-1}); (**c,d**) 0.25 equiv. (black curve), 0.375 equiv. (red curve) and 0.5 equiv. (blue curve) of Zn$_4$O(BC)$_6$ template precursor (equiv. means the equivalent of Zn$_4$O(BC)$_6$ with respect to ZrCl$_4$)).

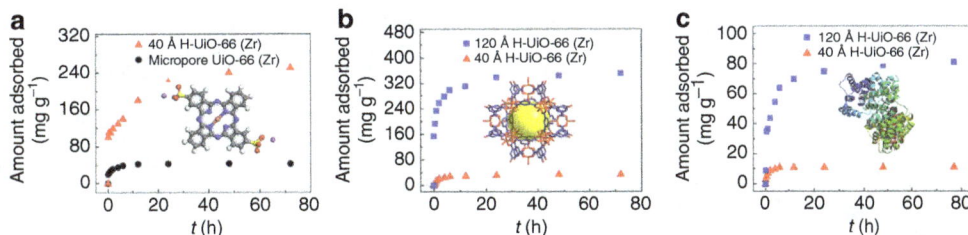

Figure 4 | Large-molecule adsorption in H-UiO-66(Zr). (**a**) Adsorption kinetics of dye DB 86 in microporous UiO-66(Zr) and 40 Å (means mesopore size in the H-MOF) H-UiO-66(Zr), (**b**) adsorption kinetics of MOP-OH in 40 and 120 Å H-UiO-66(Zr), and (**c**) adsorption kinetics of BSA in 40 and 120 Å H-UiO-66(Zr).

molecules. Here, taking H-UiO-66(Zr) as a representative of prepared H-MOFs in this work, we performed liquid-phase adsorption experiments to explore the accessibility of their mesopores towards large molecules with different sizes. For this purpose, organic dye molecule (Direct Blue 86, DB 86, about $4 \times 12 \times 14$ Å in size), MOPs (MOP-OH, about $40 \times 40 \times 40$ Å in size) and biological protein molecule (bovine serum albumin (BSA), about $140 \times 40 \times 40$ Å in size) were selected as the probe molecules. As shown in Fig. 4, microporous UiO-66(Zr) almost can not adsorb DB 86 molecules from aqueous solution, but H-UiO-66(Zr) with about 40 Å mesopores can efficiently capture this large molecule. With regard to recognizing different meso-pore sizes of H-HiO-66(Zr), larger MOP-OH and BSA were further checked. It was found that the H-UiO-66(Zr) with 40 Å mesopores can hardly accommodate the two types of molecules, but the one with 120 Å mesopores can. Although the length of BAS is a little bit longer than the mesopore size of the H-UiO-66(Zr), this material can still accommodate the protein, which can be attributed to the slender geometry of BAS and the irregular shape of mesopore in it. In addition, white H-UiO-66(Zr) powder exhibited a colour change after DB 86 or MOP-OH adsorption as shown in Supplementary Fig. 44. Clearly, these stable H-MOFs with easily tuned mesopores can find potential applications in large-molecule adsorption/separation.

Discussion

Foregoing results demonstrate that using an *in situ* self-assembly template method, by taking advantage of relative stability of different metal-organic coordination assemblies, stable H-MOFs with tailorable pore sizes can be easily prepared based on their stable parent microporous MOFs. Although the exact nature of the template can not be identified at this stage, this approach is quite feasible and seems to be versatile in preparing H-MOFs. Such types of hierarchical-pore materials may provide promising applications in large-molecule adsorption/separation, heterogeneous catalysis and drug release due to their several distinct merits, such as having co-existed micropores and mesopores, large pore size, good stability and improved mass transfer in their pores.

Methods

Synthesis. Experimental procedures for preparing all the H-MOFs achieved in this work are provided in the Supplementary Information. Here we take H-UiO-66(Zr) preparation with MOP-5 precursor as the template source, as an example to describe the preparation method. The synthesis started by mixing parent MOF precursors and template precursors, including 0.120 g of ZrCl₄, 0.160 g of H₂BDC, 0.297 g of Zn(NO₃)₂·6H₂O, with 20 ml of DMF in a 100-ml Teflon liner. After sonication for 10 min the Teflon liner vessel was sealed and placed in a preheated oven at 120 °C for 24 h. After cooling to room temperature, the resulted powder was separated by centrifugation. The supernatant was discarded and the solid was washed several times with DMF to remove the unreacted precursors, to give as-prepared template@H-UiO-66(Zr). To remove the acid-sensitive MOA templates, the solid was dispersed in 10 ml of diluted HCl solution (pH 1) and stirred for about 10 min. After discarding the supernatant, the obtained solid was re-dispersed

in DMF and then centrifuged. This process was repeated three times so that all decomposed template fragments were removed. Next, a similar process was performed by using acetone as the solvent, instead of DMF, afterwhich the solid was heated at 150 °C for 12 h at vacuum, to get activated sample for N₂ adsorption measurement.

Characterization. The characterizations of MOF-5 particle size, EA, ICP, SEM, transmission electron microscope, N₂ adsorption–desorption isotherms at 77 K, PXRD, TGA, FT-IR, XPS and NMR spectra are given in Supplementary Information. Particle size distributions and SEM micrograph of MOF-5 particles are shown in Supplementary Figs 1 and 2. N₂ adsorption–desorption isotherms and pore size distributions of H-MOFs are included in Supplementary Figs 3–15. PXRD patterns of H-MOFs are given in Supplementary Figs 16–23. TGA curves of H-MOFs are given in Supplementary Figs 24–30. FT-IR spectra of H-MOFs are given in Supplementary Figs 31–35. XPS spectra of H-MOFs are included in Supplementary Figs 36 and 37. ¹³C-NMR spectra of H-UiO-66(Zr) are given in Supplementary Figs 38 and 39. HAADF-STEM image and corresponding elemental maps in template@H-UiO-66(Zr) are shown in Supplementary Figs 40–42. PXRD patterns of H-UiO-66(Zr) before and after soaking in water for 24 h are given in Supplementary Fig. 43. Photographs of H-UiO-66(Zr) before and after large molecular adsorption are shown in Supplementary Fig. 44. Porosity properties of various H-MOFs and their parent MOFs are included in Supplementary Table 1. Element mole ratios in template@H-MOFs and H-MOFs are given in Supplementary Table 2.

References

1. Bruce, D. W., O., Hare, D. & Walton, R. I. *Porous Materials* (Wiley, 2010).
2. Zhou, H.-C., Long, J. R. & Yaghi, O. M. Introduction to metal-organic frameworks. *Chem. Rev.* **112**, 673–674 (2012).
3. Long, J. R. & Yaghi, O. M. The pervasive chemistry of metal-organic frameworks. *Chem. Soc. Rev.* **38**, 1213–1214 (2009).
4. Sumida, K. *et al.* Carbon dioxide capture in metal-organic frameworks. *Chem. Rev.* **112**, 724–781 (2012).
5. Li, J.-R., Sculley, J. & Zhou, H.-C. Metal-organic frameworks for separations. *Chem. Rev.* **112**, 869–932 (2012).
6. Yang, Q., Liu, D., Zhong, C. & Li, J. -R. Development of computational methodologies for metal-organic frameworks and their application in gas separations. *Chem. Rev.* **113**, 8261–8323 (2013).
7. Cavka, J. H. *et al.* A new zirconium inorganic building brick forming metal-organic frameworks with exceptional stability. *J. Am. Chem. Soc.* **130**, 13850–13851 (2008).
8. Park, K. S. *et al.* Exceptional chemical and thermal stability of zeolitic imidazolate frameworks. *Proc. Natl Acad. Sci. USA* **103**, 10186–10191 (2006).
9. Loiseau, T. *et al.* A rationale for the large breathing of the porous aluminum terephthalate (MIL-53) upon hydration. *Chem. Eur. J.* **10**, 1373–13823 (2004).
10. Xuan, W., Zhu, C., Liu, Y. & Cui, Y. Mesoporous metal-organic framework materials. *Chem. Soc. Rev.* **41**, 1677–1695 (2012).
11. Feng, D. *et al.* A highly stable zeotype mesoporous zirconium metal-organic framework with ultralarge pores. *Angew. Chem. Int. Ed.* **54**, 149–154 (2015).
12. Wang, K. *et al.* A series of highly stable mesoporous metalloporphyrin Fe-MOFs. *J. Am. Chem. Soc.* **136**, 13983–13986 (2014).
13. Senkovska, I. & Kaskel, S. Ultrahigh porosity in mesoporous MOFs: promises and limitations. *Chem. Commun.* **50**, 7089–7098 (2014).
14. Wang, T. C. *et al.* Ultrahigh surface area zirconium MOFs and insights into the applicability of the BET theory. *J. Am. Chem. Soc.* **137**, 3585–3591 (2015).
15. Song, L. *et al.* Mesoporous metal-organic frameworks: design and applications. *Energy Environ. Sci.* **5**, 7508–7520 (2012).
16. Eddaoudi, M. *et al.* Systematic design of pore size and functionality in isoreticular MOFs and their application in methane storage. *Science* **295**, 469–472 (2002).

17. Deng, H. *et al*. Large-pore apertures in a series of metal-organic frameworks. *Science* **336**, 1018–1023 (2012).

18. Farha, O. K. *et al*. Metal-organic framework materials with ultrahigh surface areas: is the sky the limit. *J. Am. Chem. Soc.* **134**, 15016–15021 (2012).

19. Yan, Y., Yang, S., Blake, A. J. & Schröder, M. Studies on metal-organic frameworks of Cu(II) with isophthalate linkers for hydrogen storage. *Acc. Chem. Res.* **47**, 296–307 (2014).

20. Qiu, L. *et al*. Hierarchically micro- and meso-porous metal-organic frameworks with tunable porosity. *Angew. Chem. Int. Ed.* **47**, 9487–9491 (2008).

21. Feng, S. *et al*. Synthesis of nitrogen-doped hollow carbon nanospheres for CO_2 capture. *Chem. Commun.* **50**, 329–331 (2014).

22. Zhao, Y. *et al*. Metal-organic framework nanospheres with well-ordered mesopores synthesized in an ionic liquid/CO_2/surfactant system. *Angew. Chem. Int. Ed.* **50**, 636–639 (2011).

23. Wee, L. H. *et al*. Copper benzene tricarboxylate metal-organic framework with wide permanent mesopores stabilized by keggin polyoxometallate ions. *J. Am. Chem. Soc.* **134**, 10911–10919 (2012).

24. Sun, L., Li, J. -R., Park, J. & Zhou, H. -C. Cooperative template-directed assembly of mesoporous metal-organic frameworks. *J. Am. Chem. Soc.* **134**, 126–129 (2012).

25. Peng, L. *et al*. Surfactant-directed assembly of mesoporous metal-organic framework nanoplates in ionic liquids. *Chem. Commun.* **48**, 8688–8690 (2012).

26. Wu, Y. *et al*. Amino acid assisted templating synthesis of hierarchical zeolitic imidazolate framework-8 for efficient arsenate removal. *Nanoscale* **6**, 1105–1112 (2014).

27. Pham, M., Vuong, G., Fontaine, F. & Do, T. A route to bimodal micro-mesoporous metal-organic frameworks nanocrystals. *Cryst. Growth. Des.* **12**, 1008–1013 (2012).

28. Do, X., Hoang, V. & Kaliaguine, S. MIL-53(Al) mesostructured metal-organic frameworks. *Micropor. Mesopor. Mat.* **141**, 135–139 (2011).

29. Cao, S. *et al*. Hierarchical bicontinuous porosity in metal-organic frameworks templated from functional block co-oligomer micelles. *Chem. Sci.* **4**, 3573–3577 (2013).

30. Ma, T. *et al*. Ordered mesoporous metal-organic frameworks consisting of metal disulfonates. *Chem. Mater.* **24**, 2253–2255 (2012).

31. Xue, Z. *et al*. Poly(ethylene glycol) stabilized mesoporous metal-organic framework nanocrystals: efficient and durable catalysts for the oxidation of benzyl Alcohol. *ChemPhysChem* **15**, 85–89 (2014).

32. Li, L. *et al*. A synthetic route to ultralight hierarchically micro/mesoporous Al(III)-carboxylate metal-organic aerogels. *Nat. Commun.* **4**, 1774–1783 (2013).

33. Peng, L. *et al*. Highly mesoporous metal-organic framework assembled in a switchable solvent. *Nat. Commun.* **5**, 4456–4463 (2014).

34. Reboul, J. *et al*. Mesoscopic architectures of porous coordination polymers fabricated by pseudomorphic replication. *Nat. Mater.* **11**, 717–723 (2012).

35. Morris, R. E. & Čejka, J. Exploiting chemically selective weakness in solids as a route to new porous materials. *Nat. Chem.* **7**, 381–388 (2015).

36. Schröck, K., Schröder, F., Heyden, M., Fischer, R. A. & Havenith, M. Characterization of interfacial water in MOF-5 ($Zn_4(O)(BDC)_3$), a combined spectroscopic and theoretical study. *Phys. Chem. Chem. Phys.* **10**, 4732–4739 (2008).

37. Greathouse, J. A. & Allendorf, M. D. The interaction of water with MOF-5 simulated by molecular dynamics. *J. Am. Chem. Soc.* **128**, 10678–10679 (2006).

38. DeCoste, J. B. *et al*. Stability and degradation mechanisms of metal-organic frameworks containing the $Zr_6O_4(OH)_4$ secondary building unit. *J. Mater. Chem. A* **1**, 5642–5650 (2013).

39. Han, Y., Li, J.-R., Xie, Y. & Guo, G. Substitution reactions in metal-organic frameworks and metal-organic polyhedra. *Chem. Soc. Rev.* **43**, 5952–5981 (2014).

40. Deria, P. *et al*. Beyond post-synthesis modification: evolution of metal-organic frameworks via building block replacement. *Chem. Soc. Rev.* **43**, 5896–5912 (2014).

41. Abney, C. W. *et al*. Topotactic transformations of metal-organic frameworks to highly porous and stable inorganic sorbents for efficient radionuclide sequestration. *Chem. Mater.* **26**, 5231–5243 (2014).

42. Perry, J. T., Perman, J. A. & Zaworotko, M. J. Design and synthesis of metal-organic frameworks using metal-organic polyhedra as supermolecular building blocks. *Chem. Soc. Rev.* **38**, 1400–1417 (2009).

43. Li, J. -R., Timmons, D. J. & Zhou, H. -C. Interconversion between molecular polyhedra and metal-organic frameworks. *J. Am. Chem. Soc.* **131**, 6368–6369 (2009).

44. Li, J. -R. & Zhou, H. -C. Bridging-ligand-substitution strategy for the preparation of metal-organic polyhedra. *Nat. Chem.* **2**, 893–898 (2010).

45. Kim, M. *et al*. Postsynthetic ligand and cation exchange in robust metal-organic frameworks. *J. Am. Chem. Soc.* **134**, 18082–18088 (2012).

46. Karagiaridi, O. *et al*. Opening ZIF-8: a catalytically active zeolitic imidazolate framework of sodalite topology with unsubstituted linkers. *J. Am. Chem. Soc.* **134**, 18790–18796 (2012).

47. Schaate, A. *et al*. Modulated synthesis of Zr-based metal-organic frameworks: from nano to single crystals. *Chem. Eur. J.* **17**, 6643–6651 (2011).

48. Wu, H. *et al*. Unusual and highly tunable missing-linker defects in zirconium metal-organic framework UiO-66 and their important effects on gas adsorption. *J. Am. Chem. Soc.* **135**, 10525–10532 (2013).

49. Vermoortele, F. *et al*. Synthesis modulation as a tool to increase the catalytic activity of metal-organic frameworks: the unique case of UiO-66(Zr). *J. Am. Chem. Soc.* **135**, 11465–11468 (2013).

50. Shevchenko, E. V. *et al*. Study of nucleation and growth in the organometallic synthesis of magnetic alloy nanocrystals: the role of nucleation rate in size control of $CoPt_3$ nanocrystals. *J. Am. Chem. Soc.* **125**, 9090–9101 (2003).

Acknowledgements

This work is supported by the National Key Basic Research Program of China ('973') (number 2013CB733503), the Natural Science Foundation of China (numbers 21322601, 21271015, 21136001 and 21322603), the Program for New Century Excellent Talents in University (numbers NCET-13-0647 and NCET-12-0755) and the Beijing Municipal Natural Science Foundation (number 2132013).

Author contributions

J.R.L., C.Z. and H.H. conceived and designed the experiments, and co-wrote the paper. H.H. performed most of experiments and analysed data. L.L. and K.W. participated in the N_2 adsorption test. T.H. and M.T. performed the liquid adsorption experiment under the guidance of Q.Y. and D.L. Y.X. performed the PXRD and TGA test. All authors discussed the results and commented on the manuscript.

Additional information

13

Assembling an alkyl rotor to access abrupt and reversible crystalline deformation of a cobalt(II) complex

Sheng-Qun Su[1], Takashi Kamachi[1], Zi-Shuo Yao[1], You-Gui Huang[1], Yoshihito Shiota[1], Kazunari Yoshizawa[1], Nobuaki Azuma[2], Yuji Miyazaki[2], Motohiro Nakano[2], Goro Maruta[3], Sadamu Takeda[3], Soonchul Kang[1], Shinji Kanegawa[1] & Osamu Sato[1]

Harnessing molecular motion to reversibly control macroscopic properties, such as shape and size, is a fascinating and challenging subject in materials science. Here we design a crystalline cobalt(II) complex with an *n*-butyl group on its ligands, which exhibits a reversible crystal deformation at a structural phase transition temperature. In the low-temperature phase, the molecular motion of the *n*-butyl group freezes. On heating, the *n*-butyl group rotates ca. 100° around the C–C bond resulting in 6–7% expansion of the crystal size along the molecular packing direction. Importantly, crystal deformation is repeatedly observed without breaking the single-crystal state even though the shape change is considerable. Detailed structural analysis allows us to elucidate the underlying mechanism of this deformation. This work may mark a step towards converting the alkyl rotation to the macroscopic deformation in crystalline solids.

[1] Institute for Materials Chemistry and Engineering, Kyushu University, 744 Motooka, Nishi-ku, Fukuoka 819-0395, Japan. [2] Research Center for Structural Thermodynamics, Graduate School of Science, Osaka University, Toyonaka, Osaka 560-0043, Japan. [3] Department of Chemistry, Faculty of Science, Hokkaido University, Sapporo 060-0810, Japan. Correspondence and requests for materials should be addressed to O.S. (email: sato@cm.kyushu-u.ac.jp).

The development of new materials that exhibit reversible macroscopic changes in response to external stimuli has attracted significant attention for their potential applications in actuators and their mimicry of muscle cells[1-5]. To date, numerous molecular systems that undergo reversible molecular-level motions have been developed; however, it remains extremely challenging to amplify these motions to the macroscopic regime attained through synergy[6-10]. The lack of research in this area can possibly be attributed to the subtleties of molecular motion in solids, particularly in crystals. In fact, the crystal is still an ideal medium to study the relationship of molecular movements and macroscopic properties because of its close packing potentially, enabling macroscopic changes via molecular cooperation[11-13].

As one of the typical molecular motions, rotational isomerization around the carbon–carbon bond has been widely studied[14]. Alkanes and alkyl derivatives, such as polyethylene, liquid crystals, biological membranes, and variously modified proteins are classical examples that undergo this type of isomerization, which is a major factor in the dynamics and activity of their molecular structures[15-17]. Recently, molecular crystals based on alkyl derivatives that exhibit conformational polymorphism have also been studied[18,19]. However, studies on converting alkyl rotation to reversible and large motion of the macroscopic crystal have never been reported.

With the above in mind, we turned our focus to alkyl derivatives—the potential rotors. Herein, we report a crystalline Co(II) complex containing an n-butyl derivative, which undergoes an entropy-driven structural phase transition and exhibits reversible macroscale deformation that changes with temperature. At the high-temperature phase the frozen conformation of the alkyl chain melts and undergoes the rotational isomerization accompanied by an increase in repulsion between adjacent molecules. This results in significant expansion along the long axis (c axis) of the crystal attained by synergy with π–π interactions.

Results

Preparation of crystalline complex 1.
The cobalt(II) complex with a n-butyl group in its ligand, $[Co(NO_3)_2(L)]$ (complex **1**), was synthesized by layering an acetone solution of a planar tridentate N-containing ligand (L = n-butyl-2,6-di(1H-pyrazol-1-yl) isonicotinate) on an acetone solution of $Co(NO_3)_2 \cdot 4H_2O$ in a glass tube; the complex was successfully obtained as prismatic purple crystals at room temperature (see Methods).

Characterization of the phase transition.
The occurrence of a thermal phase transition in **1** was confirmed by differential scanning calorimetry (DSC) measurements (Fig. 1a and Supplementary Fig. 1). The DSC curves for the polycrystalline **1** sample exhibit a single exothermic peak at 231 K during the cooling process; this peak corresponds to the transition to the low-temperature phase. In contrast, the DSC traces of **1** exhibit an endothermic peak at 248 K during the heating process. The relatively large thermal hysteresis of ca. 17 K and the distinct peak indicate a first-order phase transition.

Structural characterization.
The crystal analyses of **1** were performed at 123 K (low-temperature phase) and at 303 K (high-temperature phase; Supplementary Table 1). The complex at 123 K crystallizes in a monoclinic space group $P2_1/c$, with unit-cell parameters of $a = 9.794(3)$ Å, $b = 27.475(9)$ Å, $c = 7.548(3)$ Å and $\beta = 100.433(8)°$; the asymmetric unit cell consists of one complex molecule. The cobalt ion is coordinated in a distorted pentagonal bipyramid geometry by three nitrogen atoms of the L

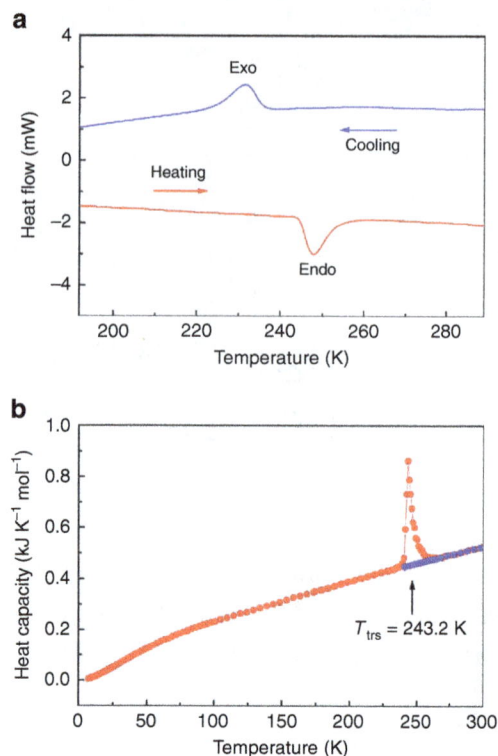

Figure 1 | Temperature dependence of differential scanning calorimetry and heat capacity for 1. (a) DSC curves of crystal **1** recorded at the rate of $10\,K\,min^{-1}$ during a cooling–heating (blue–red lines) cycle. Exo: exothermic peak; Endo: endothermic peak. **(b)** The phase transition temperature (T_{trs}) is determined to be 243.2 K on heating. Filled red and blue circles represent the data obtained during heating mode after the samples were cooled to 7.6 and 240.9 K, respectively.

ligand and four oxygen atoms of two NO_3^- anions. The two NO_3^- anions are located on both sides of the plane defined by the cobalt ion and the L ligand. The ligand L with a n-butyl group exhibits an *anti* conformation, without disorder. When the temperature was increased, a structural phase change occurred. The crystal structure at 303 K shows that **1** has the same space group as that at 123 K but different unit-cell parameters; the cell parameters at 303 K are $a = 9.8533(13)$ Å, $b = 25.995(3)$ Å, $c = 8.1782(10)$ Å and $\beta = 100.741(2)°$. The ligand L now exhibits a *gauche* conformation and disorder, suggesting the occurrence of a *gauche–anti* transition through rotational isomerization[14,20] (Fig. 2a and Supplementary Figs 2a and 3a). The *gauche* conformation, which is denoted by *gauche* **1**, is shown in Supplementary Fig. 3a. The variable-temperature infrared absorbance spectra of **1** support the induction of a structural phase transition (Supplementary Fig. 4) involving dynamic disorder of the n-butyl group[20,21]. The dynamic disorder of the n-butyl group in **1** is supported by the solid-state ^{13}C CP/MAS NMR (magic-angle-spinning nuclear magnetic resonance) spectrum of $[Zn(NO_3)_2(L)]$ (complex **2**; Supplementary Figs 5–7 and Supplementary Table 2) as an analogue to **1**, where the Co^{2+} is replaced with diamagnetic Zn^{2+}, and by the solid-state ^2H NMR spectrum of $[Zn(NO_3)_2(L\text{-}d_9)]$ (complex **2'**; Supplementary Figs 8–11 and Supplementary Table 3)[22,23].

To investigate the structural change in detail, we measured the temperature-dependent variation in length of the three crystallographic axes and variation of the angle β (Fig. 3 and Supplementary Table 1). The unit-cell parameters undergo abrupt changes at temperatures below 238 K and above 243 K during cooling and heating, respectively; these changes correspond to the

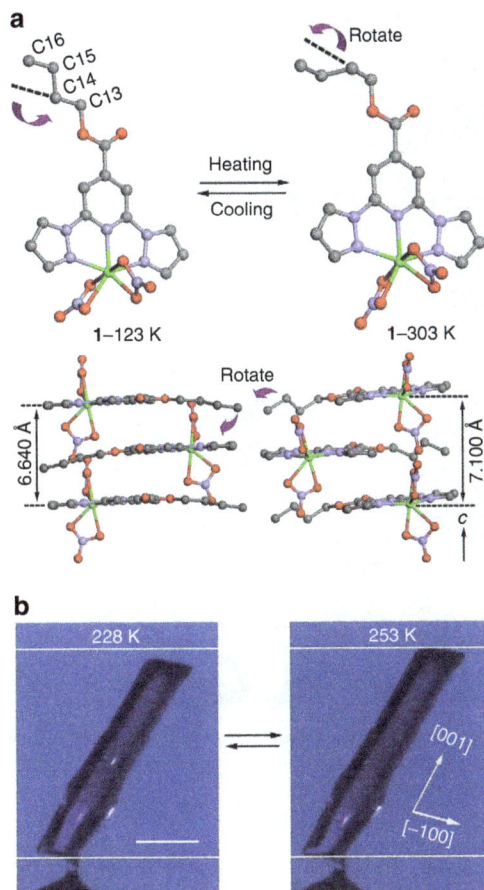

Figure 2 | Molecular structures of crystal 1 in the low- and high-temperature phase. (a) The cobalt ions are coordinated by one tripodal L ligand and two nitrate anions in both phases. The molecules stack in parallel along the c axis with different distances between molecules in the low- and high-temperature phases. The n-butyl group of the ligand rotates ca. 100° around the C13-C14 bond following phase transition. Green, Co; grey, C; blue, N; red, O. Hydrogen and disordered carbon atoms (C14' and C15') are omitted for clarity. **(b)** The crystal length changes from 1.78 to 1.90 mm along [001] from 228 to 253 K and switches back to the original state upon cooling, scale bar represents 500 μm.

occurrence of phase transition, which is consistent with the DSC experimental results. As confirmed by the unit cells measured in the temperature range 123–303 K, we established that the transition between the two phases is reversible, with hysteresis. Whereas thermal expansion along the a axis and c axis is positive, a negative thermal expansion occurs along the b axis[24].

Thermal-induced deformation. Because of the structural phase transition, the prismatic crystal significantly contracts and expands reversibly during cooling and heating. The change is as large as ~7% along the c axis and 5% along the b axis (Figs 2b and 4). The reversible crystal deformation was recorded as two videos (Supplementary Movies 1 and 2). As shown in the Supplementary Movie 1, the transformation was complete in <1 s when the cooling rate was 5 K min^{-1}; the crystal appears to bend slightly like a snake during the contraction process. Moreover, as shown in Fig. 4, a crystal up to 1.9 mm in size exhibits no evidence of appreciable fatigue after 10 cycles, which is distinguishable from the behaviour of normal molecular crystals[11,12,25–27]. Reproducible crystal deformation is an essential property for practical applications and gaining insight into the structure–property relationship.

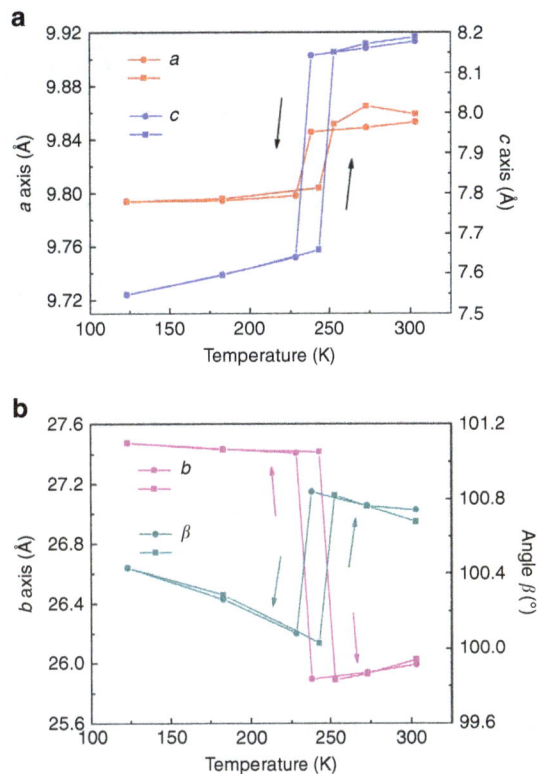

Figure 3 | Cell metrics as a function of temperature. Variable-temperature single-crystal X-ray diffraction data of complex **1** were repeatedly collected for the same crystal in the range 123–303 K (heating, filled squares; cooling, filled circles). **(a)** Changes in the crystallographic a and c axes. **(b)** Variations of the angle β and the crystallographic b axis. The length of the b and c axes exhibit a reversible and abrupt change at ~240 K and the percentage change is ~5, respectively.

Magnetic properties. Notably, the structural phase transition involves changes not only in crystal shape but also in magnetic properties. The magnetic susceptibility of microcrystal **1** was measured under heating and cooling modes in the temperature range 2–300 K; a step is observed at ~245 K (Supplementary Fig. 12a). The change in magnetization is consistent with the phase transition results. The $\chi_m T$ values below and above the transition temperature are 2.30 and 2.36 cm^3 K mol^{-1}, respectively. Magnetic anomaly can be explained as a result of the modulation of the quenching of the orbital angular momentum, which is related to the coordination environment of the cobalt ion[28]. For complex **1**, two significant changes occur in the environment of the cobalt ion before and after the phase transition: a nitrate ion is twisted along the dihedral angle N3–Co1–O6–N7 (α) from 143.43 to 150.40°, and the bond length Co1–O3 changes from 2.198 to 2.342 Å (Supplementary Fig. 12b and Supplementary Table 4).

Discussion
We rationalize the origin of the extraordinarily large thermal deformation for crystal **1** at the molecular level by systematically comparing the crystal structures before and after the phase transition. In the low-temperature phase (123 K), the molecules stack in columns along the crystallographic c axis, as shown in Supplementary Fig. 2a; within each column, they are tilted such that their molecular planes (The plane is defined by the atoms N1, N3 and N5.) form an angle φ of 61.65° relative to the stacking direction [001], which is the long axis of the prismatic crystal. The adjacent molecules in each column are parallel, with an

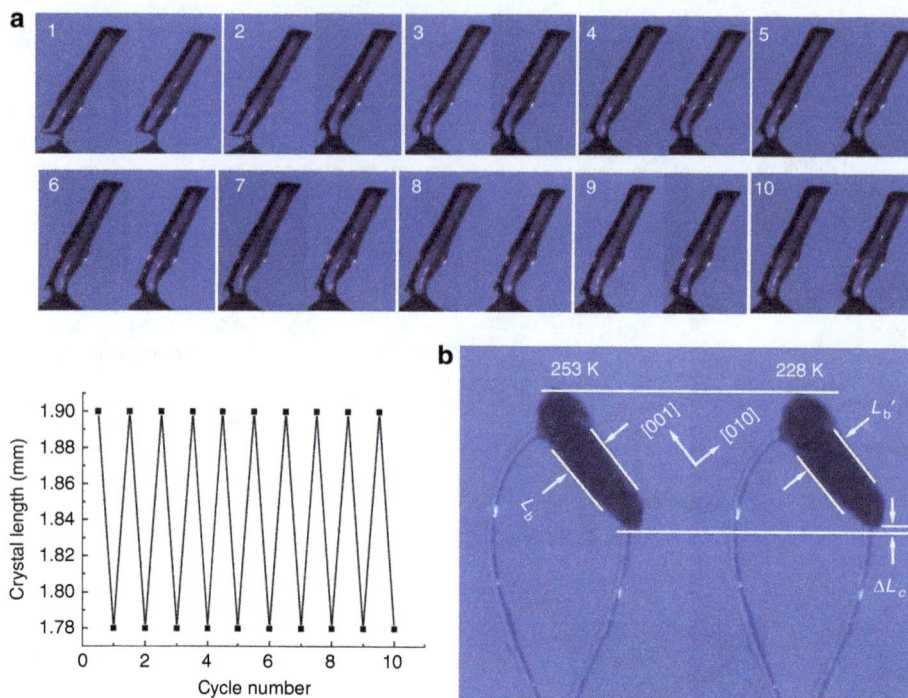

Figure 4 | Reversible shrinkage and expansion of crystal 1. (**a**) Reversible shrinkage and expansion of the prismatic crystal on alternate heating and cooling at a rate of 5 K min^{-1} in the range 228-253 K. The crystal was glued onto a nylon loop and enveloped in a temperature-controlled stream of dry nitrogen gas. The crystal exhibits shrinkage and expansion along the long axis of the crystal in response to temperature. The reversible changes were recorded for 10 cycles without breaking the single-crystal state. (**b**) The increase in crystal size from ~ 0.20 (L_b) to 0.21 mm (L_b') along the [010] direction, and the contraction by ~ 0.05 mm (ΔL_c) (from about 0.83 to 0.78 mm) during cooling from 253 to 228 K.

average distance d_c of 3.320 Å (Fig. 5a), and arrange in the scissor-crossover mode with a crossing angle ω of 74.93° (Supplementary Fig. 13). The column structure can be stabilized by π–π interactions and shape-complementary van der Waals interactions between interdigitated neighbouring π-conjugated molecules (Supplementary Fig. 14a). The three-dimensional molecular packing is primarily dictated by C–H···O nonconventional hydrogen bonds (Supplementary Fig. 2b), which play an important role in the phase transition and stabilization of the overall structure[28]; the hydrogen bond distance ranges from 3.201 to 3.281 Å.

In the high-temperature phase (303 K), the molecules stack in the same modes as in the low-temperature phase (Fig. 5a and Supplementary Fig. 3b), with a tilt angle φ' of 60.25° relative to the stacking direction and a crossing angle ω' of 65.73°; these angles are smaller than the corresponding angles (φ and ω) in the low-temperature phase. The average distance between the neighbouring molecules increases from $d_c = 3.320$ Å in the low-temperature phase to $d_c' = 3.550$ Å in the high-temperature phase on heating to reduce the repulsive interactions from the adjacent molecules along the c axis because the disordered n-butyl groups induced by rotational isomerization occupy more space. The repulsive interactions are mainly from the interactions of the n-butyl group with the molecular planes and coordinated nitrate ions of adjacent molecules (Supplementary Fig. 14b). Actually, the size of the n-butyl group in the direction perpendicular to the molecular plane increased from $d_m = 2.324$ to $d_m' = 2.724$ Å. The changes in the angle φ and in the distance between adjacent molecules result in an increase in the length of the crystal c axis from $\iota_c = 7.548$ in the low-temperature phase to $\iota_c' = 8.178$ Å in the high-temperature phase, and the crystal correspondingly extends $\sim 7\%$ ($\approx \{(d_c' \sin \varphi)/(d_c \sin \varphi') - 1\} \times 100\%$) along the long axis. In this transition process, the crossing angle ω decreases, which can be accounted for by a reduction of the

repulsive interactions that arise from a decrease in the distance between the neighbouring nitrate ions and n-butyl groups after the rotational isomerization. As the rotational isomerization of n-butyl groups occurs, the size of each molecular column along the b axis decreases from $\iota_b = 13.130$ to $\iota_b' = 12.440$ Å, corresponding to the change in the crystal b axis from 27.475 to 25.995 Å (Fig. 5b), and the crystal contracts by $\sim 5\%$ ($\approx \{1-(\iota_b'/\iota_b)\} \times 100\%$), which is not easily observed because of the small crystal size in this direction. Because of the synergetic effect between the angles φ and β, the a axis crystal parameter does not significantly change between the two phases. After the phase transition, the hydrogen bonds change slightly and the π–π interactions are weakened by an increase in d_c to 3.550 Å (Supplementary Figs 3c and 14a).

To obtain information about the phase transition, we performed heat capacity measurements using an adiabatic calorimeter[29] (Fig. 1b). The molar heat capacities under constant pressure (C_p) showed a sharp peak accompanying a latent heat at 243.2 K, which is due to a phase transition, and also exhibited a supercooling phenomenon. These facts provide evidence of a first-order phase transition. Cooperative interaction plays an essential role in the phase transition. The transition enthalpy and entropy estimated from the heat capacity are $\Delta H = 2.579 \pm 0.028$ kJ mol^{-1} and $\Delta S = 10.50 \pm 0.11$ J K^{-1} mol^{-1}, respectively. From Boltzmann's equation $\Delta S = R \ln N$, where N represents the ratio of possible conformations and R is the gas constant, the N value is ~ 4 ($N = 3.54 \approx 4$). Careful investigation of the structure shows that the n-butyl group in **1** is completely ordered in the low-temperature phase. However, two methylene carbon atoms, that is, C14 and C15, are disordered in the high-temperature phase, in which each carbon atom has two possible sites ($\Delta S = R \ln 2^2$). These results suggest that the disordering of the two methylene carbon atoms is mainly responsible for the entropy gain in the phase transition.

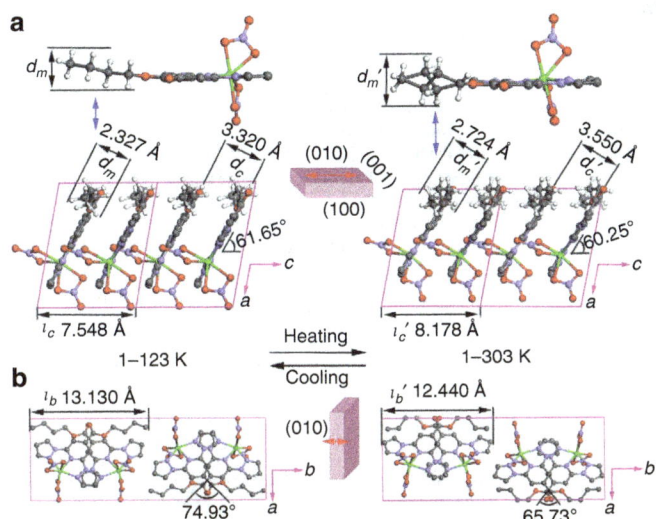

Figure 5 | Molecular packing of a crystal of 1. Crystalline packing of molecules of **1** in the low- and high-temperature phase (**1**—123 K and **1**—303 K) as viewed from (**a**) the (010) face and (**b**) the (001) face. From the low-temperature phase to the high-temperature phase (**a**) the size of the n-butyl group in the direction perpendicular to the molecular plane (d_m), the average distance between the molecules along the c axis (d_c) and the length of the crystal c axis (ι_c) increase to 2.724 Å (d_m'), 3.550 Å (d_c') and 8.178 Å (ι_c'), respectively; the tilt angle φ (Supplementary Fig. 2) decreases from 61.65 to 60.25° (φ', Supplementary Fig. 3); (**b**) the size of each molecular column decreases from 13.130 to 12.440 Å ($\iota_b \rightarrow \iota_b'$) along the b axis; and the crossing angle ω (Supplementary Fig. 2) decreases to 65.73° (ω', Supplementary Fig. 3). The outlines of the crystal morphology and red arrows in the middle of the diagrams show the direction of contraction and expansion of the crystal, respectively. Green, Co; grey, C; white, H; blue, N; red, O.

The N value of 4 suggests that the nature of the dynamic disorder motion is reorientational rather than free rotational. Furthermore, the entropy derived from crystal expansion should also contribute to the entropy gain[30].

As previously discussed, the n-butyl group of the cobalt(II) complex is in the *anti* conformation in the low-temperature phase, and in the *gauche* conformation in the high-temperature phase. We performed periodic density functional theory calculations to investigate the shrinkage and expansion of the crystal. First, we optimized the low-temperature crystal structure; the lattice parameters as well as atomic coordinates were relaxed without symmetry constraints. The calculated unit-cell parameters are in good agreement with the experimental data (Supplementary Figs 15 and 16). We manually altered the structure of the n-butyl group from the *anti* to the *gauche* **1** conformation, and the system was optimized while the unit-cell parameters were kept fixed. In the optimized structure, the n-butyl group experiences a significant steric interaction with an adjacent Co(II) complex, which is likely to increase the distance between the neighbouring Co(II) complexes. The relative energy of 24.5 kJ mol^{-1} decreased to 7.5 kJ mol^{-1} when the unit-cell parameters measured from the *anti* conformation were relaxed. The steric interaction between the n-butyl group in the *gauche* conformation and the neighbouring Co(II) complexes is reduced in the course of the unit-cell-parameter optimization (Supplementary Fig. 17), which indicates that the repulsion is a driving force of the change of the crystal size. The obtained unit cells show an expansion of 7.7 % along the c axis and a shrinkage of 5.4 % along the b axis in the *anti–gauche* transition, which is also in good agreement with the experimental data.

Consequently, compound **1** exhibits a thermally induced structural phase transition at ~240 K; the high-temperature phase has relatively larger volume per molecule, which results in a decrease of the stabilization energy derived from non-covalent interactions between constituent molecules (electro-static and π–π interactions) after the transition from the low-temperature to the high-temperature phase. The large increment of the enthalpy of the high-temperature phase relative to that of the low-temperature phase in the first-order phase transition is compensated by the gain in entropy that originates from the conformational change of the alkyl chain. These changes in orientation at the molecular level are amplified to the deformation at the micrometre scale in the crystal because of the collective motion of molecules in the whole crystal through cooperative interactions.

An important characteristic of **1** is that the thermally driven contraction and expansion of the crystal could be repeatedly observed. It has been reported that the organic compounds exhibiting crystal bending has anisotropic packing; there should be one strong interaction in a direction and a weaker interaction in a nearly perpendicular direction[31]. Careful investigation of the molecular interactions suggests that compound **1** has a one-dimensional nature in structure and that one set of interactions (π–π stacking) along the c axis is substantially stronger than those in a nearly perpendicular direction. The characteristic anisotropic molecular interaction is thought to effectively release the mechanical stress in **1** during the phase transition, resulting in the successful observation of a repeatable crystal deformation.

As described above, the Zn(II) analogue (complex **2**) exhibits the very similar structural change to **1** (Supplementary Figs 5 and 6 and Supplementary Table 2), though the phase transition temperature of **2** is different from that of **1**. On the other hand, when the pentyl, hexyl and decyl groups were used to replace the butyl group in **1**, the reversible transition was not observed in the Co(II) complexes (complexes **3–5**). The crystal structures of these complexes are different from that of **1**, including the conformation of the alkyl chain and molecular stacking in the crystals (Supplementary Fig. 18 and Supplementary Table 2). These results suggest that although the introduction of an alkyl group into the ligand is a good method to induce crystal deformation, the use of the alkyl chain does not always result in the induction of the crystal deformation.

We have reported a crystalline Co(II) complex that exhibits thermally induced, abrupt and reversible macroscopic shrinkage and expansion. In this complex, the n-butyls rotate reversibly in response to temperature as expected. Structural change at the molecular level expands to a macroscopic, abrupt crystal deformation through cooperative interaction in the crystal. The change in crystal size is 6–7% along the c axis, which is one of the largest reversible changes observed among metal–organic molecular crystals. Crystal structure analysis indicated that the π–π stacking pattern of rigid aromatic units enables the rotation of the n-butyl groups. The presence of strong and weak interactions in nearly orthogonal directions is responsible for the excellent resistance to fatigue. Thus, an alkyl chain can be used as a potential rotor for constructing thermally responsive crystalline materials, allowing for the potential use of this crystalline Co(II) complex as an actuator. This work provides a suitable model to understand the structure–property relationship between crystal deformation and alkyl rotation.

Methods

Materials. All reagents were obtained from commercial suppliers and were used without further purification. The ligands, n-butyl-2,6-di(1H-pyrazol-1-yl) isonicotinate (L), n-butyl-d_9 2,6-di(1H-pyrazol-1-yl)isonicotinate (L-d_9), *n*-pentyl 2,6-di(1*H*-pyrazol-1-yl)isonicotinate (L1), *n*-hexyl 2,6-di(1*H*-pyrazol-1-yl)isonicotinate (L2) and *n*-decyl 2,6-di(1*H*-pyrazol-1-yl)isonicotinate (L3) were synthesized according to the literature methods, with minor modifications[32].

Synthesis of complex 1. The target complex $[Co(NO_3)_2(L)]$ was prepared by layering an acetone solution of L (0.03 M, 5 cm^3) on an acetone solution of $Co(NO_3)_2 \cdot 4H_2O$ (0.03 M, 5 cm^3) in a tube. The tube was sealed and left undisturbed at room temperature. X-ray-quality purple prismatic crystals appeared 2 days later in 66% yield. Analysed (calculated) (%) for $C_{16}H_{17}CoN_7O_8$: C, 38.98 (38.88); H, 3.46 (3.47); and N, 19.73 (19.84).

Synthesis of complexes 2 and 2′. The analogue $[Zn(NO_3)_2(L)]$ (2) was prepared by mixing an acetone solution of L (0.02 M, 5 cm^3) with an acetone solution of $Zn(NO_3)_2 \cdot 6H_2O$ (0.02M, 5 cm^3) in a beaker under stirring for 2 min. The solution was filtered and allowed to slowly evaporate at room temperature to isolate crystalline solids with 52% yield. Analysed (calculated) (%) for $C_{16}H_{17}ZnN_7O_8$: C, 38.38 (37.98); H, 3.42 (3.43); and N, 19.58 (19.33). $[Zn(NO_3)_2(L\text{-}d_9)]$ (2′) was obtained in 55% yield via the same method used to prepare complex 2. Analysed (calculated) (%) for $C_{16}H_8D_9ZnN_7O_8$: C, 37.54 (37.70); and N, 19.13 (19.24).

Synthesis of complexes 3–5. Complexes 3–5 were synthesized following the same synthetic procedure as that for complex 2 by mixing an acetone solution of L1/L2/L3 (0.02 M, 5 cm^3) with an acetone solution of $Co(NO_3)_2 \cdot 4H_2O$ (0.02 M, 5 cm^3). Purple prismatic crystals were obtained (yield 68% for 3, 55% for 4 and 57% for 5). Analysed (calculated) (%) for 3 ($C_{17}H_{19}CoN_7O_8$): C, 40.17 (39.97); H, 3.77 (3.72); and N, 19.29 (19.25); for 4 ($C_{18}H_{23}CoN_7O_9$): C, 40.01 (40.02); H, 4.29 (4.34); and N, 18.15 (18.11); for 5 ($C_{22}H_{29}CoN_7O_8$): C, 45.68 (44.34); H, 5.05 (5.10); and N, 16.95 (16.25).

Single-crystal X-ray diffraction. Single-crystal X-ray data were collected on a Rigaku charge-coupled device diffractometer. A crystal was glued onto a nylon loop and enveloped in a temperature-controlled stream of dry nitrogen gas during data collection. The variable-temperature single-crystal data of complex 1 were repeatedly collected for the same crystal during cooling and heating. During cooling, single-crystal data were recorded at 303, 273, 238, 228, 183 and 123 K, however, during heating, single-crystal data were recorded at 183, 243, 253, 273 and 303 K (Supplementary Data 1). The single-crystal data of complexes 2–5 and 2′ were also collected (Supplementary Data 2 and 3). The structures were solved and refined by full-matrix least squares on F^2 using the SHELX programme[33] with anisotropic thermal parameters for all nonhydrogen atoms. Hydrogen atoms were added geometrically and refined using the riding model.

Measurements. DSC measurements of the polycrystalline sample were performed on a Seiko EXSTAR 6000 instrument using cooling and heating rates of 10 K min^{-1}. Heat capacity measurements of the polycrystalline sample were carried out in the temperature range between 7 and 300 K with a laboratory-made low-temperature adiabatic calorimeter[29]. The sample of 0.14947 g after buoyancy correction was loaded into a gold-plated copper cell and sealed together with helium gas at ambient pressure using an indium gasket. The helium gas functions as a heat exchange medium. Thermometry was performed with a rhodium–iron alloy resistance thermometer (nominal 27 Ω, Oxford Instruments) calibrated on the basis of the international temperature scale of 1990 (ITS-90). Infrared spectra were recorded using fine powder adhered to a CaF$_2$ plate on a JASCO FT/IR-600 Plus spectrometer in the 400–4,000 cm^{-1} region. ^{13}C-NMR spectra were measured on a JNM-LA400 NMR spectrometer (JEOL, Japan). Solid-state ^{13}C-NMR spectra were measured with a CP/MAS probe. The sample (ca. 60 mg) was contained in a ceramic cylindrical rotor that was spun at 15 kHz. Solid-state ^{2}H-NMR spectra were measured by a quadrupole echo pulse sequence $\pi/2x$–τ–$\pi/2y$ on a Bruker DSX 300 spectrometer; the values for the $\pi/2$ pulse width and τ were and 2.0 and 20 µs and the repetition time was varied between 2 and 5 s. Direct current (DC) magnetic susceptibility measurements were performed on a MPMS-5S SQUID magnetometer in an applied field of 5,000 G over the whole temperature range.

Computational methods. All calculations were performed with the DMol3 programme[34,35] in Material Studio (Accelrys, Inc.). The Perdew–Burke–Ernzerhof generalized gradient functional was employed for the exchange-correlation energy. The wave functions were expanded in terms of numerical basis sets. We employed the DND basis set (double numerical basis set with the *d*-type polarization functions) for geometry optimization. The Brillouin zone was sampled with a (3 × 1 × 3) Monkhorst–Pack[36] mesh of *k*-points. To reasonably describe weak interactions between the Co complexes, we used the dispersion correction method developed by Tkatchenko and Scheffler[37]. In addition to the atomic coordinates, the unit-cell parameters were optimized without symmetry constraints unless otherwise noted.

References

1. Kobatake, S., Takami, S., Muto, H., Ishikawa, T. & Irie, M. Rapid and reversible shape changes of molecular crystals on photoirradiation. *Nature* **446**, 778–781 (2007).
2. Pei, Z. *et al.* Mouldable liquid-crystalline elastomer actuators with exchangeable covalent bonds. *Nat. Mater.* **13**, 36–41 (2014).
3. Das, D., Jacobs, T. & Barbour, L. J. Exceptionally large positive and negative anisotropic thermal expansion of an organic crystalline material. *Nat. Mater.* **9**, 36–39 (2010).
4. Ma, M., Guo, L., Anderson, D. G. & Langer, R. Bio-inspired polymer composite actuator and generator driven by water gradients. *Science* **339**, 186–189 (2013).
5. Badjić, J. D., Balzani, V., Credi, A., Silvi, S. & Stoddart, J. F. A molecular elevator. *Science* **303**, 1845–1849 (2004).
6. Delius, M. V., Geertsema, Edzard, M. & Leigh, D. A. A synthetic small molecule that can walk down a track. *Nat. Chem.* **2**, 96–101 (2010).
7. Du, G. *et al.* Muscle-like supramolecular polymers: integrated motion from thousands of molecular machines. *Angew. Chem. Int. Ed.* **51**, 12504–12508 (2012).
8. Bruns, C. J. & Stoddart, J. F. Molecular machines muscle up. *Nat. Nanotech.* **8**, 9–10 (2013).
9. Browne, W. R. & Feringa, B. L. Making molecular machines work. *Nat. Nanotech.* **1**, 25–35 (2006).
10. Vogelsberg, C. S. & Garcia-Garibay, M. A. Crystalline molecular machines: function, phase order, dimensionality, and composition. *Chem. Soc. Rev.* **41**, 1892–1910 (2012).
11. Yao, Z.-S. *et al.* Molecular motor-driven abrupt anisotropic shape change in a single crystal of a Ni complex. *Nat. Chem.* **6**, 1079–1083 (2014).
12. Sahoo, S. C., Panda, M. K., Nath, N. K. & Naumov, P. Biomimetic crystalline actuators: structure – kinematic aspects of the self-actuation and motility of thermosalient crystals. *J. Am. Chem. Soc.* **135**, 12241–12251 (2013).
13. Shima, T. *et al.* Thermally driven polymorphic transition prompting a naked-eye-detectable bending and straightening motion of single crystals. *Angew. Chem. Int. Ed.* **53**, 7173–7178 (2014).
14. Zheng, J., Kwak, K., Xie, J. & Fayer, M. D. Ultrafast carbon-carbon single-bond rotational isomerization in room-temperature solution. *Science* **313**, 1951–1955 (2006).
15. Bassett, D. C. & Turner, B. New high-pressure phase in chain-extended crystallization of polythene. *Nat. Phys. Sci.* **240**, 146–148 (1972).
16. Sorai, M. & Saito, K. Alkyl chain acting as entropy reservoir in liquid crystalline materials. *Chem. Rec.* **3**, 29–39 (2003).
17. Nagle, J. F. Theory of the main lipid bilayer phase transition. *Ann. Rev. Phys. Chem.* **31**, 157–195 (1980).
18. Sokolov, A. N., Swenson, D. C. & MacGillivray, L. R. Conformational polymorphism in a heteromolecular single crystal leads to concerted movement akin to collective rack-and-pinion gears at the molecular level. *Proc. Natl Acad. Sci. USA* **105**, 1794–1797 (2008).
19. Ellena, J. *et al.* Temperature-driven isosymmetric reversible phase transition of the hormone estradiol 17β valerate. *Cryst. Growth Des.* **14**, 5700–5709 (2014).
20. Maroncelli, M., Qi, S. P., Strauss, H. L. & Snyder, R. G. Nonplanar conformers and the phase behavior of solid n- alkanes. *J. Am. Chem. Soc.* **104**, 6237–6247 (1982).
21. Sorai, M., Tsuji, K., Suga, H. & Seki, S. Studies on disc-like molecules I. Heat capacity of benzene hexa-n- hexanoate from 13 to 393 K. *Mol. Cryst. Liq. Cryst.* **59**, 33–58 (1980).
22. Facey, G. A., Connolly, T. J., Bensimon, C. & Durst, T. A solid state NMR and X-ray crystallographic investigation of dynamic disorder in solid tetrahydronaphalene derivatives. *Can. J. Chem.* **74**, 1844–1851 (1996).
23. McDowell, C. A. *et al.* The conformational analysis of tetrahydronaphthoquinones using high resolution solid state ^{13}C NMR spectroscopy. *Tetrahedron Lett.* **22**, 4779–4782 (1981).
24. Goodwin, A. L., Chapman, K. W. & Kepert, C. J. Guest-dependent negative thermal expansion in nanoporous prussian blue analogues $M^{II}Pt^{IV}(CN)_6$â · $x\{H_2O\}$ ($0 \leq x \leq 2$; M) Zn, Cd. *J. Am. Chem. Soc.* **127**, 17980–17981 (2005).
25. Skoko, Ž., Zamir, S., Naumov, P. & Bernstein, J. The Thermosalient phenomenon. 'jumping crystals' and crystal chemistry of the anticholinergic agent oxitropium bromide. *J. Am. Chem. Soc.* **132**, 14191–14202 (2010).
26. Steiner, T., Hinrichs, W., Saenger, W. & Gigg, R. 'Jumping crystals': X-ray structures of the three crystalline phases of (±)-3,4-di-O-acetyl-1,2,5,6-tetra-O-benzyl-myo-inositol. *Acta Crystallogr.* **B49**, 708–718 (1993).
27. Centore, R. *et al.* A series of compounds forming polar crystals and showing single-crystal-to-single-crystal transitions between polar phases. *CrystEngComm* **14**, 2645–2653 (2012).
28. Juhász, G. *et al.* Bistability of magnetization without spin-transition in a high-spin cobalt(II) complex due to angular Momentum quenching. *J. Am. Chem. Soc.* **131**, 4560–4561 (2009).
29. Kume, Y., Miyazaki, Y., Matsuo, T. & Suga, H. Low-temperature heat-capacities of ammonium hexachlorotellurate and its deuterated analog. *J. Phys. Chem. Solids* **53**, 1297–1304 (1992).

30. Hoffman, J. D. Hindered intermolecular rotation in the solid state: thermal and dielectric phenomena in long chain compounds. *J. Chem. Phys.* **20,** 541–549 (1952).

31. Reddy, C. M. *et al.* Structural basis for bending of organic crystals. *Chem. Commun.* 3945–3947 (2005).

32. Klein, C., Baranoff, E., Grätzel, M. & Nazeeruddin, M. K. Convenient synthesis of tridentate 2,6-di(pyrazol-1- yl)-4-carboxypyridine and tetradentate 6,6'-di(pyrazol-1-yl)-4,4'-dicarboxy-2,2'- bipyridine ligands. *Tetrahedron Lett.* **52,** 584–587 (2011).

33. Sheldrick, G. M. *SHELXL-97 Program for Refinement of Crystal Structures* (University of Göttingen, 1997).

34. Delley, B. An all-electron numerical-method for solving the local density functional for polyatomic-molecules. *J. Chem. Phys.* **92,** 508–517 (1990).

35. Delley, B. From molecules to solids with the DMol(3) approach. *J. Chem. Phys.* **113,** 7756–7764 (2000).

36. Monkhorst, H. J. & Pack, J. D. Special points for brillouin-zone integrations. *Phys. Rev. B* **13,** 5188–5192 (1976).

37. Tkatchenko, A. & Scheffler, M. Accurate molecular van der waals interactions from ground-state electron density and free-atom reference data. *Phys. Rev. Lett.* **102,** 073005 (2009).

Acknowledgements

This work was partly supported by JSPS KAKENHI Grant Numbers 26104528, 15H01018, 25410071, 24109014 and 15K13710; by Network Joint Research Center for Materials and Devices; and by Nanotechnology Platform Program (Molecule and Material Synthesis) of the Ministry of Education, Culture, Sports, Science and Technology (MEXT).

Author contributions

S.-Q.S and O.S designed this study, implemented experiments and wrote the manuscript; Z.-S.Y. and Y.-G.H. contributed to measuring the crystal deformation and diffraction; T.K., Y.S. and K.Y. performed the calculations and wrote the related discussion; N.A., Y.M. and M.N. performed the heat capacity measurements; G.M. and S.T. contributed to the measurement and analysis of the solid NMR data; S. Kanegawa and S. Kang assisted in measuring the magnetic properties; all authors discussed the results and commented on the manuscript.

Additional information

Accession codes: The X-ray crystallographic coordinates for structures reported in this article have been deposited at the Cambridge Crystallographic Data Centre (CCDC), under deposition numbers CCDC 1024026-1024036, CCDC 1407262-1407266, CCDC 1415773 and CCDC 1415774. These data can be obtained free of charge from The Cambridge Crystallographic Data Centre via www.ccdc.cam.ac.uk/data_request/cif.

Competing financial interests: The authors declare no competing financial interests.

14

Structural isomerism in gold nanoparticles revealed by X-ray crystallography

Shubo Tian[1], Yi-Zhi Li[2], Man-Bo Li[1], Jinyun Yuan[3], Jinlong Yang[3], Zhikun Wu[1] & Rongchao Jin[4]

Revealing structural isomerism in nanoparticles using single-crystal X-ray crystallography remains a largely unresolved task, although it has been theoretically predicted with some experimental clues. Here we report a pair of structural isomers, Au_{38T} and Au_{38Q}, as evidenced using electrospray ionization mass spectrometry, X-ray photoelectron spectroscopy, thermogravimetric analysis and indisputable single-crystal X-ray crystallography. The two isomers show different optical and catalytic properties, and differences in stability. In addition, the less stable Au_{38T} can be irreversibly transformed to the more stable Au_{38Q} at 50 °C in toluene. This work may represent an important advance in revealing structural isomerism at the nanoscale.

[1] Key Laboratory of Materials Physics, Anhui Key Laboratory of Nanomaterials and Nanotechnology, Institute of Solid State Physics, Chinese Academy of Sciences, Hefei, Anhui 230031, China. [2] State Key Laboratory of Coordination Chemistry, School of Chemistry and Chemical Engineering, Nanjing University, Nanjing 210093, China. [3] Hefei National Laboratory for Physical Sciences at the Microscale and Synergetic Innovation Center of Quantum Information and Quantum Physics, University of Science and Technology of China, Hefei, Anhui 230026, China. [4] Department of Chemistry, Carnegie Mellon University, Pittsburgh, Pennsylvania 15213, United States. Correspondence and requests for materials should be addressed to Z.W. (email: zkwu@issp.ac.cn) or to R.J. (email: rongchao@andrew.cmu.edu).

Structural isomerism in organic molecules is a common occurrence due to the bonding diversity of carbon. However, for nanoscale or even larger scale materials, experimental observation of structural isomerism has been largely impeded by the challenge of unravelling the intrinsic structure at the atomic level[1]. Nevertheless, theoretical and experimental efforts[2–8] in searching for structural isomerism in such materials continue, because such a finding would provide precise and insightful structure–property correlations and meaningful guidance for designing and synthesizing unique functional materials. The recently developed ultrasmall, thiolated metal nanoparticles (also called nanoclusters) provide opportunities for investigating structural isomerism, as they can now be controlled with atomic precision[9–20] and their structures can be resolved by single-crystal X-ray crystallography (SCXC) as well. To date, the structures of a series of thiolated metal nanoparticles with various sizes have been elucidated experimentally and theoretically[21–28]; however, to the best of our knowledge, no structural isomerism in thiolated metal nanoparticles has been reported, albeit $Au_{24}(SCH_2Ph-^tBu)_{20}$ and $Au_{24}(SePh)_{20}$ were revealed to have different Au_{24} core structures[29,30]. In a strict sense, $Au_{24}(SCH_2Ph-^tBu)_{20}$ and $Au_{24}(SePh)_{20}$ are not structural isomers, since their ligands are different. Thus, structural isomerism in thiolated nanoparticles remains a mystery.

In the current work, using a modified synthesis method for Au_{25}, we synthesize a nanocluster, whose composition is determined to be the same as that of the previously reported $Au_{38}(PET)_{24}$ (refs 31–33) (PET, phenylethanethiolate), as evidenced by electrospray ionization mass spectrometry (ESI–MS) in combination with X-ray photoelectron spectroscopy (XPS) and thermogravimetric analysis (TGA). SCXC reveals that the structure of this nanocluster is different from that of the previously reported structure[24]. To differentiate the two structures, the previous Au_{38} is denoted as Au_{38Q} and our nanocluster is denoted as Au_{38T} (where Q and T are the surname initial of the first author of the previous and current work, respectively). Au_{38T} and Au_{38Q} are therefore structural isomers and they represent the first pair of structural isomers in nanoparticles as revealed by SCXC, to the best of our knowledge. The two isomers exhibit distinctly different optical, stability and catalytic properties, and the less stable Au_{38T} can be irreversibly transformed to the more stable Au_{38Q} at 50 °C.

Results

Characterization. Au_{38T} was synthesized using a modified one-pot method[34] and isolated using preparative thin-layer chromatography (PTLC)[35,36]. ESI–MS was employed to determine the exact molecular mass and formula of the novel nanoparticle (note: caesium acetate was added to form positively charged adducts). Three distinct peaks centred at m/z 10910.186, 7317.312 and 5522.217 were observed in the mass spectrum (Fig. 1a). The peaks at m/z 10910.186 and 5522.217 (almost half of 10910.186) can be readily assigned to $[Au_{38}(PET)_{24}Cs]^+$ (theoretical m/z value: 10910.658; deviation: 0.472) and $[Au_{38}(PET)_{24}Cs_2]^{2+}$ (theoretical m/z value: 5522.772; deviation: 0.555), respectively. The peak at m/z 7317.312 can be assigned to $Au_{26}(PET)_{16}$ (theoretical m/z value: 7316.818; deviation: 0.494), which could be a fragment of $Au_{38}(PET)_{24}$, because the nanoparticles are monodisperse, as demonstrated by TLC, and it is also observed in the ESI spectrum of Au_{38Q} (see below). Based on the ESI–MS results, it is concluded that the as-prepared nanoparticle is neutral, and that its composition is $Au_{38}(PET)_{24}$, which is also corroborated by the TGA and XPS. TGA shows a weight loss of 30.39 wt% (Fig. 1b), corresponding to the theoretical loss of 30.55 wt% according to the formula. No other

elements (including Cl, Br, N and Na) was detected by XPS (Fig. 1c), which excludes the possibility of existence of potential counterions such as Cl^-, Br^-, $[N(C_8H_{17})_4]^+$ and Na^+; thus, the as-prepared nanoparticle is neutral. Quantitative measurement reveals that the Au/S atomic ratio is 38.0:24.3 (Supplementary Figs 3 and 4), in good agreement with the expected ratio (38.0:24.0) for the composition of $Au_{38}(PET)_{24}$. Thus, the formula is identical to that of the nanoparticle previously reported in ref. 31; however, the absorption spectrum of our nanoparticle distinctly differs from that of the previous nanoparticle. The ultraviolet–visible–near-infrared spectrum of the novel $Au_{38}(PET)_{24}$ (abbreviated as Au_{38T}) shows six absorption peaks at 505 nm (ε: 3.86 gcm l^{-1}), 540 nm (ε: 3.22 gcm l^{-1}), 610 nm (ε: 1.46 gcm l^{-1}), 700 nm (ε: 0.69 gcm l^{-1}), 880 nm and 1,090 nm (Fig. 1d and Supplementary Fig. 2). The previous $Au_{38}(PET)_{24}$ (abbreviated as Au_{38Q}) shows six absorption peaks centred at 480 nm (ε: 4.62 gcm l^{-1}), 520 nm (ε: 3.72 gcm l^{-1}), 570 nm (ε: 2.86 gcm l^{-1}), 627 nm (ε: 2.59 gcm l^{-1}), 740 nm (ε: 0.58 gcm l^{-1}) and 1,035 nm (Fig. 1d and Supplementary Fig.1). TLC also indicates that they are not the same nanoparticle (Fig. 1d, inset). Indeed, they are a pair of structural isomers (vide infra).

Atomic structure. The structure of the previous Au_{38Q} was determined by SCXC and it has a core-shell structure consisting of a face-fused bi-icosahedral Au_{23} core, which is capped by a second shell composed of the remaining 15 gold atoms (Fig. 2f). To confirm that our nanoparticle (Au_{38T}) is an isomer of Au_{38Q}, we grew high-quality single crystals and successfully elucidated the structure via SCXC. Briefly, the new structure of Au_{38T} is composed of one Au_{23} core and one mixed capping layer of thiolate ligands and gold–thiolate complex units. The Au_{23} core consists of one icosahedral Au_{13} and one Au_{10} unit, and the mixed surface layer contains two $Au_3(SR)_4$ staple units, three $Au_2(SR)_3$ staple units, three $Au_1(SR)_2$ staple units and one bridging thiolate SR ligand. The anatomy of the Au_{38T} structure starts with the central Au_{23} core (Fig. 2b), which can be viewed as one Au_{12} cap and one Au_{13} icosahedron (Fig. 2a) fused together via sharing two gold atoms (Fig. 2b, dark green gold atoms), which is in distinct contrast with the case of Au_{38Q}; for the latter, the two Au_{13} icosahedra are fused together via sharing a face (three gold atoms) to form a bi-icosahedral Au_{23} core. The Au_{12} cap is composed of three tetrahedra and the Au–Au bond lengths in each tetrahedron range from 2.71 to 2.88 Å. In the Au_{13} icosahedron, the Au–Au bond lengths between the central atom and the shell Au atoms (except for the two shared gold atoms) vary from 2.71 to 2.82 Å. The bond lengths between the two shared gold atoms and the central atom of Au_{13} icosahedron are 2.77 and 2.78 Å, respectively. The different Au_{23} core in Au_{38T} (in contrast to the biicosahedral Au_{23} core of Au_{38Q}) leads to various surface-binding structures. The Au_{23} core in our case was capped by two $Au_3(SR)_4$ units and two $Au(SR)_2$ units, and the average Au–S bond lengths/Au–S–Au bond angles were 2.33 Å/96.47° and 2.32 Å/94.43° in the $Au_3(SR)_4$ and $Au(SR)_2$ staple units, respectively (Fig. 2c). Interestingly, in addition to the two $Au_3(SR)_4$ units and two $Au(SR)_2$ units, one bridging thiolate (SR) is also found to link the Au_{13} icosahedron and the Au_{12} cap (Fig. 2c, the sulfur atom is marked in red), the two Au–S bond lengths are 2.33 and 2.30 Å, respectively, and the Au–S–Au bond angle is 92.66°. It is noteworthy that in Au_{38Q}, the Au_{23} core was protected by six $Au_2(SR)_3$ and three $Au(SR)_2$ staple units; no comparable $Au_3(SR)_4$ staple units and bridging thiolate (SR) was observed. The Au_{13} icosahedron in Au_{38T} is exclusively capped by two $Au_2(SR)_3$ staple units (the average Au–S bond lengths in the $Au_2(SR)_3$ staple units are 2.35 and 2.34 Å, respectively, and the

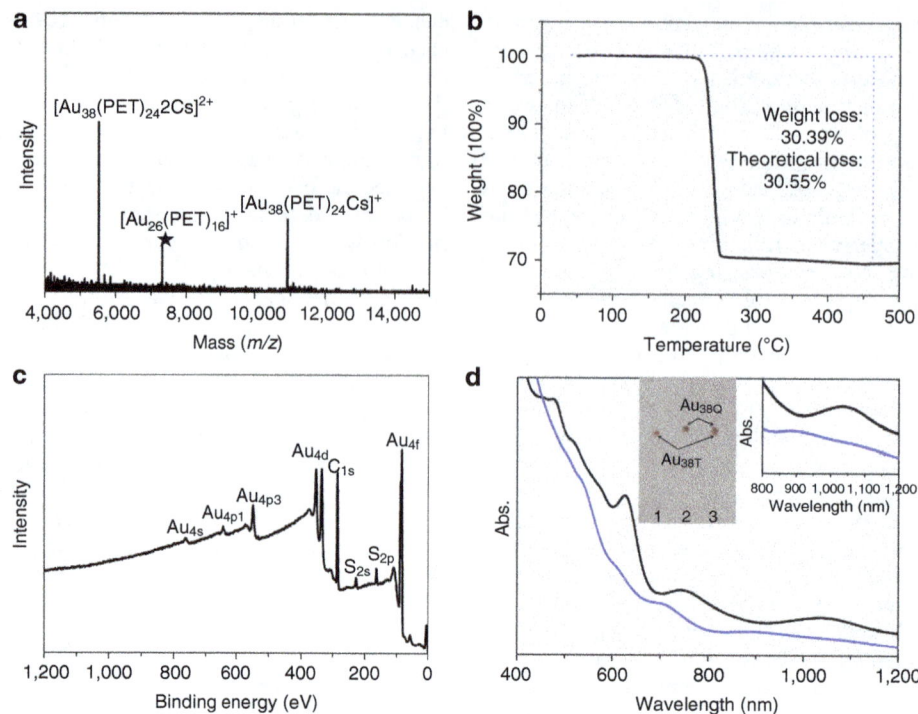

Figure 1 | Characteriztion of Au$_{38T}$. (**a**) ESI mass spectrum of the Au$_{38T}$. (**b**) TGA of Au$_{38T}$. (**c**) XPS spectrum of Au$_{38T}$. (**d**) Ultraviolet-visible-near-infrared absorption spectra of Au$_{38T}$ (blue) and Au$_{38Q}$ (black) in toluene (measurement temperature: 0 °C). Insets are the photo of thin-layer chromatography, and enlarged absorption spectra in the range from 800 to 1,200 nm of Au$_{38T}$ and Au$_{38Q}$.

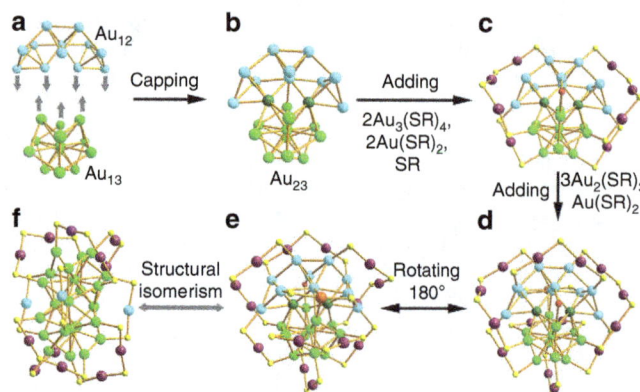

Figure 2 | Structures of Au$_{38T}$ and Au$_{38Q}$. (**a**) Anatomy of the Au$_{23}$ core, which consists of a Au$_{12}$ cap unit and Au$_{13}$ icosahedral unit. (**b**) Au$_{23}$ core, which is constructed by one Au$_{12}$ unit and one Au$_{13}$ unit sharing two gold atoms. (**c**) Two Au$_3$(SR)$_4$, two Au(SR)$_2$ and one SR linking the Au$_{12}$ cap and Au$_{13}$ icosahedron. (**d**) Three Au$_2$(SR)$_3$ and one Au(SR)$_2$ protecting the Au$_{23}$ core. (**e**) Back view of Au$_{38T}$. (**f**) The Au$_{38Q}$ structure.

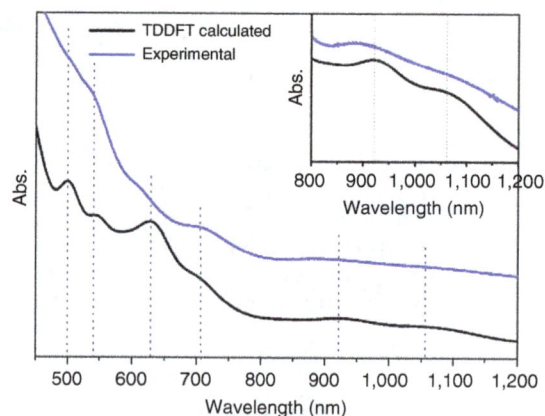

Figure 3 | Comparison of ultraviolet-visible-near-infrared absorption spectra of Au$_{38T}$. Blue: experimental; black: calculated by time-dependent density function (TDDFT) method. Inset is the enlarged spectra in the range from 800 to 1,200 nm.

average Au–S–Au bond angles in the Au$_2$(SR)$_3$ staple units are 92.20° and 90.33°, respectively) and the Au$_{12}$ cap is capped by one Au$_2$(SR)$_3$ staple unit (the average Au–S bond length is 2.33 Å and the average Au–S–Au bond angle is 97.03°). In addition, the Au$_{12}$ cap is also capped by one Au(SR)$_2$ staple unit, and the average Au–S bond length/Au–S–Au bond angle are 2.33 Å and 99.91° (Fig. 2d). The structure resolved by X-ray diffraction was further analysed by computations: the simulated ultraviolet-visible–near-infrared spectrum is close to the experimental one (Fig. 3).

As discussed above, the structure of Au$_{38T}$ is remarkably different from that of Au$_{38Q}$ and the main differences between the two structures lie in the type of Au$_{23}$ core and the surface capping

mode of the Au$_{23}$ core. Au$_{38T}$ and Au$_{38Q}$ have an identical composition but completely different structures; thus, they are literally a pair of structural isomers. Notably, the structure of Au$_{38T}$ reported in this work is novel and also differs from those theoretical structures predicted by Hakkinen et al.[37], Tsukuda and colleagues[38], Jiang et al.[39] and Zeng and colleagues[40], among others.

Transformation. Au$_{38T}$ exhibits relatively high stability at low temperatures, as no obvious spectral changes was detected when a solution of Au$_{38T}$ was stored at − 10 °C for as long as 1 month in toluene (Fig. 4a). However, the absorption spectrum of Au$_{38T}$ gradually changed to that of Au$_{38Q}$ at 50 °C in toluene (Fig. 4b), which indicates that Au$_{38T}$ can transform to Au$_{38Q}$ at elevated

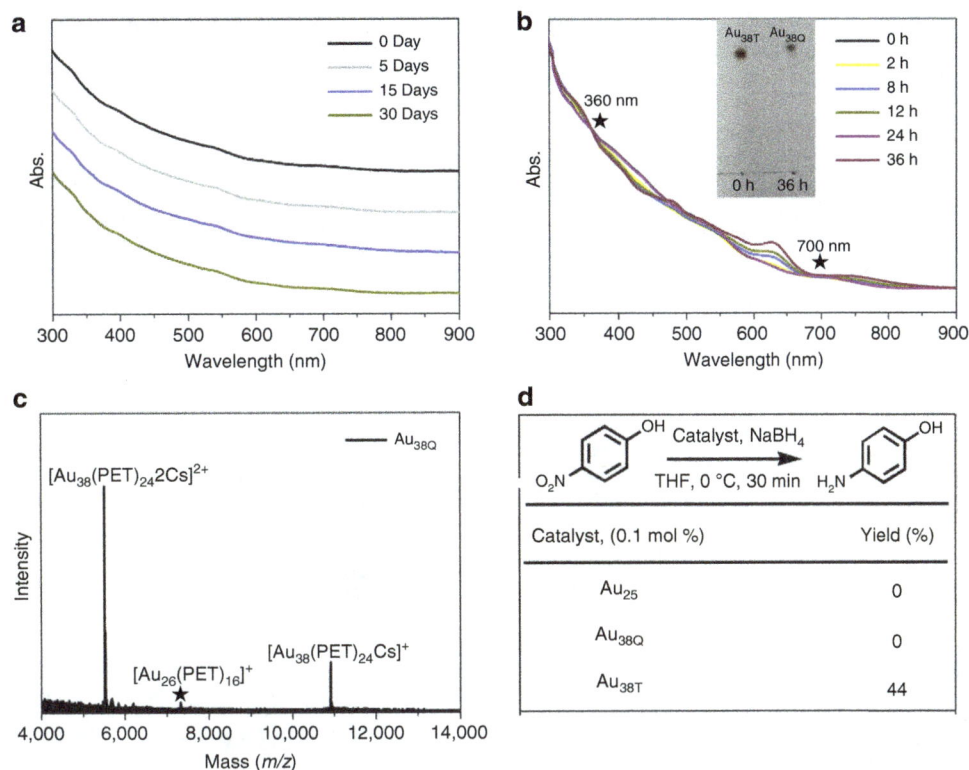

Figure 4 | Difference in stability and catalysis between Au₃₈T and Au₃₈Q. (**a**) Time-dependent ultraviolet-visible-near-infrared absorption spectra of Au$_{38T}$ at −10 °C in toluene. (**b**) Ultraviolet-visible-near-infrared absorption spectral transformation at 50 °C in toluene (the isosbestic points are at 360 and 700 nm). Inset: thin-layer chromatography of Au$_{38T}$ before and after the transformation. (**c**) ESI mass spectrum of the transformed product. (**d**) Catalytic activities of Au$_{25}$, Au$_{38T}$ and Au$_{38Q}$.

temperatures. TLC and ESI–MS further support this transformation (Fig. 4b (inset) and Fig. 4c). However, the reverse transformation (that is, from Au$_{38Q}$ to Au$_{38T}$) was not successful under various investigated conditions. These results indicate that Au$_{38T}$ is less stable than Au$_{38Q}$, and that Au$_{38T}$ can only be irreversibly transformed to Au$_{38Q}$. The reason for why the relatively unstable Au$_{38T}$ is formed rather than the stable Au$_{38Q}$ during the synthesis is probably because the former is kinetically favourable in our reaction conditions, similar to some previous reports[41,42].

Catalysis. Au$_{38T}$ exhibits remarkably higher catalytic activity than Au$_{38Q}$ at low temperature (for example, 0 °C) in reduction reactions. For example, 4-nitrophenol can be reduced to 4-aminophenol in 44% yield with 0.1 mol% Au$_{38T}$ catalyst in half an hour, whereas no reduction occurred when Au$_{25}$(PET)$_{18}^{-}$ TOA^{+} (Au$_{25}$ for short, TOA^{+}: tetra-*n*-octylammonium) or Au$_{38Q}$ was used as the catalyst (Fig. 4d and Supplementary Fig. 5) under the same reaction conditions. It is noteworthy that in other cases, Au$_{25}$ was reported to exhibit good catalytic reduction activity[43,44]. The high catalytic activity of Au$_{38T}$ may be due to its surface being not as densely protected as the surfaces of Au$_{25}$ and Au$_{38Q}$; further investigation is underway. A previous work[44] implied that the catalytic properties of gold nanoclusters are not only size dependent but also structure sensitive. However, the structure dependence of catalytic properties was unclear at that time, because the ligands were different in Au$_{44}$(PET)$_{32}$ and Au$_{44}$(TBBT)$_{28}$ (TBBT: 4-tert-butylbenzenethiolate), and the ligand effect should be considered. Herein, it is unambiguously demonstrated that the structure effect indeed exists, because Au$_{38T}$ and Au$_{38Q}$ exhibit remarkably different catalytic performance. Au$_{38T}$ is relatively robust and can retain its ultraviolet–visible–near-infrared spectrum even after 18 catalytic

cycles (Supplementary Fig. 6). It is noteworthy that the gradual decrease in the yield is primarily due to the unavoidable mass loss of the catalyst during the isolation by column chromatography. However, after 21 cycles, the catalyst transformed to more stable Au$_{38Q}$ demonstrated by the ultraviolet–visible–near-infrared spectra and accordingly the loss of catalytic activity (see Supplementary Table 1). The high catalytic activity at low temperatures indicates the potential application of Au$_{38T}$ in some catalytic processes.

Discussion
In summary, we have discovered a pair of structural isomers Au$_{38T}$ and Au$_{38Q}$, which were identified using ESI–MS, TGA, XPS and SCXC. Although both species have the same composition (that is, Au$_{38}$(PET)$_{24}$), they have distinctly different structures, which results in differences in their optical and catalytic properties, as well as structural stability. The less stable Au$_{38T}$ can be irreversibly transformed to the more stable Au$_{38Q}$ at high temperatures. The structure of Au$_{38T}$ is very interesting: it is composed of a Au$_{23}$ core (fused by one Au$_{13}$ icosahedron and one Au$_{12}$ cap by sharing two atoms) and a mixed layer of thiolate ligands and gold–thiolate complex units for surface protection. This structure is unique (that is, not found in other reported gold nanoclusters). In particular, the diversity of staple units and the bridging thiolate found in Au$_{38T}$ provide a new direction for structural studies of metal nanoclusters. The significance and novelty of this work are as follows. (i) A novel synthesis method is developed, with which a novel gold nanoparticle is readily synthesized, and the composition of the as-prepared nanoparticle is precisely determined using ESI–MS in conjunction with XPS and TGA. (ii) The structure of Au$_{38T}$ is resolved using SCXC and the unique structural features provide important implications for

nanocluster structural studies. (iii) Significantly, structural isomerism is observed in nanoparticles for the first time. (iv) The distinctly different properties (in particular the catalytic properties) of the two structural isomers indicate a structure–property correlation and this will have important implications for future catalytic studies. It is expected that our work may motivate more studies on structural isomerism and structure–property correlations in nanoscale or even larger scale materials.

Methods

Reagents. All chemicals and reagents are commercially available and were used as received. Tetraoctylammonium bromide (TOAB, 98.0%) and 4-nitrophenol (99.0%) were obtained from Aladdin; 2-phenylethanethiol (PhC_2H_4SH, 99.0%) was purchased from Sigma-Aldrich; Au_{38Q} and $Au_{25}(PET)_{18}^- TOA^+$ were synthesized following reported methods[31,34].

Synthesis. Au_{38T} was synthesized using a modified one-pot method and separated using PTLC. Briefly, $HAuCl_4 \cdot 4H_2O$ (0.20 g, 0.48 mmol) was mixed with 1.03 equivalents of TOAB (0.27 g, 0.49 mmol) in CH_2Cl_2 (40 ml). Then, 9 equivalents of phenylethanethiol (0.61 ml, 4.50 mmol) were added to this solution and the solution was stirred for ~ 2 h until it became colourless. To this solution, 5.3 equivalents of $NaBH_4$ (0.11 g, 2.80 mmol) in cold water (5 ml) was added in one shot under vigorous stirring and the reaction was allowed to proceed under constant stirring for 6 h. CH_2Cl_2 was removed via rotary evaporation at 20 °C to isolate the crude product. For purification, the crude product was extracted with a small amount of tetrahydrofuran (THF) and washed with ice water three times and with CH_3OH two times; during this procedure, traces of inorganic salt, excess TOAB and phenylethanethiol were thoroughly removed. Next, the as-obtained crude products were separated using PTLC (dichloromethane: petroleum ether = 3:4) and finally the target product was isolated from the reddish brown band of PTLC after extraction with CH_2Cl_2, with a yield of 5%.

Characterization. The ultraviolet–visible–near-infrared absorption spectrum was measured on a UV-3600 spectrophotometer (Shimadzu, Japan) at room temperature. TGA analysis was conducted under a N_2 atmosphere (~ 3 mg sample used, flow rate ~ 50 ml min^{-1}) on a TG/DTA 6300 analyzer (Seiko Instruments, Inc.) and the heating rate was 10 °C min^{-1}. XPS measurements were performed on an ESCALAB 250Xi XPS spectrometer (Thermo Scientific, USA), using a mono-chromated Al Kα source and equipped with an Ar$^+$ ion sputtering gun. All binding energies were calibrated using the C (1 s) carbon peak (284.8 eV). ESI–MS data were acquired on a Waters Q-TOF mass spectrometer equipped with a Z-spray source. The sample was dissolved in toluene (~ 1 mg ml^{-1}) and diluted 1:1 in dry ethanol (5 mM CsOAc). The sample was directly infused at 5 μl min^{-1}. The source temperature was fixed at 70 °C. The spray voltage was set at 2.20 kV and the cone voltage was set at 60 V.

Single-crystal growth and analysis. Black crystals were formed from a CH_2Cl_2/hexane solution of the nanoclusters at 4 °C after 5 days. The diffraction data for $Au_{38}(PET)_{24}$ were collected at 173 K on a Bruker APEX DUO X-ray diffractometer using Cu Kα radiation ($\lambda = 1.54184$ Å).

Theoretical methods. All calculations were performed using density functional theory with the pure functional Perdew-Burke-Ernzerhof[45,46] and the all electron basis set 6–31 g (d, p) for H, S, pseudopotential basis set LANL2DZ for Au, as implemented in the Gaussian 09 program package[47]. Time-dependent density functional calculations[48] were performed to reproduce the experimental ultraviolet–visible spectrum and –R group was replaced by –H to minimize computational work[31]. The Gaussian half-width at half-height of 0.15 eV in the Multiwfn software[49] was used to simulate the ultraviolet–visible spectrum.

General procedure for the catalyses. 4-Nitrophenol (69.50 mg, 0.500 mmol), Au_{38Q}, Au_{25} or Au_{38T} (0.100 mol%, not adsorbed on a support or calcined) and THF (5 ml) were mixed in a reaction tube at 0 °C. The mixture was stirred at this temperature for 5 min. $NaBH_4$ (189.00 mg, 5.000 mmol) dissolved in 1.0 ml of H_2O was added slowly to the mixture. After stirring at 0 °C for 30 min, a large amount of water was added to quench the reaction. The mixture was extracted with dichloromethane twice (2 × 10 ml) and then the organic layers were collected and concentrated. The reduction product (4-aminophenol) was purified by column chromatography on silica gel, with ethyl acetate and petroleum ether (ethyl acetate/petroleum ether = 1/1) as the eluant.

General procedure for the recovery of Au_{38T}. When the reduction was completed, the reaction mixture was quenched with water. Au_{38T} and other organic compounds were extracted with dichloromethane. The extract was collected and concentrated. After the other organic compounds were isolated by column

chromatography with ethyl acetate and petroleum ether, Au_{38T} was recovered using dichloromethane as the eluant. The dichloromethane was evaporated under reduced pressure and then Au_{38T} was re-used in the next cycle without further treatment.

References

1. Billinge, S. J. & Levin, I. The problem with determining atomic structure at the nanoscale. *Science* **316**, 561–565 (2007).
2. Akola, J., Walter, M., Whetten, R. L., Hakkinen, H. & Gronbeck, H. On the structure of thiolate-protected Au_{25}. *J. Am. Chem. Soc.* **130**, 3756–3757 (2008).
3. Weissker, H. C., Lopez-Acevedo, O., Whetten, R. L. & López-Lozano, X. optical spectra of the special Au_{144} gold-cluster compounds: sensitivity to structure and symmetry. *J. Phys. Chem. C.* **119**, 11250–11259 (2015).
4. Gruene, P. *et al.* Structures of neutral Au_7, Au_{19}, and Au_{20} clusters in the gas phase. *Science* **321**, 674–676 (2008).
5. Olson, R. M. & Gordon, M. S. Isomers of Au 8. *J. Chem. Phys.* **126**, 214310–214316 (2007).
6. Pyykko, P. Theoretical chemistry of gold. III. *Chem. Soc. Rev.* **37**, 1967–1997 (2008).
7. Huang, W., Pal, R., Wang, L. -M., Zeng, X. C. & Wang, L. -S. Isomer identification and resolution in small gold clusters. *J. Chem. Phys.* **132**, 054305–054305 (2010).
8. Schaefer, B. *et al.* Isomerism and structural fluxionality in the Au_{26} and Au_{26}^- nanoclusters. *ACS Nano* **8**, 7413–7422 (2014).
9. Brust, M., Walker, M., Bethell, D., Schiffrin, D. J. & Whyman, R. Synthesis of thiol-derivatized gold nanoparticles in a 2-phase liquid-liquid system. *Chem. Commun.* **7**, 801–802 (1994).
10. Qian, H. & Jin, R. Controlling nanoparticles with atomic precision: the case of $Au_{144}(SCH_2CH_2Ph)_{60}$. *Nano Lett.* **9**, 4083–4087 (2009).
11. Ackerson, C. J., Jadzinsky, P. D. & Kornberg, R. D. Thiolate ligands for synthesis of water-soluble gold clusters. *J. Am. Chem. Soc.* **127**, 6550–6551 (2005).
12. Fields-Zinna, C. A., Sardar, R., Beasley, C. A. & Murray, R. W. Electrospray ionization mass spectrometry of intrinsically cationized nanoparticles, $Au_{144/146}(SC_{11}H_{22}N(CH_2CH_3)_3^+)_x(S(CH_2)_5CH_3)_y^{x+}$. *J. Am. Chem. Soc.* **131**, 16266–16271 (2009).
13. Sardar, R., Funston, A. M., Mulvaney, P. & Murray, R. W. Gold nanoparticles: past, present, and future. *Langmuir.* **25**, 13840–13851 (2009).
14. Lopez-Acevedo, O., Kacprzak, K. A., Akola, J. & Hakkinen, H. Quantum size effects in ambient CO oxidation catalysed by ligand-protected gold clusters. *Nat. Chem.* **2**, 329–334 (2010).
15. Hakkinen, H. The gold-sulfur interface at the nanoscale. *Nat. Chem.* **4**, 443–455 (2012).
16. Lu, Y. Z. & Chen, W. Sub-nanometre sized metal clusters: from synthetic challenges to the unique property discoveries. *Chem. Soc. Rev.* **41**, 3594–3623 (2012).
17. Qian, H. F., Zhu, Y. & Jin, R. C. Atomically precise gold nanocrystal molecules with surface plasmon resonance. *Proc. Natl Acad. Sci. USA* **109**, 696–700 (2012).
18. Yau, S. H., Varnavski, O. & Goodson, T. An ultrafast look at Au nanoclusters. *Acc. Chem. Res.* **46**, 1506–1516 (2013).
19. Weissker, H. C. *et al.* Information on quantum states pervades the visible spectrum of the ubiquitous $Au_{144}(SR)_{60}$ gold nanocluster. *Nat. Commun.* **5**, 3785 (2014).
20. Yu, Y. *et al.* Solvent controls the formation of $Au_{29}(SR)_{20}$ nanoclusters in the CO-reduction method. *Part. Part. Syst. Char.* **31**, 652–656 (2014).
21. Jadzinsky, P. D., Calero, G., Ackerson, C. J., Bushnell, D. A. & Kornberg, R. D. Structure of a thiol monolayer-protected gold nanoparticle at 1.1A resolution. *Science* **318**, 430–433 (2007).
22. Heaven, M. W., Dass, A., White, P. S., Holt, K. M. & Murray, R. W. Crystal structure of the gold nanoparticle $[N(C_8H_{17})_4][Au_{25}(SCH_2CH_2Ph)_{18}]$. *J. Am. Chem. Soc.* **130**, 3754–3755 (2008).
23. Zhu, M., Aikens, C. M., Hollander, F. J., Schatz, G. C. & Jin, R. Correlating the crystal structure of A thiol-protected Au_{25} cluster and optical properties. *J. Am. Chem. Soc.* **130**, 5883–5885 (2008).
24. Qian, H., Eckenhoff, W. T., Zhu, Y., Pintauer, T. & Jin, R. Total structure determination of thiolate-protected Au_{38} nanoparticles. *J. Am. Chem. Soc.* **132**, 8280–8281 (2010).
25. Zeng, C. *et al.* Total structure and electronic properties of the gold nanocrystal $Au_{36}(SR)_{24}$. *Angew. Chem. Int. Ed.* **51**, 13114–13118 (2012).
26. Malola, S. *et al.* $Au_{40}(SR)_{24}$ cluster as a chiral dimer of 8-electron superatoms: structure and optical properties. *J. Am. Chem. Soc.* **134**, 19560–19563 (2012).
27. Jiang, D. E., Overbury, S. H. & Dai, S. Structure of $Au_{15}(SR)_{13}$ and its implication for the origin of the nucleus in thiolated gold nanoclusters. *J. Am. Chem. Soc.* **135**, 8786–8789 (2013).
28. Yang, H., Wang, Y., Edwards, A. J., Yan, J. & Zheng, N. High-yield synthesis and crystal structure of a green Au_{30} cluster co-capped by thiolate and sulfide. *Chem. Commun.* **50**, 14325–14327 (2014).

29. Das, A. *et al.* Crystal structure and electronic properties of a thiolate-protected Au$_{24}$ nanocluster. *Nanoscale* **6**, 6458–6462 (2014).

30. Song, Y. *et al.* Crystal structure of selenolate-protected Au$_{24}$(SeR)$_{20}$ nanocluster. *J. Am. Chem. Soc.* **136**, 2963–2965 (2014).

31. Qian, H., Zhu, Y. & Jin, R. Size-focusing synthesis, optical and electrochemical properties of monodisperse Au$_{38}$(SC$_2$H$_4$Ph)$_{24}$ nanoclusters. *ACS Nano.* **3**, 3795–3803 (2009).

32. Wang, Z. W., Toikkanen, O., Quinn, B. M. & Palmer, R. E. Real-space observation of prolate monolayer-protected Au$_{38}$ clusters using aberration-corrected scanning transmission electron microscopy. *Small* **7**, 1542–1545 (2011).

33. Dolamic, I., Knoppe, S., Dass, A. & Burgi, T. First enantioseparation and circular dichroism spectra of Au$_{38}$ clusters protected by achiral ligands. *Nat. Commun.* **3**, 798 (2012).

34. Wu, Z., Suhan, J. & Jin, R. One-pot synthesis of atomically monodisperse, thiol-functionalized Au$_{25}$ nanoclusters. *J. Mater. Chem.* **19**, 622–626 (2009).

35. Ghosh, A. *et al.* Simple and efficient separation of atomically precise noble metal clusters. *Anal. Chem.* **86**, 12185–12190 (2014).

36. Yao, C. *et al.* Adding two active silver atoms on Au nanoparticle. *Nano Lett.* **15**, 1281–1287 (2015).

37. Hakkinen, H., Walter, M. & Gronbeck, H. Divide and protect: apping gold nanoclusters with molecular gold-thiolate rings. *J. Phys. Chem. B.* **110**, 9927–9931 (2006).

38. Lopez-Acevedo, O., Tsunoyama, H., Tsukuda, T., Hakkinen, H. & Aikens, C. M. Chirality and electronic structure of the thiolate-protected Au$_{38}$ nanocluster. *J. Am. Chem. Soc.* **132**, 8210–8218 (2010).

39. Jiang, D. E., Luo, W., Tiago, M. L. & Dai, S. In search of a structural model for a thiolate-protected Au$_{38}$ cluster. *J. Phys. Chem. C.* **112**, 13905–13910 (2008).

40. Pei, Y., Gao, Y. & Zeng, X. C. Structural prediction of thiolate-protected Au$_{38}$: a face-fused bi-icosahedral Au core. *J. Am. Chem. Soc.* **130**, 7830–7832 (2008).

41. Wu, Z., MacDonald, M. A., Chen, J., Zhang, P. & Jin, R. Kinetic control and thermodynamic selection in the synthesis of atomically precise gold nanoclusters. *J. Am. Chem. Soc.* **133**, 9670–9673 (2011).

42. Yuan, X. *et al.* Balancing the rate of cluster growth and etching for gram-scale synthesis of thiolate-protected Au$_{25}$ nanoclusters with atomic precision. *Angew. Chem. Int. Ed.* **53**, 4623–4627 (2014).

43. Shivhare, A., Ambrose, S. J., Zhang, H., Purves, R. W. & Scott, R. W. Stable and recyclable Au$_{25}$ clusters for the reduction of 4-nitrophenol. *Chem. Commun.* **49**, 276–278 (2013).

44. Li, M. -B., Tian, S. -K., Wu, Z. & Jin, R. Cu^{2+} induced formation of Au$_{44}$(SC$_2$H$_4$Ph)$_{32}$ and its high catalytic activity for the reduction of 4-nitrophenol at low temperature. *Chem. Commun.* **51**, 4433–4436 (2015).

45. Perdew, J. P., Burke, K. & Ernzerhof, M. Generalized gradient approximation made simple. *Phys. Rev. Lett.* **77**, 3865–3868 (1996).

46. Perdew, J. P., Burke, K. & Ernzerhof, M. Generalized gradient approximation made simple (vol 77, pg 3865, 1996). *Phys. Rev. Lett.* **78**, 1396–1396 (1997).

47. Frisch, M. J. *et al. Gaussian 09, Revision B.01* (Gaussian, Inc., 2010).

48. Perdew, J. P. *et al.* Prescription for the design and selection of density functional approximations: more constraint satisfaction with fewer fits. *J. Chem. Phys.* **123**, 062001–062009 (2005).

49. Lu, T. & Chen, F. Multiwfn: a multifunctional wavefunction analyzer. *J. Comput. Chem.* **33**, 580–592 (2012).

Acknowledgements

Z.W. thank the National Basic Research Program of China (grant number 2013CB934302), the Natural Science Foundation of China (numbers 21222301 and 21171170), the Ministry of Human Resources and Social Security of China, the Innovative Program of Development Foundation of Hefei Center for Physical Science and Technology (2014FXCX002), the CAS/SAFEA International Partnership Program for Creative Research Teams and the Hundred Talents Program of the Chinese Academy of Sciences for financial support. R.J. acknowledges financial support from the U.S. Department of Energy-Office of Basic Energy Sciences, Grant DE-FG02-12ER16354, and the Natural Science Foundation of China (Overseas, Hong Kong and Macao Scholars Collaborated Researching Fund, number 21528303). We greatly appreciate Professor Linhong Weng and Professor Yuejian Lin for the assistance in the single-crystal X-ray diffraction analysis. The calculations in this paper have been done on the supercomputing system in the Supercomputing Center of University of Science and Technology of China.

Author contributions

S.T. conceived and carried out the experiments. Y.L. resolved the structure. M.L. carried out the catalytic experiments. J. Yuan and J. Yang conducted computing. Z.W. and R.J. designed the study, supervised the project and analysed the data. All authors contributed to the preparation of the manuscript.

Additional information

Accession codes: The X-ray crystallographic coordinates for structures reported in this study (see Supplementary Table 2 and Supplementary Data 1) have been deposited at the Cambridge Crystallographic Data Centre (CCDC), under deposition number CCDC 1423153. These data can be obtained free of charge from The Cambridge Crystallographic Data Centre via www.ccdc.cam.ac.uk/data_request/cif.

15

Inward lithium-ion breathing of hierarchically porous silicon anodes

Qiangfeng Xiao[1,*], Meng Gu[2,*], Hui Yang[3,*], Bing Li[4], Cunman Zhang[4], Yang Liu[5], Fang Liu[5], Fang Dai[1], Li Yang[1], Zhongyi Liu[1], Xingcheng Xiao[1], Gao Liu[6], Peng Zhao[3], Sulin Zhang[3], Chongmin Wang[2], Yunfeng Lu[5] & Mei Cai[1]

Silicon has been identified as a highly promising anode for next-generation lithium-ion batteries (LIBs). The key challenge for Si anodes is large volume change during the lithiation/delithiation cycle that results in chemomechanical degradation and subsequent rapid capacity fading. Here we report a novel fabrication method for hierarchically porous Si nanospheres (hp-SiNSs), which consist of a porous shell and a hollow core. On charge/discharge cycling, the hp-SiNSs accommodate the volume change through reversible inward Li breathing with negligible particle-level outward expansion. Our mechanics analysis revealed that such inward expansion is enabled by the much stiffer lithiated layer than the unlithiated porous layer. LIBs assembled with the hp-SiNSs exhibit high capacity, high power and long cycle life, which is superior to the current commercial Si-based anode materials. The low-cost synthesis approach provides a new avenue for the rational design of hierarchically porous structures with unique materials properties.

[1] Gereral Motors Research and Development Center, 30500 Mound Road, Warren, Michigan 48090, USA. [2] Environmental Molecular Sciences Laboratory, Pacific Northwest National Laboratory, Richland, Washington 99352, USA. [3] Department of Engineering Science & Mechanics, Pennsylvania State University, University Park, Pennsylvania 16802, USA. [4] Clean Energy Automotive Engineering Center, Tongji University, Shanghai 201804, China. [5] Department of Chemical and Biomolecular Engineering, The University of California, Los Angeles, California 90095, USA. [6] Environmental Energy Technologies Division, Lawrence Berkeley National Laboratory, Berkeley, California 94720, USA. * These authors contributed equally to this work. Correspondence and requests for materials should be addressed to S.Z. (email: suz10@psu.edu) or to C.W. (email: Chongmin.Wang@pnnl.gov) or to Y.L. (email: luucla@ucla.edu) or to M.C. (email: mei.cai@gm.com).

Lithium-ion batteries (LIBs) have emerged as the main power sources for microelectronics and are considered as the technology of choice for the vehicle electrification. Similar to other batteries, LIBs involve phase transformation accompanied by ion and electron transport. Maintaining effective and robust transport pathways, as well as minimizing reactions between electrodes and electrolytes, is the key to ensure batteries with high power and long cycling life. Silicon (Si), with a theoretical capacity of 4,200 mAh g^{-1}, has been identified as one of the most promising anode candidates for the next-generation high-energy-density LIBs. However, Si anodes generally exhibit significant volume change during electrochemical cycling, resulting in pulverization of the particles, loss of the electrical contact, rupture of the solid-electrolyte interphase (SEI)[1-3], and consequently, rapidly deteriorated storage performance.

Various strategies have been explored to mitigate these limitations, mainly by structural engineering of silicon particles, compositing with carbon (C) and adapting suitable binders. It has been demonstrated theoretically that Si materials in nanometre range exhibit alleviated lithiation-induced mechanical stress and enhanced resilience to fracture and decrepitation, which in turn result in enhanced ionic and electrical conductivities, and electrochemical performance[4,5]. Various Si materials with low-dimensional structures (for example, nanoparticles[6,7], hollow spheres[8], nest-like Si nanospheres[9], nanowires[10] and nanotubes[11]) or porous structures[12] have been fabricated by mechanical ball milling, laser pyrolysis, chemical vapour deposition, solvothermal method and two-step thermal annealing/acid etching process. These materials have shown improved cycle life, capacity retention and rate performance. Similarly, the Si/C composites, wherein the carbon moieties serve as a cushion for the volume change and provide effective electronic pathways, also exhibit enhanced electrochemical performance compared with their pure silicon counterparts[13-15]. In the aspect of binders, sodium alginate[16], poly(acrylic acid)[17], conductive polymers[18] and self-healing polymers[2] have been employed to improve the inter-particle interactions, maintain or repair the electronic contacts during the cycling. However, the outward volume expansion of the above Si materials during lithiation and potential side reactions between binders and electrolytes are detrimental to prolonged cycling. Therefore, fabrication of high-performance Si anodes with minimal outward volume expansion remains challenging.

Herein we report a size-dependent chemical transformation method to fabricate hierarchically porous Si nanospheres (hp-SiNSs), which uniquely accommodate the volume change through reversible inward Li breathing on charge/discharge cycling.

Results

Synthesis of hp-SiNSs. As illustrated in Fig. 1a, we started from silica (SiO$_2$) spheres with a mesoporous shell and solid core. The SiO$_2$ spheres were produced by simultaneous hydrolysation and condensation of tetraethoxysilane and octadecyltrimethoxysilane, followed by the removal of the organic species[19]. The shell can be visualized to be composed of nanoparticles of ~3 nm in diameter, whereas the solid core has a diameter of hundreds of nanometres, as shown in Fig. 1a (i; step I). In the presence of Mg vapour, the SiO$_2$ nanoparticles in the porous shell were converted to Si, while the solid SiO$_2$ core remained nearly intact owing to the size-dependent transformation rate (step II). It's noteworthy to point out that possible structure reorganization in the shell can occur during this step, as indicated by the broader pore size distribution in Fig. 1a (ii) than that in Fig. 1a (i). Finally, hp-SiNSs were obtained by removing MgO (Supplementary

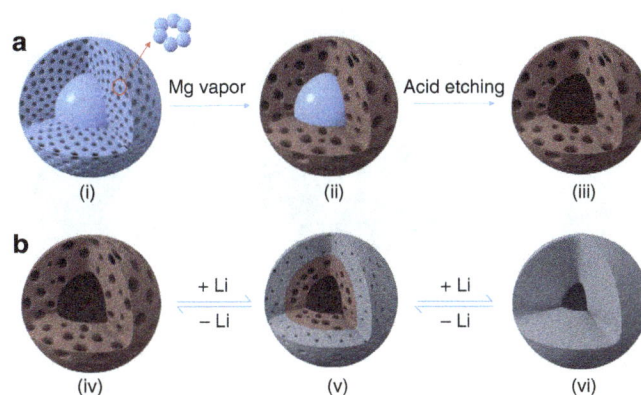

Figure 1 | Schematic of synthesis method and lithiation/delithiation process for hp-SiNSs. (**a**) Schematic of synthesis of hp-SiNSs via size-dependent reduction. The method involves three steps: (i) synthesis of solid core/mesoporous shell SiO$_2$ spheres by simultaneous hydrolysation and condensation of tetraethoxysilane and octadecyltrimethoxysilane, followed by the removal of organic species; (ii) conversion of the primary SiO$_2$ nanoparticles mesoporous shell to Si nanoparticles by Mg vapour due to size-dependent reaction; (iii) acid etching to remove residual MgO and the solid SiO$_2$ core to obtain the final hp-SiNSs with a mesoporous shell and hollow core. (**b**) Schematic of lithiation/delithiation process of the hp-SiNSs showing that the mesoporous shell directs the volume expansion towards the inner hollow core during the lithiation and recovers the morphology during the delithiation.

Fig. 1a) and residual SiO$_2$ through acid etching (step III). As schematically shown in Fig. 1b, the mesoporous shell directs the volume expansion towards the inner hollow core during the lithiation, which is enabled by the much stiffer lithiated layer than the unlithiated porous layer. During the delithiation, the morphology recovers. Opposite to the outward expansion in solid Si during lithiation, such inward Li breathing prevents the hp-SiNSs from fracture, and maintains stable SEI and ion and electron diffusion pathways. The resulting Si anodes are thus expected to possess high capacity, high power, long cycle life and high Coulombic efficiency. Previous studies demonstrated that double-walled Si nanotubes exhibit similar inward expansion and stable SEI formation, and hence impressively long cycle life[20]. However, the high cost and incompatibility with the current slurry coating electrode fabrication process prevent the nanotubes from commercialization.

Characterization of hp-SiNSs. Figure 2a shows the transmission electron microscope (TEM) images of monodisperse core/mesoporous shell SiO$_2$ spheres with a shell thickness of 75 nm and a solid core of 350 nm in diameter. After the transformation, hp-SiNSs with a mesoporous shell and hollow core were obtained, as indicated by the contrast between the darker peripheral and the lighter central regions in Fig. 2b. A high-resolution TEM image indicates that the porous shell is mostly amorphous, scattered with only a few nanocrystalline domains with a (111) interplanar spacing of 3.13 Å (Fig. 2c)[21]. The hollow core was further confirmed by the scanning TEM image in the high-angle annular dark field imaging mode for which the image contrast is proportional to mass thickness and the square of the atomic number of the element. As shown in Supplementary Fig. 2, the outer layer of each hp-SiNS is much denser than its center. Furthermore, the mesoporous shell is demonstrated by homogeneously distributed black and white spots. In a control experiment, Si products could not be obtained when solid SiO$_2$ spheres with the diameter of 350 nm were treated with the identical transformation steps (Supplementary Fig. 3).

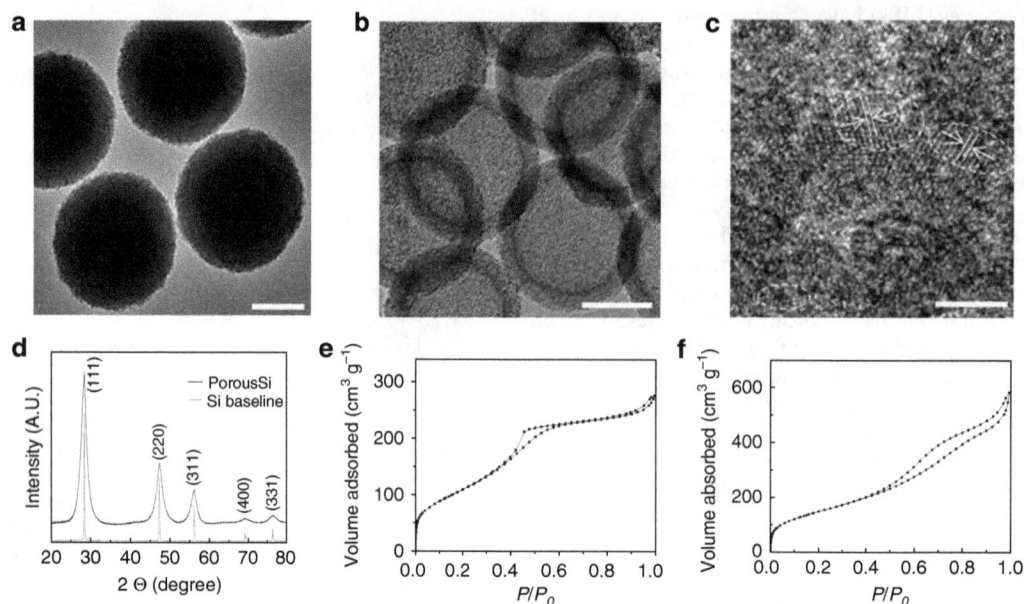

Figure 2 | Structure characterizaion of the hp-SiNSs. (a) A TEM image of solid core/mesoporous shell SiO$_2$ particles. (b) A low magnification TEM image and (c) a high-resolution TEM image of hp-SiNSs, showing the Si particles are mostly amorphous, with scattered nanocrystalline domains with a (111) interplanar spacing of 3.13 Å. (d) X-ray diffraction patterns of the hp-SiNSs. (e,f) N$_2$ isotherms for solid core/mesoporous shell SiO$_2$ particles and the hp-SiNSs, respectively. Scale bar, 200 nm (a,b) 4 nm (c).

This further indicates that the size of SiO$_2$ particles dictates the reaction rate. It is worthy to point out that similar hierarchically porous structures were obtained when changing shell thickness and core size of SiO$_2$ (Supplementary Fig. 4).

Energy dispersive X-ray spectroscopy analysis showed the expected primary Si signal with a trace amount of oxygen, while no Mg signal was observed. As excessive HF solution was used to etch inner SiO$_2$ core and the large-surface-area hp-SiNSs have high reactivity with O, we attribute the O to the air exposure during handling the sample. The Cu signal can be ascribed to TEM copper grid (Supplementary Fig. 1b). As quantified, The hp-SiNSs were constituted of 91 wt% Si and 9 wt% O. Consistently, X-ray diffraction exhibited characteristics of diamond cubic phase of Si (JCPDS card no. 27–1,402; Fig. 2d)[22]. The size of the crystallites was estimated to be 4.8 nm by the Scherrer equation. These results imply limited structure reorganization in the course of chemical transformation from silica precursor to the Si product. Raman spectrum was further used to verify the success of the conversion (Supplementary Fig. 1c). As compared with that of Si wafer reference (520 cm^{-1}), the peak of the hp-SiNSs shifted to 508 cm^{-1}. This shift was likely caused by the phonon confinement effect from the primary nanoparticles, whose first-order scattering peak shifts towards a lower energy as particle size decreases[23,24].

Nitrogen adsorption–desorption isotherms show that the structures are preserved on the conversion with a framework reconstruction. Figure 2e shows that the N$_2$ adsorption–desorption isotherms of solid core/porous shell SiO$_2$ particles. As expected, they exhibit typical type-IV features of absorbents with a H2 hysteresis. According to the Brunauer–Emmett–Teller method, the specific surface area was estimated to be 450 m^2 g^{-1}. The pore size distribution in the SiO$_2$ particles calculated by the Barrett–Joyner–Halenda method from adsorption branch shows that the pores are uniform with an average of 3.2 nm (Supplementary Fig. 5a). After converted to hp-SiNSs, the absorption–desorption hysteresis transits from H2 to H3 (Fig. 2f) and thus the pore size distribution becomes broader (Supplementary Fig. 5b)[25]. Such changes occur owing to the

reorganization of building blocks during the transformation, as reported for Si thin film by Tolbert and colleagues[26]. Consistently, Barrett–Joyner–Halenda calculation shows the pore size ranges from two to tens of nanometres. A high Brunauer–Emmett–Teller surface area of 550 m^2 g^{-1} was achieved, which was among the largest surface area ever reported[27,28]. The surface area of solid core is much smaller than that of porous shell in the starting solid core/porous shell SiO$_2$ particles. In view of the similar density of silica (2.2 g cm^{-3}) and silicon (2.3 g cm^{-3}), the increase of surface area after conversion can be mainly ascribed to the removal of the solid core.

Electrochemical performance. The hp-SiNSs show remarkable performance as an anode for LIBs, as described below. The charge/discharge capacity versus cycle number is presented in Fig. 3a. The reversible specific capacity reaches 1,850 mAh g^{-1} at 0.1C (1C = 3.6 A g^{-1}). Such a high specific capacity value indicates that most of Si is active owing to the high Li$^+$ accessibility of the hierarchically porous structures. After the two-cycle formation step at C/20, the capacity is maintained above 1,800 mAh g^{-1} after 200 cycles, demonstrating excellent capacity retention of the hp-SiNSs. The Coulombic efficiency increases from 52% at the first cycle to above 99.0% after twentieth cycle. The irreversible capacity loss at the first cycle can be compensated by prelithiation through either chemical or electrochemical methods or by using stabilized lithium metal powder[29]. In the following cycles, a stable SEI is formed and the Coulombic efficiency reached up to 99.4%. Under similar conditions, the commercial Si nanoparticles show fast capacity decay, from 2,000 mAh g^{-1} at initial cycles to <1,000 mAh g^{-1} after 100 cycles (Fig. 3a). Such comparative results indicate that the enhanced cyclability of the hp-SiNSs mainly stems from their unique porous structure rather than the binder or the conductive additives. In consistent with cyclic voltammetry results (Supplementary Fig. 6), the voltage profile in Fig. 3b shows the electrochemical behaviour of amorphous Si. The rate performance tests were carried out at various rates from C/10

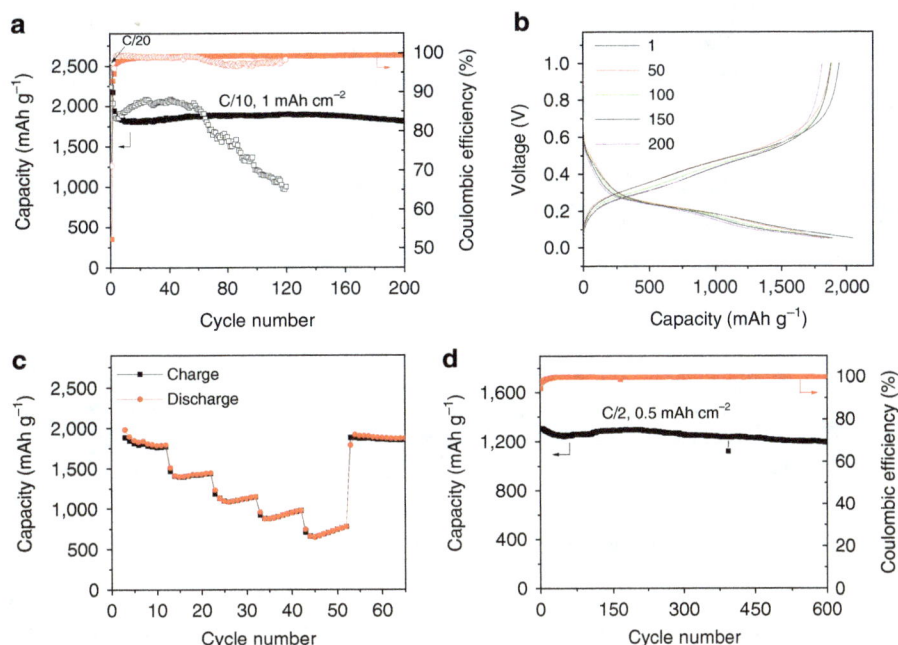

Figure 3 | Electrochemical performance of the hp-SiNSs. (**a**) Lithiation capacity and Coulombic efficiency of the hp-SiNSs and commercial solid 100 nm Si particle electrode cycled between 1 V and 0.05 V at 0.1C with the loading of 1 mAh cm^{-2}. (**b**) Galvanostatic charge–discharge profiles during cycling. (**c**) Lithiation capacity of the hp-SiNS electrode at various rates from 0.1C to 2C. (**d**) Lithiation capacity and Coulombic efficiency of the hp-SiNS electrode cycled between 1 V and 0.05 V at 0.1C with the loading of 0.5 mAh cm^{-2}.

to 2C, as shown in Fig. 3c. The discharge capacities of 1,850, 1,430, 1,125, 920 and 700 mAh g^{-1} were obtained at the rates of C/10, C/5, C/2, C and 2C, respectively. The cyclability can be improved to 600 cycles at a rate of C/2 when the loading is decreased from 1 to 0.5 mAh cm^{-2} (Fig. 3d). The average Coulombic efficiency is as high as 99.91%. It is worthy to point out that the current electrodes show moderate improvement in volumetric capacity ($\sim$760 mAh cm^{-3}) against graphitic anodes ($\sim$620 mAh cm^{-3})[13]. As compared with the currently available Si materials, our hp-SiNSs are among the most promising candidates as anode for high-performance LIBs in light of both cycling performance and rate capability[7–12].

***In situ* TEM observation of lithiation/delithiation.** The remarkable performance of this material is attributed to its unique volume accommodation mechanism, as revealed by the *in situ* TEM imaging under a dynamic operating condition of the half-cell nano battery. Figure 4a–g shows the first lithiation process of a hp-SiNS (Supplementary Movie 1). The projected area of the pristine hp-SiNS at the initial state is 182,867 nm^2, as circled in blue in Fig. 4a. On bringing the Li source to the close contract of the sphere, Li$^+$ ions diffuse quickly from the contact point via surface diffusion into the hp-SiNS through a wave-propagation-like motion, as shown in Fig. 4b–f. As lithiation proceeded, the mesopores in the shell shrunk and gradually disappeared, and the shell thickness of the hp-SiNS increased from 113 nm at 0 s to 189 nm at the end of the lithiation at 16,813 s. However, the increase in the outer diameter of the hp-SiNS was insignificant since the lithiation-induced volume increase was largely accommodated by the inward expansion of the hp-SiNS to fill the hollow pore. The total projected area of the fully lithiated hp-SiNS was increased to $\sim$233,000 nm^2. The total area increase was 27% after full lithiation based on TEM projection images. Assuming isotropic volume expansion of the amorphous Si, the total volume expansion was calculated to be around 44% (which is around 1.27$^{3/2}$–1). Considering the 300% volume expansion for

solid Si anodes[30], the porosity of the original hp-SiNS is estimated to be 0.64. In addition, the wave-propagation lithiation in our samples is in distinct contrast to the core-shell lithiation of the solid Si nanostructures, as reported in previous studies[31,32]. This different lithiation kinetics is most likely due to the extremely high surface areas in our samples. The core-shell lithiation in solid Si structures tends to generate large hoop tension at the lithiated shell, leading to surface fracture or pulverization of the Si structures[31–35]. Furthermore, the compressive stress at the reaction front retards further lithiation, limiting the rate performance and loading efficiency of the Si anodes[30,31,36]. The ultrafast diffusion through pore surfaces and wave-propagation-like lithiation exhibited by our hp-SiNSs simultaneously enhance the lithiation rate and alleviate the lithiation-induced stress, thereby giving rise to better rate performance and longer cycle life of the Si anodes.

Figure 4h–n shows the delithiation process of the hp-SiNS (Supplementary Movie 2). The starting state of the fully lithiated hp-SiNS is the same as Fig. 4g, but rotated to a different orientation. As shown in Fig. 4j–k, delithiation proceeded by the similar wave-propagation-like motion as seen in lithiation and the delithiated regions recovered to the original porous structure, where the green dashed lines mark the delithiation front. The hollow core expanded gradually as delithiation continued, as marked by the red circles in Fig. 4l,m. The fully delithiated state in Fig. 4n exhibited the same hollow center/porous shell structure as the pristine hp-SiNS in Fig. 4a. Therefore, the lithiation/delithiation processes of the hp-SiNS are highly reversible. In addition, the Si nanoparticles that constitute the hp-SiNS returned to the similar size and morphology. The total area of the fully delithiated state (around 176,000 nm^2), as circled by blue dashed line in Fig. 4n, is similar to that in the initial state shown in Fig. 4a, while the measured area of the hollow core circled by green dashed line is of 70,872 nm^2. Figure 4o–u show that the morphological and structural evolutions of the hp-SiNS during the second lithiation (Supplementary Movie 3) are very similar to the first lithiation. The highly reversible Li-breathing morphology

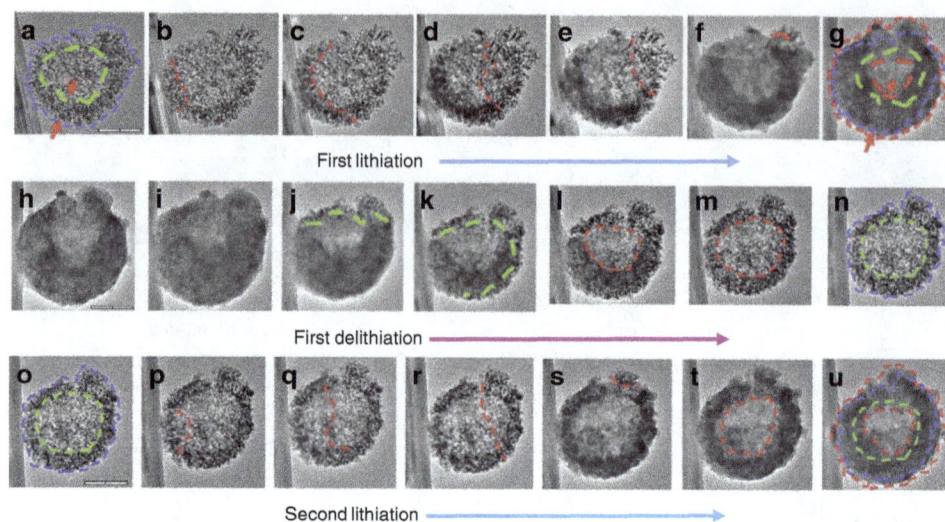

→ First lithiation →

→ First delithiation →

→ Second lithiation →

Figure 4 | *In situ* TEM characterization of the lithiation/delithiation behavior of a hp-SiNS. *In situ* TEM images of the first lithiation process at (**a**) 0 s, (**b**) 889 s, (**c**) 1,364 s, (**d**) 2,000 s, (**e**) 2,708 s, (**f**) 3,946 s and (**g**) 16,813 s; the delithiation process at (**h**) 0 s, (**i**) 6,000 s, (**j**) 6,420 s, (**k**) 6,960 s, (**l**) 7,154 s, (**m**) 7,860 s and (**n**) 9,060 s; the second lithiation process at (**o**) 0 s, (**p**) 40 s, (**q**) 480 s, (**r**) 1,104 s, (**s**) 2,065 s, (**t**) 2,840 s and (**u**) 5,694 s. (The hollow pore is circled in green and the total size is circled in blue in panels **a** and **o** the red lines in panels **b–f** shows the interface between the lithiated and unreacted regions; in panel **g** both the hollow pore and the total size are circled in red and the red arrows indicate the thickness of the shell in panels **a**,**g**. The green dashed lines indicate the interface between delithiated and remaining regions in panels **j**,**k**, the hollow pore is circled in red in panels **l**,**m** and the hollow pore is circled in green and total sphere is circled in blue in panels **n**,**o**. The red lines in panels **p–s** divide the lithiated and unlithiated regions, the hollow core is circled in red in panel **u** and the hollow core and total size of the Si sphere are circled by two red circles. The green and blue circles in panels **a**,**o** are overlaid on top of panels **g**,**u**, respectively). Scale bar, 200 nm (**a**,**h**,**o**).

during electrochemical cycling in our hp-SiNSs enables long cycle life of the Si anodes.

Chemomechanical modelling. To further appreciate the inward breathing during lithiation/delithiation cycles of the hp-SiNS, we extend a recently developed chemomechanical model[37,38] to simulate the concurrent processes of phase transformation, stress generation and morphological evolution of the three-dimensional hp-SiNS (Supplementary Note 1 and Supplementary Fig. 11). In the model, the hp-SiNS is simulated as an elasto-plastic material. The total strain ε_{ij} is composed of three parts, $\varepsilon_{ij} = \varepsilon_{ij}^{e} + \varepsilon_{ij}^{p} + \varepsilon_{ij}^{c}$, where ε_{ij}^{e} is the elastic strain, ε_{ij}^{p} is the plastic strain and $\varepsilon_{ij}^{c} = \beta c \delta_{ij}$ is the chemical strain. Here c is the local Li concentration that varies from 0 (representing the unlithiated phase) and 1 (representing the fully lithiated phase), β is the expansion coefficient. Both the Young's modulus and yield strength of the hp-SiNS are dependent on the Li concentration. The Young's modulus of porous Si can be estimated by $Y(P) = A(1 - P)^{3}$, where $A = 169$ GPa (ref. 39) is a constant and P is the porosity[40]. For $P = 0.64$, the Young's modulus of is ~ 7.9 GPa. The fully lithiated Si phase possesses a Young's modulus of ~ 40 GPa (refs 41–43). We adopt a yield stress of 0.5 GPa for both the unlithiated porous Si and the fully lithiated Si. The relatively low yield stress of the unlithiated porous Si is in accordance to the ease of generating plastic flow observed in the experiments. Both the Young's modulus and the yield strength are set to be nonlinearly dependent on Li concentration. Li insertion into porous Si induces two competing effects: a weakening effect on the Si–Si bonds in the presence of Li and a strengthening effect due to Li-insertion-induced volume expansion and subsequent pore shrinkage[44,45]. It can be anticipated that both the Young's modulus and yield strength first decrease as the Li concentration increases since the weakening effect of Li dominates. This trend reverses beyond a critical Li concentration at which the strengthening effect becomes dominant. With given materials properties of the pristine porous structure and the fully lithiated

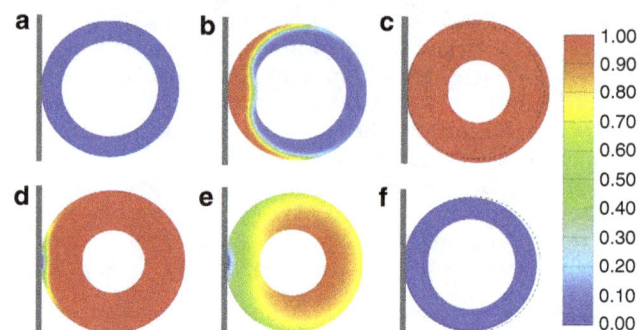

Figure 5 | Chemomechanical modelling of the lithiation/delithiation processes of a hp-SiNS. For better visualizing the processes, only a cross-section is shown. Colours denote the Li concentration, with red being fully lithiated and blue unlithiated. (**a**) The Li source, the rigid plate on the left, is brought to contact with the hp-SiNS. (**b**) Li diffuses into the hp-SiNS in a wave-propagation manner. The newly lithiated product at the reaction front pushes inward more than outward because of the lower stiffness of the inner unlithiated porous layer than that of the outer lithiated layer. (**c**) On fully lithiation, the Si sphere is slightly distorted, and the inward volume expansion is significantly larger than the outward volume expansion. (**d–f**) Delithiation also proceeds by a wave-propagation-like motion. Inner materials are dynamically pulled to the outer surface of the Si sphere due to the delithiation-induced tensile stress.

phase, we nonlinearly interpolate the Young's modulus and the yield stress, with the lowest values set at a Li concentration of 72% (corresponding to the $Li_{2.6}Si$ phase). On the basis of the experimental observations, we set the surface diffusivity of Li to be two orders of magnitude larger than its bulk diffusivity. The diffusion equations and the mechanics equilibrium equations are solved simultaneously to obtain the deformation morphology and the Li concentration profile at any given lithiation/delithiation state.

Figure 5 shows the snapshots of cross-section morphologies for both the lithiation and delithiation processes simulated by our chemomechanical modelling. Li source is placed at the contact point between a rigid plate and a hollow porous sphere. Lithiation proceeds by a wave-propagation manner from the Li source (Fig. 5b and Supplementary Movie 4), successfully mimicking the lithiation kinetics observed in the experiments. For our continuum model to take into account of the porosity of the Si sphere, the mesoporous structure undergoes phase transformation but no volume expansion until lithiation to $Li_{2.6}Si$. Further lithiation to $Li_{3.75}Si$ (the fully lithiated stage) gives rise to the apparent 44% volume expansion. At the reaction front, high incompatible strain is generated. To relax the high strain energy, the newly generated volume by lithiation pushes against both the lithiated outer layer and unlithiated porous layer[37,38,46,47]. The outward pushing effect generates hoop tension in the lithiated layer, while the inward pushing effect generates compression in the unlithiated porous layer. Owing to the lower stiffness of the inner unlithiated porous layer than that of the outer lithiated layer, the mechanical resistance for the reaction front to push inward is much smaller than outward, resulting in significantly stronger inward than outward expansion, as shown in Fig. 5c in the fully lithiated state, where the dashed lines mark the initial inner and outer surfaces in the original configuration (Fig. 5a). The fully lithiated configuration is slightly distorted, owing to the asymmetric lithiation starting from a point source. It should be noted that the inward volume expansion is precisely opposite to the volume expansion accommodation mechanism during the lithiation of a solid Si structure, where the unlithiated core is much stiffer than the lithiated shell, leading to negligible inward expansion but significant outward expansion[34,46,48,49]. Such core-shell lithiation has been widely known to generate large hoop tension in the surface layer of the lithiated Si, causing surface fracture[34,35,46]. During delithiation, Li is drawn back to the plate that acts as a Li sink, creating a Li concentration gradient. Our simulations show that delithiation also proceeds by the wave-propagation-like motion in Fig. 5d–f and Supplementary Movie 5. Owing to a lower Li concentration and hence a larger shrinkage in the outermost region, tensile stress is generated in the regions with higher Li concentration during delithiation. The inner materials are thus dynamically pulled towards the outer region of the hollow porous sphere, thereby almost completely recovering the pristine hollow porous structure.

Discussion

We have developed a novel chemical transformation method for the synthesis of hierarchically porous Si. Our starting material consists of submicron solid core/mesoporous shell SiO_2 spheres, which can be as an assembly of submicron and nanometre building blocks. The fast surface reaction in the nanoporous shell and slow bulk reaction in the solid SiO_2 core lead to porous Si/solid SiO_2 core on magnesiothermal reduction. Further acid etching of the solid SiO_2 core gives rise to the hp-SiNSs. In contrast, conventional transformation is commonly applied to nanostructure building blocks at a single length scale[50], generally leading to complete conversion of chemical composition. Our chemical transformation method innovatively exploits the nanostructure size-dependent reaction kinetics, which opens a new avenue for the synthesis of hierarchically structures materials.

Our in situ TEM characterization has demonstrated that the hp-SiNS features inward volume expansion and contraction on lithiation and delithiation, in distinct contrast to the huge outward volume changes of solid Si. Our chemomechanical modelling reveals that both the mesoporous shell and hollow internal void are indispensable for the inward expansion

and contraction during lithiation and delithiation. The inward lithium breathing facilitates stable SEI formation, enhances the capacity retention and cycling life as compared with previously reported hollow nanospheres[8,9], or porous Si particles[12]. Those nanostructures had either dense shell or not well-defined pores, or only uniformly sized pores, and consequently exhibited inferior capacity retention and limited cycling life. The hp-SiNSs have also shown improved capacity retention and Coulombic efficiency relative to other nanostructured Si materials, including nanoparticles[17], nano-Si/carbon black composite[7], nanowires[10] and nanotubes[11]. Overall, the hp-SiNSs have demonstrated potential applications as high-performance anodes. For practical usage, it is critically important in future work to increase areal loading to pair with various cathodes through engineering of electrodes and utilization of improved electrolytes.

In conclusion, the hp-SiNSs exhibit ultrahigh Li diffusion and reversible inward volume expansion/contraction, which overcome the technical barriers for using solid Si as high-capacity anodes. The design and synthesis of such hp-SiNSs constitute a generic strategy to simultaneously enhance rate performance and structural stability of high-capacity electrode materials. The size-dependent reaction kinetics provides exceptional potential in rational design of hierarchically nanostructured materials and can be broadly applied to other materials such as SiC and Si_3N_4.

Methods

Synthesis of solid core/mesoporous shell SiO_2 spheres. The solid core/mesoporous shell SiO_2 spheres produced by simultaneous hydrolysation and condensation of tetraethoxysilane and octadecyltrimethoxysilane followed by removal of organic species. Absolute ethanol (117 g, 2.54 mol), deionised water (20 g, 1.12 mol) and aqueous ammonia (32 wt%, 5.64 g and 0.34 mol) were mixed in a 250 ml flask. After heating to 30 °C, tetraethyl orthosilicate (11.2 g, 0.052 mol) was added rapidly under stirring. After 1 h, a mixture of tetraethyl orthosilicate (9.34 g, 0.044 mol) and n-octadecyltrimethoxysilane (3.536 g, 0.00944 mol) was added drop by drop over a period of 20 min. After the mixture was added, the solution was kept at ambient temperature for 12 h. The resulting white powder was obtained by centrifuge and then calcined in air for a period of 6 h at 550 °C (1 °C min^{-1}).

Synthesis of hp-SiNSs. SiO_2 powers (0.5 g) were dispersed in ethanol and cast onto a ceramic plate that was 1 cm away from 1 g of Mg powder layer dispersed at the bottom of a standard steal chamber. After the chamber was transferred into the tubular furnace, temperature was ramped to 680 °C at the rate of 10 °C and soaked for 2 h under Ar. The product was washed with dilute mixed acid (acetic acid: hydrochloric acid = 4:1), followed by etching in hydrofluoric acid solution. Finally a Si brown powder was obtained under vacuum drying at room temperature and stored in glove box.

Electrode fabrication, cell assembly and testing. The hp-SiNSs, conductive poly (9,9-dioctylfluorene-co-fluorenone-co-methyl-benzoic ester), and XG graphene nanoplatelets (Supplementary Fig. 7) with a weight ratio of 70:20:10 were dispersed in tetrahydrofuran. The slurry was cast on a copper foil and dried at room temperature. The electrode loading (including binder and carbon) is about 0.8 mg cm^{-2}. Later the electrodes were punched into a circular disc with a diameter of 0.5″ and further treated by ramping the temperature from room temperature to 500 °C at the rate of 10 °C and soaking for 1 h under Ar. The heat treatment for the electrode can improve the performance by the removal of the CHx species (Supplementary Fig. 8). After heat treatment the Si content increases to 76 wt% based on the mass change. The silicon loading is about 1 mAh cm^{-2}. For comparison, the commercial solid 100 nm Si nanoparticles were used to fabricate benchmark electrodes under the similar conditions (Supplementary Fig. 9). The coin cells, composed of a Si electrode, a microporous polyethylene separator, a lithium counter electrode were assembled in an argon-filled glove box. The electrolyte was a 1.0 M LiPF$_6$ solution in ethylene carbonate/diethyl carbonate (2/1 vol%) with 10 wt% fluorinated ethylene carbonate as an additive. The galvanostatic charge and discharge measurements were taken on a Maccor testing system. The Cyclic voltammetry was obtained using BioLogic workstation in a home-made three-electrode configure.

Characterizations. TEM, scanning TEM images and the energy dispersive X-ray spectroscopy are taken on JEOL JEM 2,100 F at 200 kV. The X-ray diffraction patterns are obtained on Bruker D8 Advance with Cu Kα radiation ($\lambda = 1.5418$ Å). In situ TEM is carried out on a FEI Titan microscope at 300 kV. To elucidate the structural advantage of the hp-SiNSs on electrochemical performance, *in situ* observation of the structural changes of the hp-SiNSs is conducted by assembling a nano battery inside of the TEM column as shown in Supplementary Fig. 10. A hp-SiNS is loaded onto a Si nanowire (SiNW). The SiNW is grown on Si substrate, which is further connected to the Au rod using conductive epoxy. During the *in situ* lithiation, the SiNW is connected to the Li/Li$_2$O end. With external bias applied, Li ions diffuse through the Li$_2$O solid electrolyte and react with the SiNW first. In the same time, the Li ions diffuse quickly along the SiNW to the hp-SiNS and react with it.

References

1. Wu, H. & Cui, Y. Designing nanostructured Si anodes for high energy lithium ion batteries. *Nano Today* **7**, 414–429 (2012).
2. Wang, C. *et al.* Self-healing chemistry enables the stable operation of silicon microparticle anodes for high-energy lithium-ion batteries. *Nat. Chem.* **5**, 1042–1048 (2013).
3. Wu, M. Y. *et al.* Toward an ideal polymer binder design for high-capacity battery anodes. *J. Am. Chem. Soc.* **135**, 12048–12056 (2013).
4. Cheng, Y. T. & Verbrugge, M. W. The influence of surface mechanics on diffusion induced stresses within spherical nanoparticles. *J. Appl. Phys.* **104**, 083521 (2008).
5. Verbrugge, M. W. & Cheng, Y. T. Stress distribution within spherical particles undergoing electrochemical insertion and extraction. *ECS Trans.* **13**, 127–139 (2008).
6. Kasavajjula, U., Wang, C. & Appleby, A. J. Nano- and bulk-silicon-based insertion anodes for lithium-ion secondary cells. *J. Power Sources* **163**, 1003–1039 (2007).
7. Li, H., Huang, X., Chen, L., Wu, Z. & Liang, Y. A high capacity nano-Si composite anode material for lithium rechargeable batteries. *Electrochem. Solid State Lett.* **2**, 547–549 (1999).
8. Yao, Y. *et al.* Interconnected silicon hollow nanospheres for lithium-ion battery anodes with long cycle life. *Nano Lett.* **11**, 2949–2954 (2011).
9. Ma, H. *et al.* Nest-like silicon nanospheres for high-capacity lithium storage. *Adv. Mater.* **19**, 4067–4070 (2007).
10. Chan, C. K. *et al.* High-performance lithium battery anodes using silicon nanowires. *Nat. Nanotechnol.* **3**, 31–35 (2008).
11. Park, M. H. *et al.* Silicon nanotube battery anodes. *Nano Lett.* **9**, 3844–3847 (2009).
12. Kim, H., Han, B., Choo, J. & Cho, J. Three-dimensional porous silicon particles for use in high-performance lithium secondary batteries. *Angew. Chem. Int. Ed.* **47**, 10151–10154 (2008).
13. Magasinski, A. *et al.* High-performance lithium-ion anodes using a hierarchical bottom-up approach. *Nat. Mater.* **9**, 353–358 (2010).
14. Chen, S. R. *et al.* Silicon core–hollow carbon shell nanocomposites with tunable buffer voids for high capacity anodes of lithium-ion batteries. *Phys. Chem. Chem. Phys.* **14**, 12741–12745 (2012).
15. Ng, S.-H. *et al.* Highly reversible lithium storage in spheroidal carbon-coated silicon nanocomposites as anodes for lithium-ion batteries. *Angew. Chem. Int. Ed.* **45**, 6896–6899 (2006).
16. Kovalenko, I. *et al.* A major constituent of brown algae for use in high-capacity Li-ion batteries. *Science* **334**, 75–79 (2011).
17. Erk, C., Brezesinski, T., Sommer, H., Schneider, R. & Janek, J. Toward silicon anodes for next-generation lithium ion batteries: a comparative performance study of various polymer binders and silicon nanopowders. *ACS Appl. Mater. Interfaces* **5**, 7299–7307 (2013).
18. Liu, G. *et al.* Polymers with tailored electronic structure for high capacity lithium battery electrodes. *Adv. Mater.* **23**, 4679–4683 (2011).
19. Buchel, G., Unger, K. K., Matsumoto, A. & Tsutsumi, K. A novel pathway for synthesis of submicrometer-size solid core/mesoporous shell silica spheres. *Adv. Mater.* **10**, 1036–1038 (1998).
20. Wu, H. *et al.* Stable cycling of double-walled silicon nanotube battery anodes through solid-electrolyte interphase control. *Nat. Nanotechnol.* **7**, 310–315 (2012).
21. Mannix, A. J., Kiraly, B., Fisher, B. L., Hersam, M. C. & Guisinger, N. P. Silicon growth at the two-dimensional limit on Ag(111). *ACS Nano* **8**, 7538–7547 (2014).
22. Ge, M. Y. *et al.* Scalable preparation of porous silicon nanoparticles and their application for lithium-ion battery anodes. *Nano Res.* **6**, 174–181 (2013).
23. Meier, C. *et al.* Raman properties of silicon nanoparticles. *Phys. E Low Dimensional Syst. Nanostructures* **32**, 155–158 (2006).
24. Ehbrecht, M., Kohn, B., Huisken, F., Laguna, M. A. & Paillard, V. Photoluminescence and resonant Raman spectra of silicon films produced by size-selected cluster beam deposition. *Phys. Rev. B* **56**, 6958–6964 (1997).
25. Thommes, M. Physical adsorption characterization of nanoporous materials. *Chem. Ing. Tech.* **82**, 1059–1073 (2010).
26. Richman, E. K., Kang, C. B., Brezesinski, T. & Tolbert, S. H. Ordered mesoporous silicon through magnesium reduction of polymer templated silica thin films. *Nano Lett.* **8**, 3075–3079 (2008).
27. Bao, Z. H. *et al.* Chemical reduction of three-dimensional silica micro-assemblies into microporous silicon replicas. *Nature* **446**, 172–175 (2007).
28. Dai, F. *et al.* Bottom-up synthesis of high surface area mesoporous crystalline silicon and evaluation of its hydrogen evolution performance. *Nat. Commun.* **5**, 3605 (2014).
29. Zhao, J. *et al.* Dry-air-stable lithium silicide-lithium oxide core-shell nanoparticles as high-capacity prelithiation reagents. *Nat. Commun.* **5**, 5088 (2014).
30. McDowell, M. T. *et al.* Studying the kinetics of crystalline silicon nanoparticle lithiation with *in situ* transmission electron microscopy. *Adv. Mater.* **24**, 6034–6041 (2012).
31. Gu, M. *et al.* In situ TEM study of lithiation behavior of silicon nanoparticles attached to and embedded in a carbon matrix. *ACS Nano* **6**, 8439–8447 (2012).
32. Liu, X. H. *et al.* Size-dependent fracture of silicon nanoparticles during lithiation. *ACS Nano* **6**, 1522–1531 (2012).
33. Liu, N. *et al.* A pomegranate-inspired nanoscale design for large-volume-change lithium battery anodes. *Nat. Nanotechnol.* **9**, 187–192 (2014).
34. Liu, X. H. *et al.* Anisotropic swelling and fracture of silicon nanowires during lithiation. *Nano Lett.* **11**, 3312–3318 (2011).
35. Lee, S. W., McDowell, M. T., Berla, L. A., Nix, W. D. & Cui, Y. Fracture of crystalline silicon nanopillars during electrochemical lithium insertion. *Proc. Natl Acad. Sci. USA* **109**, 4080–4085 (2012).
36. Liu, X. H. *et al.* Self-limiting lithiation in silicon nanowires. *ACS Nano* **7**, 1495–1503 (2012).
37. Yang, H. *et al.* Orientation-dependent interfacial mobility governs the anisotropic swelling in lithiated silicon nanowires. *Nano Lett.* **12**, 1953–1958 (2012).
38. Yang, H. *et al.* A chemo-mechanical model of lithiation in silicon. *J. Mech. Phys. Solids* **70**, 349–361 (2014).
39. Wortman, J. J. & Evans, R. A. Young's modulus, shear modulus, and Poisson's ratio in silicon and germanium. *J. Appl. Phys.* **36**, 153–156 (1965).
40. Populaire, C. *et al.* On mechanical properties of nanostructured meso-porous silicon. *Appl. Phys. Lett.* **83**, 1370–1372 (2003).
41. Fan, F. F. *et al.* Mechanical properties of amorphous LixSi alloys: a reactive force field study. *Model. Simul. Mater. Sci. Eng.* **21**, 074002 (2013).
42. Shenoy, V. B., Johari, P. & Qi, Y. Elastic softening of amorphous and crystalline Li-Si Phases with increasing Li concentration: a first-principles study. *J. Power Sources* **195**, 6825–6830 (2010).
43. Zhao, K. *et al.* Lithium-assisted plastic deformation of silicon electrodes in lithium-ion batteries: a first-principles theoretical study. *Nano Lett.* **11**, 2962–2967 (2011).
44. Yang, H. *et al.* Self-weakening in lithiated graphene electrodes. *Chem. Phys. Lett.* **563**, 58–62 (2013).
45. Huang, X. *et al.* Lithiation induced corrosive fracture in defective carbon nanotubes. *Appl. Phys. Lett.* **103**, 153901 (2013).
46. Liang, W. *et al.* Tough germanium nanoparticles under electrochemical cycling. *ACS Nano* **7**, 3427–3433 (2013).
47. Yang, H., Liang, W. T., Guo, X., Wang, C. M. & Zhang, S. L. Strong kinetics-stress coupling in lithiation of Si and Ge anodes. *Extreme Mech. Lett.* **2**, 1–6 (2015).
48. Lee, S. W., McDowell, M. T., Choi, J. W. & Cui, Y. Anomalous shape changes of silicon nanopillars by electrochemical lithiation. *Nano Lett.* **11**, 3034–3039 (2011).
49. McDowell, M. T. *et al.* In Situ TEM of two-phase lithiation of amorphous silicon nanospheres. *Nano Lett.* **13**, 758–764 (2013).
50. Moon, G. D. *et al.* Chemical transformations of nanostructured materials. *Nano Today* **6**, 186–203 (2011).

Acknowledgements

This work is supported by GM internal funds. The *in situ* TEM work was supported by the Assistant Secretary for Energy Efficiency and Renewable Energy, Office of Vehicle Technologies of the U.S. Department of Energy under Contract No. DE_AC02-05CH11231, Subcontract No. 6951379 under the advanced Battery Materials Research program and the Chemical Imaging Initiative at Pacific Northwest National Laboratory (PNNL), which was conducted in the William R. Wiley Environmental Molecular Sciences Laboratory, a national scientific user facility sponsored by DOE's Office of Biological and Environmental Research and located at PNNL. The authors also thank XG science for providing the graphene nanoplatelets. H.Y. and S.Z. acknowledge the support by the NSF-CMMI (Grant No. 0900692).

Author contributions

Q.X. and M.C. conceived or designed the experiments. M.G. and C.W. performed the *in situ* TEM study characterization and data analysis. H.Y., P.Z. and S.Z. developed the chemomechanical model. B.L. and C.Z. did BET, FTIR and TGA. Y.L. and F.L. made the schematic. F.D. and L.Y. prepared the electrolyte. Z.L. and X.X. did TEM analysis. G.L. provided the binder. Q.X., M.G. and H.Y. prepared the manuscript. S.Z., C.W., Y.L. and M.C. revised the manuscript. All authors discussed the results and commented on the manuscript.

Additional information

16

Molecular magnetic switch for a metallofullerene

Bo Wu[1], Taishan Wang[1], Yongqiang Feng[1], Zhuxia Zhang[1], Li Jiang[1] & Chunru Wang[1]

The endohedral fullerenes lead to well-protected internal species by the fullerene cages, and even highly reactive radicals can be stabilized. However, the manipulation of the magnetic properties of these radicals from outside remains challenging. Here we report a system of a paramagnetic metallofullerene $Sc_3C_2@C_{80}$ connected to a nitroxide radical, to achieve the remote control of the magnetic properties of the metallofullerene. The remote nitroxide group serves as a magnetic switch for the electronic spin resonance (ESR) signals of $Sc_3C_2@C_{80}$ via spin-spin interactions. Briefly, the nitroxide radical group can 'switch off' the ESR signals of the $Sc_3C_2@C_{80}$ moiety. Moreover, the strength of spin-spin interactions between $Sc_3C_2@C_{80}$ and the nitroxide group can be manipulated by changing the distance between these two spin centres. In addition, the ESR signals of the $Sc_3C_2@C_{80}$ moiety can be switched on at low temperatures through weakened spin-lattice interactions.

[1] Key Laboratory of Molecular Nanostructure and Nanotechnology, Beijing National Laboratory for Molecular Sciences, Institute of Chemistry, Chinese Academy of Sciences, Beijing 100190, China. Correspondence and requests for materials should be addressed to T.W. (email: wangtais@iccas.ac.cn) or to C.W. (email: crwang@iccas.ac.cn).

Endohedral fullerenes are constructed by putting atoms or clusters inside fullerene cages, which isolates the internal species with environments, so even those high-reactive species can be well stabilized inside the fullerene cages[1–6]. For example, paramagnetic endohedral fullerenes such as $N@C_{60}$ (refs 7–10), $Sc@C_{82}$ (refs 11,12), $Y@C_{82}$ (refs 13,14), $Sc_3C_2@$ I_h-C_{80} (refs 15–17) and so on, encapsulating radicals inside the fullerene cages, show also remarkable high stability, and they can be kept in air under room temperature for a long time, especially for endohedral metallofullerenes (EMFs). Considering the paramagnetic endohedral fullerenes usually show long electron spin relaxation and coherence times, they are expected to have potential applications in many fields such as spin labelling, spintronics, quantum computing and so on[18].

For paramagnetic EMFs, the electronic spin resonance (ESR) technique is a powerful tool to detect the spin distributions and spin–nucleus couplings on internal species[19–21]. By means of ESR, it was revealed that the magnetic property of the internal species can be roughly manipulated by changing the dynamic movement of internal species. For example, under room temperature the internal Y_2 cluster in $Y_2@C_{79}N$ has a free rotation that leads to a symmetric ESR pattern, but along with the temperature decreasing, the free motion of Y_2 is hindered, leading to spin anisotropy and an asymmetric ESR pattern[22]. In addition, the spin characters and couplings in $Sc_3C_2@C_{80}$ was observed to change largely upon chemical modification of the fullerene cage due to the restricted Sc_3C_2 cluster[17].

For better applying the paramagnetic EMFs in quantum information process and molecular devices, however, it is still a challenge to finely manipulate their magnetic property. Recently, Turro et al. chemically modified the $H_2@C_{60}$ with a nitroxide radical, and observed an indirect but strong magnetic communication between the electron spin of nitroxide paramagnet and the nuclear spin of encaged H_2 (refs 23–28). This finding provides a valuable clue for us on manipulating the magnetic property of paramagnetic EMFs via a foreign paramagnet[29]. Since a strong spin–spin interaction between the paramagnetic fullerene molecule and the paramagnet is expected, thus the magnetic property of paramagnetic EMFs may be controlled by the attached paramagnet.

Herein, we report detailed studies on the fine manipulation for paramagnetic $Sc_3C_2@C_{80}$ by connecting it with a paramagnet of nitroxide radical. The target system $FSc_3C_2@C_{80}PNO^{\bullet}$ contains two kinds of spins localizing on $Sc_3C_2@C_{80}$ and nitroxide radical, respectively. The remote nitroxide group serves as a magnetic switch for the ESR signals of $Sc_3C_2@C_{80}$ through spin–spin interactions. The paramagnetic properties of the metallofullerene $Sc_3C_2@C_{80}$ can be delicately adjusted by changing the temperature, varying the distance between the two spin centres, or simply quenching the nitroxide radical.

Results

Preparation of metallofullerene and its derivatives.
Metallofullerene $Sc_3C_2@C_{80}$ was synthesized by the Kräschmer–Huffman arc-discharging method[30] and isolated by multi-stage high-performance liquid chromatography. Two $Sc_3C_2@C_{80}$ derivatives, $FSc_3C_2@C_{80}PNOH$ and $FSc_3C_2@C_{80}PNO^{\bullet}$, were first synthesized through a Prato reaction[31], respectively, as shown in Fig. 1a,b. The structures and spin density distributions of $FSc_3C_2@C_{80}PNOH$ and $FSc_3C_2@C_{80}PNO^{\bullet}$ were calculated as well, as shown in Fig. 1c,d. The $FSc_3C_2@C_{80}PNOH$ has one spin centre that is localized on $Sc_3C_2@C_{80}$, whereas the $FSc_3C_2@C_{80}PNO^{\bullet}$ has two unpaired spins localizing on the $Sc_3C_2@C_{80}$ moiety and nitroxide radical, respectively.

The ESR analysis of metallofullerene with a nitroxide radical.
$Sc_3C_2@C_{80}$ is a typical paramagnetic endohedral fullerene. As reported previously, the ESR spectrum of the pristine $Sc_3C_2@C_{80}$ shows a symmetric pattern with 21 resonant lines, however, those of $Sc_3C_2@C_{80}$ derivatives showed a distorted pattern with a greatly increased amount of resonant lines[17]. Therefore, ESR spectroscopy was first employed to reveal the electron spin characters of $FSc_3C_2@C_{80}PNOH$ and $FSc_3C_2@C_{80}PNO^{\bullet}$.

As shown in Fig. 1e, the ESR spectrum of $FSc_3C_2@C_{80}PNOH$ was measured and analysed. Because the I_h symmetry of $Sc_3C_2@C_{80}$ is broken down after chemical modification, the original three equivalent scandium nuclei ($I_{Sc} = 7/2$) are classified into two groups in $FSc_3C_2@C_{80}PNOH$, in which one group contains a single Sc nucleus ($g = 1.9948$, hyperfine coupling constants (hfcc) $= 8.5\,G$), and the other group contains two equivalent Sc nuclei (hfcc $= 5.0\,G$). In comparison, the previously studied $Sc_3C_2@C_{80}$ fulleropyrrolidine shows a similar ESR pattern with hfcc of $8.6\,G$ (one Sc nucleus) and $4.8\,G$ (two Sc nuclei), respectively[17]. These ESR results reveal that the $FSc_3C_2@C_{80}PNOH$ has a same reaction site with that of $Sc_3C_2@C_{80}$ fulleropyrrolidine[17].

However, the ESR study of $FSc_3C_2@C_{80}PNO^{\bullet}$ showed only three resonant lines ($g = 2.0026$, $a = 15.5\,G$) that are derived from nitroxide radical ($I_N = 1$), and the ESR signals of $Sc_3C_2@C_{80}$ moiety were not observed, as illustrated in Fig. 1f. The current results reveal that the ESR signals of $Sc_3C_2@C_{80}$ can be switched off by paramagnetic nitroxide radical through spin–spin interaction. Thus it is interesting that the nitroxide radical group can serve as a remote controller for the ESR signals of $Sc_3C_2@C_{80}$ moiety. Vividly, if the ESR signals of $Sc_3C_2@C_{80}$ moiety are regarded as an indicating lamp, the nitroxide radical group can switch it off.

To reveal the mechanism of how the ESR signals of $Sc_3C_2@C_{80}$ moiety are switched off by the nitroxide radical, we synthesized $FSc_3N@C_{80}PNO^{\bullet}$ in a same way for comparison. $Sc_3N@C_{80}$ is a diamagnetic molecule, and no spin–spin interaction is expected for $FSc_3N@C_{80}PNO^{\bullet}$. The signal intensity of $FSc_3N@C_{80}PNO^{\bullet}$ was observed to be stronger than that of $FSc_3C_2@C_{80}PNO^{\bullet}$ at the same concentration (0.151 p.p.m.). That is to say, the spin–spin interactions weaken both of the ESR signals of $Sc_3C_2@C_{80}$ and nitroxide radical (Supplementary Fig. 1).

For $FSc_3C_2@C_{80}PNO^{\bullet}$, the spin–spin interactions can be expressed as below:

$$E_{dip} = \frac{\mu_0}{4\pi} g^2 \beta_e^2 \left[\frac{S_1 \times S_2}{r^3} - \frac{3(S_1 \times r)(S_2 \times r)}{r^5} \right] \quad (1)$$

Where E_{dip} is the energy of dipolar coupling, and r is the dipole–dipole distance.

In fact, the spin–spin interactions between $Sc_3C_2@C_{80}$ and nitroxide radical broaden the resonance lines and lower the line intensity in the meantime, so the line width (ΔH) is adopted to represent the ESR line intensity and interpret the spin–spin interactions[32,33].

In general, for paramagnetic molecules the ESR line width is inversely proportional to the relaxation time (T), including the spin–lattice relaxation time (T_1) and the spin–spin relaxation time (T_2):

$$\Delta H = \frac{h}{2\pi g \beta} \left(\frac{1}{T_1} \right) + \frac{h}{2\pi g \beta} \left(\frac{1}{T_2} \right) \quad (2)$$

Therefore, the strong dipole–dipole interactions between nitroxide radical and $Sc_3C_2@C_{80}$ in $FSc_3C_2@C_{80}PNO^{\bullet}$ reduced the spin–spin relaxation time (T_2), resulting in decreased ESR signals of both $Sc_3C_2@C_{80}$ and nitroxide radical moieties.

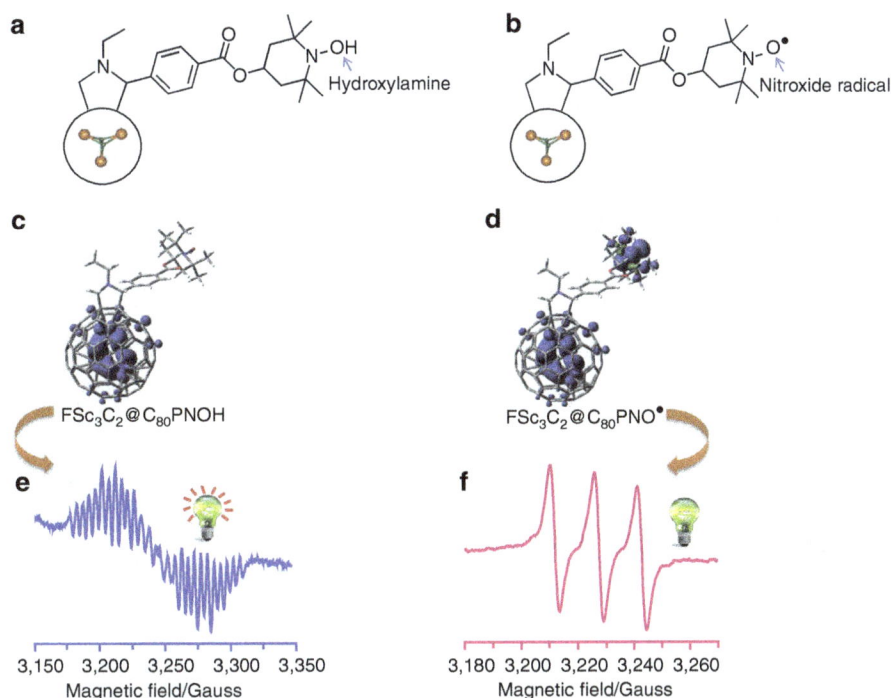

Figure 1 | Magnetic switch for the ESR signals of Sc₃C₂@C₈₀. (**a**) The structure of FSc₃C₂@C₈₀PNOH. (**b**) The structure of FSc₃C₂@C₈₀PNO•. (**c**) The calculated structure and spin density distributions of FSc₃C₂@C₈₀PNOH. (**d**) The calculated structure and spin density distributions of FSc₃C₂@C₈₀PNO•. (**e**) The ESR spectrum of FSc₃C₂@C₈₀PNOH at 293 K in toluene. (**f**) The ESR spectrum of FSc₃C₂@C₈₀PNO• at 293 K in toluene. The lamps in **e** and **f** show the 'on' and 'off' states of Sc₃C₂@C₈₀ ESR signals, respectively.

Note that the transformation between FSc₃C₂@C₈₀PNO• and FSc₃C₂@C₈₀PNOH is reversible, that is, the nitroxide radical in FSc₃C₂@C₈₀PNO• turns into the corresponding hydroxylamine derivative (FSc₃C₂@C₈₀PNOH) using p-toluenesulfonohydrazide, and the FSc₃C₂@C₈₀NOH can be back to FSc₃C₂@C₈₀PNO• by means of oxidation with copper acetate (Supplementary Fig. 2). Therefore, the magnetic property of Sc₃C₂@C₈₀ can be easily manipulated by a chemical method, and the remote nitroxide group serves as a switch in this process.

The distance-dependent ESR signals. On the basis of equation (1), the spin–spin interactions between Sc₃C₂@C₈₀ and nitroxide radical moieties can be efficiently reduced by elongating their distance, thus two other Sc₃C₂@C₈₀ and nitroxide radical derivatives, that is, FSc₃C₂@C₈₀PNO•-2 and FSc₃C₂@C₈₀PNO•-3, were synthesized with longer distances between these two spin centres. The pulsed ESR measurements on FSc₃C₂@C₈₀PNO• and FSc₃C₂@C₈₀PNO•-2 revealed that the T_2 of nitroxide radical becomes longer when the distance of these two spins increases (Supplementary Fig. 3). As shown in Fig. 2, it is obvious that the ESR signals of Sc₃C₂@C₈₀ also gradually boost up along with the distance increasing, and this process is like lighting a lamp and making it brighter.

However, the increased chain length between Sc₃C₂@C₈₀ and nitroxide radical would result in a strengthened spin–lattice interaction, which will bring another line-broadening effect for both nitroxide radical and Sc₃C₂@C₈₀ moiety. The spin–lattice relaxation time (T_1) can be expressed as below:

$$\frac{1}{T_1} = \frac{2\pi g \beta_e}{h} \left(B_x^2 + B_y^2 \right) \frac{\tau_c}{1 + \tau_c^2 \omega_s^2} \tag{3}$$

Where B_x^2 and B_y^2 are mean square amplitudes of the fluctuating fields along the x- and y-directions, and τ_c is the correlation time of the motion that causes the fluctuation. For molecules with a

spherical shape, τ_c in liquid solution corresponds to the rotational correlation time τ_r, which can be approximated by the Stokes–Einstein relation:

$$\tau_r = \frac{4\pi \eta a^3}{3 k_B T} \tag{4}$$

Where a is the rotationally effective radius of the molecule, and η is the viscosity of the solvent. In liquids with a low viscosity ($\tau_c^2 \omega_s^2 \approx 1$), the T_1 is dependent of a^3, and T_1 decreases with increasing the rotationally effective radius of the molecule. Therefore, increasement of the molecular size would shorten the T_1, leading to weaker ESR signals.

Since FSc₃C₂@C₈₀PNO•, FSc₃C₂@C₈₀PNO•-2 and FSc₃C₂@C₈₀PNO•-3 are all rigid structural molecules, the intramolecular dipolar coupling strength (D) can be estimated, in which the coupling strength of FSc₃C₂@C₈₀PNO• with $r = 1.38$ nm was estimated to be about 24.3 MHz following the classical point dipole approximation, and those of FSc₃C₂@C₈₀PNO•-2 and FSc₃C₂@C₈₀PNO•-3 were estimated to be about 11.2 and 5.48 MHz, respectively.

The temperature-dependent ESR signals. It is known that the spin–lattice relaxation time (T_1) of unpaired spin is tightly related to the temperature. As expressed in equation (3), the decreasement of temperature would reduce the B_x and B_y, and then increase the T_1 and lead to higher ESR line intensity. Therefore, the temperature-dependent ESR studies of FSc₃C₂@C₈₀PNO• were performed, as shown in Fig. 3. It can be observed that no ESR signal of Sc₃C₂@C₈₀ was observed at 293 K, but since 253 K, the ESR signals of Sc₃C₂@C₈₀ appeared together with three strong resonant lines of nitroxide radical, and continuously increased along with the temperature further decreasing. Finally at 213 K, the ESR signals of Sc₃C₂@C₈₀ moiety can be clearly observed. The electrostatic spin–phonon

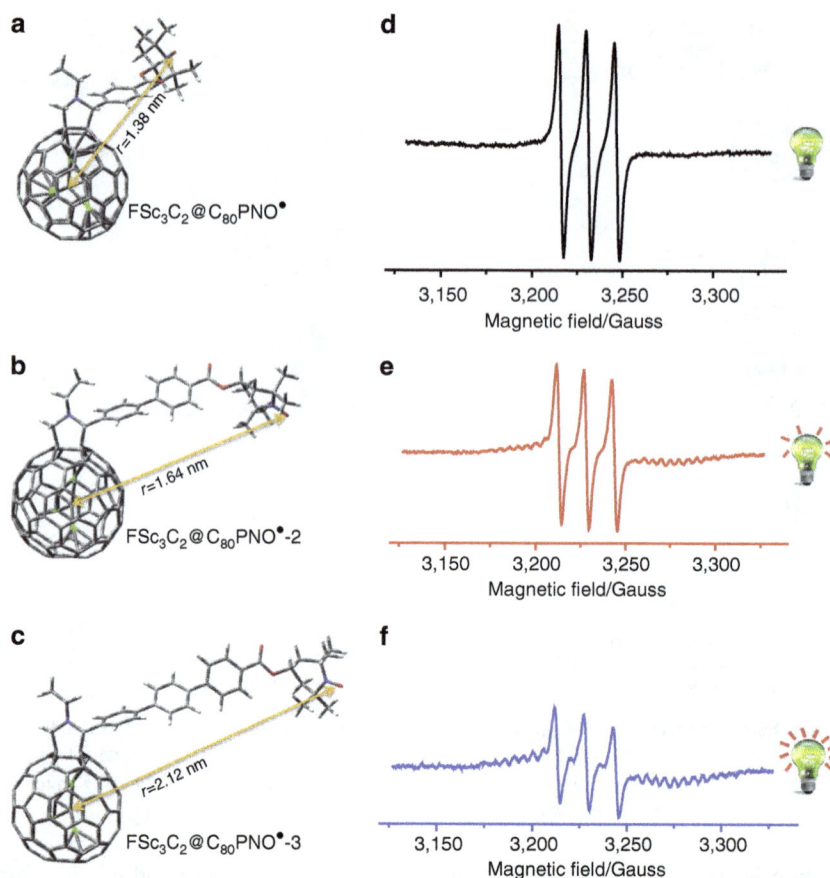

Figure 2 | Distance-dependent ESR signals of $Sc_3C_2@C_{80}$ derivatives. (a–c) Structures of $FSc_3C_2@C_{80}PNO^\bullet$, $FSc_3C_2@C_{80}PNO^\bullet$-2 and $FSc_3C_2@C_{80}PNO^\bullet$-3. (d–f) ESR signals of $FSc_3C_2@C_{80}PNO^\bullet$, $FSc_3C_2@C_{80}PNO^\bullet$-2 and $FSc_3C_2@C_{80}PNO^\bullet$-3 at 293 K. The lamps in d–f show the strengthened ESR signals of $Sc_3C_2@C_{80}$ moiety along with the enlarged distance from nitroxide radical.

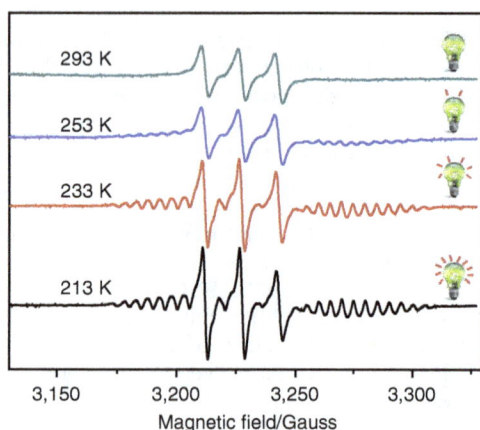

Figure 3 | Temperature-dependent ESR signals of $Sc_3C_2@C_{80}$ derivatives. The ESR spectra of $FSc_3C_2@C_{80}PNO^\bullet$ at variable temperatures in toluene solution. The lamps represent the strengthened ESR signals of $Sc_3C_2@C_{80}$ moiety along with the decreased temperatures.

interaction was also analysed for these temperature-dependent ESR spectra[34]. As the lowest temperature in our system is 213 K, under this condition the $FSc_3C_2@C_{80}PNO^\bullet$ toluene solution is still in liquid state, so the spin–phonon interaction is rather small and negligible. From the temperature-dependent ESR spectra, it can be seen that the temperature also can light the signals of $Sc_3C_2@C_{80}$ moiety in $FSc_3C_2@C_{80}PNO^\bullet$ and make it brighter.

Table 1 | ESR data.

Temperature (K)	Line width of nitroxide (G)	Line width of $Sc_3C_2@C_{80}$ (G)
293	3.17	—
253	3.08	2.63
233	2.47	2.37
213	2.35	2.13

ESR, electronic spin resonance.
The ESR spectra line width of nitroxide and $Sc_3C_2@C_{80}$ in $FSc_3C_2@C_{80}PNO^\bullet$ at variable temperatures.

Moreover, it should be noted that the ESR signals of the nitroxide were also enhanced. Therefore, along with the temperature decreasing, the prolonged T_1 enhances not only the ESR signals of $Sc_3C_2@C_{80}$ moiety, but also those of the nitroxide radical (Supplementary Fig. 4). The line widths of $Sc_3C_2@C_{80}$ moiety and nitroxide radical are listed in Table 1.

Discussion

Through connecting the paramagnetic metallofullerene $Sc_3C_2@C_{80}$ with a nitroxide radical, we have realized the manipulation of ESR signals of $Sc_3C_2@C_{80}$. The remote nitroxide group serves as a magnetic switch for ESR signals of $Sc_3C_2@C_{80}$, that is, the paramagnetic nitroxide group can 'switch off' the ESR signals of $Sc_3C_2@C_{80}$ moiety. It was revealed that the spin–spin interactions between $Sc_3C_2@C_{80}$ and nitroxide radical play a key role in realizing this kind of magnetic switch. Moreover, through increasing the distance between $Sc_3C_2@C_{80}$ and nitroxide radical, or decreasing the temperature, we can finely adjust the paramagnetic property of $Sc_3C_2@C_{80}$.

Such controllable paramagnetism and switchable ESR signals have potential applications in quantum information processing and molecular devices. For example, these magnetic molecules can be fabricated to a single molecule membrane for data storage considering their transferrable two electron spin states (0/1), which can be written and read out by means of scanning tunnelling microscope. In addition, these magnetic molecules can be utilized as a probe for the reaction transition state considering the susceptible magnetic switch of the nitroxide group.

Methods

Synthesis of $Sc_3C_2@C_{80}$ and $Sc_3N@C_{80}$ derivatives. $Sc_3C_2@C_{80}$ and $Sc_3N@C_{80}$ were heated with N-ethylglycine and 2,2,6,6 tetramethylpiperidine-1-oxyl 4-formylbenzoate (1a) (Supplementary Fig. 5), which were synthesized as literature methods[24] at 120 °C to give corresponding fulleropyrrolidines with yields of nearly 50% in toluene solution for 15 and 50 min, respectively. Pure $FSc_3C_2@C_{80}PNO^{\bullet}$ and $FSc_3N@C_{80}PNO^{\bullet}$ were isolated by HPLC using Buckyprep column (Supplementary Fig. 6). $FSc_3C_2@C_{80}PNO^{\bullet}$-2 and $FSc_3C_2@C_{80}PNO^{\bullet}$-3 were synthesized according to same procedure used for compound $FSc_3C_2@C_{80}PNO^{\bullet}$, except 2,2,6,6 tetramethylpiperidine-1-oxyl 4'-formylbiphenyl-4-carboxylate (2a) and 2,2,6,6 tetramethylpiperidine-1-oxyl 4'-p-terphenyl-4-carboxylate (3a) (Supplementary Fig. 7) were used.

Synthesis of $FSc_3C_2@C_{80}PNOH$. To a solution of ~ 0.5 mg of the nitroxide derivative $FSc_3C_2@C_{80}PNO^{\bullet}$ in ~ 2 ml toluene was added ~ 1 mg p-toluenesulfonohydrazide, and stirred under air for about 15 min.

Characterization of metallofullerene derivatives. Ultraviolet/visible–near-infrared spectra of purified metallofullerene derivatives (Supplementary Fig. 8) were collected on Lambda 950 UV/Vis/NIR Spectrometer (PerkinElmer Instruments). ^{1}H NMR spectra of $FSc_3N@C_{80}PNOH$ was measured in chloroform-d on a Bruker 600 MHz spectrometer (Supplementary Fig. 9).

ESR measurements of metallofullerene derivatives. ESR spectra were measured on a JEOL JEF FA200 X-band spectrometer (Supplementary Fig. 10). The samples were degassed and the oxygen was removed from the solutions. All of the samples are dissolved in toluene solution at the same concentration.

Calculations on $Sc_3C_2@C_{80}$ derivatives. Density functional theory calculations were investigated by Perdew, Burke and Enzerhof/double numerical plus polarization using the DMol3 code in Accelrys Materials Studio[35,36].

References

1. Taylor, R. & Walton, D. R. M. The chemistry of fullerenes. *Nature* **363**, 685–693 (1993).
2. Popov, A. A., Yang, S. & Dunsch, L. Endohedral fullerenes. *Chem. Rev.* **113**, 5989–6113 (2013).
3. Yang, S., Liu, F., Chen, C., Jiao, M. & Wei, T. Fullerenes encaging metal clusters-clusterfullerenes. *Chem. Commun.* **47**, 11822–11839 (2011).
4. Schenning, A. P. H. J. & George, S. J. Self-assembly: phases full of fullerenes. *Nat. Chem.* **6**, 658–659 (2014).
5. Hollamby, M. J. *et al.* Directed assembly of optoelectronically active alkyl-π-conjugated molecules by adding n-alkanes or π-conjugated species. *Nat. Chem.* **6**, 690–696 (2014).
6. Popov, A. A. & Dunsch, L. Structure, stability, and cluster-cage interactions in nitride clusterfullerenes $M_3N@C_{2n}$ (M = Sc, Y; 2n = 68 − 98): a density functional theory study. *J. Am. Chem. Soc.* **129**, 11835–11849 (2007).
7. Morton, J. J. L. *et al.* Electron spin relaxation of $N@C_{60}$ in CS_2. *J. Chem. Phys.* **124**, 014508 (2006).
8. Morton, J. J. L. *et al.* The $N@C_{60}$ nuclear spin qubit: bang-bang decoupling and ultrafast phase gates. *Phys. Stat. Sol.* **243**, 3028–3031 (2006).
9. Liu, G. *et al.* $N@C_{60}$–porphyrin: a dyad of two radical centers. *J. Am. Chem. Soc.* **134**, 1938–1941 (2012).
10. Morton, J. J. L. *et al.* Environmental effects on electron spin relaxation in $N@C_{60}$. *Phys. Rev. B* **76**, 085418 (2007).
11. Morley, G. W. *et al.* Hyperfine structure of $Sc@C_{82}$ from ESR and DFT. *Nanotechnology* **16**, 2469–2473 (2005).
12. Ito, Y. *et al.* Magnetic properties and crystal structure of solvent-free $Sc@C_{82}$ metallofullerene microcrystals. *Chem. Phys. Chem.* **8**, 1019–1024 (2007).
13. Kikuchi, K. *et al.* Characterization of the isolated $Y@C_{82}$. *J. Am. Chem. Soc.* **116**, 9367–9368 (1994).
14. Misochko, E. Y. *et al.* EPR spectrum of the $Y@C_{82}$ metallofullerene isolated in solid argon matrix: hyperfine structure from EPR spectroscopy and relativistic DFT calculations. *Phys. Chem. Chem. Phys.* **12**, 8863–8869 (2010).
15. Taubert, S., Straka, M., Pennanen, T. O., Sundholm, D. & Vaara, J. Dynamics and magnetic resonance properties of $Sc_3C_2@C_{80}$ and its monoanion. *Phys. Chem. Chem. Phys.* **10**, 7158–7168 (2008).
16. Wang, T. *et al.* Preparation and ESR study of $Sc_3C_2@C_{80}$ bis-addition fulleropyrrolidines. *Dalton Trans.* **41**, 2567–2570 (2012).
17. Wang, T. *et al.* Spin divergence induced by exohedral modification: ESR Study of $Sc_3C_2@C_{80}$ fulleropyrrolidine. *Angew. Chem. Int. Ed.* **49**, 1786–1789 (2010).
18. Krause, M. *et al.* Fullerene quantum gyroscope. *Phys. Rev. Lett.* **93**, 137403 (2004).
19. Wang, T. & Wang, C. Endohedral metallofullerenes based on spherical I_h-C_{80} Cage: molecular structures and paramagnetic properties. *Acc. Chem. Res.* **47**, 450–458 (2013).
20. Sato, S. *et al.* Mechanistic study of the Diels–Alder reaction of paramagnetic endohedral metallofullerene: reaction of $La@C_{82}$ with 1,2,3,4,5-pentamethylcyclopentadiene. *J. Am. Chem. Soc.* **135**, 5582–5587 (2013).
21. Elliott, B. *et al.* Spin density and cluster dynamics in $Sc_3N@C_{80}^{-}$ upon [5, 6] exohedral functionalization: an ESR and DFT study. *J. Phys. Chem. C* **117**, 2344–2348 (2013).
22. Ma, Y. *et al.* Susceptible electron spin adhering to an yttrium cluster inside an azafullerene $C_{79}N$. *Chem. Commun.* **48**, 11570–1157 (2012).
23. Li, Y. *et al.* A magnetic switch for spin-catalyzed interconversion of nuclear spin isomers. *J. Am. Chem. Soc.* **132**, 4042–4043 (2010).
24. Li, Y. *et al.* Distance-dependent paramagnet-enhanced nuclear spin relaxation of $H_2@C_{60}$ derivatives covalently linked to a nitroxide radical. *J. Phys. Chem. Lett.* **1**, 2135–2138 (2010).
25. Li, Y. *et al.* Distance-dependent para-$H_2 \rightarrow$ ortho-H_2 conversion in $H_2@C_{60}$ derivatives covalently linked to a nitroxide radical. *J. Phys. Chem. Lett.* **2**, 741–744 (2011).
26. Li, Y. *et al.* Synthesis, isomer count, and nuclear spin relaxation of $H_2O@open$-C_{60} nitroxide derivatives. *Org. Lett.* **14**, 3822–3825 (2012).
27. Turro, N. J. *et al.* The spin chemistry and magnetic resonance of $H_2@C_{60}$. From the pauli principle to trapping a long lived nuclear excited spin state inside a buckyball. *Acc. Chem. Res.* **43**, 335–345 (2009).
28. Turro, N. J. *et al.* Demonstration of a chemical transformation inside a fullerene. The reversible conversion of the allotropes of $H_2@C_{60}$. *J. Am. Chem. Soc.* **130**, 10506–10507 (2008).
29. Farrington, B. J. *et al.* Chemistry at the nanoscale: synthesis of an $N@C_{60}$–$N@C_{60}$ endohedral fullerene dimer. *Angew. Chem. Int. Ed.* **51**, 3587–3590 (2012).
30. Kroto, H. W., Heath, J. R., O'Brien, S. C., Curl, R. F. & Smalley, R. E. C_{60}: Buckminsterfullerene. *Nature* **318**, 162–163 (1985).
31. Brough, P., Klumpp, C., Bianco, A., Campidelli, S. & Prato, M. [60] Fullerene-pyrrolidine-N-oxides. *J. Org. Chem.* **71**, 2014–2020 (2006).
32. Schweiger, A. & Jeschke, G. *Principles of Pulse Electron Paramagnetic Resonance* (Oxford Univ. Press, 2001).
33. Berliner, L. J., Eaton, G. R. & Eaton, S. S. *Distance Measurements in Biological Systems by EPR* (Springer, 2002).
34. Abragam, A. & Bleaney, B. *Electron Paramagnetic Resonance of Transition Ions* (Dover Publications, 1970).
35. Perdew, J. P., Burke, K. & Ernzerhof, M. Generalized gradient approximation made simple. *Phys. Rev. Lett.* **77**, 3865–3868 (1996).
36. Delley, B. An all-electron numerical method for solving the local density functional for polyatomic molecules. *J. Chem. Phys.* **92**, 508–517 (1990).

Acknowledgements

This work was supported by the National Basic Research Program (2012CB932901), National Natural Science Foundation of China (21203205, 61227902, 51472248 and 21273006), NSAF (11179006) and the Key Research Program of the Chinese Academy of Sciences (KGZD-EW-T02). T.W. thanks the Youth Innovation Promotion Association of CAS. We thank the help from Dr Guoquan Liu in Max Planck Institute for pulsed ESR measurements.

Author contributions

T.W. conceived and designed the experiments. T.W. and C.W. wrote the paper. Experiments were carried out by B.W. The ESR data were analysed by T.W. and B.W. Calculations were carried out by Z.Z. All authors discussed the results and contributed to manuscript preparation.

Additional information

Competing financial interests: The authors declare no competing financial interests.

17

Enhancing the magnetic anisotropy of maghemite nanoparticles via the surface coordination of molecular complexes

Yoann Prado[1], Niéli Daffé[1,2,3], Aude Michel[1], Thomas Georgelin[4,5], Nader Yaacoub[6], Jean-Marc Grenèche[6], Fadi Choueikani[3], Edwige Otero[3], Philippe Ohresser[3], Marie-Anne Arrio[2], Christophe Cartier-dit-Moulin[7,8], Philippe Sainctavit[2,3], Benoit Fleury[7,8], Vincent Dupuis[1], Laurent Lisnard[7,8] & Jérôme Fresnais[1]

Superparamagnetic nanoparticles are promising objects for data storage or medical applications. In the smallest—and more attractive—systems, the properties are governed by the magnetic anisotropy. Here we report a molecule-based synthetic strategy to enhance this anisotropy in sub-10-nm nanoparticles. It consists of the fabrication of composite materials where anisotropic molecular complexes are coordinated to the surface of the nanoparticles. Reacting 5 nm γ-Fe$_2$O$_3$ nanoparticles with the [Co^{II}(TPMA)Cl$_2$] complex (TPMA: tris(2-pyridylmethyl)amine) leads to the desired composite materials and the characterization of the functionalized nanoparticles evidences the successful coordination—without nanoparticle aggregation and without complex dissociation—of the molecular complexes to the nanoparticles surface. Magnetic measurements indicate the significant enhancement of the anisotropy in the final objects. Indeed, the functionalized nanoparticles show a threefold increase of the blocking temperature and a coercive field increased by one order of magnitude.

[1] Sorbonne Universités, UPMC Univ Paris 06, UMR 8234, PHENIX, CNRS, F-75005 Paris, France. [2] Institut de Minéralogie, de Physique des Matériaux et de Cosmochimie, UMR 7590, CNRS, UPMC, IRD, MNHN, F-75005 Paris, France. [3] Synchrotron SOLEIL, L'Orme des Merisiers, Saint-Aubin—BP 48, 91192 Gif-sur-Yvette, France. [4] Sorbonne Universités, UPMC Univ Paris 06, UMR 7197, LRS, F-94200 Ivry-sur-Seine, France. [5] CNRS, UMR 7197, Laboratoire de Réactivité de Surface, F-94200 Ivry-sur-Seine, France. [6] Institut des Molécules et Matériaux du Mans CNRS UMR-6283, Université du Maine, F-72085 Le Mans, France. [7] Sorbonne Universités, UPMC Univ Paris 06, UMR 8232, IPCM, F-75005 Paris, France. [8] CNRS, UMR 8232, Institut Parisien de Chimie Moléculaire, F-75005 Paris, France. Correspondence and requests for materials should be addressed to Y.P. (email: yoann.prado@upmc.fr) or to L.L. (email: laurent.lisnard@upmc.fr) or to J.F. (email: jerome.fresnais@upmc.fr).

In single-domain superparamagnetic nanoparticles, magnetic anisotropy has a direct impact on the magnetization, its remanence, reversal and relaxation. Magnetic anisotropy is therefore a key parameter in the preparation of magnetic nanocrystals designed for high-density data storage applications or for medical applications[1–7]. For such applications, achieving a controlled modulation of the magnetic anisotropy for a given size of crystals represents thus one of the most efficient ways to improve and tune their magnetic properties. For example, the optimization of the specific loss power of nanocrystals for magnetic hyperthermia and the comprehension of the interplay between magnetic anisotropy and magnetization is of crucial importance for applications in nanomedicine[8,9]. On the other hand, the fabrication of small magnetic nanocrystals displaying high blocking temperature while maintaining magnetic bistability—that is, large coercive fields—remains a considerable challenge aimed at overtaking the so-called superparamagnetic limit and increasing data storage densities[10].

Different chemical approaches have been used so far to alter the magnetic anisotropy of single-domain magnetic nanoparticles: particle doping or formation of alloys[11–14], coordination of solvent or organic molecules to the nanoparticles surface[15–17], preparation of core-shell systems incorporating highly anisotropic components[8,9,12,18–23] and insertion of the nanoparticles into magnetic host matrices[10,24].

Herein, we report a strategy for the preparation of anisotropically enhanced magnetic nanoparticles. Our synthetic strategy is based on the direct coordination of magnetic molecular complexes to the surface of the nanoparticles. Using molecular complexes as the enhancing unit is a straightforward approach to improve the nanoparticles properties and it also represents a very advantageous functionalization tool. Indeed, coordination chemistry with its great versatility allows the design and the use of specific complexes where both the local ion anisotropy and the nature of the coordination sphere are controlled. The use of coordination complexes to achieve surface functionalization will also prevent diffusion of the ions into the particles. Decorating magnetic nanoparticles with molecular complexes is thus equivalent to building a new surface with ions whose environment and hence local anisotropy is predetermined. This represents an efficient way to not only modify but also tune the nanoparticles surface anisotropy, which is known to contribute greatly to the whole magnetic anisotropy and actually be one of its most influential components in small systems[25,26]. Furthermore, investigations motivated by the study of the possible interactions between two magnetic components in hybrid materials made from inorganic nanomaterials and coordination compounds remain scarce[27]. In another type of multi-scale hybrid system, it has been shown that the magnetization relaxation of deposited mononuclear complexes can be influenced by a magnetic substrate[28]. Our work, however, is motivated by the opposite phenomena, that is the influence of the complexes on the magnetic behaviour of the nanoparticles. We demonstrate here that the grafting of an adequate magnetic complex—even for a low quantity—on a superparamagnetic nanoparticle leads to a massive improvement of the magnetic properties.

Results

Strategy

In combining small superparamagnetic nanoparticles and magnetic molecular complexes we wished to efficiently transmit the anisotropy of the complex to the particles and thus achieve the modulation of the particles magnetic anisotropy. As the magnetic moment originating from mononuclear complexes (few Bohr magnetons) is weak compared to that of a particle (thousands of Bohr magnetons), weak interactions or electrostatic interactions between the complexes and the particles were not suitable for the occurrence of a significant magnetic effect. Therefore, we have targeted the formation of a chemical bond between the complexes and the particles as the best way to promote a strong exchange pathway, namely, a coordination bridge between the metal ions of the complexes and the ones located at the surface of the particles. To successfully coordinate complexes at the surface of nanoparticles and yield an anisotropically enhanced system, we have followed three prerequisites.

First, we have selected a magnetic complex made with a polydentate ligand and bearing labile groups. Such a ligand should warrant the stability and the geometry of the complex while the labile groups will promote the coordination of the complex to the particle surface. In this work, we have chosen the complex $[Co(TPMA)Cl_2]$ (ref. 29; TPMA, tris(2-pyridylmethyl)amine, Fig. 1a). With its intrinsic magnetocrystalline anisotropy, cobalt(II) was an obvious choice of metal ion[30–32]. The tetradentate TPMA ligand occupies four positions in the cobalt(II) coordination sphere leaving two cis-coordinated chloride groups that are easily substitutable. As the geometry of the mononuclear cobalt(II) complex allows no more than one layer of complex at the surface of the particles, this approach guarantees a low increase of the nanoparticle size.

Second, we have chosen magnetic nanoparticles possessing coordinating atoms at their surface that could be synthesized without any stabilizing organic species. The use of bare nanoparticles is indeed necessary to promote the approach of the complexes to the particles surface and then the coordination reaction. The well-known Massart procedure along with a thorough size sorting procedure allows the synthesis of maghemite nanoparticles γ-Fe_2O_3 with a relatively low polydispersity and a small size[33,34]. The colloidal solutions of the particles are stabilized by the pH-dependent surface charge (positive or negative under acidic or basic conditions, respectively) that prevents the use of any stabilizing organic species.

Finally, it was crucial to maintain colloidal stability during and after the coordination reaction of the complexes at the nanoparticles surface. We took special care to ensure that no aggregation of the particles was occurring in the solution, since it would have led to undesirable magnetic dipolar interactions. Indeed, such inter-particle interactions would have masked the actual impact of the complexes on the nanoparticles magnetic behaviour.

Synthesis and characterization

We have used small maghemite nanoparticles in acidic colloidal solution (sample 0a, $D_0 = 5.1$ nm, $\sigma = 0.12$; refs 33,34). In a first step, the $[Co(TPMA)Cl_2]$ complex is added at room temperature to the nanoparticles acidic solution. The number of added complexes has been varied from ~3 to 210 per nanoparticle. At this stage, no iono-covalent bond between the complexes and the particles surface is expected, albeit supramolecular interactions cannot be ruled out. There are furthermore no sign of increase of the hydrodynamic diameter in dynamic light scattering (DLS). A single peak is detected and it remains close to the value observed for the bare acidic particles 0a ($Z_{av} = 7.3$ nm). In the second step, the condensation of the complexes at the surface of the particles takes place by a brutal modification of the pH: from 2.4 to 11 with the addition of a concentrated solution of tetramethylammonium hydroxide (TMAOH), to give the functionalized nanoparticles (Fig. 1c). The very quick crossing of the zero point charge (pH 7–8) allows the conservation of the colloidal stability. An increase of the hydrodynamic diameter is observed after the condensation

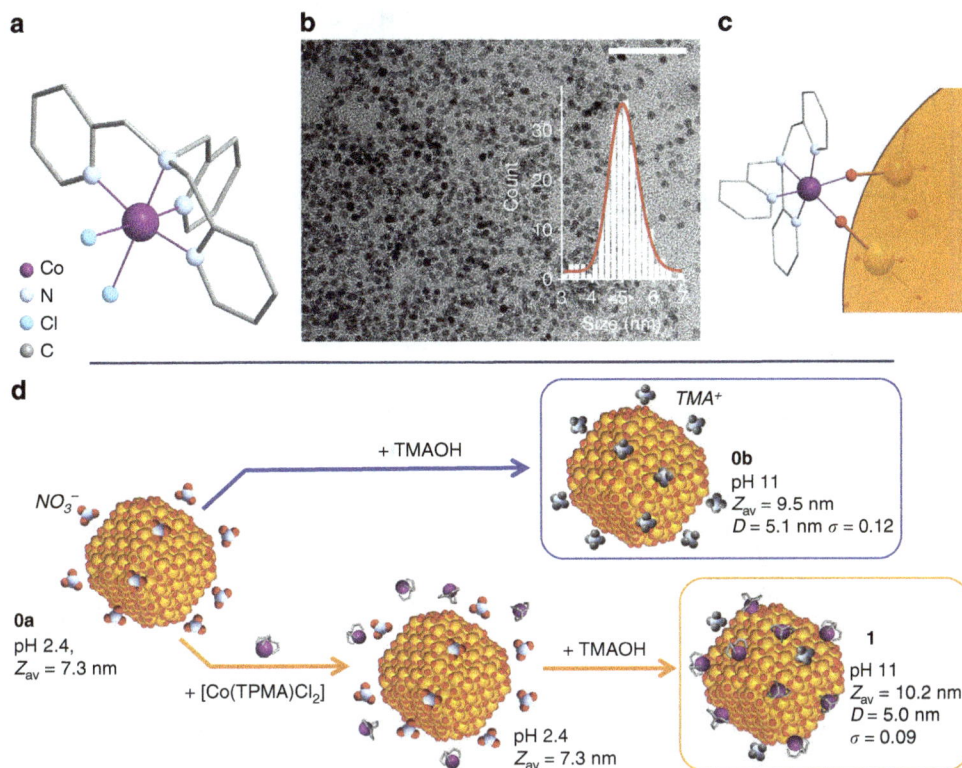

Figure 1 | Enhancing molecular complex and functionalized maghemite nanoparticles. (**a**) Representation of the [Co(TPMA)Cl$_2$] complex used to enhance the magnetic anisotropy of the γ-Fe$_2$O$_3$ nanoparticles. (**b**) TEM image of the γ-Fe$_2$O$_3$ nanoparticles functionalized with the cobalt(II) complex: **1** (5.0 nm, $\sigma = 0.09$) and (**c**) schematic view of the coordination of the complex with the iron ions. (**d**) Synthesis scheme with measured pH values, hydrodynamic diameters (Z_{av}), sizes (D) and distributions (σ).

reaction, up to $Z_{av} = 10.2$ nm when ~ 85 complexes were added per nanoparticle in the first step (Supplementary Fig. 1 and Supplementary Table 1). Sample **1**, which corresponds to the nanoparticles functionalized by the addition of ca. 60 complexes per particle in the first synthesis step, shows a similar increase. This can be ascribed to the coordination of the complexes and to the presence around the particles of TMA$^+$ counterions, which accompany the modification of the nature of the surface charge (from positively to negatively charged). Indeed, for bare nanoparticles, DLS shows a similar increase of the hydrodynamic diameter (from 7.3 to 9.5 nm) when performing the brutal pH change in the absence of complexes: passing from **0a** (pH 2.4) to the basified colloidal solution (sample **0b**, pH 11) with the addition of TMAOH. For the functionalized nanoparticles, the absence of any additional peaks in DLS indicates that there are neither aggregation of particles nor side nucleation of cobalt oxide—that could have occurred were the complexes unstable. No evolution of the single peak has been observed over weeks. The addition of > 85 complexes per particle induces a dramatic increase of the hydrodynamic diameter, followed by the flocculation of the particles. The latter is probably caused by the loss of the electrostatic repulsion-induced stabilization that should accompany the increase of the grafting rate. In the following we will focus on **0b** and **1**.

Transmission electron microscopy (TEM) indicates that very similar sizes and distributions are observed for **0b** and **1** (5.1 nm, $\sigma = 0.12$ and 5.0 nm, $\sigma = 0.09$, respectively; Fig. 1b and Supplementary Fig. 2). Along with the DLS experiments, this supports the absence of aggregation or of higher size particles. It also indicates that the functionalization has a negligible effect on the size of the objects. X-ray powder pattern analysis shows that **0b** and **1** both display the cubic structure of the maghemite

(Fd-$3m$) while the estimated crystallite sizes are in agreement with TEM imaging (Supplementary Fig. 3 and Supplementary Table 2). High-resolution TEM also confirms the cubic structure for the particles and indicates that no structural evolution has occurred during the functionalization reaction (Supplementary Fig. 4 and Supplementary Table 3). In addition, X-ray photoelectron spectroscopy (XPS) measurement at the Fe 2p edges shows an energy gap between 2p$_{1/2}$ and 2p$_{3/2}$ (13.7 eV), in agreement with the γ-Fe$_2$O$_3$ structure[35] (Supplementary Fig. 5).

XPS measurements at the N 1s edge show two peaks at 404 and 399 eV for **1** (Fig. 2). The spectra of the bare nanoparticles **0b** and of the complex display only one peak at 403 and 398 eV, respectively. As the presence of nitrogen atoms can originate from the TMA$^+$ counterions in **0b** and in **1**, and from the TPMA ligand in **1** and in [Co(TPMA)Cl$_2$], the low energy contribution at 399 eV can be assigned to the nitrogen atoms from the TPMA ligand and the high energy peak at 404 eV to the contribution from the TMA$^+$ counterion. The experimental Fe/N$_{lig}$ atomic ratio of the peaks has been found equal to 1.00 for sample **1**, which differs from the calculated one (10 accounting for the TPMA ligands only). Nevertheless, the experimental Fe/N$_{lig}$ ratio agrees well with the calculated one if only surface iron ions ($\sim 10\%$) are taken into account. Atomic absorption spectroscopy (AAS) measurements made on a precipitated sample of **1** confirm the presence of cobalt(II) ions. The found 46 ± 4 Fe/Co ratio corresponds to 52 complexes per particle (considering 2418 Fe(III) ions for a spherical 5 nm γ-Fe$_2$O$_3$ nanoparticle). This would indicate an 86% grafting rate corresponding to a surface density of 0.66 complex per nm^2.

The presence of complexes coordinated to the nanoparticles surface has been further evidenced by X-ray absorption spectroscopy (XAS) measurements at the $L_{2,3}$ edges of the iron and

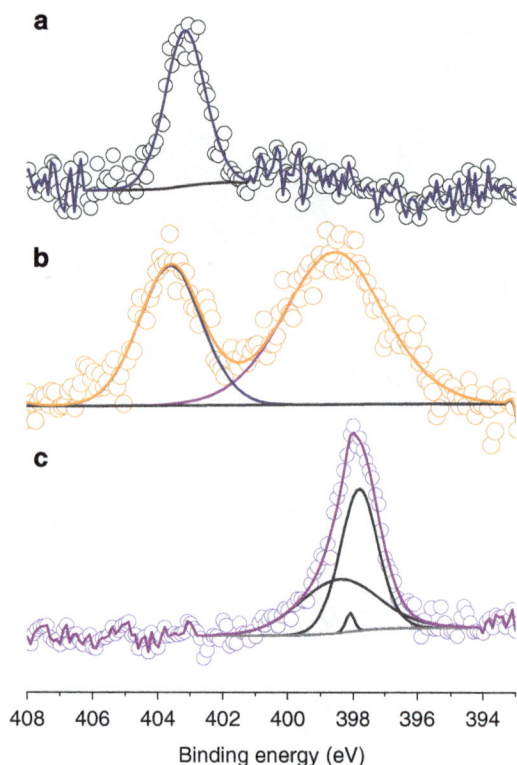

Figure 2 | Presence of the enhancing unit in the functionalized maghemite nanoparticles. XPS spectra at the N1s edge of samples **0b** (**a**), **1** (**b**) and of the [Co(TPMA)Cl$_2$] complex (**c**).

cobalt ions for sample **1** and at the cobalt $L_{2,3}$ edges for the [Co(TPMA)Cl$_2$] complex. For **1**, the spectra at the cobalt edges confirm the presence of octahedral Co(II) and the absence of Co(III) (Supplementary Fig. 6). Moreover, differences are observed between the spectra of **1** and of the 'ungrafted' complex [Co(TPMA)Cl$_2$]. They can be attributed to a change in the first coordination sphere of the cobalt ion, since we expect the replacement of chloride ions by oxo ligands through the condensation at the nanoparticle surface. Indeed chloride ligands are expected to induce a weaker ligand field than the oxo groups from the particle surface[36]. This is confirmed by ligand field multiplet calculations of Co $L_{2,3}$ edges that indicate a ligand field in **1** stronger than in the [Co(TPMA)Cl$_2$] complex (Supplementary Fig. 7 and Supplementary Methods).

In summary, the combination of XPS, AAS and XAS measurements clearly confirms the presence of the {Co^{II}(TPMA)}$^{2+}$ complex at the surface of the nanoparticles and the formation of an oxo-bridge between Co(II) and Fe(III) ions.

Magnetic characterization. To assess the influence of the molecular complex and investigate the magnetic properties of the functionalized nanoparticles we have performed d.c. magnetization measurements, as well as Mössbauer and X-ray magnetic circular dichroism (XMCD) spectroscopies. Where the former gives the macroscopic behaviour of the functionalized nanoparticles, the latter two—as local probes—give element-specific information.

In **1**, the presence of the {Co^{II}(TPMA)}$^{2+}$ complexes at the nanoparticles surface increases considerably the temperature of the maximum in the zero-field-cooled (ZFC) magnetization curve, reaching 30 K (11 K for **0b**, Fig. 3). The fit of the ZFC

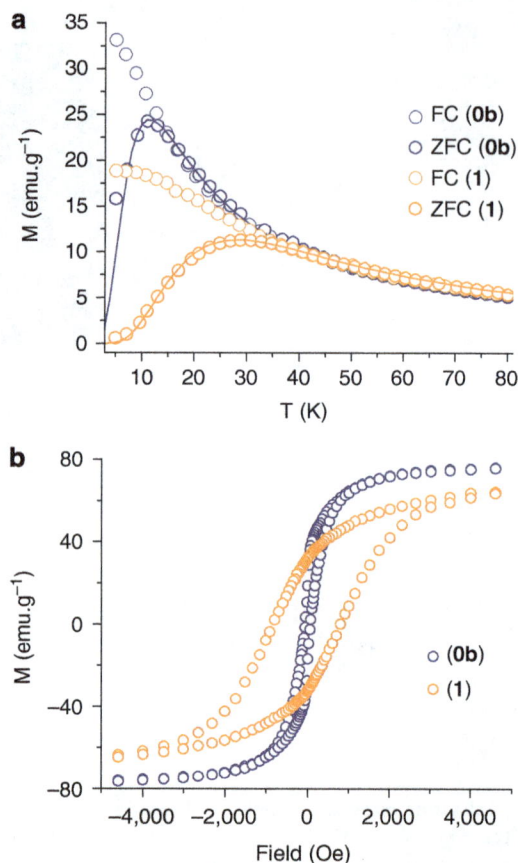

Figure 3 | Enhanced anisotropy and improved magnetic properties. (**a**) Field-cooled and zero-field-cooled (FC/ZFC) magnetization curves measured in the 5–80 K temperature range under an applied field of 50 Oe and (**b**) magnetization vs field curves measured at 5 K for **0b** and **1** in diluted solutions (%v < 0.15). Lines in the ZFC plots represent the best fit (see Methods for calculation details).

curves gives—using the same size distribution function—effective anisotropy constants of 26 and 65 kJ m^{-3} for **0b** and **1**, respectively, attesting thus the anisotropy enhancement (Fig. 3a, see methods for calculation details). This enhancement is also confirmed with the magnetization vs field curves. No break in the hysteresis curve around the remnant magnetization is observed, in agreement with a uniform reversal of the magnetization (Fig. 3b and Supplementary Fig. 8). The presence of the complexes impressively increases the coercive field of the nanoparticles, multiplying the value by 13 (from 62 Oe for **0b** to 839 Oe for **1**). In an attempt to differentiate the effect of a surface modification due to the coordination of the complexes from that of a magnetic coupling between the Co(II) complexes and the nanoparticle, a Zn(II) analogue of **1** has been prepared and measured (**2**). The same quantity of the diamagnetic {Zn^{II}(TPMA)}$^{2+}$ fragment grafted on the particle surface does not induce a comparable effect on the temperature of the maximum in the zero-field-cooled magnetization curve (from 11 to 14 K; Supplementary Fig. 9). In the magnetization vs field curve, the presence of the {Zn^{II}(TPMA)}$^{2+}$ complex has a slight effect on the remnant magnetization but an almost negligible one on the coercive field (from 62 to 73 Oe; Supplementary Fig. 10). These results indicate that the effect of the {Co^{II}(TPMA)}$^{2+}$ units on the magnetic properties does not originate from a simple modification of the environment of the iron ions located at the nanoparticles surface. Moreover, since no aggregation of the nano-objects occurs after the condensation of the complexes, the observed effect necessarily results from the magnetic interaction of the complexes with the particles, leading to an increase of the magnetic anisotropy. The transmission of the anisotropy from the complexes to the particles is possible only if there is an exchange interaction between the Co(II) and the Fe(III) ions. As the observed enhancement of the magnetic properties is important and effective at relatively high temperature, electrostatic interactions must be ruled out. Only the occurrence of a chemical bond such as an oxo-bridge between the Co(II) and the Fe(III) ions can support the effective anisotropy enhancement, source of the improved properties.

^{57}Fe Mössbauer spectrometry has been performed at 77 K on frozen solutions of **0b** and **1** (Fig. 4 and Supplementary Fig. 11) to discriminate the chemical environment and magnetic properties of the different Fe species, through the analysis of the hyperfine interactions[37]. Indeed, this local probe technique remains a powerful tool for investigating Fe-containing nanoparticles and the influence of the functionalization, thanks to its high sensitivity to electron transfer[38]. The 77 K spectra result from a minor central quadrupolar doublet and a prevailing broadened lines magnetic sextet: they have exactly the same isomer shift and their proportions are rather independent of the samples. These two

contributions are unambiguously assigned to Fe species with fast and weak superparamagnetic relaxation phenomena, due to size distributions in the samples. The lack of resolution does not allow the proportions of iron in tetrahedral and octahedral sites to be estimated but they were accurately estimated from in-8 T field Mössbauer spectra at 12 K ($Fe^{Oh}(III)/Fe^{Td}(III) = 1.70$ close to 5/3 as expected for maghemite; Supplementary Fig. 11). The mean values of isomer shift (at 77 K 0.41(2) mm s^{-1}), which probes the electronic density at the ^{57}Fe nuclei, that is the valence state, are consistent with the presence of pure ferric species for both **0b** and **1**. This excludes the presence of a ferric impurity and the occurrence of Fe^{2+} species or intermediate valence state. It further evidences that no electron transfer is induced by the presence of the Co(II) complexes. The mean hyperfine field distribution profiles, which correspond to the shape of the magnetic lines, indicate clearly that the grafting of the complexes gives rise to both a shrinkage of the distribution and a shift towards larger hyperfine fields, that is a significant increase of the mean hyperfine field (28.4(5) and 35.1(5) T, respectively). These features distinctly attest a slowdown of the relaxation phenomena of the magnetization in **1** because the attached Co(II) complexes increase the magnetic anisotropy of the Fe(III) moments, strengthening thus the magnetization of each nanoparticle, in agreement with the ZFC measurements.

The shape and intensity of the XMCD signals at the Fe $L_{2,3}$ edges for **1** are similar to those observed for previously reported maghemite nanoparticles[39]. It bears the signature of antiferromagnetic coupling between Fe(III) ions in tetrahedral sites and Fe(III) ions in octahedral sites (Fig. 5). The magnetic moment for Fe(III) ions in the sub-network of the octahedral Fe(III) is parallel to the external magnetic field. The ratio between the occupation of the tetrahedral and octahedral sites can be determined from the ligand field multiplet analysis of the XMCD shape and a $Fe^{Oh}(III)/Fe^{Td}(III)$ ratio close to 5/3 is found, as expected for maghemite. Traces of Fe(II) have also been detected. The latter are due to sample preparation (see methods). The XMCD at Co $L_{2,3}$ edges in **1** is mainly negative at the L_3 edge indicating that the Co(II) magnetic moment is, at 6 T, parallel to the octahedral Fe(III) ions and antiparallel to the tetrahedral Fe(III) ions. Element-specific magnetization curves for Fe and Co were also obtained measuring the dependence of the XMCD signal as a function of the applied magnetic field amplitude (see methods). The Co-specific magnetization curve (Fig. 6 and Supplementary Fig. 12) does not show any inversion in the sign of the XMCD when varying the magnetic field, indicating that no inversion of coupling can be expected at low magnetic field. All three curves are superimposed demonstrating that the Co(II) is magnetically coupled to the Fe(III) ions of the maghemite nanoparticle. Moreover, the Co-specific magnetization curve of **1** differs drastically from the XMCD-detected magnetization curve of the [Co(TPMA)Cl$_2$] complex. The latter shows a slow increase of the magnetization with no saturation reached at 6.5 T, as expected for a non-interacting paramagnetic Co(II) ion. For **1**, the magnetization increases abruptly and saturates above 2 T. This behaviour evidences and confirms that the Co(II) ions within the grafted complexes are magnetically coupled to the iron(III) ions at the nanoparticles surface.

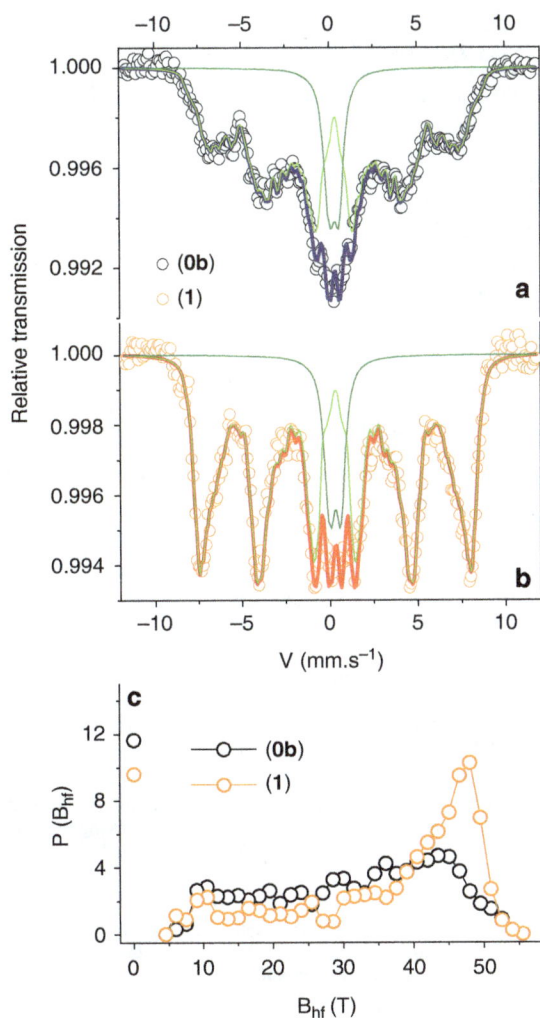

Figure 4 | Slowdown of the relaxation of the magnetization. Zero field ^{57}Fe Mössbauer spectra (circles: experimental; lines: calculated) measured at 77 K for **0b** (**a**) and **1** (**b**) and corresponding hyperfine field distributions ($P(B_{hf})$) vs hyperfine field (B_{hf}) plot (**c**).

Discussion

We have presented in this work a synthetic strategy, which, in combining molecular and nano chemistry, offers a way towards control and modulation of the magnetic anisotropy in nanoparticles. Magnetic measurements, Mössbauer spectrometry and XMCD measurements show that {Co^{II}(TPMA)}$^{2+}$ complexes grafted on the surface of maghemite nanoparticles

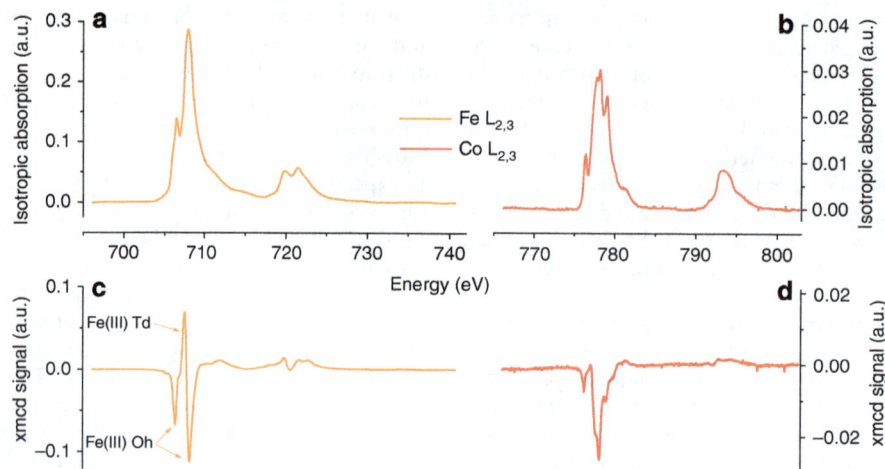

Figure 5 | Element-specific characterization of the functionalized nanoparticles. XAS and XMCD signals measured on sample **1** at the Fe (**a,c**) and Co (**b,d**) $L_{2,3}$ edges at 5 K and 6 T.

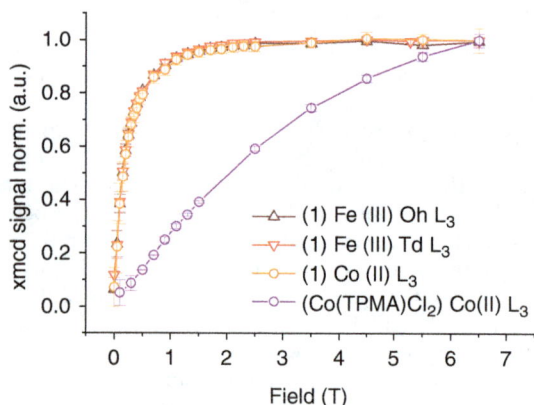

Figure 6 | Fe-specific and Co-specific XMCD-detected magnetization curves at 5 K. The XMCD curves for octahedral Fe(III) and for Co(II) were multiplied by -1 before normalization. All the curves were normalized to one at the highest field value, error bars are s.d.

massively enhance the magnetic properties of the nano-objects. Our results also indicate that the strong influence of the molecular component on the nanoparticle comes from the covalent linking of the two species through oxo-bridges and the resulting magnetic interaction.

This work may open tremendous prospects in the design of nanomagnets and of multifunctional nano-platforms. Provided that the choice of nanoparticle to functionalize allows the formation of a coordination bridge able to promote magnetic exchange, and that the particle size and the characteristics of the molecule are adequately matched, it should be possible to obtain composite nano-objects with desired blocking temperature and coercive field. Objects prepared in soft conditions, in air, in aqueous media, and in the lack of surfactant to stabilize the colloidal solution represents an important advantage for the preparation of applied materials (surface deposition, biocompatible polymers). The possibility of performing chemistry on the ligand born by the complex also represents an asset and offers many additional possibilities. The adequate choice of ligand could easily allow pre- or post- functionalization, whether to add a property (organic chromophores) or to structure the objects (polymerizable/'clickable' units) towards polyfunctional devices.

Methods

Preparation of 0a. The solution was prepared according to literature procedures[33,34] (5.1 nm, $\sigma = 0.12$, [Fe] $= 0.87$ M, %m $= 6.96$%, %v $= 1.39$%, pH $= 1.8$).

Preparation of 0b. A measure of 250 µl of **0a** were diluted 10 times with a H_2O:MeOH 50% v/v mixture. Then, 500 µl of an aqueous TMAOH solution (2.8 M) were brutally added to the solution under strong stirring leading to the sample **0b**.

Preparation of 1. A measure of 250 µl of **0a** were diluted 10 times with a H_2O:MeOH 50% v/v mixture. A volume of 0.550 ml of a [Co(TPMA)Cl_2] solution (10 mM, H_2O:MeOH 50% v/v) were added dropwise under stirring, followed by the rapid addition of 500 µl of an aqueous TMAOH solution (2.8 M) under strong stirring. Then, the solution was stirred for 4 h at 60 °C and for 24 h at room temperature leading to sample **1**.

Preparation of 2. The sample was prepared following the procedure described for **1** using [Zn(TPMA)Cl_2] instead of [Co(TPMA)Cl_2].

Precipitation of the particles. The addition of three volumes of acetone into the solutions led to the precipitation of the particles. The suspension was placed on a NdFeB magnet to settle the particles and the supernatant was removed. The obtained paste-like solid was washed with an aliquot of ethanol and dried in an oven at 40 °C for 48 h.

Atomic absorption spectroscopy. The total iron, cobalt and zinc concentration (mol l^{-1}) was determined by AAS with a Perkin–Elmer Analyst 100 apparatus after degrading the precipitated particles in HCl (37%).

Transmission electron microscopy. Images have been performed on a JEOL 100CX2 microscope with 65 keV incident electrons focused on the specimen. High-resolution TEM has been achieved on a JEOL JEM 2011 microscope with an acceleration voltage of 200 kV and a resolution of 0.18 nm.

Dynamic light scattering. The DLS measurements have been performed on a Malvern Zetasizer nanoZS model equipped with a backscattering mode on the solutions containing the particles using the intensity profile. The sizes given in the article correspond to the Z average measurements.

X-ray powder diffraction. Patterns were collected on a Philips X'pert Pro diffractometer using Co-Kα1 monochromatic radiation ($\lambda = 1.78901$ Å) and equipped with a X'celerator linear detector.

Magnetic measurements. Magnetic measurements were carried out with Quantum Design MPMS-XL and MPMS-5S magnetometers working in d.c. mode on frozen solutions of the samples. The solution was diluted in a H_2O:MeOH 50% v/v mixture before measurements. The solution volume was 150 µl and the weight concentration 0.4%. The solution is placed in a 0.2 ml eppendorf and inserted in the cryostat of the superconducting quantum interference device

(SQUID) magnetometer and frozen directly from room temperature to 100 K in zero magnetic field (lowering the rod takes a few seconds) before any measurement. The temperature sweeping rate for ZFC/FC measurements was 2 K min^{-1}.

Anisotropy constants calculations. Following Tamion et al.[40] we have used their semi-analytical model to describe the temperature dependence of the ZFC magnetization and extract an estimate of the magnetic anisotropy energy density K_{eff}.

Having defined a switching-field frequency:

$$v(T) = v_0 \exp\left[\frac{-K_{eff}V_{mag}}{k_B T}\right] \quad (1)$$

with $v_0 = 10^9$ Hz the attempt frequency and a characteristic time $\delta_t(T)$, which depends on the temperature sweeping rate (here 2 K min^{-1}), the magnetic moment measured during a ZFC protocol is given by

$$m_{ZFC}(T) = N_T \int_0^\infty M_0 V_{mag}\left[e^{-v(T)\delta t(T)} + \frac{K_{eff}V_{mag}}{k_B T}\left(1 - e^{-v(T)\delta t(T)}\right)\right]P(D_{mag})dD_{mag} \quad (2)$$

where $M_0 V_{mag} = \mu_0 m_s^2 H/(3K_{eff}V_{mag})$ (equation 3) is the initial ZFC susceptibility in the frozen low temperature state and $P(D_{mag})$ is the size distribution function taken here as a lognormal with characteristic parameter taken from the TEM characterization of the particles. In the equations above, μ_0 and k_B denote the magnetic permeability of vacuum and the Boltzmann constant, H is the magnetic field strength, m_S is the saturation magnetization of the maghemite and N_T is the number of magnetically active clusters.

Using this equation, it is quite straightforward to obtain a calculated ZFC curve and optimize the value of K_{eff} in order that make it best match the experimental data.

^{57}Fe Mössbauer spectrometry. ^{57}Fe Mössbauer spectra were performed at 77 K using a conventional constant acceleration transmission spectrometer with a ^{57}Co source (Rh matrix) and a bath cryostat and at 12 K in a 8 T external field applied parallel to the γ-beam in a cryomagnetic device. The spectra were fitted by means of the MOSFIT program and an α-Fe foil was used as the calibration sample.

X-ray photoemission spectroscopy analyses. At various times of adsorption, $t = 30$ s, 1, 3 and 30 min, the chamber was evacuated to some 10^{-10} torr, and the sample was analysed by X-ray photoemission spectroscopy using an Omicron NanoTechnology GmbH (Taunusstein, Germany) Argus hemispherical analyser and a monochromatic AlKα X-ray source (1,486.6 eV). After recording a broad range spectrum (pass energy 100 eV), high-resolution spectra were recorded for the N 1s, C 1s, O 1s and Fe 2p core levels (pass energy 20 eV). High-resolution XPS conditions have been fixed: 'sweep' analysis mode and an electron beam power of 280 W (14 kV and 20 mA). The spectra were fitted using the Casa XPS v.2.3.16 Software (Casa Software Ltd., UK) and applying a Gaussian/Lorentzian ratio G/L equal to 70/30.

XAS and XMCD measurements. XAS and XMCD spectra at Fe and Co $L_{2,3}$ edges were recorded on the soft X-ray beamline DEIMOS[41] at synchrotron SOLEIL (France). Circularly polarized photons delivered by an Apple II undulator are monochromatised by a variable groove depth (VGD) grating monochromator working in the inverse Petersen geometry. All reported spectra were measured using total electron yield detection under a 10^{-10} mbar ultra-high vacuum (UHV). The XMCD signals were recorded by both flipping the circular polarization (either left or right helicity) and the applied magnetic field (either $+6$ or -6 T). The XMCD signal is obtained as the difference $\sigma_{XMCD} = \sigma^- - \sigma^+$ where $\sigma^- = [\sigma_L(H^-) + \sigma_R(H^+)]/2$, $\sigma^+ = [\sigma_L(H^+) + \sigma_R(H^-)]/2$, σ_L (σ_R) is the cross-section with left (right) polarized X-rays, and H$^+$ (H$^-$) the magnetic field parallel (antiparallel) to the X-ray propagation vector. This procedure ensured a high signal-to-noise ratio and allowed us to discard any spurious systematic signals. XAS and XMCD spectra were measured for samples cooled to 5 K and in a 6 T applied magnetic field.

The XMCD-detected magnetization curves are the field dependence of the dichroic signal. The XMCD amplitude is recorded at the energy of its maximum amplitude (707.56 eV for Td Fe, 708.19 eV for Oh Fe and 778.19 eV for Co) by quickly switching the circular polarization thanks to the electromagnet/permanent magnet helical undulator (EMPHU)[41] available on DEIMOS beamline. Due to the presence of TMA$^+$ cations, drop-casts of the nanoparticles solution on gold-coated silicon plate yielded highly hydroscopic deposits. The solid samples were thus prepared by precipitation in acetone. The solid was suspended in ethanol, drop-casted on a slide, dried on a hot plate and fixed on carbon conductive tape to the copper sample holder. Traces of Fe(II) have been detected on the XAS spectrum of **1** and estimated at ~2% of the iron signal. This cannot be the result of an electron transfer from the Co(II) to the Fe(III) (Supplementary methods).

References

1. Magnetic recording media. *Fuji Electric Review* **57**, 30–62 (2011).
2. Rosensweig, R. E. Heating magnetic fluid with alternating magnetic field. *J. Magn. Magn. Mater.* **252**, 370–374 (2002).
3. Jun, Y., Seo, J. & Cheon, J. Nanoscaling laws of magnetic nanoparticles and their applicabilities in biomedical sciences. *Acc. Chem. Res.* **41**, 179–189 (2008).
4. Frey, N. A., Peng, S., Cheng, K. & Sun, S. Magnetic nanoparticles: synthesis, functionalization, and applications in bioimaging and magnetic energy storage. *Chem. Soc. Rev.* **38**, 2532–2542 (2009).
5. Fortin, J.-P. et al. Size-sorted anionic iron oxide nanomagnets as colloidal mediators for magnetic hyperthermia. *J. Am. Chem. Soc.* **129**, 2628–2635 (2007).
6. Mehdaoui, B. et al. Optimal size of nanoparticles for magnetic hyperthermia: a combined theoretical and experimental study. *Adv. Funct. Mater.* **21**, 4573–4581 (2011).
7. Georgelin, T., Bombard, S., Siaugue, J.-M. & Cabuil, V. Nanoparticle-mediated delivery of bleomycin. *Angew. Chem. Int. Ed.* **49**, 8897–8901 (2010).
8. Noh, S. et al. Nanoscale magnetism control via surface and exchange anisotropy for optimized ferrimagnetic hysteresis. *Nano Lett.* **12**, 3716–3721 (2012).
9. Lee, J.-H. et al. Exchange-coupled magnetic nanoparticles for efficient heat induction. *Nat. Nano* **6**, 418–422 (2011).
10. Skumryev, V. et al. Beating the superparamagnetic limit with exchange bias. *Nature* **423**, 850–853 (2003).
11. Salazar-Alvarez, G. et al. Reversible post-synthesis tuning of the superparamagnetic blocking temperature of γ-Fe2O3 nanoparticles by adsorption and desorption of Co(II) ions. *J. Mater. Chem.* **17**, 322–328 (2007).
12. Prado, Y. et al. Tuning the magnetic anisotropy in coordination nanoparticles. Random distribution versus core-shell architecture. *Chem. Commun.* **48**, 11455–11457 (2012).
13. Fantechi, E. et al. Exploring the effect of Co doping in fine maghemite nanoparticles. *J. Phys. Chem. C* **116**, 8261–8270 (2012).
14. Vichery, C. et al. Introduction of cobalt Ions in γ-Fe2O3 nanoparticles by direct coprecipitation or postsynthesis adsorption: dopant localization and magnetic anisotropy. *J. Phys. Chem. C* **117**, 19672–19683 (2013).
15. Vestal, C. R. & Zhang, Z. J. Effects of surface coordination chemistry on the magnetic properties of MnFe2O4 spinel ferrite nanoparticles. *J. Am. Chem. Soc.* **125**, 9828–9833 (2003).
16. Salafranca, J. et al. Surfactant organic molecules restore magnetism in metal-oxide nanoparticle surfaces. *Nano Lett.* **12**, 2499–2503 (2012).
17. Prado, Y. et al. Magnetization reversal in CsNiIICrIII(CN)6 coordination nanoparticles: unravelling surface anisotropy and dipolar interaction effects. *Adv. Funct. Mater.* **24**, 5402–5411 (2014).
18. Zeng, H., Li, J., Wang, Z. L., Liu, J. P. & Sun, S. Bimagnetic core/shell FePt/Fe3O4 nanoparticles. *Nano Lett.* **4**, 187–190 (2003).
19. Nogués, J. et al. Exchange bias in nanostructures. *Phys. Rep.* **422**, 65–117 (2005).
20. Catala, L. et al. Core-multishell magnetic coordination nanoparticles: toward multifunctionality on the nanoscale. *Angew. Chem. Int. Ed.* **48**, 183–187 (2009).
21. Salazar-Alvarez, G. et al. Two-, three-, and four-component magnetic multilayer onion nanoparticles based on iron oxides and manganese oxides. *J. Am. Chem. Soc.* **133**, 16738–16741 (2011).
22. Dia, N. et al. Synergy in photomagnetic/ferromagnetic Sub-50 nm core-multishell nanoparticles. *Inorg. Chem.* **52**, 10264–10274 (2013).
23. Estrader, M. et al. Robust antiferromagnetic coupling in hard-soft bi-magnetic core/shell nanoparticles. *Nat. Commun.* **4**, 2960 (2013).
24. Zeng, H., Li, J., Liu, J. P., Wang, Z. L. & Sun, S. Exchange-coupled nanocomposite magnets by nanoparticle self-assembly. *Nature* **420**, 395–398 (2002).
25. Poulopoulos, P. & Baberschke, K. Magnetism in thin films. *J. Phys. Condens. Matter* **11**, 9495 (1999).
26. Surface effects in magnetic nanoparticles (ed. Fiorani, D.) 1–298 (Springer US, 2005).
27. Zoppellaro, G., Tuček, J., Herchel, R., Šafářová, K. & Zbořil, R. Fe3O4 nanocrystals tune the magnetic regime of the Fe/Ni molecular magnet: a new class of magnetic superstructures. *Inorg. Chem.* **52**, 8144–8150 (2013).
28. Lodi Rizzini, A. et al. Coupling single molecule magnets to ferromagnetic substrates. *Phys. Rev. Lett.* **107**, 177205 (2011).
29. Davies, C. J., Solan, G. A. & Fawcett, J. Synthesis and structural characterisation of cobalt(II) and iron(II) chloride complexes containing bis(2-pyridylmethyl)amine and tris(2-pyridylmethyl)amine ligands. *Polyhedron* **23**, 3105–3114 (2004).
30. Batchelor, L. J. et al. Pentanuclear cyanide-bridged complexes based on highly anisotropic coii seven-coordinate building blocks: synthesis, structure, and magnetic behavior. *Inorg. Chem.* **50**, 12045–12052 (2011).
31. Mondal, A. et al. A cyanide and hydroxo-bridged nanocage: a new generation of coordination clusters. *Chem. Commun.* **49**, 1181–1183 (2013).

32. Ruamps, R. *et al.* Ising-type magnetic anisotropy and single molecule magnet behaviour in mononuclear trigonal bipyramidal Co(II) complexes. *Chem. Sci.* **5,** 3418–3424 (2014).

33. Massart, R. Preparation of aqueous ferrofluids without using surfactant; behavior as a function of pH and counterions. *C. R. Seances Acad. Sci. Ser. C* **291,** 1–3 (1980).

34. Lefebure, S., Dubois, E., Cabuil, V., Neveu, S. & Massart, R. Monodisperse magnetic nanoparticles: preparation and dispersion in water and oils. *J. Mater. Res.* **13,** 2975–2981 (1998).

35. Grosvenor, A. P., Kobe, B. A., Biesinger, M. C. & McIntyre, N. S. Investigation of multiplet splitting of Fe 2p XPS spectra and bonding in iron compounds. *Surf. Interface Anal.* **36,** 1564–1574 (2004).

36. Lever, A. B. P. *Inorganic electronic spectroscopy* (Elsevier, 1984).

37. Greneche, J.-M. in *Mössbauer Spectroscopy* (eds. Yoshida, Y. & Langouche, G.) 187–241 (Springer, 2013).

38. Fouineau, J. *et al.* Synthesis, mössbauer characterization, and ab initio modeling of iron oxide nanoparticles of medical interest functionalized by dopamine. *J. Phys. Chem. C* **117,** 14295–14302 (2013).

39. Brice-Profeta, S. *et al.* Magnetic order in γ-Fe2O3 nanoparticles: a XMCD study. *J. Magn. Magn. Mater.* **288,** 354–365 (2005).

40. Tamion, A., Hillenkamp, M., Tournus, F., Bonet, E. & Dupuis, V. Accurate determination of the magnetic anisotropy in cluster-assembled nanostructures. *Appl. Phys. Lett.* **95,** 062503 (2009).

41. Ohresser, P. *et al.* DEIMOS: a beamline dedicated to dichroism measurements in the 350–2500 eV energy range. *Rev. Sci. Instrum.* **85,** 013106 (2014).

Acknowledgements

This work was supported by the Centre National de la Recherche Scientifique (CNRS, France), the Ministère de l'Enseignement Supérieur et de la Recherche (MESR, France), the LabEx MATISSE and by the LabEx MiChem part of French state funds managed by the ANR within the Investissements d'Avenir programme under reference ANR-11-IDEX-0004-02. We acknowledge SOLEIL for provision of synchrotron radiation facilities. We thank Sandra Casale and the UPMC Chemistry department microscopy service for high-resolution TEM images. Y.P. and L.L. would like to thank Laure Catala and Talal Mallah for their constant support and encouragements.

Author contributions

J.F., L.L., V.D. and B.F. conceived and supervised the project. Y.P., J.F., L.L. and B.F. planned and implemented the synthetic and analytical experiments. A.M. and J.F. performed the TEM imaging. N.D. and L.L. performed the powder X-ray diffraction measurements, N.Y. and J.-M.G. analysed the X-ray data. T.G performed and analysed the XPS measurements. Y.P., L.L. and V.D. performed and analysed the SQUID measurements. N.Y. and J.-M.G. performed and analysed the Mössbauer spectrometry measurements. M.-A.A., N.D., C.C.-.d.-M., L.L., P.S., F.C., E.O. and P.O. performed and analysed the XAS and XMCD measurements. Y.P. and L.L. wrote the manuscript with the help of all authors.

Additional information

18

Monolayer-to-bilayer transformation of silicenes and their structural analysis

Ritsuko Yaokawa[1], Tetsu Ohsuna[1], Tetsuya Morishita[2], Yuichiro Hayasaka[3], Michelle J.S. Spencer[4] & Hideyuki Nakano[1,5]

Silicene, a two-dimensional honeycomb network of silicon atoms like graphene, holds great potential as a key material in the next generation of electronics; however, its use in more demanding applications is prevented because of its instability under ambient conditions. Here we report three types of bilayer silicenes that form after treating calcium-intercalated monolayer silicene ($CaSi_2$) with a BF_4^- -based ionic liquid. The bilayer silicenes that are obtained are sandwiched between planar crystals of CaF_2 and/or $CaSi_2$, with one of the bilayer silicenes being a new allotrope of silicon, containing four-, five- and six-membered sp^3 silicon rings. The number of unsaturated silicon bonds in the structure is reduced compared with monolayer silicene. Additionally, the bandgap opens to 1.08 eV and is indirect; this is in contrast to monolayer silicene which is a zero-gap semiconductor.

[1] TOYOTA Central R&D Labs, Inc., 41-1, Yokomichi, Nagakute, Aichi 480-1192, Japan. [2] CD-FMat, National Institute of Advanced Industrial Science and Technology (AIST), Central 2, 1-1-1 Umezono, Tsukuba, Ibaraki 305-8568, Japan. [3] The Electron Microscopy Center, Tohoku University, Katahira 2-1-1, Aoba-ku, Sendai 980-8577, Japan. [4] School of Science, RMIT University, GPO Box 2476, Melbourne, Victoria 3001, Australia. [5] JST Presto, Kawaguchi 332-0012, Japan. Correspondence and requests for materials should be addressed to R.Y. (email: e4777@mosk.tytlabs.co.jp) or to H.N. (email: hnakano@mosk.tytlabs.co.jp).

A frenzy of interest in graphene has spawned many theoretical and experimental studies[1-4]. After calculating the structures of two-dimensional (2D) crystals of silicon (silicene)[5-7], researchers have speculated that silicon atoms might form graphene-like sheets and have attempted to produce such silicene structures[8-12]. Very recently, Tao et al.[13] succeeded in fabricating the first silicene transistor, although the device's performance was modest. Nonetheless, the development of much more facile and practical processing methods has remained a challenging issue. The most difficult problem is that silicene grows on specific substrates and is stable only under vacuum conditions[8,9,14,15]. Another issue is that the influence of the substrate cannot be removed; the strong hybridization between Si and the substrate may stabilize silicene grown on specific substrates[8,14-16].

In a previous report on calcium-intercalated silicene ($CaSi_2$), we observed a massless Dirac-cone band dispersion at the k-point in the Brillouin zone, which was located far from the Fermi level because of the substantial charge transfer from the Ca atoms to the silicene layers[17]. This result is similar to the previously reported band structures of silicenes deposited on specific substrates[9] because $CaSi_2$ is a type of Zintl silicide, in which the formal charge is rewritten as Ca^{2+} and Si^- (ref. 18). Therefore, the intrinsic electronic structure of silicene has never been observed. In the calculated results, a van der Waals bonded silicene layer has been deposited on an intact multi-CaF_2 layer[19]. If the Ca layer of $CaSi_2$ had been exchanged with a CaF_2 layer, the influence of the substrate would have been almost completely suppressed. To reduce the influence of external factors on the electronic structure of silicene (for example, from substrates or counter ions) and to increase the stability under ambient condition, we replaced monolayer silicene with bilayer silicene.

The existence of a bilayer silicene structure, whose density of unsaturated silicon bonds is reduced in comparison with monolayer silicene, has been predicted by molecular dynamics (MD) calculations[20-27]. If we could experimentally prepare a similar bilayer silicene, we could then investigate its intrinsic electronic structure. Because of the electron transfer from the calcium cation, the monolayer silicene in $CaSi_2$ is a formally anionic layer[17]: when the calcium cation becomes electrically neutral, the silicene will not retain its honeycomb structure and will reconstruct to form a more stable structure. Under this supposition, we attempted to segregate the Ca and Si phases while maintaining the layer structures by diffusing fluoride (F) atoms, which are more electronegative than Si, into $CaSi_2$; the goal was to form an ionic bond (or interaction) between Ca and F. In this study, BF_4 anion based ionic liquid was used for the origin of fluoride anion.

Results

Fluoride diffusion into $CaSi_2$. When the $CaSi_2$ crystal (Supplementary Fig. 1) was annealed in [BMIM][BF_4] ionic liquid at 250–300 °C, it was changed to a $CaSi_2F_X$ ($0 \leq X \leq 2.3$) compound through diffusion of F^-, in which the local F^- concentration gradually decreased from the crystal edge to the interior (Fig. 1a,b and Supplementary Fig. 2). As a result, three types of bilayer Si in a $CaSi_2$ single crystal were obtained by diffusion of F^-. Figure 1c, which displays a high-angle annular dark field scanning transmission electron microscopy (HAADF-STEM) image taken of the $CaSi_2F_{1.8}$ compound, shows the alternate stacking of planar crystal domains with layer thicknesses of 1–2 nm. The HAADF-STEM imaging provided an atomic-scale Z-contrast image (Z: atomic number) to distinguish the heavier constituent elements[28-30]. STEM-energy-dispersive X-ray spectroscopy (STEM-EDX) elemental mapping identified

the bright-contrast crystal domains, which were identified as the CaF_2 phase and the dark domains, which were identified as Si phases (Fig. 1f–j). We determined the crystal structures of the entire planar region in the images of the $CaSi_2F_{1.8}$ and $CaSi_2F_{2.0}$ compounds shown in Fig. 1c,d, respectively. These planar domains were identified as trilayer CaF_2, trilayer Si, bilayer CaF_2 and a novel bilayer silicene (denoted as w-BLSi in Fig. 1c,d) that has not been previously predicted by MD calculations[20-27]. Furthermore, two types of bilayer silicenes, one with inversion symmetry (i-BLSi) and one with mirror symmetry (m-BLSi), were recognized in the $CaSi_2F_{0.6-1.0}$ composition area (Fig. 1e and Supplementary Fig. 3). The formation of m-BLSi is in accordance with predictions from a previous MD study[22]. The i- and m-BLSi must be adjacent to a pair of CaF_2 and $CaSi_2$ crystal layers. The abundance ratio of i-BLSi to m-BLSi was 124:3 in the observed HAADF-STEM images. Because the calculated energy of i-BLSi was 0.03 eV per atom lower than that of m-BLSi under vacuum, the abundance ratio is qualitatively reasonable. The average size of w-BLSi is ~30 nm, and that of m-BLSi is ~10 nm. The size of i-BLSi is greater than 51 nm, which is the maximum size that can be observed by STEM imaging.

Structural determination of w-BLSi. The atomic structure of the bilayer silicene was determined from HAADF-STEM images that were taken with different incident electron beam directions (Fig. 2a–c, Supplementary Fig. 4 and Supplementary Note 1). As shown in Fig. 2d, the bilayer silicene structure had a 2D translation symmetry and a wavy morphology (hereafter, we refer to the structure as w-BLSi). The w-BLSi structure consists of two silicenes, with alternating chair and boat conformations, that are vertically connected via four-, five- and six-membered rings. Because w-BLSi consists of only Si atoms exhibiting tetrahedral coordination, the top atom of the five-membered silicon ring possesses unsaturated silicon bonds (dangling bonds). Therefore, compared with those in monolayer silicene and i- (or m-) BLSi, the density of unsaturated silicon bonds in w-BLSi decreased to 25 and 50%, respectively (Supplementary Fig. 5).

We determined the atomic positions of w-BLSi from high-resolution transmission electron microscopy and HAADF-STEM images as accurately as possible (Supplementary Figs 6 and 7, Supplementary Table 1 and Supplementary Note 2). The 2D translation periods of w-BLSi were $a = 0.661(2)$ nm and $b = 0.382(3)$ nm, and the two translation axes were normal to each other (Supplementary Fig. 8). The a period of w-BLSi is similar to the triple lattice spacing of d_{11-2} in CaF_2 (0.223 nm), and the b period is similar to d_{-110} in CaF_2 (0.386 nm); that is, the difference between w-BLSi and CaF_2 (111) is less than the observation error (Supplementary Fig. 9). Because the atomic arrangement of the (111) plane of the CaF_2 crystal exhibited threefold symmetry, three equivalent relative rotation angles were observed between w-BLSi and the CaF_2 (111) plane (Supplementary Figs 10 and 11). In addition, the angle between the $[01]_{w-BLSi}$ and $[11]_{w-BLSi}$ directions was almost 60° (Supplementary Fig. 10, w-BLSi is described in 2D notation, because 2D can be expressed more simply than three dimensions). Therefore, Figs 1c and 2a show the contrast of two different arrangements of bright dots—specifically, the [01] and [11] direction images (Fig. 2e,g) in the w-BLSi regions. In almost all of the observed HAADF-STEM images, w-BLSi always faced the (111) plane of CaF_2, and the F vacancies (red arrows in Fig. 1d) on the CaF_2 (111) surface were recognized at special positions associated with the wavy structure of w-BLSi. A w-BLSi was observed to be sandwiched between two CaF_2 layers with an F-site surface vacancy of ~0.5 at the interface (Fig. 1d, Supplementary Figs 7,12–15 and Supplementary Note 3).

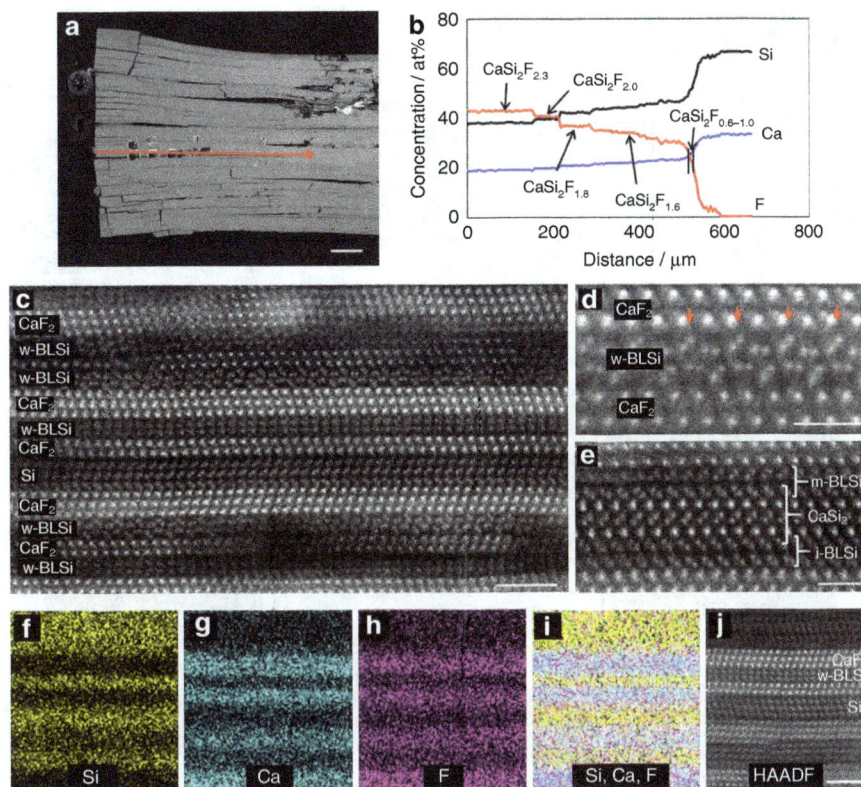

Figure 1 | Visualization of fluoride diffusion. (**a**) Cross-sectional BSE image of the crystal grain including $CaSi_2F_X$ compound. (**b**) EPMA quantitative line analysis result along the red arrow in **a**. (**c**) HAADF-STEM image taken from a region with $CaSi_2F_{1.8}$ in **b**; the strip contrast corresponds to Si (dark domain) and CaF_2 (bright domain) planar crystals. (**d**) An enlarged HAADF-STEM image taken from a region with $CaSi_2F_2$ in **b**; red arrows indicate an F-vacancy site. (**e**) HAADF-STEM image taken from a region with $CaSi_2F_{0.6-1.0}$ in **b**; bright dots, corresponding to the projected atomic positions of m-and i-BLSi, can be observed in the image. (**f-i**) STEM-EDX elemental mapping results of the $CaSi_2F_2$ composition region. One-element mapping (**f**: Si; **g**: Ca; and **h**: F). (**i**) Overlapped-mapping of Si, Ca and F. (**j**) HAADF-STEM image of the STEM-EDX elemental mapping area. The scale bars in **a**; **c** and **j**; and **d** and **e**; 100 μm, 2 nm and 1 nm.

DFT and *ab initio* MD calculations and optical properties. The w-BLSi structure appears to resemble re-BLSi[20] in appearance; however, its atomic arrangement is clearly different (Supplementary Fig. 16). An *ab initio* MD calculation was performed for BLSi under the conditions corresponding to the experimentally observed structure, that is, BLSi was sandwiched between two CaF_2 layers with an F-site surface vacancy of 0.5 at the interface. The MD calculation was started with the i-BLSi structure, but it was immediately transformed to another BLSi structure. The system was then equilibrated, and the resultant BLSi structure was found to perfectly agree with the experimentally observed w-BLSi structure in Fig. 3a (Supplementary Tables 2–5, Supplementary Fig. 17 and Supplementary Note 4). The electronic density of states (DOS) for w-BLSi was calculated by using the structure in Fig. 3a, and the decomposed DOSs for Si, Ca, and F are shown in Fig. 3b. The Ca and F bands are located far below the Fermi level, and the valence bands consist of only Si bands. An ionic rather than a covalent interaction is thus expected between Si and Ca or F. We also observe that the bandgap opens to $\sim 0.65\,eV$, in contrast to monolayer silicene, which is a zero-gap semiconductor[31]. Interestingly, however, the gap closes when w-BLSi is isolated without geometry optimization under vacuum (Supplementary Fig. 18b). This result indicates that, in the $CaSi_2F_X$ compound, charge transfer from Ca to Si occurs, filling the energy levels that are unoccupied under vacuum (Supplementary Discussion). Thus, the electronic properties of w-BLSi appear to be sensitive to its environmental conditions.

The presence of the F vacancies allows the electrons on Ca to transfer to Si, which enhances the stability of the w-BLSi structure (Fig. 3a) by saturating the dangling bonds. The CaF_{2-X} domains (specifically, ionic crystalline domains) surrounding the Si layers are key to the formation of the w-BLSi structure.

The optical bandgap can be calculated from the absorption spectrum. The diffuse reflectance spectrum of the powder sample with $CaSi_2F_{1.8-2.3}$ composition was measured, and the obtained reflectance spectrum data (Supplementary Fig. 19) were converted to a Kubelka–Munk function (K/S), which is proportional to the absorption coefficient (α). The sample was a mixture of w-BLSi, two types of trilayer silicene (with dangling bonds and terminated with F atoms, as shown in Supplementary Fig. 20) and a CaF_2 layer (Supplementary Note 5). The relationship between the absorption coefficient (α) and the bandgap energy (Eg) can be described by two types of equations: $\alpha h\nu = const$ (direct gap) and $\alpha h\nu = A\,(h\nu - Eg)$ (indirect gap), where the DOS for 2D crystals is constant as a function of energy[32–35] (Supplementary Note 5). Here, h, ν and A are Planck's constant, light frequency and proportional constant, respectively. From two linear fittings of the spectrum, the latter equation was found to be suitable for the sample. The absorption edges of the $CaSi_2F_{1.8-2.3}$ compound were observed at 1.08 and 1.78 eV (Fig. 3c), assuming indirect transitions.

Freestanding trilayer silicene is semi-metallic, as shown by density functional theory (DFT) calculations[36]. It has been suggested that the bandgap of trilayer silicene with dangling bonds in $CaSi_2F_{1.8-2.3}$ is nearly zero if charge transfer between the

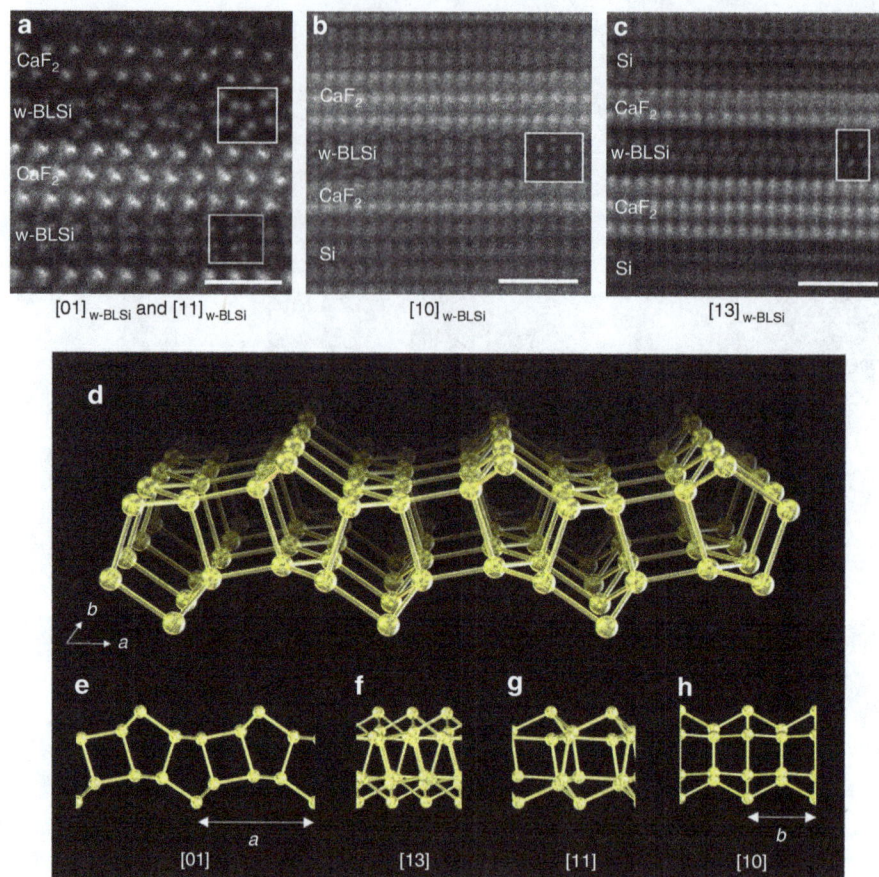

Figure 2 | Structural determination. (**a–c**) HAADF-STEM and simulation (insets) images of w-BLSi. (**a**) the [01]$_{w\text{-BLSi}}$ and [11]$_{w\text{-BLSi}}$ incident directions ([1-10]$_{CaF2}$), (**b**) the [10]$_{w\text{-BLSi}}$ and [11-2]$_{Si\ and\ CaF2}$ directions and (**c**) the [13]$_{w\text{-BLSi}}$ and [11-2]$_{Si\ and\ CaF2}$ directions. (**d**) Schematic illustration of the w-BLSi atomic structure. (**e–h**) Schematic structures projected in each direction in **e** [01], **f** [13], **g** [11] and **h** [10] directions. All scale bars in (**a–c**), 1 nm.

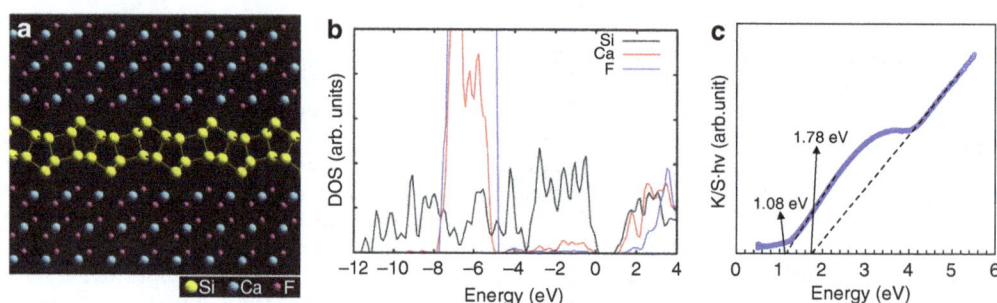

Figure 3 | DFT and *ab initio* MD results and optical property. (**a**) Structure of w-BLSi sandwiched between two CaF$_2$ crystals, with vacancies at half of the F sites on the interface; this structure was used to calculate the DOS and was obtained from the transformation of i-BLSi in the *ab initio* MD simulation and the subsequent quenching process (Supplementary Method). (**b**) Decomposed DOS for Si, Ca and F in w-BLSi displayed in **a**. (**c**) Plot of multiplication of the K/S and energy as a function of energy for CaSi$_2$F$_{1.8-2.3}$ consisting of w-BLSi, trilayer silicene with dangling bonds and F-terminated trilayer silicene. The absorption spectrum suggests two indirect gaps with values of 1.08 and 1.78 eV.

trilayer silicene and the CaF$_2$ layer is inhibited[19]. From previous DFT results of monolayer and multilayer silicene terminated with atoms[37,38], it is conjectured that the bandgap of F-terminated trilayer silicene would be ∼1 eV within the framework of the DFT and Perdew, Burke and Ernzerhof (PBE) technique. It should be noted that DFT calculations using a standard generalized gradient approximation functional tend to underestimate the bandgap (roughly ∼2/3 in crystal Si). This indicates that the bandgap experimentally measured for the trilayer silicene should be ∼1.5 eV. Meanwhile, the bandgap for w-BLSi, which is estimated to be ∼0.65 eV in the DFT–PBE

calculation, is expected to be ∼1 eV in the experimental measurement. Therefore, the measured gaps were estimated such that the gaps of w-BLSi and F-terminated trilayer silicene were 1.08 and 1.78 eV, respectively.

Transformation process from monolayer silicene to w-BLSi. On the basis of the HAADF-STEM data, we discussed a model for the transformation process from a monolayer silicene in CaSi$_2$ (Fig. 4a) to w-BLSi (Fig. 4f). When F$^-$ ions diffuse from the surface of a CaSi$_2$ crystallite into the crystal along the Ca layer,

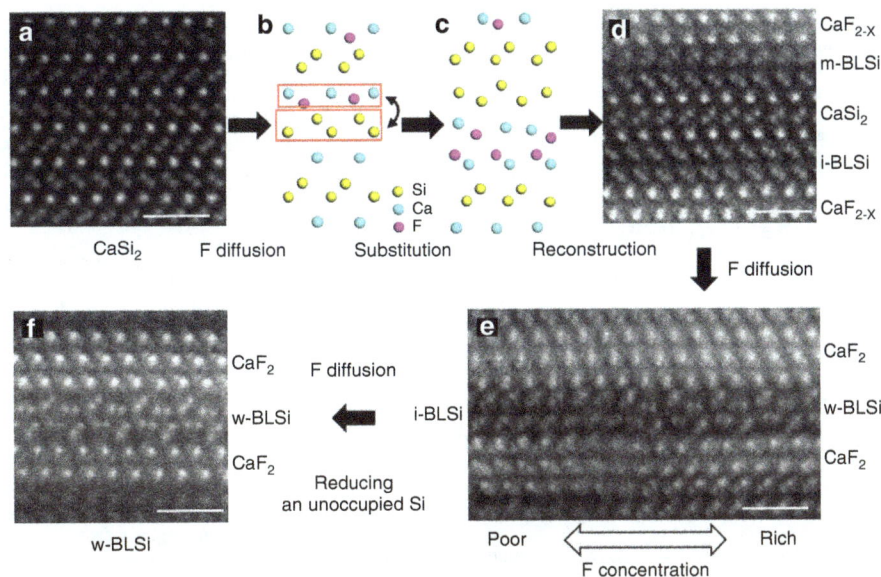

Figure 4 | A model for the transformation process from monolayer Si to w-BLSi. (a) and (d-f) HAADF-STEM image. (b,c) A schematic model. (a) Raw tr6 $CaSi_2$. (b) F diffusion into $CaSi_2$. (c) A random arrangement of i-BLSi and bilayer CaF_{2-x} in $CaSi_2$. (d) i-BLSi, CaF_{2-x} and $CaSi_2$ in a region with $CaSi_2F_{0.6-1.0}$. (e) i-BLSi and w-BLSi formed within the same layers in $CaSi_2F_{0.6-1.0}$. (f) w-BLSi in $CaSi_2F_{2.0}$. All scale bars, 1 nm.

thin CaF_{2-x} planar crystals are formed; as a result, anionic silicene layers assemble to reduce the number of unsaturated bonds beyond the Ca layer (Fig. 4b). During this movement, the Si covalent bonding network with honeycomb symmetry is broken and its arrangement consequently becomes random (Fig. 4c). As shown in Fig. 4d, two types of bilayer silicenes, i-BLSi and m-BLSi, which formed in the slit-like regions, as predicted by the MD calculation[22], co-exist with $CaSi_2$ in the low F-concentration region. Both of these structures are stabilized as a result of charge transferred from the Ca atoms which saturate the silicon dangling bonds.

We analysed more than 200 STEM images of BLSi; w-BLSi was recognized at F concentrations surpassing that of $CaSi_2F_{1.8}$. With increasing F concentration, the site occupancy of F concentration at the interface of the CaF_2 planar crystal reached ~0.5, then w-BLSi was formed by the change of ionic interactions among Si, Ca and F (Fig. 3e,f). In this process, negatively charged Si atoms tend to lose their electrons, which makes i- (or m-) BLSi less stable because the 'capping' of the dangling bonds by extra electrons from Ca is reduced and the dangling bonds destabilize the sp^3 tetrahedral configuration. Thus, the anionic honeycomb structure of i- (or m-) BLSi is transformed to w-BLSi, which is approximately neutral because of the fluorination of the Ca cation.

Discussion

We focused on calcium-intercalated silicene ($CaSi_2$) and discovered a strategy for transforming monolayer silicene into a novel bilayer silicene (w-BLSi). From HAADF-STEM images, we observed that w-BLSi was formed between the planar crystals of CaF_2 and contained four-, five- and six-membered silicon rings, although w-BLSi consists of only Si atoms exhibiting tetrahedral coordination. Compared with monolayer silicene, the number of unsaturated silicon bonds in w-BLSi decreased to 25% of the unit cell. The transformation process from monolayer silicene in $CaSi_2$ to w-BLSi was estimated from HAADF-STEM data. When F^- ions diffuse into the $CaSi_2$ crystal along the Ca layer, thin CaF_{2-x} planar crystals and two types of bilayer silicenes (i-BLSi and m-BLSi) are formed, following breakage of the Si covalent bonding monolayer network. Both of these Si structures were stabilized as a result of charge transferred from the Ca atoms

which saturate the silicon dangling bonds. With increasing F content, i- (or m-) BLSi is transformed to w-BLSi. Additionally, the structure possesses an indirect bandgap of 1.08 eV in contrast to monolayer silicene, which is a zero-gap semiconductor.

Methods

Synthesis of $CaSi_2F_X$ compound. $CaSi_2$ single-crystal grains (0.1 g) were reacted with 5 ml of ionic liquid [BMIM][BF_4] (1-butyl-3-methylimidazolium tetrafluoroborate) at 300 °C for 15 h. BF_4^- decomposed into F^- during annealing, and the $CaSi_2$ crystal was changed to $CaSi_2F_X$ compounds ($0 \leq X \leq 2.3$) through the diffusion of F^- (Fig. 1a,b). More details are given in Supplementary Method.

Chemical composition analysis. The chemical compositions of the $CaSi_2F_X$ domains were determined by electron probe microanalyser (EPMA) with a wave dispersion system (JEOL JXA-8200), an accelerating voltage of 10 kV, a specimen current of 50 nA, and an electron irradiation area of 5 μmφ. Single-phase CaF_2 and Si crystals were used as the standard for quantitative composition analysis of Ca, F and Si. EPMA line analyses were performed with 5 μm steps from the edge to the inside of the $CaSi_2F_X$ crystallites cross-sectioned parallel to the $CaSi_2$ [001] direction.

TEM/STEM analysis. HAADF-STEM observations[28-30] and STEM energy-dispersive X-ray spectroscopy (EDX) analyses were performed with a Titan[3]. G2 60–300 electron microscope (FEI, Cs = 156 nm) operated at 300 kV. HAADF-STEM imaging was capable of providing an atomic-scale Z-contrast image associated with the heavier constituent elements. The annular detector was set to collect the electrons scattered at angles between 50.5 and 200 mrad. High-resolution transmission electron microscopy observations were obtained with a JEM-2000EX electron microscope (JEOL, Cs = 0.7 mm) operating at 200 kV. TEM specimens of $CaSi_2F_X$ were detected with five different F concentration ranges ($CaSi_2F_{0.6-1.0}$, $CaSi_2F_{1.6}$, $CaSi_2F_{1.8}$, $CaSi_2F_{2.0}$ and $CaSi_2F_{2.3}$) by using the FIB micro-sampling method[39]. The atomic positions in the w-BLSi crystal and the interface structure were characterized by comparing the HAADF-STEM image contrasts with simulated contrasts calculated by the multi-slice method using MacTempasX.

Computational method. DFT and *ab initio* MD calculations were performed to calculate the DOS and to examine the structural stability of BLSi using the Vienna *Ab initio* Simulation Package (ref. 40). The projector augmented wave method[41] and generalized gradient approximation with the exchange and correlation functions of PBE were employed[42]. A plane-wave basis set with an energy cutoff of 400 eV was used with Γ-point sampling in the Brillouin zone. To model the BLSi systems observed in our experiments, two-layer Si structures were sandwiched by CaF_2 crystal domains, each consisting of three sets of CaF_2 layers, with or without the F-site vacancy at the Si/CaF_2 interfaces. The DOS for the w-BLSi was calculated for the structure obtained after the quenching process (shown in Fig. 3a) following the 300 K run. More details are given in Supplementary Method.

Optical reflectivity. Diffuse reflectance spectra were obtained for the $CaSi_2F_{1.8-2.3}$ composition powder sample using a spectrophotometer (JASCO V-670).

The diffuse reflectance spectra were processed under the Kubelka–Munk formalism, and the bandgaps were determined using a plot of the multiplication of the K/S and energy. More details are given in Supplementary Methods.

References

1. Novoselov, K. S. et al. Electric field effect in atomically thin carbon films. *Science* **306**, 666–669 (2004).
2. Novoselov, K. S. et al. Two-dimensional gas of massless dirac fermions in graphene. *Nature* **438**, 197–200 (2005).
3. Zhang, Y., Tan, Y. W., Stormer, H. L. & Kim, P. Experimental observation of the quantum hall effect and berry's phase in graphene. *Nature* **438**, 201–204 (2005).
4. Novoselov, K. S. et al. Room-temperature quantum Hall Effect in Graphene. *Science* **315**, 1379 (2007).
5. Takeda, K. & Shiraishi, K. Theoretical possibility of stage corrugation in Si and Ge analogs of graphite. *Phy. Rev. B* **50**, 14916 (1994).
6. Cahangirov, S., Topsakal, M., Aktürk, E., Şahin, H. & Ciraci, S. Two- and one-dimensional honeycomb structures of silicon and germanium. *Phys. Rev. Lett.* **102**, 236804 (2009).
7. Lebègue, S. & Eriksson, O. Electronic structure of two-dimensional crystals from *ab initio* theory. *Phys. Rev. B* **79**, 115409 (2009).
8. Vogt, P. et al. Silicene: Compelling experimental evidence for graphenelike two-dimensional silicon. *Phys. Rev. Lett.* **108**, 155501 (2012).
9. Fleurence, A. et al. Experimental evidence for epitaxial silicene on diboride thin films. *Phys. Rev. Lett.* **108**, 245501 (2012).
10. Okamoto, H. et al. Silicon nanosheets and their self-assembled regular stacking structure. *J. Am. Chem. Soc.* **132**, 2710–2718 (2010).
11. Sugiyama, Y. et al. Synthesis and optical properties of monolayer organosilicon nanosheets. *J. Am. Chem. Soc.* **132**, 5946–5947 (2010).
12. Okamoto, H., Sugiyama, Y. & Nakano, H. Synthesis and modification of silicon nanosheets and other silicon nanomaterials. *Chem. Eur. J.* **17**, 9864–9887 (2011).
13. Tao, L. et al. Silicene field-effect transistors operating at room temperature. *Nat. Nanotechnol.* **10**, 227–231 (2015).
14. Morishita, T., Spencer, M. J. S., Kawamoto, S. & Snook, I. K. A new surface and structure for silicene: polygonal silicene formation on the Al(111) surface. *J. Phys. Chem. C* **117**, 22142–22148 (2013).
15. Gao, J. & Zhao, J. Initial geometries, interaction mechanism and high stability of silicene on Ag(111) surface. *Sci. Rep.* **2**, 861 (2012).
16. Cahangirov, S. et al. Electronic structure of silicene on Ag(111): strong hybridization effects. *Phys. Rev. B* **88**, 035432 (2013).
17. Noguchi, E. et al. Direct observation of dirac cone in multilayer silicene intercalation compound $CaSi_2$. *Adv. Mater.* **27**, 856–860 (2015).
18. Yaokawa, R., Nakano, H. & Ohashi, M. Growth of $CaSi_2$ single phase polycrystalline ingots using the phase relationship between $CaSi_2$ and associated phases. *Acta Mater.* **81**, 41–49 (2014).
19. Kokott, S., Pflugradt, P., Matthes, L & Bechstedt, F. Nonmetallic substrates for growth of silicene: an *ab initio* prediction. *J. Phys. Condens. Matter* **26**, 185002 (2014).
20. Morishita, T., Spencer, M. J. S., Russo, S. P., Snook, I. K. & Mikami, M. Surface reconstruction of ultrathin silicon nanosheets. *Chem. Phys. Lett.* **506**, 221–225 (2011).
21. Sakai, Y. & Oshiyama, A. Structural stability and energy-gap modulation through atomic protrusion in freestanding bilayer silicene. *Phy. Rev. B* **91**, 201405(R) (2015).
22. Morishita, T., Nishio, K. & Mikami, M. Formation of single- and double-layer silicon in slit pores. *Phy. Rev. B* **77**, 081401(R) (2008).
23. Bai, J., Tanaka, H. & Zeng, X. C. Graphene-like bilayer hexagonal silicon polymorph. *Nano Res* **3**, 694–700 (2010).
24. Johnston, J. C., Phippen, S. & Molinero, V. A single-component silicon quasicrystal. *J. Phys. Chem. Lett.* **2**, 384–388 (2011).
25. Pflugradt, P., Matthes, L. & Bechstedt, F. Unexpected symmetry and AA stacking of bilayer silicene on Ag(111). *Phys. Rev. B* **89**, 205428 (2014).
26. Guo, Z.-X. & Oshiyama, A. Structural tristability and deep Dirac states in bilayer silicene on Ag(111) surfaces. *Phys. Rev. B* **89**, 155418 (2014).
27. Cahangirov, S. et al. Atomic structure of the $\sqrt{3}\times\sqrt{3}$ phase of silicene on Ag(111). *Phys. Rev. B* **90**, 035448 (2014).
28. Pennycook, S. J. & Jesson, D. E. High-resolution incoherent imaging of crystals. *Phys. Rev. Lett* **64**, 938–941 (1990).
29. Pennycook, S. J. & Jesson, D. E. High-resolution z-contrast imaging of crystals. *Ultromicroscopy* **37**, 14–38 (1991).
30. Pennycook, S. J. & Jesson, D. E. Atomic resolution Z-contrast imaging of interfaces. *Acta Mater.* **40**, S149–S159 (1992).
31. Huang, S., Kang, W. & Yang, L. Electronic structure and quasiparticle bandgap of silicene structures. *Appl. Phys. Lett.* **102**, 133106 (2013).
32. Lee, P. A., Said, G., Davis, R. & Lim, T. H. On the optical properties of some layer compounds. *J. Phys. Chem. Solids* **30**, 2719–2729 (1969).
33. Mak, K. F., Lee, C., Hone, J., Shan, J. & Heinz, T. F. Atomically thin MoS_2: a new direct-gap semiconductor. *Phys. Rev. Lett.* **105**, 136805 (2010).
34. Gaiser, C. et al. Band-gap engineering with $HfS_XSe_{2\square X}$. *Phys. Rev. B* **69**, 075205 (2004).
35. Bianco, E. et al. Stability and exfoliation of germanane: a germanium graphane analogue. *ACS Nano* **7**, 4414–4421 (2013).
36. Kamal, C., Chakrabarti, A., Banerjee, A. & Deb, S. K. Silicene beyond mono-layers —different stacking configurations and their properties. *J. Phys. Condens. Matter* **25**, 085508 (2013).
37. Gao, N., Zheng, W. T. & Jiang, Q. Density functional theory calculations for two-dimensional silicene with halogen functionalization. *Phys. Chem. Chem. Phys.* **14**, 257–261 (2012).
38. Morishita, T. et al. First-principles study of structural and electronics properties of ultrathin silicon nanosheets. *Phys. Rev. B* **82**, 045419 (2010).
39. Kirk, E. C. G., Williams, D. A. & Ahmed, H. Cross-sectional transmission electron microscopy of precisely selected regions from semiconductor devices. *Inst. Phys. Conf. Ser.* **100**, 501–506 (1989).
40. Kresse, G. & Furthmüller, J. Efficiency of *ab-initio* total energy calculations for metals and semiconductors using a plane-wave basis set. *Comput. Mater. Sci.* **6**, 15–50 (1996).
41. Blöchl, P. E. Projector augmented-wave method. *Phys. Rev. B* **50**, 17953 (1994).
42. Perdew, J. P., Burke, K. & Ernzerhof, M. Generalized gradient approximation made simple. *Phys. Rev. Lett.* **77**, 3865 (1996).

Acknowledgements

This work was supported in part by PRESTO, the Japan Science and Technology Agency, and by a Grant-in-Aid for Scientific Research from the Ministry of Education, Culture, Sports, Science and Technology (MEXT), Japan. HAADF-STEM observations in this work were supported by the 'Nanotechnology Platform' of MEXT, Japan, at the Center for Integrated Nanotechnology Support, Tohoku University. The computations were undertaken with the assistance of resources from the National Computational Infrastructure (NCI), which is supported by the Australian Government, the Pawsey Supercomputing Centre with funding from the Australian Government and the Government of Western Australia, the Multi-modal Australian ScienceS Imaging and Visualisation Environment (MASSIVE) and the Victorian Partnership for Advanced Computing Limited (VPAC Ltd) through the V3 Alliance, Australia, and at the computational facilities at the Research Center for Computational Science, National Institute of Natural Sciences, and at the Research Institute for Information Technology, Kyushu University, Japan. We thank Mr Y. Yagi for the EPMA measurements, Prof. S. Yamanaka and Dr Y. Takeda for fruitful discussion.

Author contributions

R.Y. and H.N. conceived the idea. R.Y. and T.O. designed the experiments. R.Y. synthesized the $CaSi_2$ single crystals and $CaSi_2F_X$ compounds. Y.H. performed the HAADF-STEM observations and the EDX analyses. T.M. and M.J.S.S. performed the theoretical work. R.Y. and T.O. characterized the w-BLSi. R.Y., T.O., T.M. and H.N. wrote the manuscript. All the authors have read the manuscript and agree with its content.

Additional information

19

Electrochemical oxygen reduction catalysed by Ni$_3$(hexaiminotriphenylene)$_2$

Elise M. Miner[1], Tomohiro Fukushima[1], Dennis Sheberla[1], Lei Sun[1], Yogesh Surendranath[1] & Mircea Dincă[1]

Control over the architectural and electronic properties of heterogeneous catalysts poses a major obstacle in the targeted design of active and stable non-platinum group metal electrocatalysts for the oxygen reduction reaction. Here we introduce Ni$_3$(HITP)$_2$ (HITP = 2, 3, 6, 7, 10, 11-hexaiminotriphenylene) as an intrinsically conductive metal-organic framework which functions as a well-defined, tunable oxygen reduction electrocatalyst in alkaline solution. Ni$_3$(HITP)$_2$ exhibits oxygen reduction activity competitive with the most active non-platinum group metal electrocatalysts and stability during extended polarization. The square planar Ni-N$_4$ sites are structurally reminiscent of the highly active and widely studied non-platinum group metal electrocatalysts containing M-N$_4$ units. Ni$_3$(HITP)$_2$ and analogues thereof combine the high crystallinity of metal-organic frameworks, the physical durability and electrical conductivity of graphitic materials, and the diverse yet well-controlled synthetic accessibility of molecular species. Such properties may enable the targeted synthesis and systematic optimization of oxygen reduction electrocatalysts as components of fuel cells and electrolysers for renewable energy applications.

[1]Department of Chemistry, Massachusetts Institute of Technology, 77 Massachusetts Avenue, Cambridge, Massachusetts 02139, USA. Correspondence and requests for materials should be addressed to M.D. (email: mdinca@mit.edu).

The development of heterogeneous oxygen reduction reaction (ORR) electrocatalysts for implementation into fuel cell and electrolyser cathodes is a major research thrust in the arena of renewable fuel development. Achieving desired architectural and electronic properties of such catalysts remains difficult, however, because several variables must be optimized simultaneously, requiring synthetic tunability that is rarely available in the solid state. Desirable characteristics of an ORR electrocatalyst include: high active site density, reproducible synthesis and catalytic activity, stability in the electrolyte and in oxygen and peroxide, and low overpotential relative to the thermodynamic $4e^-$ oxygen-to-water reduction potential of 1.23 V (versus the reversible hydrogen electrode, RHE). One structural motif that has proven successful in catalysing ORR with high activity and physical robustness is the M-N_x unit, where M = a non-platinum group metal (for example, Fe, Co, Ni, Cu) chelated in a nitrogenous environment. These structures were popularized after the 1964 report by Jasinski[1] that detailed the high ORR activity of cobalt phthalocyanine complexes blended with electrically conductive acetylene black. The ability for oxygen to chemisorb onto these M-N_x sites without degrading the material fuelled extensive investigations of ORR on M-N_x-containing catalysts[2-5]. Though active towards ORR, M-N_x complexes have shown inconsistent stability in various electrolytes, motivating high-temperature treatment of the materials to enhance catalyst longevity and electrical conductivity[3,5]. Thermal treatment indeed increased the stability of the materials, but introduced new challenges in maintaining synthetic control over structure formation, identifying the catalytic active sites, and establishing structure–function relationships useful for catalyst optimization and mechanistic understanding. Thus, the search for active, intrinsically conductive, and chemically and electrochemically stable ORR electrocatalysts possessing well-defined and tunable active sites continues.

One class of materials that could answer these challenges is metal-organic frameworks (MOFs). These materials are compelling choices for electrocatalytic applications because their high surface area maximizes active site density, and their tunable chemical structure affords tailor-made microenvironments for controllable reaction conditions within the pores. Despite their promising features, MOFs have rarely been used for electrocatalytic applications because they are typically electrical insulators[6-11]. Recently, synthetic advances have given rise to conductive MOFs, some of which exhibit encouraging properties as electrocatalysts[12-16], but to our knowledge none have been experimentally shown to mediate ORR electrocatalysis.

Here we introduce $Ni_3(HITP)_2$ (HITP = 2, 3, 6, 7, 10, 11-hexaiminotriphenylene), a conductive two-dimensionally layered material structurally reminiscent of the long-studied M-N_x ORR electrocatalysts (Fig. 1)[17], as a representative of a new class of highly ordered ORR electrocatalysts exhibiting ORR activity and electrical conductivity ($\sigma = 40\,\mathrm{S\,cm^{-1}}$)[17] with no post-synthetic treatment or modification. In addition to possessing ORR activity competitive with the most active non-platinum group metal (nPGM) electrocatalysts to date, $Ni_3(HITP)_2$ retains 88% of its current density and undergoes no visible morphological degradation during prolonged electrochemical cycling. This study highlights conductive MOFs as a powerful platform for the development of tunable, designer electrocatalysts. It is noted that MOFs have been used as scaffolds for ORR electrocatalysts formed from high-temperature (>600 °C) pyrolysis[18-37] as well as incorporated into composites containing graphene oxide and porphyrin additives[7,8]. Whereas such materials indeed exhibit competitive ORR activity, the pyrolysis involved in their preparation eliminates the crystallinity and synthetic control inherent to MOFs. Our aim herein is to introduce a multi-faceted handle on imposing in a controlled manner structural, chemical and electronic properties on our material for reaction-targeted, MOF-based electrocatalyst design.

Results

Synthesis and quantification of $Ni_3(HITP)_2$. $Ni_3(HITP)_2$ can be grown solvothermally as a thin film on a variety of electrode surfaces using synthetic conditions mimicking those employed for the synthesis of bulk material[17]. Glassy carbon disk electrodes (5 mm diameter) served as the working electrodes for all investigations described herein unless otherwise noted, and all potentials are referenced to RHE. Deposition of $Ni_3(HITP)_2$ onto the glassy carbon electrodes typically afforded loadings of $\sim 5\,\mu g$ of MOF. The loadings were determined precisely in each case by atomic absorption spectroscopy (AAS) and verified by inductively coupled plasma-mass spectrometric (ICP-MS) measurements. The thickness of the film was analysed by atomic force microscopy. $Ni_3(HITP)_2$ films grown on glassy carbon electrodes have a thickness of $\sim 120\,\mathrm{nm}$, whereas films grown on indium tin oxide exhibit a similar morphology with a thickness of $\sim 300\,\mathrm{nm}$ (Supplementary Fig. 1).

ORR activity of $Ni_3(HITP)_2$. Cyclic voltammograms of $Ni_3(HITP)_2$ thin films on glassy carbon rotating disk electrodes recorded in the absence of O_2 revealed a significant double layer capacitance that increased with increasing scan rate (Supplementary Fig. 2), reflecting the high surface area of the modified electrodes[38]. Indeed, $Ni_3(HITP)_2$ exhibits a Brunauer–Emmett–Teller-specific surface area of $629.9 \pm 0.7\,\mathrm{m^2\,g^{-1}}$, as calculated from its nitrogen adsorption isotherm (Supplementary Fig. 3). Under O_2 atmosphere, the material reduces oxygen with an onset potential ($j = -50\,\mu\mathrm{A\,cm^{-2}}$) of 0.82 V in a 0.10 M aqueous solution of KOH (pH = 13.0; Fig. 2). The measured ORR onset potential is competitive with the most active nPGM ORR electrocatalysts reported thus far[39] and sits at an overpotential of 0.18 V relative to Pt ($E_{onset} = 1.00\,\mathrm{V}$).

Notably, cyclic voltammetry of the film on indium tin oxide electrodes shows the same ORR activity as the films on the glassy carbon electrodes (Supplementary Fig. 4), verifying that the MOF does not simply enhance the ORR activity of the glassy carbon electrode but rather acts as a stand-alone ORR electrocatalyst regardless of the substrate.

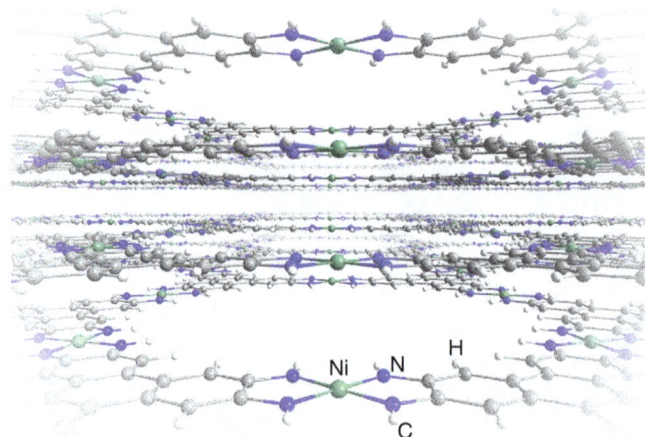

Figure 1 | $Ni_3(HITP)_2$ structure. Perspective view of the two-dimensional layered structure of $Ni_3(HITP)_2$ (ref. 17).

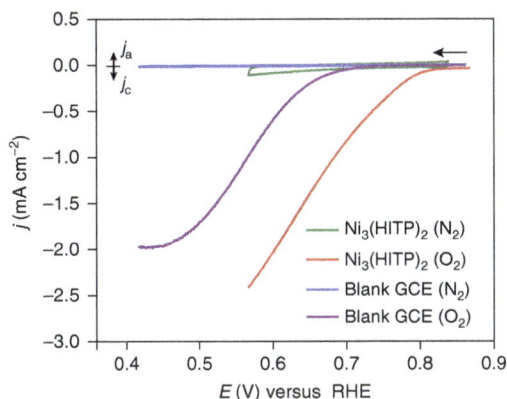

Figure 2 | ORR performance. Polarization curves of $Ni_3(HITP)_2$ under N_2 (green) versus O_2 atmosphere (red) as well as of the blank glassy carbon electrode under N_2 versus O_2 atmosphere (blue and purple, respectively). Scan rate $= 5\,mV\,s^{-1}$, rotation rate $= 2,000$ r.p.m., electrolyte $= 0.10\,M$ aqueous KOH, counter electrode $=$ Pt mesh, reference electrode $=$ Hg/HgO (1.00 M KOH), working electrode $=$ glassy carbon electrode (GCE).

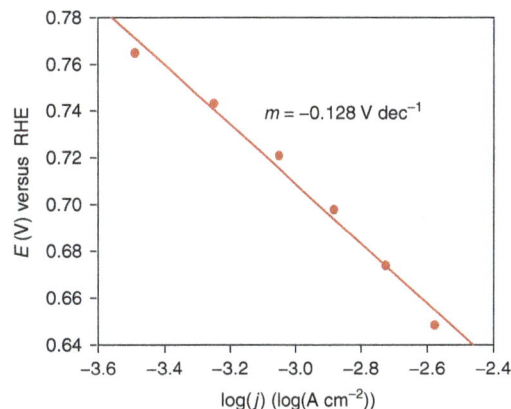

Figure 3 | ORR Tafel plot. Activation-controlled Tafel plot for $Ni_3(HITP)_2$-electrocatalyzed ORR, derived from the Koutecky–Levich plots (Supplementary Fig. 13).

Stability of $Ni_3(HITP)_2$ during ORR. Steady-state potentiostatic measurements at $E = 0.77\,V$ showed that 88% of the initial current density is retained over 8 h (Supplementary Fig. 5), in-line with other nPGM ORR catalysts[5,40–44]. Cyclic voltammetry of the modified electrode after the durability study showed no shift in the diffusion-limited region of the polarization curve, indicating that any alterations to the material during electrocatalysis were not significant enough to decrease the mass transport properties of the $Ni_3(HITP)_2$ film (Supplementary Fig. 6). Moreover, cyclic voltammetry of the electrolyte after these 8 h, using a fresh unmodified glassy carbon electrode, indicated no leaching from the $Ni_3(HITP)_2$ films, evidencing the heterogeneous nature of the catalyst. Additionally, ICP-MS and AAS analyses of films before and post electrolysis indicated the same quantity of Ni, suggesting that no part of the catalyst, homogeneous or heterogeneous, is lost from the films during catalysis (Supplementary Tables 1 and 4).

Characterization of $Ni_3(HITP)_2$ before and after ORR. Spectroscopic, microscopic and diffractometric techniques enabled analysis of the film before and after ORR catalysis. X-ray photoelectron spectroscopy (XPS) of catalyst films before and after catalysis revealed an increase in binding energy of the Ni_{2p} envelope region (850–885 eV) by $+1.0\,eV$ (Supplementary Fig. 7). Also visible by XPS was a shoulder peak that was present in the N_{1s} region before ORR (399 eV) which disappears after catalysis (Supplementary Fig. 8). Importantly, though the catalyst may undergo minor structural rearrangement during ORR, the high activity is largely retained over extended steady-state measurements, that is, neither access to the active sites nor the integrity of the active sites themselves is severely compromised during prolonged electrocatalysis. The structural robustness was supported by Raman spectroscopy conducted on the unused film, the film after submersion in the 0.10 M KOH electrolyte, and the film after electrochemical cycling under either N_2 or O_2. There were no missing or additional Raman bands for any of the altered films compared with the spectrum of the unused film (Supplementary Fig. 9). Additional evidence supporting the stability of the film was observed in scanning electron micrographs (SEMs) of the film taken before and after ORR catalysis (Supplementary Fig. 10). No perturbations in the morphology of the film were observed upon electrochemical cycling under O_2. Finally, grazing incidence X-ray diffraction of

the $Ni_3(HITP)_2$ film before and after ORR catalysis showed retention of the long-range order in the ab plane of $Ni_3(HITP)_2$ during ORR, further highlighting the structural stability of this catalyst during electrochemical cycling under O_2 (Supplementary Fig. 11).

ORR kinetics on $Ni_3(HITP)_2$. Using standard rotating ring-disk electrode experiments (Supplementary Fig. 12) and assuming that catalytically competent sites within $Ni_3(HITP)_2$ are distributed homogeneously throughout the film and not just on the surface, lower limit turnover frequencies (TOFs), determined by AAS, were found to be 0.042 electrons $[Ni_3(HITP)_2]^{-1}\,s^{-1}$ and 0.052 electrons $[Ni_3(HITP)_2]^{-1}\,s^{-1}$ for H_2O_2 and H_2O production, respectively, at $E = 0.79\,V$. Quantifying the Ni content in the same films by ICP-MS gave lower limit TOF values of 0.046 electrons $[Ni_3(HITP)_2]^{-1}\,s^{-1}$ and 0.056 electrons $[Ni_3(HITP)_2]^{-1}\,s^{-1}$ for H_2O_2 and H_2O production, respectively, also at $E = 0.79\,V$ (Supplementary Tables 1–6). The TOF values for H_2O_2 and H_2O production increase by one order of magnitude to 0.491 electrons $[Ni_3(HITP)_2]^{-1}\,s^{-1}$ and 0.466 electrons $[Ni_3(HITP)_2]^{-1}\,s^{-1}$, respectively, at 0.67 V. If the active sites in $Ni_3(HITP)_2$ are the Ni atoms, the lower limit TOFs derived from AAS quantification of Ni are 0.014 electrons $[Ni]^{-1}\,s^{-1}$ and 0.017 electrons $[Ni]^{-1}\,s^{-1}$ for H_2O_2 and H_2O production, respectively, at $E = 0.79\,V$. Minimum TOF values calculated from the ICP-MS quantification of Ni were 0.015 electrons $[Ni]^{-1}\,s^{-1}$ and 0.019 electrons $[Ni]^{-1}\,s^{-1}$ for H_2O_2 and H_2O production, respectively, also at $E = 0.79\,V$. Notably, the intrinsic ORR turnover frequencies for $Ni_3(HITP)_2$ could exceed the values reported here because the Ni quantification methods do not distinguish exclusively electroactive Ni sites; if some fraction of the potentially active sites are not catalytically competent because of mass transport limitations within the films, the ORR current-to-active-site ratio would increase, consequently increasing the TOF.

Mechanistic insight into ORR on $Ni_3(HITP)_2$. The activation-controlled Tafel plot generated from Koutecky–Levich (K–L) data (Supplementary Fig. 13) revealed a Tafel slope of $-128\,mV\,dec^{-1}$ (Fig. 3). This Tafel slope corresponds to an irreversible one-electron pre-equilibrium process, likely indicating the formation of the superoxide anion as the rate-limiting step (theoretical Tafel slope $= -120\,mV\,dec^{-1}$). The total number of electrons transferred during ORR was determined using the inverse of the slope of the K–L plots, termed the B factor (equation (1)):

$$B = 0.62nFD_{O_2}^{2/3}v^{-1/6}c_{O_2}\left(\frac{2\pi}{60}\right)^{1/2} \qquad (1)$$

where $n =$ number of electrons transferred, $F =$ Faraday's constant, $D_{O_2} = O_2$ diffusion coefficient in the electrolyte, $v =$ kinematic viscosity of the electrolyte, and $c_{O_2} =$ the saturation concentration of O_2 in the electrolyte at 1 atm O_2 pressure. Assuming the typical values in 0.10 M KOH: $D_{O_2} = 1.9 \times 10^{-5}$ cm s^{-1}, $v = 0.1$ m^2 s^{-1} (ref. 7), and $c_{O_2} = 1.26 \times 10^{-6}$ mol cm^{-3}, and using $B = 0.0831$ mA cm^{-2} r.p.m.$^{-1/2}$ from the K–L plot at $E = 0.767$ V, the number of transferred electrons in our system was calculated to be $n = 2.25$ (Supplementary Table 7). This electron transfer number is consistent with predominant (87.5%) production of H_2O_2 (more accurately, production of HO_2^- in 0.10 M KOH given the pK_a of $H_2O_2 = 11.63$)[45], with the remaining activity ascribed to 4e$^-$ reduction to H_2O.

The Faradaic efficiency for H_2O_2 production was determined by measuring the ratio of the ring current to the disk current in rotating ring-disk electrochemical experiments (see Methods section). In the 0.82–0.54 V potential range, the Faradaic efficiency for H_2O_2 production decreases from 100 to 63% (Fig. 4) as formation of H_2O increases with increasing overpotential before reaching a plateau at ~ 0.75 V.

The H$^+$ order for ORR catalysis was probed galvanostatically at $I = -5.0$ µA in a 0.1 M NaClO$_4$/0.1 M NaOH aqueous electrolyte titrated from pH 12.89 to 11.54 with 1.0 M HClO$_4$. These studies revealed a slope of zero for $\delta E/\delta$pH above pH 12.80, suggesting a zeroth order dependence on [H$^+$] for the kinetic rate law (Supplementary Fig. 14). However, a non-zero slope was observed below pH 12.80, indicating a change in mechanism that involves proton-coupled electron transfer or proton-dependent chemical steps before or during the rate-limiting step (Supplementary Note 1). Detailed mechanistic investigations of ORR with our catalyst are currently underway.

Discussion

Direct adhesion of the Ni$_3$(HITP)$_2$ film onto the electrode surface eliminates the need for binders or conductive additives that may block access to active sites by pore filling. This direct contact between the parent material and the electrode allowed for investigation of the inherent electrocatalytic behaviour of pure Ni$_3$(HITP)$_2$. The high surface area and porosity inherent to Ni$_3$(HITP)$_2$ may increase the density of and facilitate easy access to the catalytic active sites on Ni$_3$(HITP)$_2$, contributing to the notable ORR activity. Given that this high ORR activity is observed after the film purification procedure which involves

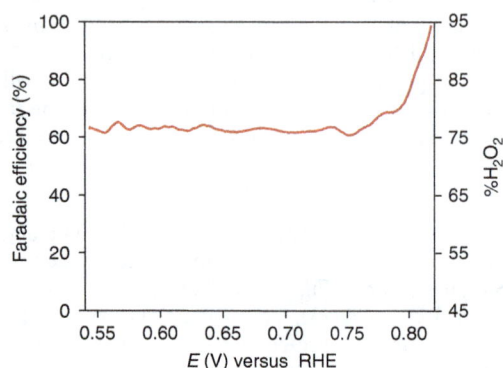

Figure 4 | Faradaic efficiency for H$_2$O$_2$ and %H$_2$O$_2$. Potential-dependent Faradaic efficiency for H$_2$O$_2$ production and %H$_2$O$_2$ production during ORR catalysed by Ni$_3$(HITP)$_2$ at pH 13.

heating the modified electrode in methanol at 65 °C for 20 h, Ni$_3$(HITP)$_2$ and related materials may be strong candidates for implementation into direct methanol fuel cells where methanol tolerance of the anodic and cathodic catalysts is a necessity[46]. High stability in the presence of methanol is not observed for Pt-based electrocatalysts, a major hurdle currently slowing direct methanol fuel cell development[47].

Further insight into the robustness of Ni$_3$(HITP)$_2$ during ORR was achieved using several spectroscopic and microscopic techniques to probe the catalyst structure before and after catalysis. The +1 eV shift in the Ni$_{2p}$ XPS after ORR catalysis could be indicative of a Ni–O interaction[48,49], or alternatively a strengthening of the ligand field as electron density around the imine decreases. The loss of asymmetry in the N$_{1s}$ region of the XPS after catalysis is consistent with an alteration of the ligand field during ORR. Though some minor changes in the film structure may take place during ORR, retention of the majority of ORR activity over the steady-state potentiostatic measurements provides encouraging evidence that neither access to the active sites nor the integrity of the active sites themselves is severely compromised during prolonged electrocatalysis. Furthermore, any subtle alterations of the film affected neither the film's microstructure nor the polarizability of the Ni$_3$(HITP)$_2$ bonds as shown by SEM and Raman spectroscopy, respectively. The stability of the catalyst in aqueous media is industrially advantageous given the lower cost of water-based electrolytes.

To the best of our knowledge, the foregoing results demonstrate for the first time electrocatalytic ORR activity in a well-defined, intrinsically conductive MOF. Clearly, the faradaic efficiency for water production should be increased for maximizing energy density in industrial settings, but such a goal may be more tractable with MOFs, whose well-defined structures provide the ability to systematically investigate a number of variables including the metal centre identity, valency and coordination environment. Structure–function and mechanistic studies will facilitate understanding, development, and diversification of this material into a platform structure primed for the targeted design of other ORR electrocatalysts.

Methods

Characterization of the Ni$_3$(HITP)$_2$ film. Samples were prepared for ICP-MS and AAS analysis by sonication of the modified electrode buttons in concentrated ICP (Omnitrace purity, 67–70% w/w; EMD) grade nitric acid for 4 h. The electrode buttons were removed from the acid, and the acid was diluted to 2% v/v with Milli-Q water.

ICP-MS was conducted on an Agilent 7900 at the MIT Center for Environmental and Health Sciences (Cambridge, MA, USA). An external calibration curve was generated with a nickel standard (1,000 p.p.m. in 2% HNO$_3$; Ultra Scientific) diluted to 0, 15, 30, 60 and 120 p.p.b. in 2% ICP grade nitric acid. Argon flowing at 1.06 l min^{-1} was used as the carrier gas. The ICP-MS data was analysed by MassHunter 4.1 software.

Graphite furnace AAS was conducted on a Perkin Elmer AAnalyst 600 GFAAS (property of the Lippard Group, MIT, Cambridge, MA, USA). An ICP grade Ni standard (1,000 p.p.m. in 2% HNO$_3$) (Ultra Scientific) was diluted to 100 p.p.b. in 2% HNO$_3$ in Milli-Q water. The AAS performed a serial dilution to generate a nickel calibration curve with 0, 25, 50, 75 and 100 p.p.b. nickel calibration points. The nickel content was probed by monitoring the optical absorption at $\lambda = 232.0$ nm. The graphite furnace temperature was ramped from 110 to 2,500 °C during AAS analysis. The AAS results were analysed by WinLab32 for AA, version 6.5.0.0266.

Atomic force microscopy was conducted at the MIT Institute for Soldier Nanotechnologies (Cambridge, MA, USA) using a Veeco Dimension 3,100 scanning probe microscope (Veeco Digital Instruments by Bruker) equipped with a Nanoscope V controller. Images were recorded in tapping mode in the air at room temperature (23–25 °C) using an Al reflex coated silicon micro cantilever (AC240TS-R3, Asylum Research). The scan rate was set at 1.0 Hz. The atomic force microscopy results were analysed by Gwyddion 2.43 software.

XPS was conducted at the Harvard Center for Nanoscale Systems (Cambridge, MA, USA) on a Thermo Scientific K-Alpha XPS. A survey scan was taken and C, N, O and Ni were probed with a pass energy = 50 eV, beam width = 400 µm. Data analysis was executed with the Advantage 5.938 software programme.

Raman spectroscopy was conducted on a Horiba Raman spectrophotometer (property of the Myerson Group, MIT, Cambridge, MA, USA) operated at 457 nm with a hole diameter of 500 μm, a slit size of 100 μm, a range of 100–3,000 cm^{-1}, a 100 × magnification lens, a laser intensity of 39 A, and 2 s runs with three accumulations per sample.

Scanning electron microscopy was conducted at the Harvard Center for Nanoscale Systems (Cambridge, MA, USA) on a Zeiss Ultra Plus FE-SEM with an InLens detector, a voltage of 10 kV, and 200 k × magnification. Data analysis was executed with SmartSEM V05.04.02.00 software.

Grazing incidence X-ray diffraction was conducted at the MIT Center for Materials Science and Engineering (Cambridge, MA, USA) on a Bruker D8 Discover Diffractometer with a Vantec 2,000 two-dimensional detector, a Cu K_{α} X-ray source (1.5409 Å), and a tube voltage and current of 40 kV and 40 mA, respectively. The diffraction patterns were collected in a grazing incidence geometry with a grazing incidence angle of 3.6°. The blank indium tin oxide slide and the indium tin oxide slides modified with the $Ni_3(HITP)_2$ film were secured onto the diffractometer stage with double-sided tape during data collection. The data for each sample was collected in a single exposure with an exposure time of 10 min per sample. The two-dimensional data were reduced by azimuth averaging over 180° of the Debye Scherrer ring. It is noted that the remaining 180° of the Debye Scherrer ring was blocked by the sample due to the grazing incidence geometry.

Electrochemistry with the $Ni_3(HITP)_2$ film. KOH (99.99% trace metals) was purchased from Sigma-Aldrich. Oxygen gas was purchased from Airgas (99.8% purity). Reference and glassy carbon working electrodes were purchased from CH Instruments. Pt gauze (100 mesh, 99.9% metal basis) and wires ($\phi = 0.404$ mm, annealed, 99.9% metal basis, and $\phi = 0.5$ mm dia., hard, 99.95% metal basis) comprising the auxiliary electrode were purchased from Alfa Aesar. The auxiliary electrode was cleaned by submersion in concentrated HCl followed by sonication for 5 min, washing with Milli-Q water, and drying under a stream of air before each experiment. Working electrodes were cleaned by submersion in concentrated HCl followed by sonication for 5 min, washing with Milli-Q water, and drying under a stream of air. The working electrodes were then sequentially polished with 100, 30 and 5 μm diameter alumina powder from BASI. Unless otherwise noted, all electrochemical experiments were executed with a Bio-Logic SP200 potentiostat/galvanostat in a custom 2-compartment electrochemical cell. Rotating disk electrode and rotating ring-disk electrode studies were conducted with a Bio-Logic VMP3 potentiostat/galvanostat Pine Research Instrumentation Modulated Speed Rotator. Unless otherwise specified, internal resistance of the electrolyte was measured with the Bio-Logic SP200 potentiostat/galvanostat, and iR drop correction was applied. Generally, the resistance of 0.10 M KOH was measured to be ~40 Ω.

Synthesis of the $Ni_3(HITP)_2$ film on glassy carbon electrode. 2, 3, 6, 7, 10, 11-hexaaminotriphenylene hexahydrochloric acid (HATP · 6HCl) salt (10.4 mg) was dissolved in Milli-Q water (6 ml) and heated to 65 °C with stirring in a 20 ml capped glass vial (Vial A). In a second glass reaction vial (Vial B), nickel(II) chloride hexahydrate (4.6 mg) was dissolved in Milli-Q water (4 ml) and to this was added concentrated aqueous ammonium hydroxide (0.4 ml, 25% aqueous solution). The heated HATP solution in Vial A was added to the $NiCl_2/NH_4OH$ solution (Vial B) and two alumina micropolished glassy carbon electrodes (5 mm diameter) were placed in the reaction so that the polished faces of the glassy carbon buttons were parallel to the bottom of the reaction vial. Each button was inserted into an NMR tube cap so that only the polished face of the glassy carbon was exposed for modification. The vial was capped and the reaction was heated without stirring at 65 °C for 15 h. The next day, the reaction afforded a translucent film on the glassy carbon electrode buttons. Additionally, a translucent black film was visible on the reaction vial walls and a black flaky solid had settled at the bottom of the reaction vial. The electrode film and the reaction mixture solid were purified separately.

The electrode was removed from the reaction mixture and heated in Milli-Q water (20 ml) at 65 °C for 4 h in a capped vial, rinsed with Milli-Q water, then heated again in water at 65 °C for 15 h in a capped vial. The electrode was rinsed with CH_3OH and then heated in fresh CH_3OH in a capped vial at 65 °C for 5 h. The CH_3OH was removed and the electrode was heated at 65 °C for 15 h in fresh CH_3OH. The next day after drying under dynamic vacuum, a black translucent film coating the polished side of the glassy carbon button was visible. The electrode button was stored under dynamic vacuum.

For purification of the black powder, the remaining reaction mixture was centrifuged, the supernatant was removed, and the remaining solid was sonicated in Milli-Q water (15 ml) for 5 min then heated in a capped vial with stirring at 65 °C for 4 h. The same procedure was repeated once more, with the final heating step duration of 15 h. The powder was once again centrifuged, followed by removal of the supernatant, and then the powder was sonicated in CH_3OH (15 ml) for 5 min, then heated in the capped vial in CH_3OH at 65 °C for 5 h. The CH_3OH wash procedure was also repeated one more time, then the powder was centrifuged, the supernatant was removed, and the black solid was dried under vacuum for 15 h.

Determination of the 2-and-4-electron ORR TOFs. The background-subtracted ring current (Supplementary Fig. 12) was taken for each potential probed during potentiostatic measurements ($E_{disk} = 0.807$, 0.787, 0.767, 0.747, 0.727, 0.707, 0.687, and 0.667 V) then divided by 1,000 to calculate current passed in $A = C s^{-1}$ ($I = Q/t$). That current was divided by 0.2 to account for the 20% ring collection efficiency, then divided by Faraday's constant (96,485.3365 $C mol^{-1}$) and multiplied by Avogadro's number (6.022×10^{23} electrons per mol) to determine the number of electrons transferred to O_2 when reducing O_2 to H_2O_2 (2-electron ORR) per second. By ICP-MS the electrode was calculated to have an average of 1.1015×10^{16} nickel sites deposited (see Supplementary Table 1 for s.d.). By AAS, the electrode was calculated to have an average of 1.26199×10^{16} nickel sites deposited (see Supplementary Table 4 for s.d.). The number of electrons transferred per second was divided by the number of nickel sites as determined by AAS and ICP-MS, respectively, to convert the 2-electron ORR TOF to electrons nickel per site per second according to the two nickel quantification methods (that is, AAS and ICP-MS) (Supplementary Tables 2 and 5). Alternatively, the number of electrons transferred per second was divided by the number of $Ni_3(HITP)_2$ units to convert the 2-electron ORR TOF to electrons per $Ni_3(HITP)_2$ formula unit per second. The number of $Ni_3(HITP)_2$ formula units was directly calculated from the number of nickel sites derived from the two nickel quantification methods. In the main text, the TOF is expressed as a range defined by the values calculated using the two employed nickel quantification methods.

To determine the TOF for 4-electron ORR, the background and collection efficiency-corrected ring current was subtracted from the disk current (A) (Supplementary Fig. 12) to obtain the current passed during 4-electron ORR. The current (A) was divided by Faraday's constant then by number of nickel or $Ni_3(HITP)_2$ sites to calculate the TOF (electrons per nickel site per second, or electrons per $Ni_3(HITP)_2$ formula unit per second, respectively) during 4-electron ORR according to the two employed nickel quantification methods (Supplementary Tables 3 and 6). In the main text, the TOF is expressed as a range defined by the values calculated using the two employed nickel quantification methods.

Rotating disk and rotating ring-disk electrode investigations. Experiments were conducted in a two-compartment cell with a glass frit separating the auxiliary electrode from the working electrode; electrolyte = 0.10 M KOH; auxiliary electrode = Pt mesh; reference electrode = Hg/HgO (1.00 M KOH), working electrode = blank glassy carbon button (5 mm diameter) or glassy carbon button modified with $Ni_3(HITP)_2$ film and inserted in a polyarylether-ether ketone rotating ring-disk electrode (RRDE) tip with a platinum ring; rotation rate = 2,000 r.p.m.; scan speed = 5 $mV s^{-1}$; atmosphere = N_2 or O_2 sparged for 10 min through a fritted sparge tube before data collection, with continuous sparging during data collection. All voltammograms were collected by scanning cathodically from $E = 0$ V versus open circuit potential (OCP) to −0.300 V versus Hg/HgO (0.567 V versus RHE). The electrolyte solvent window was established by cycling a blank glassy carbon button under N_2 atmosphere. ORR activity of the unmodified glassy carbon was observed by cycling the unmodified glassy carbon electrode under O_2 from $E = 0$ V versus OCP to $E = 0.400$ V versus RHE. These controls preceded data collection of $Ni_3(HITP)_2$-modified glassy carbon under N_2 and O_2 atm. When relevant, a potential of $E = 1.23$ V versus RHE was applied to the Pt ring disk for oxidation of the ORR products. A 20% collection efficiency was applied for quantification of the ORR products using the current measured at the Pt ring disk.

Potentiostatic steady-state durability test. While rotating at 2,000 r.p.m., cyclic voltammetry (CV) of the $Ni_3(HITP)_2$-modified glassy carbon electrode was conducted at 5 $mV s^{-1}$ from $E = 0$ V versus OCP .o $E = -0.3$ V versus Hg/HgO ($E = 0.567$ V versus RHE) under sparging O_2 atmosphere to measure the ORR activity. In this experiment, a titanium plate auxiliary electrode was used. O_2 sparged throughout the entirety of the experiment. The potential was held at $E = 0.767$ V versus RHE for 8 h with O_2 sparging continuously. The current response was monitored with data points collected every 60 s. After the potentiostatic stability test was completed, CV was conducted again under O_2 atmosphere at 5 $mV s^{-1}$ from $E = 0$ V versus OCP to $E = -0.3$ V versus Hg/HgO to compare the mass transport of the used material to that of the material before the stability test.

Koutecky-Levich and Tafel studies. CV (5 $mV s^{-1}$) under N_2 atmosphere was conducted from 0 V versus OCP to −0.3 V versus Hg/HgO (ORR potential range for $Ni_3(HITP)_2$). CV (5 $mV s^{-1}$) under O_2 atmosphere was conducted from 0 V versus OCP to −0.3 V versus Hg/HgO (ORR potential range for $Ni_3(HITP)_2$). Galvanostatic measurements were conducted with $I = -1$, −10 and −100 μA to identify the reliable potential range for potentiostatic measurements. Potentiostatic measurements were conducted from −20 to −200 mV versus Hg/HgO in increments of 20 mV. Each potential was held for 1 min. This was conducted five times, with altering rotation speeds to extrapolate the diffusion coefficient. The electrode was rotated at 2,000, 625, 816, 550 and 1,189 r.p.m., respectively. This allowed for elimination of mass transport limitations when analysing Tafel behaviour via generation of the activation-controlled Tafel plot. CV (5 $mV s^{-1}$) under O_2 atmosphere was conducted from 0 V versus OCP to −0.3 V versus Hg/HgO (ORR potential range for $Ni_3(HITP)_2$). Chronoamperometry at $E = -0.2$ V versus Hg/HgO was run for 8 min under N_2 sparging atmosphere to eliminate O_2. CV (5 $mV s^{-1}$) under N_2 atmosphere was conducted from 0 V versus OCP to −0.3 V versus Hg/HgO (ORR potential range for $Ni_3(HITP)_2$) to recheck the double layer capacitance as an indicator of potential catalyst decomposition. Ohmic drop was measured at $I = -0.1$ mA for iR correction.

ORR proton order study. Potential was measured over 25 min at a constant current $I = -5\,\mu A$ while varying the pH from 12.89 to 11.54 in the 0.10 M KOH electrolyte titrated with 1.0 M $HClO_4$.

References

1. Jasinski, R. Cobalt phthalocyanine as a fuel cell cathode. *J. Electrochem. Soc.* **112**, 526–528 (1965).
2. Tang, H. *et al.* Molecular architecture of cobalt porphyrin multilayers on reduced graphene oxide sheets for high-performance oxygen reduction reaction. *Angew. Chem. Int. Ed.* **52**, 5585–5589 (2013).
3. Schafer, F. P. *et al. Physical and Chemical Applications of Dyestuffs* (Springer, 1976).
4. Masa, J., Xia, W., Muhler, M. & Schuhmann, W. On the role of metals in nitrogen-doped carbon electrocatalysts for oxygen reduction. *Angew. Chem. Int. Ed.* **54**, 10102–10120 (2015).
5. Chen, Z., Higgins, D., Yu, A., Zhang, L. & Zhang, J. A review on non-precious metal electrocatalysts for PEM fuel cells. *Energy Environ. Sci.* **4**, 3167–3192 (2011).
6. Mao, J., Yang, L., Yu, P., Wei, X. & Mao, L. Electrocatalytic four-electron reduction of oxygen with copper(II)-based metal-organic frameworks. *Electrochem. Commun.* **19**, 29–31 (2012).
7. Jahan, M., Bao, Q. & Loh, K. P. Electrocatalytically active graphene-porphyrin MOF composite for oxygen reduction reaction. *J. Am. Chem. Soc.* **134**, 6707–6713 (2012).
8. Jahan, M., Liu, Z. & Loh, K. P. A graphene oxide and copper-centered metal organic framework composite as a tri-functional catalyst for HER, OER, and ORR. *Adv. Funct. Mater.* **23**, 5363–5372 (2013).
9. Jiang, M., Li, L., Zhu, D., Zhang, H. & Zhao, X. Oxygen reduction in the nanocage of metal–organic frameworks with an electron transfer mediator. *J. Mater. Chem. A* **2**, 5323–5329 (2014).
10. Wang, H., Yin, F., Chen, B. & Li, G. Synthesis of an ε-MnO_2/metal–organic-framework composite and its electrocatalysis towards oxygen reduction reaction in an alkaline solution. *J. Mater. Chem. A* **3**, 16168–16176 (2015).
11. Barkholtz, H., Chong, L., Kaiser, Z., Xu, T. & Liu, D.-J. Highly active non-PGM catalysts prepared from metal organic frameworks. *Catalysts* **5**, 955–965 (2015).
12. Clough, A. J., Yoo, J. W., Mecklenburg, M. H. & Marinescu, S. C. Two-dimensional metal-organic surfaces for efficient hydrogen evolution from water. *J. Am. Chem. Soc.* **137**, 118–121 (2015).
13. Hinogami, R. *et al.* Electrochemical reduction of carbon dioxide using a copper rubeanate metal organic framework. *ECS Electrochem. Lett.* **1**, H17–H19 (2012).
14. Lin, S. *et al.* Covalent organic frameworks comprising cobalt porphyrins for catalytic CO_2 reduction in water. *Science* **349**, 1208–1213 (2015).
15. Dong, R. *et al.* Large-area, free-standing, two-dimensional supramolecular polymer single-layer sheet for highly efficient electrocatalytic hydrogen evolution. *Angew. Chem. Int. Ed.* **54**, 12058–12063 (2015).
16. Zhang, P., Hou, X., Liu, L., Mi, J. & Dong, M. Two-dimensional π-conjugated metal bis(dithiolene) complex nanosheets as selective catalysts for oxygen reduction reaction. *J. Phys. Chem. C* **119**, 28028–28037 (2015).
17. Sheberla, D. *et al.* High electrical conductivity in $Ni_3(2,3,6,7,10,11$-hexaiminotriphenylene)$_2$, a semiconducting metal-organic graphene analogue. *J. Am. Chem. Soc.* **136**, 8859–8862 (2014).
18. Afsahi, F. & Kaliaguine, S. Non-precious electrocatalysts synthesized from metal-organic frameworks. *J. Mater. Chem. A* **2**, 12270–12279 (2014).
19. Aijaz, A., Fujiwara, N. & Xu, Q. From metal-organic framework to nitrogen-decorated nanoporous carbons: high CO_2 uptake and efficient catalytic oxygen reduction. *J. Am. Chem. Soc.* **136**, 6790–6793 (2014).
20. Li, J. *et al.* Metal-organic framework templated nitrogen and sulfur co-doped porous carbons as highly efficient metal-free electrocatalysts for oxygen reduction reactions. *J. Mater. Chem. A* **2**, 6316–6319 (2014).
21. Li, Q. *et al.* Graphene/graphene-tube nanocomposites templated from cage-containing metal-organic frameworks for oxygen reduction in Li–O_2 batteries. *Adv. Mater.* **26**, 1378–1386 (2014).
22. Li, Q. *et al.* Metal-organic framework-derived bamboo-like nitrogen-doped graphene tubes as an active matrix for hybrid oxygen-reduction electrocatalysts. *Small* **11**, 1443–1452 (2014).
23. Kong, A. *et al.* From cage-in-cage MOF to N-doped and Co-nanoparticle-embedded carbon for oxygen reduction reaction. *Dalton Trans.* **44**, 6748–6754 (2015).
24. Ge, L. *et al.* High activity electrocatalysts from metal–organic framework-carbon nanotube templates for the oxygen reduction reaction. *Carbon* **82**, 417–424 (2015).
25. Palaniselvam, T., Biswal, B. P., Banerjee, R. & Kurungot, S. Zeolitic imidazolate framework (ZIF)-derived, hollow-core, nitrogen-doped carbon nanostructures for oxygen-reduction reactions in PEFCs. *Chemistry* **19**, 9335–9342 (2013).
26. Pandiaraj, S., Aiyappa, H. B., Banerjee, R. & Kurungot, S. Post modification of MOF derived carbon via g-C_3N_4 entrapment for an efficient metal-free oxygen reduction reaction. *Chem. Commun.* **50**, 3363–3366 (2014).
27. Strickland, K. *et al.* Highly active oxygen reduction non-platinum group metal electrocatalyst without direct metal–nitrogen coordination. *Nat. Commun.* **6**, 7343–7351 (2015).
28. Kung, C.-W. *et al.* Metal–organic framework thin films composed of free-standing acicular nanorods exhibiting reversible electrochromism. *Chem. Mater.* **25**, 5012–5017 (2013).
29. Xia, W. *et al.* Well-defined carbon polyhedrons prepared from nano metal-organic frameworks for oxygen reduction. *J. Mater. Chem. A* **2**, 11606–11613 (2014).
30. Wang, X. *et al.* MOF derived catalysts for electrochemical oxygen reduction. *J. Mater. Chem. A* **2**, 14064–14070 (2014).
31. Zhang, G. *et al.* One-step conversion from metal–organic frameworks to Co_3O_4@N-doped carbon nanocomposites towards highly efficient oxygen reduction catalysts. *J. Mater. Chem. A* **2**, 8184–8189 (2014).
32. Zhang, L. *et al.* Highly graphitized nitrogen-doped porous carbon nanopolyhedra derived from ZIF-8 nanocrystals as efficient electrocatalysts for oxygen reduction reactions. *Nanoscale* **6**, 6590–6602 (2014).
33. Zhao, D. *et al.* Iron imidazolate framework as precursor for electrocatalysts in polymer electrolyte membrane fuel cells. *Chem. Sci.* **3**, 3200–3205 (2012).
34. Zhao, D. *et al.* Highly efficient non-precious metal electrocatalysts prepared from one-pot synthesized zeolitic imidazolate frameworks. *Adv. Mater.* **26**, 1093–1097 (2014).
35. Zhao, S. *et al.* Carbonized nanoscale metal-organic frameworks as high performance electrocatalyst for oxygen reduction reaction. *ACS Nano* **8**, 12660–12668 (2014).
36. Zhao, X. *et al.* One-step synthesis of nitrogen-doped microporous carbon materials as metal-free electrocatalysts for oxygen reduction reaction. *J. Mater. Chem. A* **2**, 11666–11671 (2014).
37. Zhu, D. *et al.* Nitrogen-doped porous carbons from bipyridine-based metal-organic frameworks: electrocatalysis for oxygen reduction reaction and Pt-catalyst support for methanol electrooxidation. *Carbon* **79**, 544–553 (2014).
38. Gileadi, E. *Physical Electrochemistry* (Wiley-VCH Verlag GmbH & Co, 2011).
39. Shi, H. *et al.* Recent advances of doped carbon as non-precious catalysts for oxygen reduction reaction. *J. Mater. Chem. A* **2**, 15704–15716 (2014).
40. Serov, A. *et al.* Nano-structured non-platinum catalysts for automotive fuel cell application. *Nano Energy* **16**, 293–300 (2015).
41. Xie, Y. *et al.* Carbonization of self-assembled nanoporous hemin with a significantly enhanced activity for the oxygen reduction reaction. *Faraday Discuss.* **176**, 393–408 (2014).
42. Lin, Q. *et al.* Heterometal-embedded organic conjugate frameworks from alternating monomeric iron and cobalt metalloporphyrins and their application in design of porous carbon catalysts. *Adv. Mater.* **27**, 3431–3436 (2015).
43. Wang, Y., Kong, A., Chen, X., Lin, Q. & Feng, P. Efficient oxygen electroreduction: hierarchical porous Fe–N-doped hollow carbon nanoshells. *ACS Catal.* **5**, 3887–3893 (2015).
44. Liu, Y. *et al.* Iron(II) phthalocyanine covalently functionalized graphene as a highly efficient non-precious-metal catalyst for the oxygen reduction reaction in alkaline media. *Electrochim. Acta* **112**, 269–278 (2013).
45. Blizanac, B. B., Ross, P. N. & Markovic, N. M. Oxygen electroreduction on Ag(111): the pH effect. *Electrochim. Acta* **52**, 2264–2271 (2007).
46. Zhao, S. *et al.* Three dimensional N-doped graphene/PtRu nanoparticle hybrids as high performance anode for direct methanol fuel cells. *J. Mater. Chem. A* **2**, 3719–3724 (2014).
47. Alonso-Vante, N. Platinum and non-platinum nanomaterials for the molecular oxygen reduction reaction. *Chemphyschem.* **11**, 2732–2744 (2010).
48. Grosvenor, A. P., Biesinger, M. C., Smart, R. S. C. & McIntyre, N. S. New interpretations of XPS spectra of nickel metal and oxides. *Surf. Sci.* **600**, 1771–1779 (2006).
49. Biesinger, M. C., Payne, B. P., Lau, L. W. M., Gerson, A. & Smart, R. S. C. X-ray photoelectron spectroscopic chemical state quantification of mixed nickel metal, oxide and hydroxide systems. *Surf. Interface Anal.* **41**, 324–332 (2009).

Acknowledgements

This work was supported by the US Department of Energy, Office of Science, Office of Basic Energy Sciences (Award DESC0006937). We thank Dr M. Li for valuable discussions and assistance with XPS, Dr I. Riddell (Lippard Group, MIT) for assistance with AAS, Dr K. Taghizadeh (MIT Center for Environmental and Health Sciences) for assistance with ICP-MS, Dr S. Perala (Myerson Group, MIT) for assistance with Raman spectroscopy, Mr D. Lange (Harvard Center for Nanoscale Systems) for assistance with SEM, and Dr C. Settens (MIT Center for Materials Science and Engineering) for assistance with the grazing incidence X-ray diffraction.

Author contributions

T.F. assisted with the RRDE experimental setup and design of electrochemical procedures. Y.S. provided electrochemical equipment. D.S. conducted N_2 isotherms and Brunauer–Emmett–Teller surface area analysis of $Ni_3(HITP)_2$. L.S. conducted atomic force microscopy measurements. E.M.M. prepared all samples for electrochemical investigations and all other analyses, and conducted the reported experiments. E.M.M., T.F., Y.S. and M.D. contributed to data analysis. E.M.M. and M.D. wrote the manuscript.

Additional information

Competing financial interests: The authors declare no competing financial interests.

20

Making hybrid [n]-rotaxanes as supramolecular arrays of molecular electron spin qubits

Antonio Fernandez[1], Jesus Ferrando-Soria[1], Eufemio Moreno Pineda[1], Floriana Tuna[1], Iñigo J. Vitorica-Yrezabal[1], Christiane Knappke[2], Jakub Ujma[1,3], Christopher A. Muryn[1], Grigore A. Timco[1], Perdita E. Barran[1,3], Arzhang Ardavan[4] & Richard E.P. Winpenny[1]

Quantum information processing (QIP) would require that the individual units involved—qubits—communicate to other qubits while retaining their identity. In many ways this resembles the way supramolecular chemistry brings together individual molecules into interlocked structures, where the assembly has one identity but where the individual components are still recognizable. Here a fully modular supramolecular strategy has been to link hybrid organic–inorganic [2]- and [3]-rotaxanes into still larger [4]-, [5]- and [7]-rotaxanes. The ring components are heterometallic octanuclear $[Cr_7NiF_8(O_2C^tBu)_{16}]^-$ coordination cages and the thread components template the formation of the ring about the organic axle, and are further functionalized to act as a ligand, which leads to large supramolecular arrays of these heterometallic rings. As the rings have been proposed as qubits for QIP, the strategy provides a possible route towards scalable molecular electron spin devices for QIP. Double electron–electron resonance experiments demonstrate inter-qubit interactions suitable for mediating two-qubit quantum logic gates.

[1] School of Chemistry and Photon Science Institute, The University of Manchester, Oxford Road, Manchester M13 9PL, UK. [2] Department of Chemistry, University of Oxford, Oxford OX1 3TA, UK. [3] The Michael Barber Centre for Collaborative Mass Spectrometry, Manchester Institute of Biotechnology, The University of Manchester, Oxford Road, Manchester M13 9PL, UK. [4] Department of Physics, Centre for Advanced Electron Spin Resonance, The Clarendon Laboratory, University of Oxford, Parks Road, Oxford OX1 3PU, UK. Correspondence and requests for materials should be addressed to R.W. (email: richard.winpenny@manchester.ac.uk).

A marked trend during the last decades has been the extreme miniaturization of components in information technologies aiming to produce devices with higher speed and storage capacity. As we reach features of ~ 10 nm classical physical laws cease to be dominant, and quantum devices will result as we enter the nanometre scale, with quantum mechanics governing the behaviour of the systems. Owing to the intrinsic nature of quantum systems, quantum information processing offers the possibility of performing some computational tasks far more quickly than is possible using conventional computers. These tasks include searching of unsorted directories, using the Grover algorithm[1], and factoring large numbers in primes, using the Shor algorithm[2]. Furthermore, Lloyd proved the correctness of the conjecture made by Feynman, stating that quantum systems are most successfully simulated using other quantum systems[3,4]. Therefore, the motivation to produce quantum computers is considerable, and has attracted a great deal of interest from scientist working in materials science, chemistry, physics and nano-fabrication technologies. For example, the company D-Wave has reported a quantum annealer[5] that performs certain calculations sufficiently rapid that a consortium involving Google and NASA has invested staggering sums of money in one such device. Many implementation strategies have been proposed for creating quantum computers including superconducting qubits[6], quantum dots[7], photons and trapped atoms[8]. Particularly relevant here are experiments involving nuclear spins, for example recent astonishing work on terbium phthalocyanine complexes[9], where it is possible to drive quantum oscillations within single nuclei using microwaves.

A major challenge in developing devices for quantum computation is bringing together sufficient interacting units (qubits) to carry out useful algorithms. This is the major difficulty for quantum systems such as ^{13}C sites near nitrogen vacancies in diamond; the individual sites have excellent performance[10], but cannot be linked controllably into useful arrays. In an alternative approach, molecular nanomagnets have been proposed as qubits. The first molecular magnet to be proposed for quantum computing was the famous polymetallic cage {Mn$_{12}$} (ref. 11). Since then other molecular systems have been proposed, chiefly the idea of bringing together two two-level systems to make universal quantum gates[12–20]. Individually, they do not have the very long phase memory times associated with defect sites, but in supramolecular chemistry[21] a methodology exists to link together molecular qubits into entangled arrays with great control. Supramolecular chemistry brings together molecules with interactions that are weaker than covalent bonds, such that the individual units retain their individual character. In many ways, this is reminiscent of the idea for obtaining an entanglement equilibrium state with quantum mechanics; where individual components retain their identities, but an action at one unit influences another.

We have proposed using heterometallic rings as a qubit[22], specifically looking at the {Cr$_7$Ni} rings that have an $S = \frac{1}{2}$ ground state as two-level systems that meet criteria required for qubits, that is, it is possible to initialize them and the energy scales are correct[22]. They also have sufficient phase memory times T_m to allow gate operation before state degradation can occur[23], and we can control the interaction between rings to create entangled states[24]. We have also shown these rings have significant T_m values when doped into diamagnetic lattices[25], which in the longer term brings the possibility of addressing individual transitions in orientated single crystals. Other groups have reported still longer T_m values for simpler paramagnets[26,27]. The chemical advantage of using the heterometallic ring is that there are many methods available, due to their vast chemical versatility, to link them together or to other chemical entities[28–31]. Here we propose a fully modular design strategy to create large arrays of molecular qubits and show the first steps in delivering supramolecular arrays of qubits.

Results

Design and synthesis of hybrid rotaxanes. Conceptually what we intend involves growing a large array through combining two steps, building from our reports of hybrid organic–inorganic rotaxanes[32,33]. Hybrid rotaxanes contain an organic thread to bring together two or more heterometallic rings. In a first step, by means of basic concepts of organic and supramolecular chemistry, we build rotaxanes of increasing complexity, where we functionalise one end of a thread with a pyridine group. In a second step, we make use of single metal sites and coordination complexes with open coordination positions, to act as a central core and bring together very large number of qubits within a supramolecular assembly (Fig. 1).

To show the general validity and versatility of our design strategy, firstly we synthesized the [2]-rotaxane [AH$_2$\{Cr$_7$Ni(μ-F)$_8$(O$_2$C^{t}Bu)$_{16}$\}$_2$] (1), that contains a pyridine group at the edge of the thread, from the reaction of 2-phenyl-N-\{[4′-(pyridin-4-yl)-(1,1′-biphenyl)-4-yl]methyl\}ethanamine A (ref. 34), chromium fluoride (CrF$_3 \cdot$4H$_2$O) and a nickel pivalate salt [Ni$_2$(μ-OH$_2$)(O$_2$C^{t}Bu)$_4$(HO$_2$C^{t}Bu)$_4$] (ref. 35) (1:6:2 molar ratio) using pivalic acid as solvent with a moderate yield (10%). Subsequently reaction of the [2]-rotaxane 1 with hydrated [Cu(hfacac)$_2$]$\cdot$$nH_2$O (2:1 molar ratio) yields the [3]-rotaxane [cis-\{Cu(hfacac)$_2$\}AH$_2$\{Cr$_7$Ni(μ-F)$_8$(O$_2$C^{t}Bu)$_{16}$\}$_2$] (2) (Fig. 2a), with three potential qubits (the single Cu^{II} ion is also a two-level system), which contains two heterometallic rings above the central copper, with a cis geometry (see Supplementary Methods for synthetic detail).

To increase the number of qubits in an assembly further we can bind the [2]-rotaxane to a coordination complex that possesses more than two open coordination positions. To this end, we prepared thread B, 4-phenyl-N-(4-(pyridin-4-yl)benzyl)butan-1-amine[36], and make the [2]-rotaxane [BH$_2$\{Cr$_7$Ni(μ-F)$_8$(O$_2$C^{t}Bu)$_{16}$\}] (3). Reaction of 3 with [Fe$_2$Co(μ_3-O)(O$_2$C^{t}Bu)$_6$(H$_2$O)$_3$] (ref. 37) (3:1 molar ratio) gives a triangular [4]-rotaxane [Fe$_2$Co(μ_3-O)(O$_2$C^{t}Bu)$_6$\{BH$_2$\{Cr$_7$Ni(μ-F)$_8$(O$_2$C^{t}Bu)$_{16}$\}\}$_3$] (4), with three qubits, where the terminal water molecules of the metal carboxylate triangle have been replaced by pyridine-nitrogen atoms of three [2]-rotaxane (Fig. 2b).

Another route to increase the number of qubits in the supramolecular array is to increase the number of heterometallic rings in each individual rotaxane. With this in mind, we synthesized threads C, N^1-(4-(methylthio)benzyl)-N^{12}-(4-(pyridin-4-yl)benzyl)dodecane-1,12-diamine and D, N^1-(4-(methylthio)benzyl)-N^{12}-(pyridin-4-ylmethyl)dodecane-1,12-diamine[38], in good yield (Supplementary Methods). Reaction of C with CrF$_3 \cdot$4H$_2$O and [Ni$_2$(μ-OH$_2$)(O$_2$C^{t}Bu)$_4$(HO$_2$C^{t}Bu)$_4$] (1:14:4.5 molar ratio) with pivalic acid as solvent, leads to a [3]-rotaxane with formula [CH$_2$\{Cr$_7$Ni(μ-F)$_8$(O$_2$C^{t}Bu)$_{16}$\}$_2$] (5) in 23% yield, while reaction with D gives [3]-rotaxane [DH$_2$\{Cr$_7$Ni(μ-F)$_8$(O$_2$C^{t}Bu)$_{16}$\}$_2$] (6) in 60% yield (6). These yields are good given the complex assembly process involved. The reactions in each case protonate both secondary amines and a {Cr$_7$Ni} ring grows around each, leaving the pyridine group available for further coordination. Addition of 6 or 5 with [Cu$_2$(O$_2$C^{t}Bu)$_4$(HO$_2$C^{t}Bu)$_2$] and Cu(NO$_3$)$_2 \cdot$6H$_2$O, respectively (2:1 molar ratio) in hot toluene, allow us to obtain [\{Cu(O$_2$C^{t}Bu)$_2$\}DH$_2$\{Cr$_7$Ni(μ-F)$_8$(O$_2$C^{t}Bu)$_{16}$\}$_2$]$_2$ (7) and [\{Cu(NO$_3$)$_2$\}CH$_2$\{Cr$_7$Ni(μ-F)$_8$(O$_2$C^{t}Bu)$_{16}$\}$_2$]$_2$ (8) (Fig. 2c,d). Compounds 7 and 8 are [5]-rotaxanes that are almost 7 nm long; they are formed in an overall yield of 70 (7) and 75% (8) based on thread C and D,

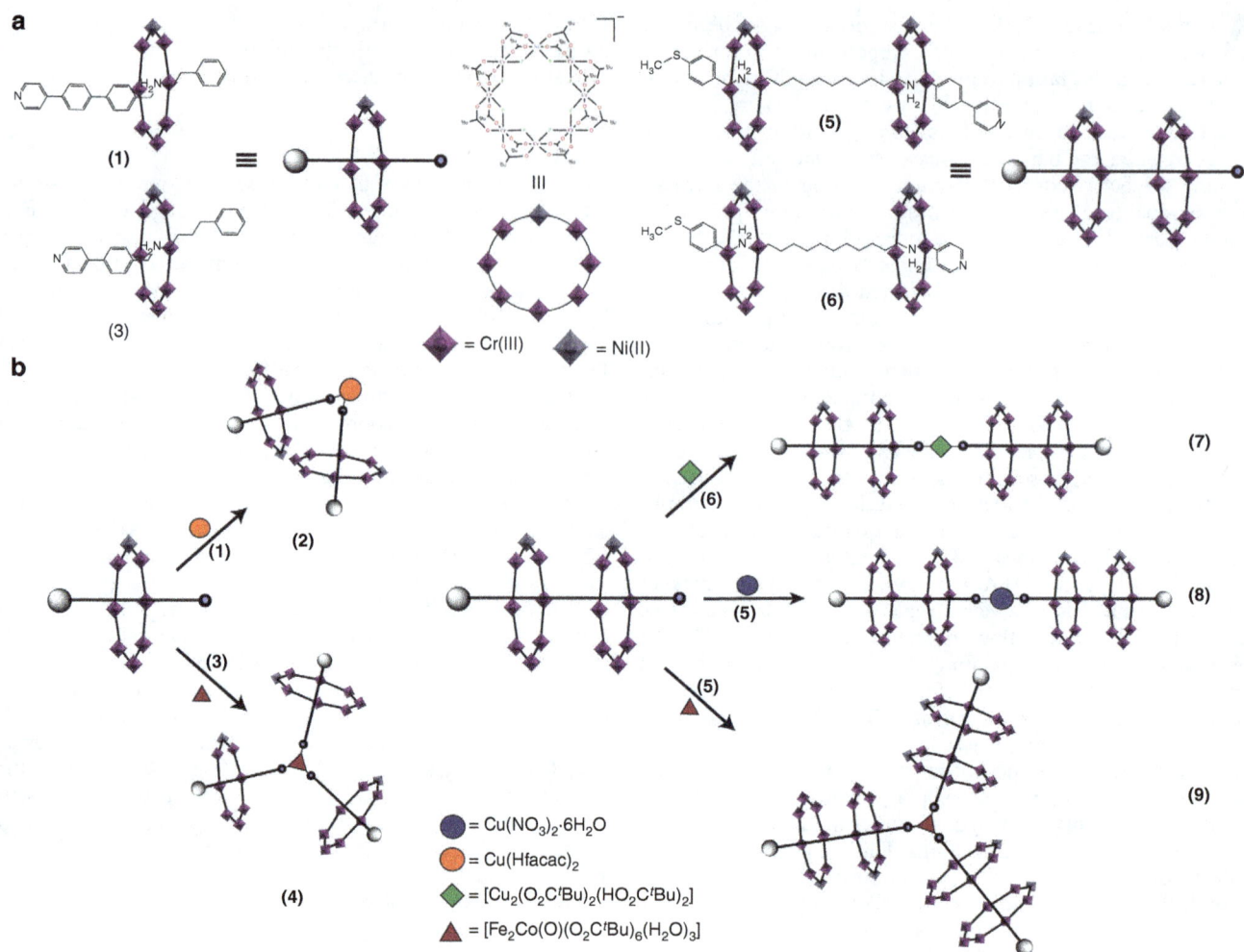

Figure 1 | Supramolecular strategy for constructing large arrays of qubits using rotaxanes as ligands. (a) Scheme for the components of the larger arrays. **(b)** Reactions to give larger [n]-rotaxanes. The numbers in parentheses refer to compounds discussed in the text.

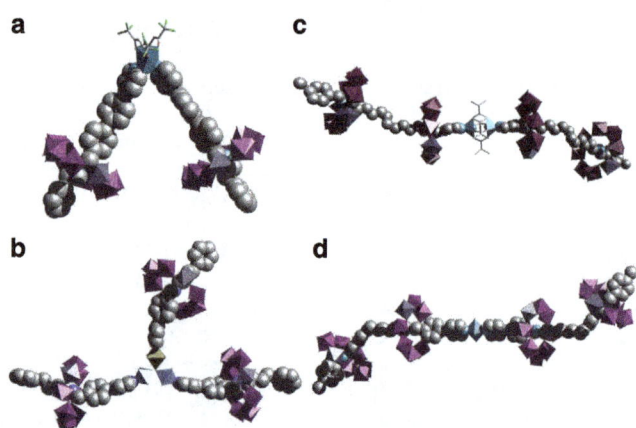

Figure 2 | Single crystal structures of polyrotaxanes. (a) [3]-rotaxane (**2**); **(b)** [4]-rotaxane (**4**); **(c)** [5]-rotaxane (**7**) and **(d)** [5]-rotaxane (**8**). Hydrogen atoms and pivalate groups omitted for clarity. Metal sites shown as polyhedra. Colour code: Cu, light blue; Cr, purple; Ni, teal; N, blue; O, red; C, grey; F, green; H omitted for clarity. Crystal and refinement parameters given in Supplementary Table 1.

respectively. These [5]-rotaxanes contain four {Cr$_7$Ni} rings; in **8** the presence of a paramagnetic Cu^{II} ion within the adds a further qubit, producing a [5]-rotaxane containing five qubits.

The final step of the strategy is to combine these two routes to increase the number of heterometallic rings per assembly. In this sense, **5** was reacted with the central core [Fe$_2$Co(μ_3-O)(O$_2$C^{t}Bu)$_6$(H$_2$O)$_3$] (3:1 molar ratio) in hot acetone, with the aim to get a [7]-rotaxane of formula [{Fe$_2$Co(μ_3-O)(O$_2$C^{t}Bu)$_6$}{CH$_2${Cr$_7$Ni(μ-F)(O$_2$C^{t}Bu)$_{16}$}$_2$}$_3$] (**9**). Unfortunately, while compounds **2**, **4**, **7** and **8** yield X-ray quality crystals, we were unable to get suitable crystals of **9** for single-crystal X-ray diffraction experiment.

Characterization of a [7]-rotaxane. In the absence of a crystal structure, and because of the large number of paramagnetic centres present in **9** (that makes NMR spectroscopy less useful than for larger organic molecules), we used small angle X-ray scattering (SAXS) to demonstrate that the [7]-rotaxane has been made. SAXS has been used to characterize porphyrin arrays[39–41], which present the same problems as **9**. Owing to poor sample solubility, saturated solutions were employed in the SAXS collections. To ensure no aggregation had occurred, diluted collections were also performed and compared (see Supplementary Figures 11–17). We began by examining compound **4**, where we have a full crystal structure determination and which can be considered a small version of **9**. Here SAXS data shows two broad maxima at 12 and 22 Å, which match perfectly with the distance obtained by X-ray crystallography in an individual [2]-rotaxane

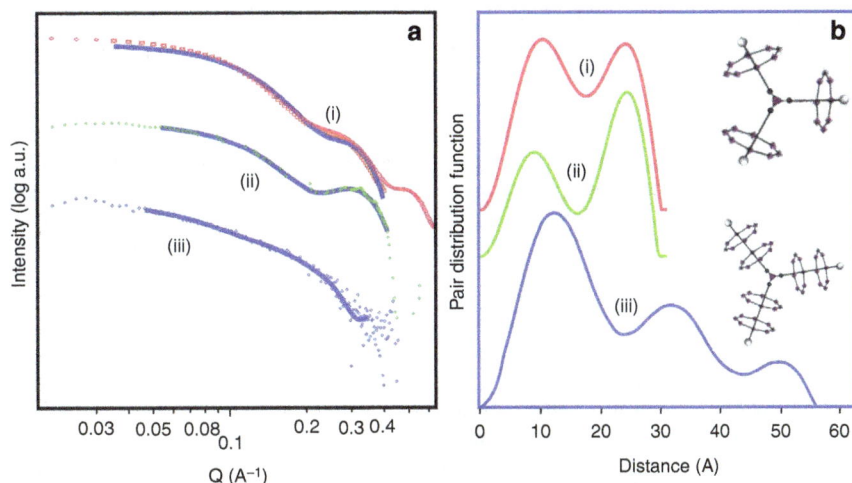

Figure 3 | SAXS on [4]- and [7]-rotaxanes. (a) SAXS data; (i) calculated diffraction from crystallographically determined structure of **4**, (ii) experimental data from **4**, (iii) experimental data from **9**. The solid line in each spectra is the fit that is associated with the pair distribution function plots shown in **b**. **(b)** Pair distance distribution function; (i) fit to calculated SAXS data for **4**, (ii) fit to experimental data collected for **4** and (iii) fit to experimental data collected for **9**. Inset: schematic representation of **4** (top) and **9** (bottom). The pair distance distribution function curves presented were also correlated with Guinier plots and gyration radius calculations to ensure consistency within the data analysis.

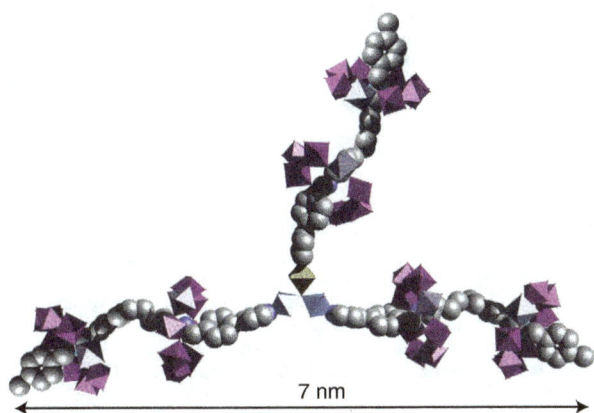

Figure 4 | Proposed structure for [7]-rotaxane 9. The proposed formula for this compound is $[Fe_2Co(\mu_3\text{-}O)(O_2C^tBu)_6\{CH_2\{Cr_7Ni(\mu\text{-}F)(O_2C^tBu)_{16}\}_2\}_3]$; colour code as Fig. 2.

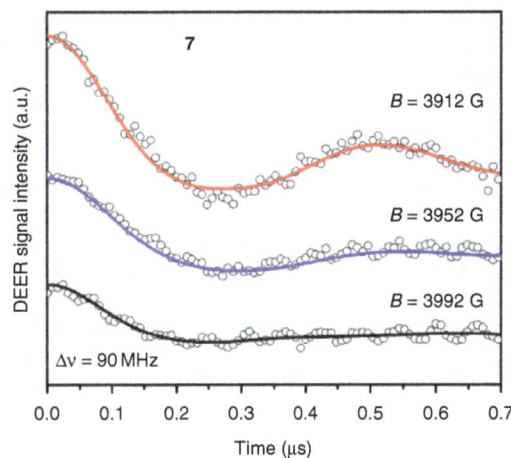

Figure 5 | DEER data for compound 7. Samples were measured in toluene at 2.6 K for three values of the magnetic field within the resonance line (3,912, 3,952 and 3,992 G). The probe and pump frequencies were $v_1 = 9.7648$ GHz and $v_2 = v_1 - 90$ MHz, respectively. See Supplementary Methods for more experimental details.

from the central core to the centre of the heterometallic ring and the size of the assembly, respectively (Fig. 3).

Then we move to compound **9**, where we have no structure. The first remarkable feature that we noticed was a significant change in the X-ray scattering compare with **4**, as we expected owing to the larger number of heterometallic rings in this supramolecular array. Interestingly, the analysis of the diffraction data with pair distance distribution function shows pairs of particles to much larger distances, with the maximum extent being ca. 55 Å and three maxima at 12, 32 and 50 Å. It can be speculated that the 12 Å distance is related to contacts within a single ring and is consistent with the data from the [4]-rotaxane. The 32 and 50 Å could then be related to distances between one inner ring with an outer ring on a different arm and an outer ring to outer ring distance. The model is therefore consistent with six {Cr7Ni} rings about a central {Fe2Co} triangle, and matches the proposed structure for **9** (Fig. 4).

We attempted to derive further evidence for the formation of **9** from electrospray mass spectrometry, however this [7]-rotaxane did not fly as an intact species under mass spectrometry conditions.

However, we prepared an analogous version of **9** with Zn(II) $[\{Fe_2Co(\mu_3\text{-}O)(O_2C^tBu)_6\}\{CH_2\{Cr_7Zn(\mu\text{-}F)(O_2C^tBu)_{16}\}_2\}_3]$ (**10**) and this shows a peak for **10** for a trication at $m/z = 5,188$; matching well with the calculated mass of 15,424 (see Supplementary Methods and Supplementary Figures 6–10 for further details).

Hybrid rotaxanes in quantum logic gates. A key question remains whether it is possible to produce and control interactions between qubits that would allow operation of quantum logic gates. The heterometallic ring monomers discussed here can exhibit phase memory times (T_m) exceeding 10 μs (ref. 23), although more typically they are a little below 1 μs (refs 25,42). A single-qubit manipulation using commercial electron paramagnetic resonance (EPR) apparatus takes of the order of 10 ns. Any two-qubit interaction capable of, for example, generating controlled entanglement, should do so on a timescale intermediate between the single-qubit manipulation time and the phase memory time, that is, a gate time of 100–300 ns.

A major concern has been that in complex molecules featuring multiple weakly interacting spins, this will introduce a source of decoherence, and therefore significantly decrease the value of T_m. However, a recent study of a molecule containing twenty-four qubits suggests this is not a significant problem[31]. To confirm this, we have studied T_m for compounds 2, 4, 5 and 7 using pulsed EPR spectroscopy (see Supplementary Methods for experimental details); in 2, 5 and 7 we can fit the measurements to a stretched exponential with $T_m = 795$ ns and a stretch parameter of 1.7. These values are similar to many other heterometallic rings containing pivalate ligands[23,25]. In compound 4 $T_m = 700$ ns with a similar stretch parameter. These results confirm that bringing together multi-qubit assemblies will not have a deleterious effect on the phase memory.

To examine the interaction between the qubits we have used double electron–electron resonance (DEER) spectroscopy[43], which is established in structural biology as a way of measuring distances through the strength of the dipolar interaction between spins. Here we are interested in the two-qubit gate time, which comes directly out of the DEER experiment (see SI for experimental details, Supplementary Figures 1–5 and Supplementary Table 2). The experiment involves flipping the spin on one centre and examining the resulting precession of a second spin under the change in its effective local magnetic field. The time taken for half of a complete precession corresponds directly to the duration of a two-qubit conditional phase gate (one of the fundamental entangling gates)[44].

Figure 5 shows typical DEER data from compound 7, for several different magnetic fields within the EPR absorption spectrum. Compound 7 was studied because the central ring…ring interaction involves a rigid thread with the two rings co-parallel. This makes interpretation of the DEER data more straightforward than in cases where the thread is flexible or where the rings are not co-parallel. A large and oscillatory modulation of the DEER signal as a function of the time position of the pump pulse is visible for a field of 3,912 G; this oscillation represents the precession of the probe spin in the effective magnetic field of the pump spin, yielding a two-qubit gate time of ~260 ns consistent with a dipolar interaction between the qubits. Weaker modulations occur for the higher magnetic fields of 3,952 G and 3,992 G. The field dependence arises because different orientational subpopulations are excited at each magnetic field, owing to the anisotropy of the monomers' g factors[44].

This two-qubit gate time falls between the operation time for a single qubit and the phase memory time T_m, matching observations we have made in simpler [3]-rotaxanes[44]. This establishes the presence of two-qubit interactions that are of the right strength for implementing quantum logic gates within assemblies of heterometallic rings. Studies of the other rotaxanes reported here are progressing.

Discussion

In summary, we have shown the general validity of our modular design strategy to overcome one of the major challenges that face quantum computation to bring together many potential qubits. Using simple concepts from coordination and supramolecular chemistry we have shown a gradual increase of the complexity in the arrays synthesized. These results are the first steps of this new strategy to increase the number of qubits in a supramolecular array.

The longer perspective is how such supramolecular compounds could be used as multi-qubit arrays. First, an array of qubits could be prepared in an initial state by cooling the sample in the presence of a moderate magnetic field. This is as the relevant energy scale for preparation of an initial state is the Zeeman energy of the individual electron spins. The inter-qubit interaction energy scale is much smaller, therefore if we can prepare one qubit by cooling in field, we can prepare a multi-qubit array in the same way. This differs from nuclear spins, where the Zeeman energy is much smaller. The major remaining challenge is then to achieve local control to address individual pairs of electron spins to perform selected qubit gates. The DEER results reported above and elsewhere[44] suggest this should be possible. The EPR spectra of these compounds are strongly anisotropic and hence it is only at specific orientations where the spin on one qubit interacts with the spin on a second. Studies of orientated single crystals of multi-qubit arrays are planned for the future where we will attempt to demonstrate how such selectivity can be used to perform simple two-qubit gates.

Methods

Synthesis. Unless stated otherwise, all reagents and solvents were used without further purification. The syntheses of the hybrid organic–inorganic rotaxanes were carried out in Erlenmeyer Teflon FEP flasks supplied by Fisher. Column chromatography was carried out using Silica 60A (particle size 35–70 μm, Fisher, UK) as the stationary phase, and TLC was performed on precoated silica gel plates (0.25 mm thick, 60 F254, Merck, Germany) and observed under ultraviolet light. NMR spectra were recorded on Bruker AV 400 and Bruker DMX 500 instruments. Chemical shifts are reported in p.p.m. from low to high frequency and referenced to the residual solvent resonance. electrospray ionization mass spectrometry, matrix-assisted laser desorption ionization-time of flight spectrometry and microanalysis were carried out by the services at the University of Manchester. Details of the synthetic procedures are given in the Supplementary Methods.

Details of the physical characterization and DEER studies are also given in the Supplementary Methods.

References

1. Grover, L. K. Quantum computers can search arbitrarily large databases by a single query. *Phys. Rev. Lett.* **79**, 4709–4712 (1997).
2. Shor, P. W. Polynomial-time algorithms for prime factorization and discrete logarithms on a quantum computer. *SIAM J. Comput.* **26**, 1484–1509 (1997).
3. Lloyd, S. Universal quantum simulators. *Science* **273**, 1073–1078 (1996).
4. Feynman, R. P. Simulating physics with computers. *Int. J. Theor. Phys.* **21**, 467–488 (1982).
5. Johnson, M. W. *et al.* Quantum annealing with manufactured spins. *Nature* **473**, 194–198 (2011).
6. DiCarlo, L. *et al.* Preparation and measurement of three-qubit entanglement in a superconducting qubit. *Nature* **467**, 547–578 (2010).
7. Elzerman, J. M. *et al.* Single-shot read-out of an individual electron spin in a quantum dot. *Nature* **430**, 431–435 (2004).
8. Ladd, T. D. *et al.* Quantum computers. *Nature* **464**, 45–53 (2010).
9. Thiele, S. *et al.* Electrically driven nuclear spin resonance in single-molecule magnets. *Science* **344**, 1135–1138 (2014).
10. Maurer, P. C. *et al.* Room-temperature quantum bit memory exceeding one second. *Science* **336**, 1283–1286 (2012).
11. Leuenberger, M. & Loss, D. Quantum computing in molecular magnets. *Nature* **410**, 789–793 (2001).
12. Meier, F., Levy, J. & Loss, D. Quantum computing with spin cluster qubits. *Phys. Rev. Lett.* **90**, 47901–47904 (2003).
13. Lehmann, J., Gaita-Ariño, A., Coronado, E. & Loss, D. Spin qubits with electrically gated polyoxometalate molecules. *Nat. Nanotechnol.* **2**, 312–317 (2007).
14. Sañudo, E. C. *et al.* Molecules composed of two weakly magnetically coupled [Mn$_4^{III}$] clusters. *Inorg. Chem.* **46**, 9045–9047 (2007).
15. Santini, P., Carretta, S., Troiani, F. & Amoretti, G. Molecular nanomagnets as quantum simulators. *Phys. Rev. Lett.* **107**, 230502 (2011).
16. Troiani, F. & Affronte, M. Molecular spins for quantum information technologies. *Chem. Soc. Rev.* **40**, 3119–3129 (2011).
17. Nakazawa, S. *et al.* Synthetic two-spin quantum bit: g-engineered exchange-coupled biradical designed for controlled-NOT gate operations. *Angew. Chem. Int. Ed.* **51**, 9860–9864 (2012).
18. Aromí, G., Aguilà, D., Gamez, P., Luis, F. & Roubeau, O. Design of magnetic coordination complexes for quantum computing. *Chem. Soc. Rev.* **41**, 537–546 (2012).
19. Plant, S. R. *et al.* A two-step approach to the synthesis of N@C$_{60}$ fullerene dimers for molecular qubits. *Chem. Sci.* **4**, 2971–2975 (2013).

20. Aguilà, D. *et al.* Heterodimetallic [LnLn'] lanthanide complexes: toward a chemical design of two-qubit molecular spin quantum gates. *J. Am. Chem. Soc.* **136**, 14215–14222 (2014).

21. Lehn, J.-M. *Supramolecular Chemistry: Concepts and Perspectives* (Wiley-VCH, 1995).

22. Troiani, F. *et al.* Molecular engineering of antiferromagnetic rings for quantum computation. *Phys. Rev. Lett.* **94**, 207208 (2005).

23. Wedge, C. J. *et al.* Chemical engineering of molecular qubits. *Phys. Rev. Lett.* **108**, 107204 (2012).

24. Candini, A. *et al.* Entanglement in supramolecular spin systems. *Phys. Rev. Lett.* **104**, 037203 (2010).

25. Moro, F. *et al.* Coherent electron spin manipulation in a dilute oriented ensemble of molecular nanomagnets: pulsed EPR on doped single crystals. *Chem. Commun.* **50**, 91–93 (2014).

26. Warner, W. *et al.* Potential for spin-based information processing in a thin-film molecular semiconductor. *Nature* **503**, 504–508 (2013).

27. Bader, K. *et al.* Room temperature quantum coherence in a potential molecular qubit. *Nat. Commun.* **5**, 5304 (2014).

28. Timco, G. A. *et al.* Engineering the coupling between molecular spin qubits by coordination chemistry. *Nat. Nanotechnol.* **4**, 173–178 (2008).

29. Whitehead, G. F. S. *et al.* A ring of rings and other multicomponent assemblies of cages. *Angew. Chem. Int. Ed.* **52**, 9932–9935 (2013).

30. Whitehead, G. F. S. *et al.* The acid test: the chemistry of carboxylic acid functionalised {Cr$_7$Ni} rings. *Chem. Sci.* **5**, 235–239 (2014).

31. Ferrando-Soria, J. *et al.* Controlled synthesis of nanoscopic metal cages. *J. Am. Chem. Soc.* **137**, 7644–7647 (2015).

32. Lee, C.-F. *et al.* Hybrid organic-inorganic rotaxanes and molecular shuttles. *Nature* **458**, 314–318 (2009).

33. Ballesteros, B. *et al.* Synthesis, structure and dynamic properties of hybrid organic-inorganic rotaxanes. *J. Am. Chem. Soc.* **132**, 15435–15444 (2010).

34. Cunningham, P. D. & Hayden, L. M. Carrier dynamics resulting from above and below gap excitation of P3HT and P3HT/PCBM investigated by optical-pump terahertz-probe spectroscopy. *J. Phys. Chem. C* **112**, 7928–7935 (2008).

35. Chaboussant, G. *et al.* Nickel pivalate complexes: structural variations and magnetic susceptibility and ineslastic neutron scattering studies. *Dalton Trans.* 2758–2766 (2004).

36. You, Y. *et al.* Micromolding of a highly fluorescent reticular coordination polymer: solvent-mediated reconfigurable polymerization in a soft lithographic mold. *Angew. Chem. Int. Ed.* **49**, 3757–3761 (2010).

37. Abdulwahab, K. O. *et al.* A one pot synthesis of monodispersed cobalt and manganese ferrite nanoparticles from bimetallic pivalate clusters. *Chem. Mater.* **26**, 999–1013 (2014).

38. Rath, H. *et al.* Studies of hybrid organic–inorganic [2]- and [3]-rotaxanes bound to Au surfaces. *Chem. Commun.* **49**, 3404–3406 (2013).

39. Sprafke, J. K. *et al.* Belt-shaped π-systems: relating geometry to electronic structure in a six-porphyrin nanoring. *J. Am. Chem. Soc.* **133**, 17262–17273 (2011).

40. Kelley, R. F. *et al.* Intramolecular energy transfer within butadiyne-linked chlorophyll and porphyrin dimer-faced, self-assembled prisms. *J. Am. Chem. Soc.* **130**, 4277–4284 (2008).

41. Tiede, D. M., Zhang, R., Chen, L. X., Yu, L. & Lindsey, J. S. Structural characterization of modular supramolecular architectures in solution. *J. Am. Chem. Soc.* **126**, 14054–14062 (2004).

42. Kaminski, D. *et al.* Quantum spin coherence in halogen-modified Cr$_7$Ni molecular nanomagnets. *Phys. Rev. B* **90**, 184419 (2014).

43. Schiemann, O. & Prisner, T. F. Long-range distance determinations in biomacromolecules by EPR spectroscopy. *Q. Rev. Biophys.* **40**, 1–53 (2007).

44. Ardavan, A. *et al.* Precise control of two-qubit gate times for supramolecular molecular spin qubits. *npj Quantum Inf* **1**, 15012 (2015).

Acknowledgements

This work was supported by the EPSRC(UK), the European Commission (Marie Curie Intra-European Fellowship to A.F. (300402) and J.F-.S. (622659)). E.M.P. thanks the Panamanian agency SENACYT-IFARHU for funding. R.E.P.W. thanks the Royal Society for a Wolfson Merit Award. We also thank EPSRC (UK) for funding an X-ray diffractometer (grant no. EP/K039547/1) and for access to pulsed EPR spectrometers through the National EPR Facility. We thank Diamond Light Source for access to synchrotron X-ray facilities, and specifically Drs Nuntaporn Kamonsutthipaijit, Dmitry Kondratiuk and Sophie Rousseaux for assistance with SAXS experiments and Drs Dave Allan and Harriet Nowell for help with single-crystal measurements.

Author contributions

A.F., G.A.T. and R.E.P.W. designed the research. A.F. synthesized the organic threads and rotaxanes; G.A.T. helped with crystallisations. Single-crystal X-ray data were collected and refined by E.M.P., I.J.V-.Y. and C.A.M. SAXS data were collected in Manchester by C.A.M. and at Diamond Light Source by C.K. Cryo-electrospray mass spectrometry was carried out by J.U. and P.E.B., F.T. carried out the EPR studies; use of DEER was suggested by A.A., J.F-.S. and R.E.P.W. wrote the paper with input from all other authors.

Additional information

Accession codes: The X-ray crystallographic coordinates for structures reported in this study have been deposited at the Cambridge Crystallographic Data Centre, under deposition numbers 1037544–1037547. These data can be obtained free of charge from The Cambridge Crystallographic Data Centre via www.ccdc.cam.ac.uk/data_request/cif.

21

Metal–organic framework with optimally selective xenon adsorption and separation

Debasis Banerjee[1], Cory M. Simon[2], Anna M. Plonka[3], Radha K. Motkuri[4], Jian Liu[4], Xianyin Chen[5], Berend Smit[2,6], John B. Parise[3,5,7], Maciej Haranczyk[8,9] & Praveen K. Thallapally[1]

Nuclear energy is among the most viable alternatives to our current fossil fuel-based energy economy. The mass deployment of nuclear energy as a low-emissions source requires the reprocessing of used nuclear fuel to recover fissile materials and mitigate radioactive waste. A major concern with reprocessing used nuclear fuel is the release of volatile radionuclides such as xenon and krypton that evolve into reprocessing facility off-gas in parts per million concentrations. The existing technology to remove these radioactive noble gases is a costly cryogenic distillation; alternatively, porous materials such as metal–organic frameworks have demonstrated the ability to selectively adsorb xenon and krypton at ambient conditions. Here we carry out a high-throughput computational screening of large databases of metal–organic frameworks and identify SBMOF-1 as the most selective for xenon. We affirm this prediction and report that SBMOF-1 exhibits by far the highest reported xenon adsorption capacity and a remarkable Xe/Kr selectivity under conditions pertinent to nuclear fuel reprocessing.

[1] Physical and Computational Science Directorate, Pacific Northwest National Laboratory, Richland, Washington 99352, USA. [2] Department of Chemical and Biochemical Engineering, University of California, Berkley, Berkeley, California 94720, USA. [3] Department of Geosciences, Stony Brook University, Stony Brook, New York 11794, USA. [4] Energy and Environmental Directorate, Pacific Northwest National Laboratory, Richland, Washington 99352, USA. [5] Department of Chemistry, Stony Brook University, Stony Brook, New York 11794, USA. [6] Institut des Sciences et Ingénierie Chimiques, Valais, Ecole Polytechnique Fédérale de Lausanne (EPFL), Rue de l'Industrie 17, CH-1951 Sion, Switzerland. [7] Photon Sciences, Brookhaven National Laboratory, Upton, New York 11973, USA. [8] Computational Research Division, Lawrence Berkeley National Laboratory, Berkeley, California 94720, USA. [9] IMDEA Materials Institute, C/Eric Kandel 2, 28906 Getafe, Madrid, Spain. Correspondence and requests for materials should be addressed to B.S. (email: Berend.Smit@berkeley.edu) or to M.H. (email: mharanczyk@lbl.gov) or to P.K.T. (email: Praveen.thallapally@pnnl.gov).

One of the grandest challenges of our generation is to meet our rapidly growing energy demand without further increasing the emission of greenhouse gases[1,2]. Nuclear energy is one of the cheapest alternatives to carbon-based fossil fuels that, because of its high energy density and minimal land use requirements, can be scaled up to meet global energy demands. Life cycle analyses indicate that greenhouse gas emissions of a nuclear power plant are significantly lower than fossil fuel technologies and comparable to other renewable electricity generation technologies, such as solar photovoltaics[3]. For the mass implementation of nuclear energy as a low-emissions energy source, we must also safely sequester the associated high-level radioactive waste[2]. In this, most attention is given to recovering the heavy, long-lived nuclear elements in used nuclear fuel (UNF), such as uranium and plutonium. Less discussed are the volatile radionuclides (for example, Xe, Kr) that evolve into the off-gas of UNF aqueous reprocessing facilities[4]. In these off-gases, gaseous radioactive ^{85}Kr has a long half-life ($t_{1/2} = 10.8$ years) and therefore must be captured and removed from the off-gas to prevent its uncontrolled release into the atmosphere[4]. In contrast, the radioactive Xe isotopes ($t_{1/2} \approx 36.3$ days for ^{127}Xe) have decayed by the time the fuel is reprocessed. As high purity Xe is used in many applications, including commercial lighting, propulsion, imaging, anesthesia and insulation, the recovered Xe could be sold into the chemical market to offset operating costs. At present, cryogenic distillation is the most mature technology to separate Xe and Kr from air, but it is energy- and capital-intensive and therefore expensive[5,6]. Furthermore, the radiolytic formation of ozone poses an explosion hazard during cryogenic distillation[4]. These factors incentivize the development of an alternative technology for a less energy-intensive, more cost-effective and safer process to capture Kr and Xe from UNF reprocessing facility off-gas.

A promising alternative technology for Xe/Kr removal from reprocessing off-gas is an adsorption-based process at room temperature using a selective, solid-state adsorbent. These solid-state adsorbents are found to be almost exclusively Xe-selective, and thus a dual step process whereby, first, the Xe is selectively removed from the off-gas, is a necessary requirement for a practical application[7]. In the subsequent step, the radioactive Kr can be removed from the Xe-free effluent using the same material or a different material. Adsorbents such as silver-loaded zeolites and activated carbon have been proposed[4], but these fall short compared with high surface area, crystalline metal–organic frameworks (MOFs) and porous organic cage compounds[8–22]. Among the many novel materials tested thus far, HKUST-1 (ref. 14), Co-formate[15,20] and CC3 (ref. 12) are shown to be promising for Xe/Kr separations, showing high capacity and good selectivity for Xe over Kr.

An important advantage of MOFs is their chemical tunability; by combining different linkers and metal centres that self-assemble to form ordered, pre-determined crystal structures, one can synthesize millions of possible materials[23]. MOFs can thus be tailor-made to be optimal for applications related to gas storage and separation, catalysis, chemical sensing and optics[9,11,23–34]. Our goal here is to identify an optimal MOF for selectively capturing Xe from the off-gas of UNF reprocessing facilities. In practice, however, constraints in resources allow us to synthesize and test only a small subset of chemical space. Molecular models and simulations of adsorption can rapidly and cost-effectively rank MOFs by their Xe/Kr selectivity with reasonable accuracy (Supplementary Methods). High-throughput computational screenings thus play a valuable role of elucidating design rules, determining performance limits, and predicting performance rankings of materials to focus experimental efforts on the most promising MOFs for Xe/Kr separations[35–38].

In this work, we use molecular simulations to screen over 125,000 MOF structures[39,40] for selectively adsorbing Xe over Kr at dilute conditions pertinent to UNF reprocessing. Our computational screening predicts that one of the most Xe-selective MOFs is a calcium-based nanoporous MOF, SBMOF-1 [also known as CaSDB, SDB = 4,4 -sulfonyldibenzoate], that has not yet been tested for Xe/Kr separations[41]. We affirm this prediction by synthesizing SBMOF-1 and measuring its pure-component Xe and Kr adsorption isotherms. SBMOF-1 exhibits the highest Xe Henry coefficient and thermodynamic Xe/Kr selectivity at dilute conditions among MOFs tested to date. In addition to its high thermal and chemical stability, column breakthrough experiments reveal that SBMOF-1 is a practical, near-term material for capturing Xe from reprocessing facilities.

Results

High-throughput computational screening.

For capturing Xe from nuclear reprocessing, the Xe/Kr selectivity is the most important thermodynamic property determining the performance of a MOF. We used molecular simulations to predict the Xe/Kr selectivity of 125,000 MOF structures at dilute conditions relevant to UNF reprocessing (Supplementary Methods, Fig. 1a,b). The distribution of simulated selectivities in the MOFs is shown in Fig. 1a. We partitioned this distribution into a database of existing MOFs ($\sim$5,000 structures)[39] and a database of predicted/hypothetical structures ($\sim$120,000)[40]. These distributions span a large range of selectivities, illustrating the unique tunability of MOF materials. Our simulations predict that the most selective material in the database of existing MOFs is SBMOF-1 (Fig. 1c), a three-dimensional, permanently porous MOF (Cambridge Structural Database (CSD) code: KAXQIL)[41]. Furthermore, the Xe/Kr selectivity of SBMOF-1 is ranked in the top 0.01 percentile in the database of 120,000 hypothetical MOFs (Fig. 1b). The red line in Fig. 1b illustrates the outlying Xe/Kr selectivity of SBMOF-1 predicted by our screening. While SBMOF-1 has been synthesized and considered for CO_2/N_2 separation[41], it has not been tested for Xe/Kr separations.

Synthesis and equilibrium adsorption measurements.

Encouraged by the data from our high-throughput screening, we synthesized SBMOF-1 and measured its pure-component Xe and Kr adsorption isotherms at room temperature (see synthesis section of Supplementary Methods, Supplementary Figs 17–21)[41]. Our first measurement of low pressure Xe uptake in SBMOF-1 at 298 K, when activated by the reported activation procedure[41], was much lower than predicted by molecular simulation (Supplementary Fig. 22). However, we found that activating SBMOF-1 at a lower temperature yielded low pressure Xe uptake closer to the simulation (Fig. 2a, Supplementary Fig. 23, see effect of activation temperature section of the Supplementary Methods). The Xe adsorption isotherm in SBMOF-1 saturates at a low pressure, indicative of a high affinity for Xe compared with other gases including Kr (Fig. 2a, Supplementary Figs 24 and 25). The Kr adsorption isotherm exhibits a smaller slope and does not saturate even at 1 bar, indicative of a much weaker affinity for Kr. This hints that SBMOF-1 is highly discriminatory for Xe over Kr. Indeed, identifying the Xe and Kr Henry coefficients from the pure-component adsorption isotherms, we predict SBMOF-1 to exhibit a thermodynamic Xe/Kr selectivity of 16 at dilute conditions at 298 K.

It is interesting to compare the equilibrium Xe and Kr uptake of SBMOF-1 with the reported top-performing MOFs. We collected from the literature experimentally measured Xe and Kr

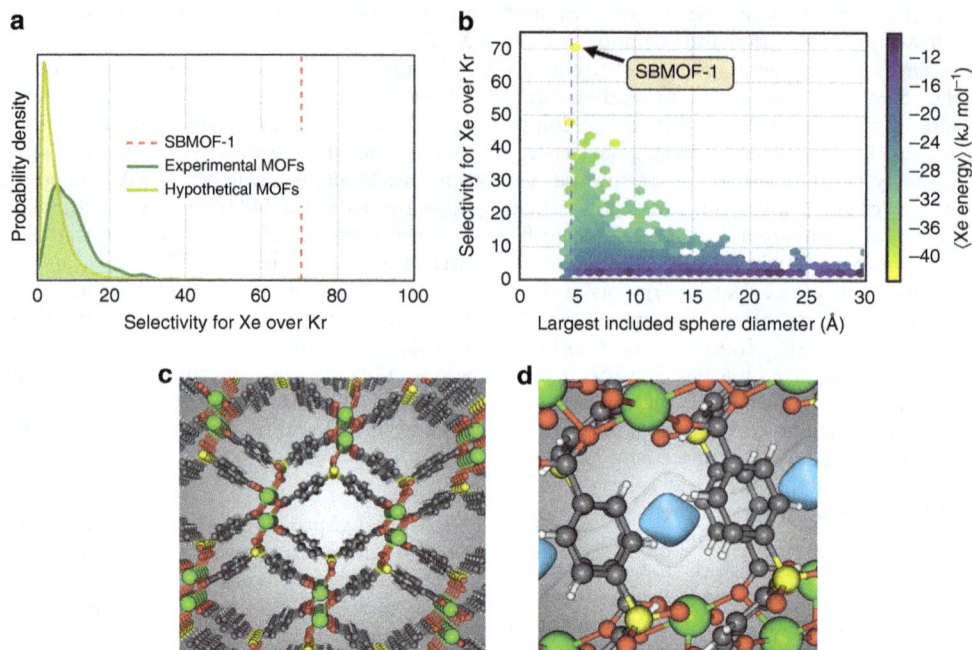

Figure 1 | Computational screening of MOFs for Xe/Kr separations at dilute conditions relevant to UNF reprocessing off-gas. We computed the Henry coefficients of Xe and Kr in ~125,000 MOF structures; the selectivity at dilute conditions is the ratio of Henry coefficients. (**a**) Distribution of simulated selectivities for experimentally synthesized (green) and hypothetical (yellow) MOF structures; vertical, dashed line is SBMOF-1 (KAXQIL in the Cambridge Structural Database (CSD)). (**b**) Histogram showing relationship between selectivity and pore size, with the largest included sphere diameter as a metric; colour shows average energy of Xe adsorption in that bin. SBMOF-1 (KAXQIL in CSD), with simulated selectivity 70.6 and largest included sphere diameter of 5.1 Å, is indicated. Vertical, dashed line is the distance that yields the minimum energy in a Xe–Xe Lennard–Jones potential. (**c**) SBMOF-1 is composed of corner sharing, octahedrally coordinated calcium chains along the crystallographic b direction, which are connected by organic linkers, forming a one-dimensional nanoporous channel. (**d**) Side view. Shown are the calculated potential energy contours of a Xe atom adsorbed in the pore (blue surface, $-32\,kJ\,mol^{-1}$; white surface, $15\,kJ\,mol^{-1}$).

adsorption isotherms in Co-formate[15,20], SBMOF-2 (ref. 13), HKUST-1 (ref. 14), MOF-505 (ref. 10), PCN-14 (ref. 19), Ni-MOF-74 (ref. 8), Zinc tetrazolate[21], IRMOF-1 (ref. 22), and FMOF-Cu[9] and identified the Xe and Kr Henry coefficients from the data in the low pressure regime (see Computational Calculation section of Supplementary Methods, Fig. 2b, Supplementary Figs 1–16). The saturation loading of Xe in SBMOF-1 is lower than observed in the majority of these materials due to the comparatively low ($\sim 145\,m^2\,g^{-1}$) surface area of SBMOF-1 (Supplementary Figs 26 and 27)[41]. However, the Henry coefficient of Xe in SBMOF-1 is a factor of two higher than in CC3, the material in our survey with the second highest Xe Henry coefficient; we thus expect SBMOF-1 to have an outstanding Xe uptake under UNF reprocessing off-gas conditions. Figure 2b shows that SBMOF-1 exhibits by far the largest Xe Henry coefficient and the highest Xe/Kr selectivity at dilute conditions among all reported Xe and Kr adsorption isotherms in our literature survey.

Adsorption kinetics and column breakthrough experiments. From a practical point of view, it is important that the kinetics of Xe adsorption/desorption are sufficiently fast and the material can undergo multiple ad-/de-sorption cycles without losing capacity. We measured the kinetics of Xe adsorption into an SBMOF-1 sample by connecting a chamber of Xe at 1 bar and 298 K to an evacuated chamber with the SBMOF-1 sample, then opening a valve to allow flow. Figure 2c shows that the rate of Xe uptake is sufficiently fast, reaching ~80% of saturation uptake within 10 min. Next, we performed 10 ad-/de-sorption cycles to test if SBMOF-1 retains its high Xe adsorption capacity after many

cycles. Figure 2d shows that SBMOF-1 retains its performance after multiple cycles. In addition, SBMOF-1 shows high thermal stability up to 500 K (Supplementary Fig. 20). To demonstrate the practical applicability of SBMOF-1 for capturing Xe from UNF reprocessing off-gas, we conducted single-column breakthrough experiments with a representative gas mixture (400 p.p.m. Xe, 40 p.p.m. Kr, 78.1% N_2, 20.9% O_2, 0.03% CO_2 and 0.9% Ar) (see breakthrough measurement section of the Supplementary Information, Supplementary Figs 28 and 29)[17]. We fed this gas mixture through a column packed with SBMOF-1 and initially purged with He. Figure 3 shows that all gases except Xe broke through the column within minutes, whereas Xe was retained in the column for more than an hour). This demonstrates that SBMOF-1 can selectively remove Xe from air at UNF reprocessing conditions. Under these conditions, SBMOF-1 adsorbed 13.2 mmol Xe per kg, higher than the reported breakthrough Xe capacities of benchmark materials, Ni-MOF-74 (4.8 mmol Xe per kg) and CC3 (11 mmol Xe per kg) (Supplementary Fig. 29)[12,17]. The experimental breakthrough capacity is close to that predicted from the Henry coefficient of the pure-component Xe isotherm ($15.4\,mmol\,kg^{-1}$), suggesting minimal diffusion limitations in the SBMOF-1 pellets. Next, we conducted column breakthrough experiments on SBMOF-1 in the presence of 42% relative humidity (Fig. 3b). Remarkably, SBMOF-1 retains a high Xe uptake ($\sim 11.5\,mmol\,kg^{-1}$) even in the presence of water vapor. These results suggest the outstanding stability of SBMOF-1 makes it a practical material for the removal of Xe from UNF reprocessing off-gas. Such stability is a desirable property, as very few metal–organic hybrid materials exhibit such properties[42–44]. We postulate the absence of open metal sites to be responsible for the stability of SBMOF-1 in the presence of water vapour[45].

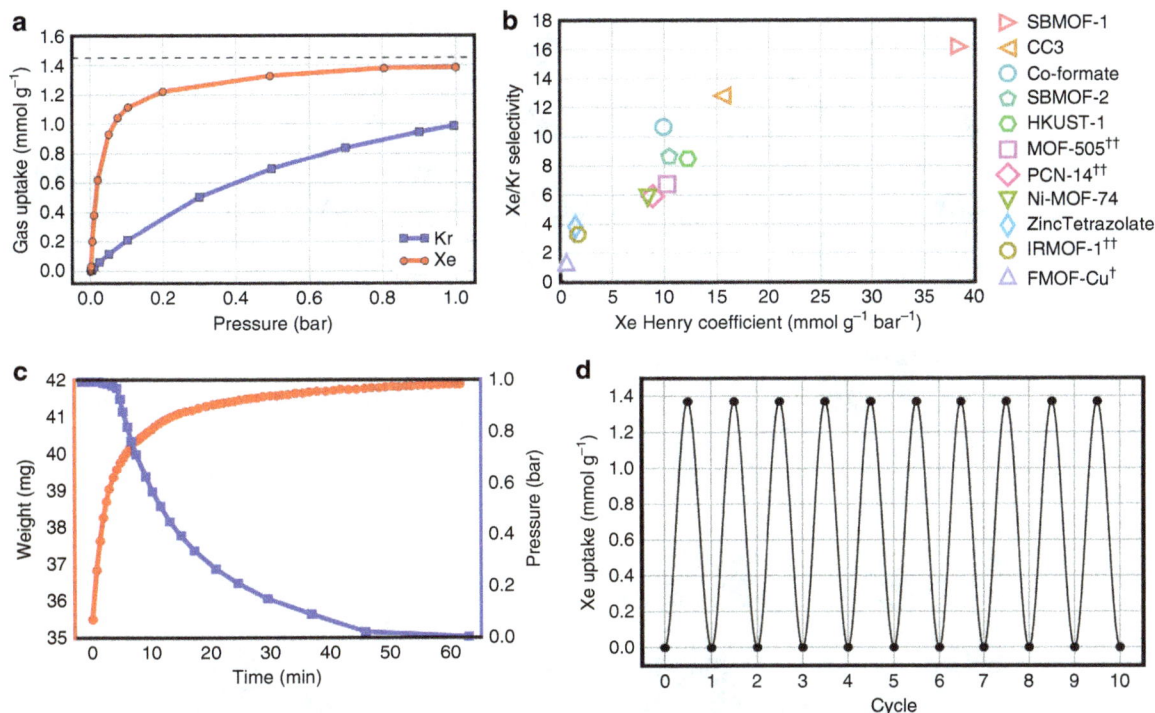

Figure 2 | Experimental characterization of Xe and Kr adsorption in SBMOF-1. (**a**) Experimental Xe and Kr adsorption isotherms. Horizontal line indicates one atom per pore segment. (**b**) Survey of thermodynamic Xe/Kr separation performance in top-performing materials. Henry coefficients are extracted from pure-component Xe and Kr adsorption isotherms reported in the literature (see Methods). Data at 298 K, exceptions denoted by a dagger (†) for 297 K and a double dagger (††) for 292 K. (**c**) Xe adsorption kinetics experiments. The blue curve shows the pressure drop in a chamber feeding Xe to an initially evacuated chamber with the SBMOF-1 sample; the red curve shows the corresponding weight increase due to Xe adsorption. (**d**) Xe adsorption/ desorption cycling data; a sinusoidal curve is superimposed on the data. (**a,c,d**) Data for SBMOF-1 at 298 K.

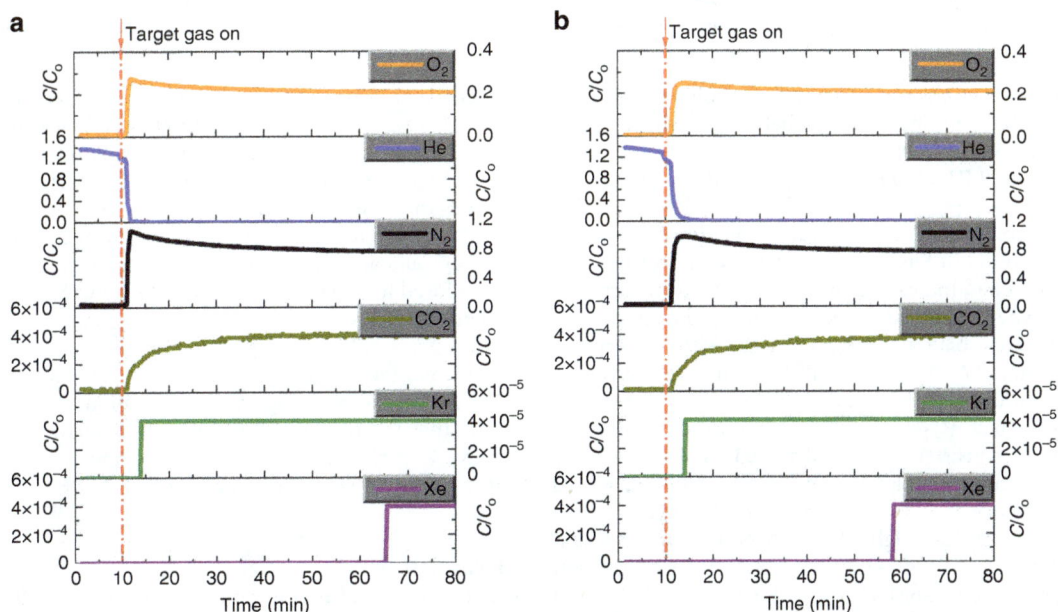

Figure 3 | Single column breakthrough experiments using SBMOF-1 at room temperature and 1 atm. Column is initially purged with He. (**a**) Inlet is a dry gas mixture with 400 p.p.m. Xe and 40 p.p.m. Kr balanced with air. (**b**) Inlet is the same gas mixture as in (**a**) with 42% relative humidity. Note that the Xe breakthrough time is only marginally decreased in the presence of water.

Revealing the Xe adsorption sites in SBMOF-1. To identify the location of adsorbed Xe and Kr, we performed single-crystal X-ray diffraction experiments on activated SBMOF-1 (Supplementary Data 1 and 2). Single-crystal analysis of

Xe-loaded SBMOF-1 reveals that Xe adsorbs at a single site, near the midpoint of the channel, interacting with the channel wall composed of aromatic rings by mainly van der Waals interactions. Due to symmetry considerations (space group

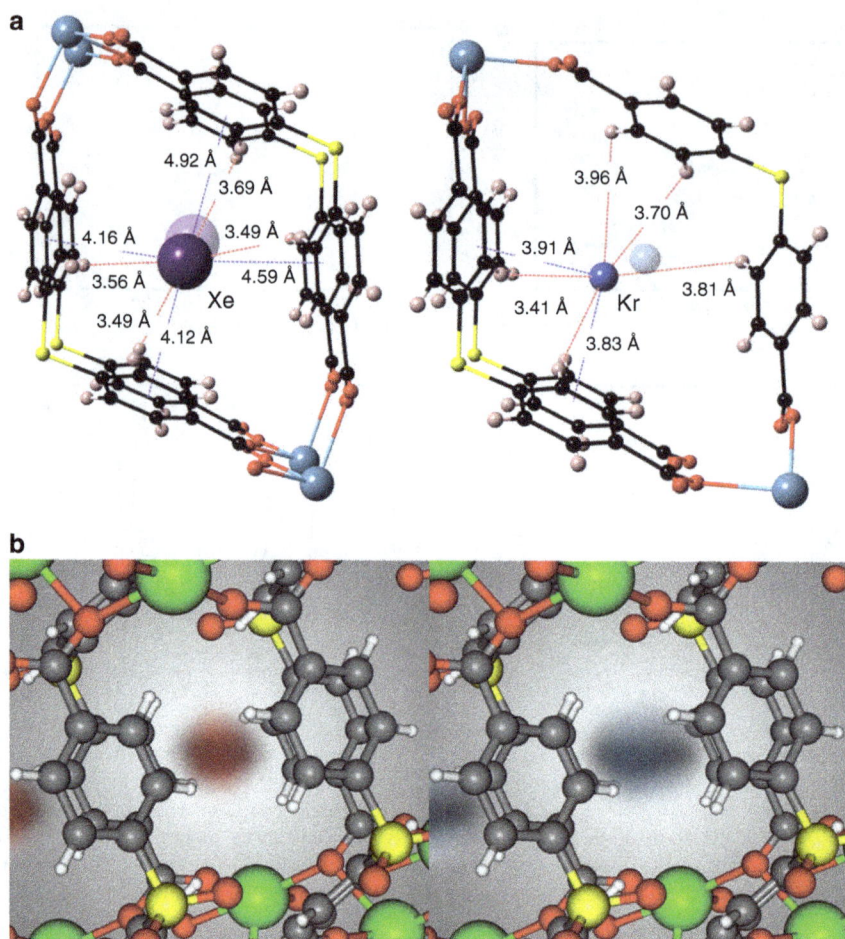

Figure 4 | Xe and Kr adsorption sites in SBMOF-1. (**a**) Xe and Kr positions determined by single-crystal X-ray diffraction. (**b**) Spatial probability densities from recording adsorbate positions during pure-component grand-canonical Monte Carlo simulations at 1 bar and 298 K (Xe, red; Kr, blue).

$P2_1/c$), each Xe atom is positioned at two possible sites (Fig. 4, Supplementary Table 1). The distance between each Xe atom along the b axis is 5.56 Å, closely matching the b axis length of the unit cell. There are 1.72 atoms of Xe per unit cell based on crystallographic analysis (~ 1.25 mmol g^{-1}), close to the loading obtained from gas-adsorption data (1.38 mmol g^{-1}). The saturation loading of Xe in Fig. 2a approaches two atoms per unit cell (see horizontal line), indicating commensurate Xe adsorption, which occurs when the adsorbed amount, location and orientation of an adsorbate are commensurate with the crystallographic symmetry of the adsorbent[46]. Such commensurate adsorption in SBMOF-1 was previously observed for small hydrocarbon molecules (C$_2$–C$_3$)[47,48]. The observed position of Xe in the pore is consistent with calculated potential energy contours and molecular simulations of Xe adsorption (Figs 1d and 4b, Supplementary Fig. 30, Supplementary Table 2).

We can rationalize the high Xe adsorption capacity and selectivity exhibited by SBMOF-1 by its optimal Xe adsorption site. First, the pore size of SBMOF-1 is tailored for Xe[35,38]. As a metric for pore size, we calculate the diameter of the largest included hard-sphere that can fit inside the pore of SBMOF-1 as 4.2 Å, slightly larger than a Xe atom, ~ 4.1 Å. Simulations of Xe/Kr adsorption in the database of experimental MOFs show that all of the most selective MOFs have pore sizes slightly larger than a Xe atom (Fig. 1b). Such a pore diameter is a prerequisite for a highly Xe-selective material, as the pore size controls the proximity and degree of overlap from multiple framework atoms contributing van der Waals interactions from multiple directions

to achieve a highly favourable host–Xe interaction. A pore of optimal size for Xe is suboptimal for Kr because of the size difference, so this forms a pore that is highly discriminatory for Xe over Kr[38]. As shown in Fig. 1c, the pore size of SBMOF-1 falls in the optimal pore size window for Xe/Kr separations[37], distinguishing it from other MOFs. Porous organic cage CC3, another outstanding Xe-selective material, also exhibits a pore size tailored for Xe (pore window 4.4 Å), but SBMOF-1 constructs a denser wall of chemical moieties than CC3 to achieve a higher Xe binding energy, enhancing its preference for Xe (Supplementary Table 3). This is the second reason why SBMOF-1 is outstanding in Xe adsorption; the colour in Fig. 1b shows that the dense wall of SBMOF-1 surrounds a Xe atom to achieve a high energy of Xe adsorption and thus a high Xe selectivity, following the trend in other MOFs.

Discussion

We demonstrated that a nanoporous MOF, SBMOF-1, identified as an outstanding Xe/Kr selective material from molecular simulations, shows exceptional Xe uptake at low pressure, selectivity for Xe, thermal and water stability, and adsorption kinetics. These attributes make SBMOF-1 potentially useful as a practical, near-term material for removal of Xe and Kr from nuclear reprocessing facilities with a far less energy requirement than cryogenic distillation. The selective adsorption of Xe from relevant gas mixtures even with $\sim 42\%$ relative humidity demonstrate practicality and offer improvements over current technologies. Our recent economic analysis showed the cost benefits of using Ni-MOF-74 for an adsorption-based separation process at room temperature in comparison with cryogenic distillation[7,11]. The discovery of the high Xe uptake and selectivity of SBMOF-1 at UNF reprocessing conditions—also in the presence of humidity—will enable an even more cost-effective process. The exceptional selectivity of SBMOF-1 is attributed to its pore size tailored to Xe and its dense wall of atoms that constructs a binding site with a high affinity for Xe, as evident by single-crystal X-ray diffraction and molecular models. As molecular simulations predicted SBMOF-1 to be among the most selective of ~5,000 experimentally reported MOFs and ~120,000 hypothetical MOF structures *a priori*, this work is a rare case of a computationally inspired materials discovery.

Methods

Synthesis and scale up. SBMOF-1 was originally synthesized using a previously published literature procedure[41]. In a typical synthesis, a mixture of 0.6 mmol of CaCl$_2$ (0.074 g) and 0.6 mmol of 4, 4'-SDB (0.198 g) were added in 10 ml of ethanol and stirred for ~2 h to achieve homogeneity (molar ratio of metal chloride: ligand:solvent = 1:1:380). The resultant solution was heated at 180 °C for 3 days. Colourless, needle-shaped crystals were recovered as product and washed with ethanol (yield: 45% based on CaCl$_2$, 0.1 g). For scale up, 1.44 g of CaCl$_2$ (13 mmol) and 3.98 g of 4,4'-SDB (13 mmol) were added to 120 ml of ethanol and stirred for ~2 h to achieve homogeneity (molar ratio: 1:1:156). The well-mixed solution was then transferred to three 100 ml Teflon-lined stainless steel Parr autoclaves and heated for 3 days at 180 °C. The product was obtained as white powder and washed by ethanol (3 times, 50 ml), followed by drying under vacuum (yield: 2.2 g, 50% based on CaCl$_2$). The as-synthesized material was then exchanged with methanol (3 ×, 50 ml) for a total period of 3 days. The product purity was confirmed by powder XRD.

Gas-adsorption and breakthrough experiments. The methanol-exchanged SBMOF-1 was activated at 100 °C for 12 h under dynamic vacuum. Single-component gas-adsorption isotherms were collected in a Quantachrome Autosorb-1 and dynamic sorption analyzer (ARBC, Hiden Analytical Ltd., Warrington, UK). The later instrument was also used to collect breakthrough measurement data. Breakthrough measurements were conducted on 20–35 mesh (500–850 μm) pellets of SBMOF-1 (1.48 g) using a gas mixture composition simulating UNF conditions (400 p.p.m. Xe, 40 p.p.m. Kr, 78.1% N$_2$, 20.9% O$_2$, 0.03% CO$_2$ and 0.9% Ar).

Computational methodologies. At dilute conditions relevant to UNF reprocessing off-gas, we modelled Xe and Kr adsorption in the MOFs with Henry's law. Let P be the pressure (units: bar) and $\sigma(P)$ be the gas uptake (units: mmol g^{-1}) as a function of pressure (the adsorption isotherm). Henry's law, only valid at low surface coverage, is then:

$$\sigma(P) = K_H P, \qquad (1)$$

where K_H is the Henry coefficient (units: mmol g^{-1} bar^{-1}) of the gas in the adsorbent. The Xe/Kr selectivity is then the ratio of the Henry coefficients. We calculated the Henry coefficient in each MOF using Widom particle insertions, a Monte Carlo integration[49]. We model the energetic interactions between Xe and Kr with the atoms of the MOFs using Lennard Jones potentials. We took parameters for Xe and Kr from Boato *et al.* and for the MOF atoms from the Universal Force Field, applying Lorentz–Berthelot mixing rules to obtain cross-interactions[50,51]. We hold the MOF structures rigid throughout the simulation and apply periodic boundary conditions to mimic an infinite crystal. For potential energy contours and spatial probability density plots for SBMOF-1, we utilized a hybrid Dreiding-TraPPE force field, as this force field produces a better match to the Xe and Kr isotherms in SBMOF-1 than the UFF[52,53]. The largest included hard sphere diameter is calculated using Zeo + + (refs 54,55). See Supplementary Methods for more details.

Literature survey for pure-component Xe and Kr adsorption isotherms. To generate Fig. 2b, we collected from the literature experimentally measured single-component Xe and Kr adsorption isotherms in MOFs and porous organic cage materials. Focusing on the low-pressure regime of the adsorption isotherm that exhibits linear behaviour—the Henry regime where Henry's law in equation (1) is valid—we fit a line with zero intercept to this data to identify K_H of Xe and Kr in the material. See Supplementary Section for the data and visualizations of the resulting fits to equation (1). Our data and code to reproduce Fig. 2b are openly available on GitHub at https://github.com/CorySimon/XeKrMOFAdsorptionSurvey.

Single-crystal X-ray diffraction. The single-crystal data on the Xe- and Kr-loaded activated SBMOF-1 were collected using a four circle kappa Oxford Gemini

diffractometre equipped with an Atlas detector ($\lambda = 0.71073$) at 100 K. The raw intensity data were collected, integrated and corrected for absorption effects using CrysAlis PRO software. Data sets were corrected for absorption using a multi-scan method, and structures were solved by direct methods using SHELXS-97 and refined by full-matrix least squares on F^2 with SHELXL-97 (ref. 56).

References

1. Hoffert, M. I. *et al.* Advanced technology paths to global climate stability: energy for a greenhouse planet. *Science* **298**, 981–987 (2002).
2. Chu, S. & Majumdar, A. Opportunities and challenges for a sustainable energy future. *Nature* **488**, 294–303 (2012).
3. Lenzen, M. Life cycle energy and greenhouse gas emissions of nuclear energy: a review. *Energ. Convers. Manage.* **49**, 2178–2199 (2008).
4. Soelberg, N. R. *et al.* Radioactive iodine and krypton control for nuclear fuel reprocessing facilities. *Sci. Technol. Nucl. Ins.* **2013**, 1–12 (2013).
5. Ying, R. T. *Gas Separation by Adsorption Processes* (Butterworth-Heinemann, 2013).
6. Kerry, F. G. *Industrial Gas Handbook: Gas Separation and Purification* (CRC Press, 2007).
7. Liu, J., Fernandez, C. A., Martin, P. F., Thallapally, P. K. & Strachan, D. M. A two-column method for the separation of Kr and Xe from process off-gases. *Ind. Eng. Chem. Res.* **53**, 12893–12899 (2014).
8. Thallapally, P. K., Grate, J. W. & Motkuri, R. K. Facile xenon capture and release at room temperature using a metal-organic framework: a comparison with activated charcoal. *Chem. Commun. (Camb)* **48**, 347–349 (2012).
9. Fernandez, C. A., Liu, J., Thallapally, P. K. & Strachan, D. M. Switching Kr/Xe selectivity with temperature in a metal-organic framework. *J. Am. Chem. Soc.* **134**, 9046–9049 (2012).
10. Bae, Y. S. *et al.* High xenon/krypton selectivity in a metal-organic framework with small pores and strong adsorption sites. *Micropor. Mesopor. Mater.* **169**, 176–179 (2013).
11. Banerjee, D. *et al.* Potential of metal-organic frameworks for separation of xenon and krypton. *Acc. Chem. Res.* **48**, 211–219 (2014).
12. Chen, L. *et al.* Separation of rare gases and chiral molecules by selective binding in porous organic cages. *Nat. Mater.* **13**, 954–960 (2014).
13. Chen, X. *et al.* Direct observation of Xe and Kr adsorption in a Xe-selective microporous metal–organic framework. *J. Am. Chem. Soc.* **137**, 7007–7010 (2015).
14. Hulvey, Z. *et al.* Noble gas adsorption in copper trimesate, HKUST-1: an experimental and computational study. *J. Phys. Chem. C* **117**, 20116–20126 (2013).
15. Lawler, K. V., Hulvey, Z. & Forster, P. M. Nanoporous metal formates for krypton/xenon separation. *Chem. Commun. (Camb)* **49**, 10959–10961 (2013).
16. Liu, J., Strachan, D. M. & Thallapally, P. K. Enhanced noble gas adsorption in Ag@MOF-74Ni. *Chem. Commun. (Camb)* **50**, 466–468 (2014).
17. Liu, J., Thallapally, P. K. & Strachan, D. Metal-organic frameworks for removal of Xe and Kr from nuclear fuel reprocessing plants. *Langmuir* **28**, 11584–11589 (2012).
18. Meek, S. T., Teich-McGoldrick, S. L., Perry, J. J., Greathouse, J. A. & Allendorf, M. D. Effects of polarizability on the adsorption of noble gases at low pressures in monohalogenated isoreticular metal-organic frameworks. *J. Phys. Chem. C* **116**, 19765–19772 (2012).
19. Perry, J. J. *et al.* Noble gas adsorption in metal-organic frameworks containing open metal sites. *J. Phys. Chem. C* **118**, 11685–11698 (2014).
20. Wang, H. *et al.* The first example of commensurate adsorption of atomic gas in a MOF and effective separation of xenon from other noble gases. *Chem. Sci.* **5**, 620–624 (2014).
21. Xiong, S. S. *et al.* A flexible zinc tetrazolate framework exhibiting breathing behaviour on xenon adsorption and selective adsorption of xenon over other noble gases. *J. Mater. Chem. A* **3**, 10747–10752 (2015).
22. Mueller, U. *et al.* Metal-organic frameworks - prospective industrial applications. *J. Mater. Chem.* **16**, 626–636 (2006).
23. Yaghi, O. M. *et al.* Reticular synthesis and the design of new materials. *Nature* **423**, 705–714 (2003).
24. Zhou, H.-C., Long, J. R. & Yaghi, O. M. Introduction to metal-organic frameworks. *Chem. Rev.* **112**, 673–674 (2012).

25. Eddaoudi, M., Sava, D. F., Eubank, J. F., Adil, K. & Guillerm, V. Zeolite-like metal-organic frameworks (ZMOFs): design, synthesis, and properties. *Chem. Soc. Rev.* **44**, 228–249 (2015).

26. Ferey, G. Hybrid porous solids: past, present, future. *Chem. Soc. Rev.* **37**, 191–214 (2008).

27. James, S. L. Metal-organic frameworks. *Chem. Soc. Rev.* **32**, 276–288 (2003).

28. Kitagawa, S., Kitaura, R. & Noro, S.-i. Functional porous coordination polymers. *Angew. Chem. Int. Ed.* **43**, 2334–2375 (2004).

29. Li, J.-R., Kuppler, R. J. & Zhou, H.-C. Selective gas adsorption and separation in metal-organic frameworks. *Chem. Soc. Rev.* **38**, 1477–1504 (2009).

30. Sumida, K. *et al.* Carbon dioxide capture in metal–organic frameworks. *Chem. Rev.* **112**, 724–781 (2012).

31. Lee, J. *et al.* Metal-organic framework materials as catalysts. *Chem. Soc. Rev.* **38**, 1450–1459 (2009).

32. Kreno, L. E. *et al.* Metal–organic framework materials as chemical sensors. *Chem. Rev.* **112**, 1105–1125 (2012).

33. Allendorf, M. D., Bauer, C. A., Bhakta, R. K. & Houk, R. J. T. Luminescent metal-organic frameworks. *Chem. Soc. Rev.* **38**, 1330–1352 (2009).

34. Motkuri, R. K. *et al.* Fluorocarbon adsorption in hierarchical porous frameworks. *Nat. Commun.* **5**, 4368 (2014).

35. Ryan, P., Farha, O. K., Broadbelt, L. J. & Snurr, R. Q. Computational screening of metal-organic frameworks for xenon/krypton separation. *Aiche J.* **57**, 1759–1766 (2011).

36. Van Heest, T., Teich-McGoldrick, S. L., Greathouse, J. A., Allendorf, M. D. & Sholl, D. S. Identification of metal-organic framework materials for adsorption separation of rare gases: applicability of ideal adsorbed solution theory (IAST) and effects of inaccessible framework regions. *J. Phys. Chem. C* **116**, 13183–13195 (2012).

37. Sikora, B. J., Wilmer, C. E., Greenfield, M. L. & Snurr, R. Q. Thermodynamic analysis of Xe/Kr selectivity in over 137 000 hypothetical metal–organic frameworks. *Chem. Sci.* **3**, 2217 (2012).

38. Simon, C. M., Mercado, R., Schnell, S. K., Smit, B. & Haranczyk, M. What are the best materials to separate a xenon/krypton mixture? *Chem. Mater.* **27**, 4459–4475 (2015).

39. Chung, Y. G. *et al.* Computation-ready, experimental metal-organic frameworks: a tool to enable high-throughput screening of nanoporous crystals. *Chem. Mater.* **26**, 6185–6192 (2014).

40. Wilmer, C. E. *et al.* Large-scale screening of hypothetical metal-organic frameworks. *Nat. Chem.* **4**, 83–89 (2012).

41. Banerjee, D., Zhang, Z. J., Plonka, A. M., Li, J. & Parise, J. B. A calcium coordination framework having permanent porosity and high CO2/N-2 selectivity. *Cryst Growth Des* **12**, 2162–2165 (2012).

42. Howarth, A. J. *et al.* Chemical, thermal and mechanical stabilities of metal–organic frameworks. *Nat. Rev. Mater.* **1**, 15018 (2016).

43. Deria, P. *et al.* Ultraporous, water stable, and breathing zirconium-based metal–organic frameworks with ftw topology. *J. Am. Chem. Soc.* **137**, 13183–13190 (2015).

44. Kalidindi, S. B. *et al.* Chemical and structural stability of zirconium-based metal–organic frameworks with large three-dimensional pores by linker engineering. *Angew. Chem. Int. Ed. Engl.* **127**, 223–228 (2015).

45. Kizzie, A. C., Wong-Foy, A. G. & Matzger, A. J. Effect of humidity on the performance of microporous coordination polymers as adsorbents for CO2 Capture. *Langmuir* **27**, 6368–6373 (2011).

46. Wu, H., Gong, Q., Olson, D. H. & Li, J. Commensurate adsorption of hydrocarbons and alcohols in microporous metal organic frameworks. *Chem. Rev.* **112**, 836–868 (2012).

47. Banerjee, D. *et al.* Direct structural evidence of commensurate-to-incommensurate transition of hydrocarbon adsorption in a microporous metal organic framework. *Chem. Sci.* **7**, 759–765 (2016).

48. Plonka, A. M. *et al.* Light hydrocarbon adsorption mechanisms in two calcium-based microporous metal organic frameworks. *Chem. Mater.* **28**, 1636–1646 (2016).

49. Frenkel, D. & Smit, B. *Understanding Molecular Simulation: From Algorithms to Applications* (Academic Press, 2001).

50. Boato, G. & Casanova, G. A self-consistent set of molecular parameters for neon, argon, krypton and xenon. *Physica* **27**, 571 -& (1961).

51. Rappe, A. K., Casewit, C. J., Colwell, K. S., Goddard, W. A. & Skiff, W. M. Uff a full periodic-table force-field for molecular mechanics and molecular-dynamics simulations. *J. Am. Chem. Soc.* **114**, 10024–10035 (1992).

52. Wick, C. D., Martin, M. G. & Siepmann, J. I. Transferable potentials for phase equilibria. 4. United-atom description of linear and branched alkenes and alkylbenzenes. *J. Phys. Chem. B* **104**, 8008–8016 (2000).

53. Mayo, S. L., Olafson, B. D. & Goddard, W. A. Dreiding - a generic force-field for molecular simulations. *J. Phys. Chem.* **94**, 8897–8909 (1990).

54. Willems, T. F., Rycroft, C., Kazi, M., Meza, J. C. & Haranczyk, M. Algorithms and tools for high-throughput geometry-based analysis of crystalline porous materials. *Micropor. Mesopor. Mater.* **149**, 134–141 (2012).

55. Pinheiro, M., Martin, R. L., Rycroft, C. H. & Haranczyk, M. High accuracy geometric analysis of crystalline porous materials. *CrystEngComm* **15**, 7531–7538 (2013).

56. Sheldrick, G. M. A short history of SHELX. *Acta Cryst. A.* **64**, 112–122 (2008).

Acknowledgements

We (PNNL) acknowledge US Department of Energy (DOE), Office of Nuclear Energy for synthesis, Xe/Kr adsorption, kinetics and breakthrough measurements. C.M.S. is supported by the US DOE, Office of Science, Office of Workforce Development for Teachers and Scientists, Office of Science Graduate Student Research (SCGSR) program. The SCGSR program is administered by the Oak Ridge Institute for Science and Education for the DOE under Contract No. DE-AC05-06OR23100. B.S. is supported by the Center for Gas Separations Relevant to Clean Energy Technologies, an Energy Frontier Research Center funded by the US DOE, Office of Science, Office of Basic Energy Sciences under Award No. DE-SC0001015. M.H. was supported by the Center for Applied Mathematics for Energy Research Applications (CAMERA), funded by the U.S. Department of Energy under Contract No. DE-AC02- 05CH11231. This research used resources of the National Energy Research Scientific Computing Center, which is supported by the Office of Science of the US DOE under Contract No. DE-AC02-05CH11231. A.M.P., X.C. and J.B.P. were supported by the National Science Foundation DMR-1231586 and CHE-0840483. P.K.T. would like to acknowledge Dr Terry Todd at Idaho National Laboratory, Dr Robert Jubin at Oakridge National Laboratory, Dr Denis Strachan, Dr John Vienna at PNNL, Kimberly Gray (DOE-NE HQ) and Jim Breese (DOE-NE HQ) for programmatic support. PNNL is a multi-program national laboratory operated for the US DOE by Battelle Memorial Institute under Contract DE-AC05-76RL01830.

Author contributions

D.B. synthesized SBMOF-1, and performed scale-up operation, activation and single-component gas-adsorption experiments. J.L. and D.B. performed breakthrough measurement. R.K.M. performed cycle studies. C.M.S., B.S. and M.H. performed computational work. A.M.P., X.C. and J.B.P. performed single-crystal X-ray diffraction studies. P.K.T. conceived and executed the project. P.K.T., D.B., C.M.S., B.S. and M.H. wrote the manuscript with input from all the co-authors.

Additional information

22

Engineering electrocatalytic activity in nanosized perovskite cobaltite through surface spin-state transition

Shiming Zhou[1], Xianbing Miao[1], Xu Zhao[1], Chao Ma[1], Yuhao Qiu[2], Zhenpeng Hu[2,3], Jiyin Zhao[1], Lei Shi[1] & Jie Zeng[1]

The activity of electrocatalysts exhibits a strongly dependence on their electronic structures. Specifically, for perovskite oxides, Shao-Horn and co-workers have reported a correlation between the oxygen evolution reaction activity and the e_g orbital occupation of transition-metal ions, which provides guidelines for the design of highly active catalysts. Here we demonstrate a facile method to engineer the e_g filling of perovskite cobaltite $LaCoO_3$ for improving the oxygen evolution reaction activity. By reducing the particle size to ~ 80 nm, the e_g filling of cobalt ions is successfully increased from unity to near the optimal configuration of 1.2 expected by Shao-Horn's principle. Consequently, the activity is significantly enhanced, comparable to those of recently reported cobalt oxides with $e_g^{\sim 1.2}$ configurations. This enhancement is ascribed to the emergence of spin-state transition from low-spin to high-spin states for cobalt ions at the surface of the nanoparticles, leading to more active sites with increased reactivity.

[1] Hefei National Laboratory for Physics Sciences at the Microscale, Hefei Science Center, University of Science and Technology of China, Hefei, Anhui 230026, China. [2] School of Physics, Nankai University, Tianjin 300071, China. [3] State Key Laboratory of Luminescent Materials and Devices, South China University of Technology, Guangzhou 510640, China. Correspondence and requests for materials should be addressed to S.Z. (email: zhousm@ustc.edu.cn) or to Z.H. (email: zphu@nankai.edu.cn) or to J.Z. (email: zengj@ustc.edu.cn).

Electrochemical water splitting is regarded as a prime approach for renewable energy conversion and storage[1,2]. One of the critical reactions of this process is oxygen evolution reaction (OER). However, this reaction is kinetically sluggish due to a complex multistep four-electron oxidation[1-5]. To reach a desirable current density of $10\,mA\,cm^{-2}$, which is a metric associated with solar fuel synthesis, a considerable overpotential (η) relative to thermodynamic potential of the reaction is required, but will hinder the large-scale electrochemical water splitting[3,4]. Currently, RuO_2 and IrO_2 are among the most active OER catalysts, however, their unacceptable cost and low abundance severely restrict their large-scale applications[5,6]. Therefore, it is of great technological and scientific significance to pursue highly efficient alternatives on the basis of earth-abundant nonprecious materials.

Transition-metal oxides and their derivatives have received much attention because of their earth-abundant reserves, low cost, environment-friendly features and remarkable OER activities[7-23]. Specially, cobalt oxides, such as layered $LiCoO_2$ (refs 11,12), Co-oxyhydroxides (refs 13,14), spinel Co_3O_4 (refs 15-17), perovskite $Ba_{0.5}Sr_{0.5}Co_{0.8}Fe_{0.2}O_{3-\delta}$ (BSCF)[18] and $PrBaCo_2O_{5+\delta}$ (ref. 19), have been explored as potential OER catalysts due to their high activities comparable to and even better than the precious metal oxides. Theoretical and experimental works have shown that the OER activity is intrinsically related to the electronic structure of Co ions, including oxidation state and spin configuration, which enlightens the rational design of optimal catalysts[7-21]. For example, Lu et al.[12] recently reported that the OER activity of $LiCoO_2$ was remarkably enhanced by tuning the Co oxidation state via electrochemical delithiation process. More interestingly, for perovskite transition-metal oxides, Shao-Horn and co-workers[18] demonstrated the direct dependence of OER activity on the e_g orbital filling of the transition-metal ions, and that the oxide BSCF with $e_g^{\sim 1.2}$ configuration exhibited the highest OER activity. The performance of the BSCF was found to be nearly identical to IrO_2 in terms of catalytic activity. However, this Co-based oxide undergoes surface amorphization after long-term potential cycles under OER conditions[22]. Therefore, substantial progress is still needed to develop perovskite-type OER catalysts with improved activity and stability. In this respect, Shao-Horn's principle highlights that the optimization of e_g filling close to 1.2 should be an alternative strategy to develop the transition-metal oxides as the effective OER catalysts. Practically, a recent study by Zhu et al.[23] supported this principle, where an improved OER activity was realized through adjusting the e_g filling to ~ 1.2 with partial substitution of Co by Nb in $SrCo_{0.8}Fe_{0.2}O_{3-\delta}$ (ref. 23).

In this work, we present a facile method to modify the electronic structure of Co ions by varying the particle size for the improvement in the OER activity. We focus on the perovskite cobaltite $LaCoO_3$ (LCO), which is well-known for its unique thermally-driven transition of Co^{3+} ions from low spin (LS: $t_{2g}^6 e_g^0$) at low temperatures to higher spin state with e_g orbital configuration of $e_g^{\sim 1.0}$ at room temperature[18,24-31]. This compound was reported to exhibit reasonable OER activity but much less than BSCF[18]. Here, by reducing the particle size, the e_g filling is successfully increased from unity to close to the optimization configuration of ~ 1.2. As a consequence, the OER activity of the 80-nm LCO is higher than those of other sized samples as well as the bulk, and comparable to those of the reported cobalt oxides with $e_g^{\sim 1.2}$ filling, which enables the nanosized LCO to be applicable as a promising OER catalyst.

Results

Crystal structures of the bulk and nanosized LCO. The LCO
samples were prepared by a sol–gel method[32,33]. The precursory powders derived from the gel were annealed at 600, 700 and 800 °C

for 6 h to produce the LCO nanoparticles with the particle size of about 60, 80 and 200 nm (ref. 32), respectively, as well as at 1,000 °C for 12 h to the bulk sample with the particle size of about 0.5–1 μm (Supplementary Fig. 1). A representative transmission electron microscopy (TEM) image for the 700 °C annealing sample is shown in Fig. 1a. The high-resolution TEM images and the selective area electron diffraction patterns reveal a single-crystal structure of small LCO particles with a high crystallinity (Supplementary Fig. 2). The X-ray diffraction patterns (Fig. 1b and Supplementary Fig. 3) reveal that all the samples take a rhombohedral structure with R$\bar{3}$c space group (Fig. 1c). The structural parameters obtained from the Rietveld refinements on the diffraction data are given in Supplementary Table 1. As the particle size is reduced, the unit cell is found to be expanded. Specially, the bond length of Co–O exhibits an obvious increase (Fig. 1d).

Spin structures of the bulk and nanosized LCO. The temperature-dependent magnetizations were measured with a magnetic field of $H=1$ kOe under field-cooling procedures for all the samples (Supplementary Fig. 4a) to study the spin structures of Co ions controlled by the particle size. Above 150 K, the susceptibilities derived from the magnetizations ($\chi = M/H$) obey a paramagnetic Curie–Weiss law: $\chi = C/(T-\Theta)$, where C is Curie constant, and Θ is Curie–Weiss temperature. From the fitting results (Fig. 2a), an effective magnetic moment μ_{eff} can be calculated through $\mu_{eff} = \sqrt{8C}\ \mu_B$ (Supplementary Fig. 4b). For the bulk LCO, although the exact nature of whether the higher spin state of Co^{3+} ions at room temperature is intermediate-spin (IS: $t_{2g}^5 e_g^1$) state or a mixture of LS and high spin (HS: $t_{2g}^4 e_g^2$) states was controversial in the past decades[24-27], a large number of recent theoretical and experimental studies reveal that the mixture of LS and HS states is more favourable[28-31]. Here, the calculated μ_{eff} of 3.48 μ_B for the bulk sample is well consistent with those values reported for the polycrystalline bulk LCO, which corresponds to the Co^{3+} ions in 50% HS + 50% LS states as well as the e_g filling of ~ 1.0 (refs 25, 28-31). For the nanosized LCO, μ_{eff} shows a gradual increase with the decrease of the particle size, suggesting that the spin state of partial Co ions transmits from LS to HS state. Using the calculated μ_{eff}, the spin states are estimated to be 55% HS + 45% LS, 60.5% HS + 39.5% LS and 63.7% HS + 36.3% LS for the 200, 80 and 60 nm samples (Supplementary Note 1 and Supplementary Table 2), meaning that about 5, 10.5 and 13.7% Co^{3+} ions in LS state change to be in HS state, respectively. As shown in Fig. 2b, the corresponding e_g filling is about 1.1, 1.2 and 1.27 for the nanosized LCO, respectively. It is worthwhile to emphasize that by reducing the particle size to about 80 nm, we successfully tune the e_g filling from unity to the optimization value of 1.2 expected by Shao-Horn's principle.

To further confirm the spin-state transition and to explore its possible origin, the electron energy loss spectroscopy (EELS) analyses on the LCO nanoparticles were performed on the scanning TEM. Figure 2c,d shows the representative EELS spectra at Co L- and O K-edges from the centre and edge of the 80 nm nanoparticle, which are sensitive to the electronic structure from the core (bulk) and shell (surface), respectively[34,35]. For the Co L-edge spectra, no noticeable changes in the L_3/L_2 ratio between the bulk and surface were found, suggesting that the oxidation state of Co ions remained unchanged[35-39]. For the O K-edge spectra, three characteristic peaks near the edge onset, labelled a, b and c, were observed. The prepeak a, b and c were assigned to the hybridization of O 2p with Co 3d, La 5d and Co 4sp orbitals, respectively[36-39]. Compared with that from the centre position, the spectrum from the edge shows an obvious reduction in the intensity of the prepeak a. Similar results are also found in the

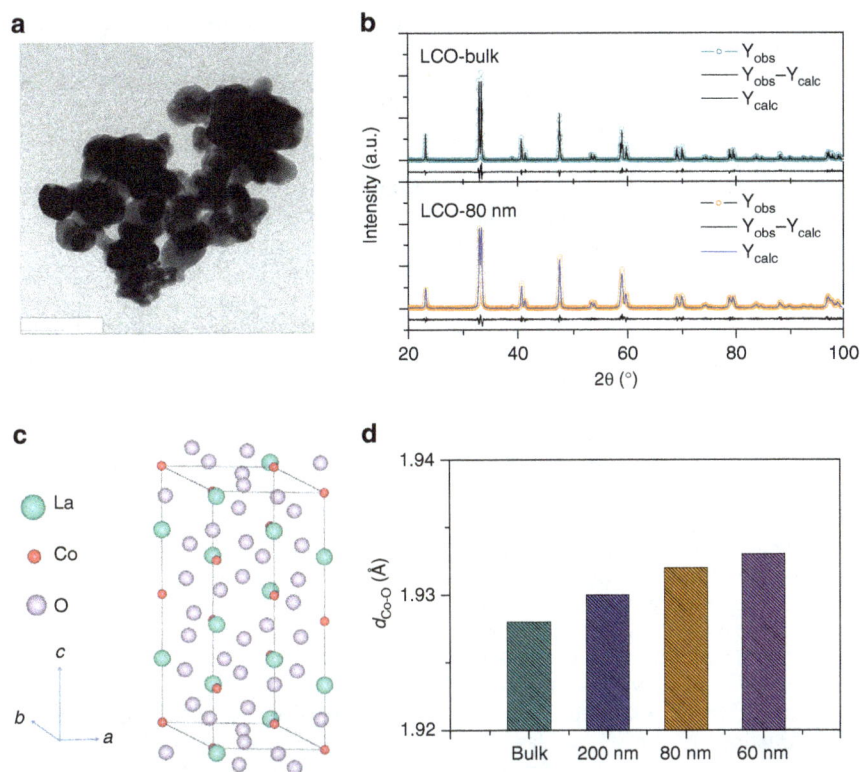

Figure 1 | TEM image and crystal structure analyses of the bulk and nanosized LCO. (**a**) TEM image for the 80 nm LCO. (**b**) X-ray diffraction patterns for bulk and 80 nm LCO together with the Rietveld refined results. (**c**) LCO crystal structure. (**d**) The length of Co-O bond for the bulk and nanosized LCO. Scale bar, 180 nm.

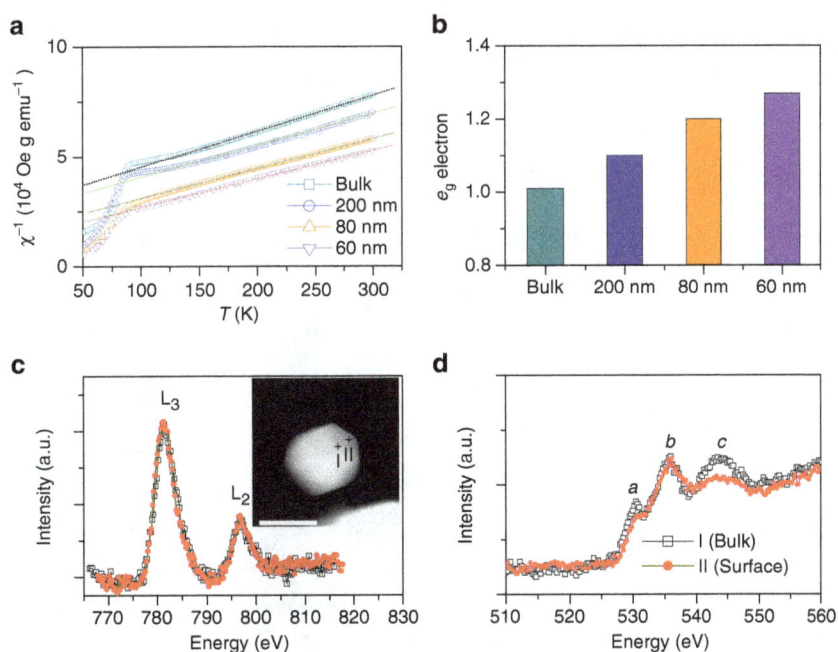

Figure 2 | Spin structure analyses of the bulk and nanosized LCO. (**a**) The temperature dependence inverse susceptibilities for all the LCO samples. The dotted lines are the fitting results by a Curie–Weiss law. (**b**) The corresponding e_g filling. (**c,d**) Representative EELS spectra of the 80 nm LCO at Co L-edge and O K-edge, respectively. The inset corresponds to the representative position of EELS acquisition. Scale bar, 50 nm.

other nanoparticles (Supplementary Fig. 5). This reduction is generally attributed to the formation of oxygen vacancy or the weakening of Co 3d–O 2p hybridization[36–39]. Since the Co L-edge spectra revealed no change in the Co oxidation state, the formation of oxygen vacancies can be excluded. Thus, the modification of O K-edge would originate from the change of the Co 3d–O 2p hybridization. Previous works have been widely reported that the spin-state transition of Co³⁺ ions in the LCO can significantly modify the hybridization of Co 3d–O 2p orbitals and then the intensity of the prepeak a (refs 35–39). In the LS configuration, the 3d e_g levels of Co^{3+} ions are completely empty, allowing the electrons from the filled O 2p levels to be shared with Co e_g orbitals, and accordingly creating O 2p holes. Thus, the hybridization of Co 3d with the O 2p states promotes electron transitions between 1s and the unfilled O 2p state, resulting in the prepeak a of the O K-edge. However, as a HS state of the Co^{3+} ions emerges, the e_g orbitals are increasingly occupied, which prevents the charge transfer and weakens the hybridization of O 2p with Co 3d orbitals, resulting in the decreased intensity of the prepeak a (refs 36–39). Therefore, the decrease in the intensity of the prepeak a at the surface confirms the existence of surface spin-state transitions in the nanosized LCO.

We proposed a mechanism to explain the presence of surface spin-state transitions of Co^{3+} ions where the modified crystal field splitting of Co 3d orbital at the surface favors the Co^{3+} ions to be in HS states, which has also been reported in nanosized stoichiometric LiCoO$_2$ (ref. 40). Assuming that Co^{3+} ions within surface layers are all transited to be in HS state, we can give a rough estimate of the e_g filling for the nanosized LCO on the basis of a simple core-shell model (Supplementary Note 2 and Supplementary Fig. 6). For nanosized perovskite oxides with the particle size ranging from tens to hundreds of nanometres, the surface layers are usually reported to be about 2–5 nm in the thickness[41–44]. Taking the thickness of 3 nm (see Supplementary Fig. 2), the estimated volume fractions of the surface layer are about 8.7, 20.8 and 27.1% for the 200, 80 and 60 nm LCO, respectively. Consequently, the increased fractions of the HS Co^{3+} ions are 4.4, 10.4 and 13.6%, which means that the e_g fillings are increased to be about 1.09, 1.21 and 1.27, respectively. These values are well consistent with those obtained from the magnetizations, further supporting that the tuning of e_g filling by the size reduction originates from the surface spin-state transition. In addition, since the radii of Co^{3+} ions increases when their spin state changes from LS to HS, the presence of this

transition is also confirmed by the crystal structure data, where the expansion of the unit cell and the increase of Co–O length under decreasing the particle size are found (Supplementary Table 1 and Fig. 1d).

OER activities of the bulk and nanosized LCO. To shed light on the role that the surface spin-state transition plays in the OER activity for the LCO, the electrochemical measurements were carried out in O$_2$-saturated 0.1 M KOH solutions using a standard three-electrode system. Figure 3a shows the iR-corrected polarization curves for all the samples, where the LCO nanoparticles exhibited smaller onset overpotentials than that of the bulk (~0.37 V). In particular, the smallest onset overpotential of ~0.32 V was observed in the 80 nm LCO. Similarly, the overpotential η required to achieve a current density of 10 mA cm^{-2} was also reduced from 0.62 V for the bulk to 0.54, 0.49 and 0.55 V as the particle size decreased to about 200, 80 and 60 nm, respectively. Figure 3b plots the dependence of mass activity at $\eta = 0.49$ V on the e_g filling. As the e_g filling approached the optimal configuration of ~1.2, the current density reached the largest value, which was about 4.3 times larger than that of the bulk. Moreover, the corresponding Tafel plots (Fig. 3d) also reveal that the 80 nm LCO possessed the smallest Tafel slope of ~69 mV dec^{-1}, much smaller than that of the bulk (~102 mV dec^{-1}). This large reduction suggests that the rate-determining step tends to change from the –OH adsorption to the O–OH formation[18,45,46]. In addition, the preliminary stability tests for the bulk and 80 nm LCO under a constant galvanostatic current of 10 A g^{-1} (Supplementary Fig. 7) also demonstrate that the 80 nm sample exhibited better stability than the bulk. As revealed by Co 2p X-ray photoelectron spectra of the 80 nm LCO before and after the electrolysis (Supplementary Fig. 8), no visible changes for the Co 2p spectra were found during the electrolysis, suggesting that the electronic state of the Co ions may remain unchanged. All of the above results clearly indicate that the OER

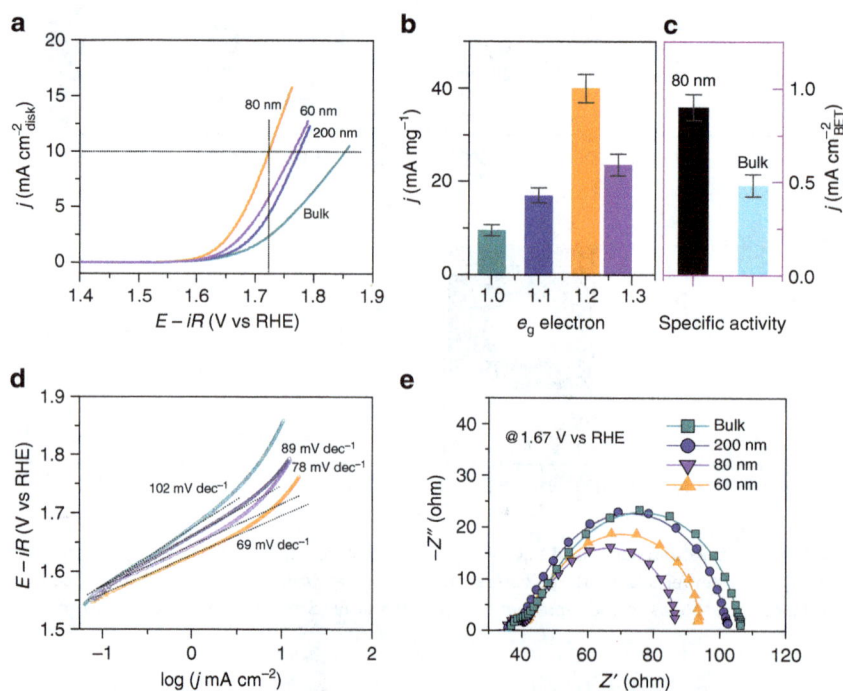

Figure 3 | OER activities of the bulk and nanosized LCO. (**a**) Polarization curves of the bulk and nanosized LCO. (**b**) Mass and (**c**) special activities at $\eta = 0.49$ V. (**d**) Tafel plots for the bulk and nanosized LCO. (**e**) Nyquist plots for the bulk and nanosized LCO. Error bars represent the s.d. from at least three independent measurements.

Table 1 | Comparison of OER activity for different catalysts.

Catalyst	Onset overpotential (V vs RHE)	η @ $j=10\,mA\,cm^{-2}$ (V)	Tafel slope (mV dec^{-1})	e_g filling	Ref.
LCO-bulk	0.37	0.62	102	$e_g^{\sim1.0}$	This work
LCO-80 nm	0.33	0.49	69	$e_g^{\sim1.2}$	This work
$Ba_{0.5}Sr_{0.5}Co_{0.8}Fe_{0.2}O_{3-\delta}$	—	0.49	84	$e_g^{\sim1.2}$	(ref. 21)
$Ba_{0.5}Sr_{0.5}Co_{0.8}Fe_{0.2}O_{3-\delta}$	0.30	0.50	94	$e_g^{\sim1.2}$	(ref. 23)
$SrCo_{0.9}Ti_{0.1}O_{3-\delta}$	—	0.51	88	$e_g^{\sim1.16}$	(ref. 21)
$SrNb_{0.1}Co_{0.7}Fe_{0.2}O_{3-\delta}$	0.30	0.50	76	$e_g^{\sim1.2}$	(ref. 23)

OER, oxygen evolution reaction.

activity of the LCO is successfully modified by controlling the particle size. As the particle size is reduced to about 80 nm with the $e_g^{\sim1.2}$ configuration, the activity is significantly enhanced.

The origin of the enhanced OER activity. As the sample size is reduced, the enhanced mass catalytic activity towards the OER is largely ascribed to the increase of the surface area. Such scenarios were reported in various Co-based oxides such as BSCF[18], $SrNd_{0.1}Co_{0.7}Fe_{0.2}O_{3-\delta}$ (ref. 23) and $NiCo_2O_4$ (ref. 47). However, in those cases, the size reduction leaded to a large decrease in the specific OER activities, that is, the normalized activities by the surface area. Since the specific activity reflects the intrinsic activity of the catalysis, this decrease indicates that intrinsic OER activity was deteriorated by reducing the sample size for those oxides. To clarify whether the enhancement of the OER activity in our LCO nanoparticles is intrinsic, the specific activities are further calculated on the basis of two types of surface areas, the Brunauer-Emmett-Teller (BET) surface areas and the electrochemically active surface areas, obtained by means of the gas desorption (Supplementary Fig. 9) and electrochemical double-layer capacitance measurements (Supplementary Fig. 10 and Supplementary Note 3), respectively. The specific activities at $\eta = 0.49\,V$ normalized by the surface areas exhibit similar dependences on the e_g filling to the mass activity (Supplementary Fig. 11). Compared with the bulk, the 80 nm sample is still 1.8 times more active in the specific activity normalized by the BET area (Fig. 3c), which strongly suggests that the increased number of active sites from the surface areas may be not the main contribution to the significant enhancement of the OER activity. The improved performance would be mainly attributed to the increased reactivity of the active sites due to the spin-state transition of Co^{3+} ions at surfaces. When the e_g filling of Co^{3+} ions increases from about 1.0 to 1.2, the electron occupancy of the Co 3d–O 2p σ^* band increases with the elongation of the length of Co–O bond as shown in Figs 1d and 2d. Thus, the hybridization of Co 3d–O 2p orbitals and the strength of Co–O bond become weaker, which leads to a less surface coverage by –OH groups on the active sites and thereby facilitates the formation of –OOH species[45,46]. As a result, the Tafel slope is reduced and the OER activity is improved. On the other hand, it has been generally demonstrated that H_2O molecules are initially adsorbed onto the surface of catalysts during the OER process[13,18,45]. Consequently, the adsorption energy of H_2O onto the active site plays a crucial role in the OER activity. Our density functional theory (DFT) calculations on the adsorption energy of H_2O onto the surface Co ions in different spin states reveal that the surface Co ions being in HS state are more favourable for adsorbing H_2O molecular (Supplementary Fig. 12 and Supplementary Table 3), well consistent with the improvement of the OER activity by reducing the particle size to 80 nm. However, it is worthwhile to note that as the particle size is further reduced to about 60 nm the activity decreases again, which cannot be explained by the above factors. We propose that the excessive e_g occupancy (>1.2) of the Co^{3+} ions in this sample would make the charge transfer ability lower. As such, when the two neighboured Co ions in Co–O–Co network are both in

HS state, the half-filling of e_g orbitals tends to prevent the charge transfer. To confirm this point, the electrochemical impedance spectroscopy experiments have been carried out. As shown in Fig. 3e, the Nyquist plots reveal that the charge transfer resistance gradually decreases with the reduction of the particle size to 80 nm, while increases again as the size is further reduced to 60 nm. Therefore, we conclude that the modifications of the Co–O binding strength and the charge transfer ability associated with the surface spin-state transition in the LCO are responsible for the size-dependent OER activity.

Finally, we compared the OER activity of our LCO samples with those of the recently reported Co-based perovskite oxides with the optimal configuration of $e_g^{\sim1.2}$. As illustrated in Table 1, it is interesting to find that the 80 nm LCO exhibits a well comparable activity with those well-known catalysts, which further consolidates Shao-Horn's principle and suggests that tuning the spin state can provide an effective strategy to improve the OER activity.

Discussion
In summary, we highlight an effective strategy to engineer the electronic configuration of perovskite cobalt oxide for the development of high active electrocatalysts. By reducing the particle size to about 80 nm, the spin filling of Co ions in LCO is successfully tuned from unity (bulk) to near the optimization configuration of ~1.2 expected by Shao-Horn's principle. Through X-ray diffraction, magnetic measurements and EELS analysis, we confirm that this modification originates from the size-induced spin-state transition of Co^{3+} ions from LS to HS state. Consequently, the nanosized sample exhibits an improved OER activity with lower overpotential, smaller Tafel slope and better stability compared with the bulk. More interestingly, the performance of the 80 nm LCO can be comparable with those of the reported cobalt oxides with $e_g^{\sim1.2}$ filling, suggesting that the LCO in this nanosized form can serve as a promising OER catalyst. Our work paves the way for the rational design of high-efficient OER catalysts.

Methods
Synthesis and characterization. $La(NO_3)_3 \cdot 6H_2O$ and $Co(NO_3)_2 \cdot 6H_2O$ were dissolved in deionized water, followed by the addition of a mixture of citric acid and ethylene glycol. Subsequently, the obtained transparent solution was slowly evaporated to get a gel, which was decomposed at about 400 °C for 4 h to result in dark brown powders. The precursor powders were further annealed at 600, 700 and 800 °C for 6 h to produce LCO nanoparticles with different particle sizes, and at 1,000 °C for 12 h to the bulk sample. The phase purity and crystal structure of the samples were determined by X-ray diffraction at room temperature on a Rigaku TTR-III diffractometer using Cu K_a radiation ($\lambda = 1.5418$ Å). The field emission SEM and TEM images were obtained on a JEOL-2010 SEM and a JEM-2100F TEM, respectively. The HRTEM images and the EELS analyses were performed on a JEOL JEM-ARM200F TEM/scanning TEM with a spherical aberration corrector. The magnetic measurements were carried out with a MPMS SQUID magnetometer. The nitrogen adsorption – desorption isotherms were conducted on a Micromeritics ASAP 2000 system at 77 K. X-ray photoelectron spectra were carried out on an ESCALAB 250 X-ray photoelectron spectrometer with Al $K\alpha$ as the excitation source.

Electrochemical measurements. The electrochemical tests were performed in O_2-saturated 0.1 M KOH with a conventional three-electrode on the CHI660B electrochemical station. Saturated Ag/AgCl and platinum wires were used as the reference and the counter electrodes, respectively. The reference electrode was calibrated with respect to the reversible hydrogen electrode (RHE), which was carried out in the high-purity hydrogen saturated electrolyte with a Pt wire as the working electrode. Cyclic voltammetry was run at a sweep rate of $1\,mV\,s^{-1}$. The average of the two potentials at which the current crossed zero was taken to be the thermodynamic potential for the hydrogen electrode reactions. In 0.1 M KOH, $E_{RHE} = E_{Ag/AgCl} + 0.964\,V$. To prepare the working electrode, 3.5 mg of electro-catalyst and $20\,\mu l$ of 5 wt% Nafion solutions were dispersed in 1 ml ethanol with sonication for at least 30 min to form a mixted ink. Then, $5\,\mu l$ of this solution was drop-casted onto a 3 mm in diameter glassy carbon electrode and dried naturally, yielding a catalyst loading of $0.25\,mg\,cm^{-2}$. Linear sweeping voltammograms were obtained at a scan rate of $5\,mV\,s^{-1}$. The potentials are corrected to compensate for the effect of solution resistance, which were calculated by the following equation: $E_{iR-corrected} = E - iR$, where i is the current, and R is the uncompensated ohmic electrolyte resistance ($\sim 36\,\Omega$) measured via high frequency ac impedance in O_2-saturated 0.1 M KOH. The polarization curves were replotted as overpotential (η) vs log current (log j) to get Tafel plots for quantification of the OER activities of investigated catalysts. Electrochemical impedance spectroscopy were conducted with AC voltage with 5 mV amplitude at the potential of 1.67 V vs RHE within the frequency range from 100 KHz to 100 mHz. Durablity test was performed at room temperature under a constant galvanostatic current of $10\,A\,g^{-1}$. Error bars represented s.d. from at least three independent measurements.

DFT calculations. DFT+U calculations with the Vienna *ab initio* simulation package[48] for a water molecule adsorbed on (001) surface of LCO with LS and HS state on Co^{3+} ions were performed to study the influence on adsorption energy of water molecule with different spin states of Co^{3+} ions. A slab model consisting of eight atom-layers ($La_{16}Co_{16}O_{48}$) was used to simulate the (001) surface with two terminations. In calculations, an effective U of 3.4 eV was added on Co $3d$ orbital, the plane wave energy cut-off was set to 400 eV, and a $2 \times 2 \times 1$ Monkhorst–Pack k-point mesh was used. During geometry optimization, converge criteria were $10^{-5}\,eV$ for energy and $0.05\,eV\,Å^{-1}$ for force. The LS of Co^{3+} was not obtained on CoO_2 terminated (001) surface, since surface Co^{3+} was in a five-coordinated structure and turns into HS during the calculation even though it was set to LS initially.

References

1. Gray, H. B. Powering the planet with solar fuel. *Nat. Chem.* **1**, 7 (2009).
2. Lewis, N. S. & Nocera, D. G. Powering the planet: chemical challenges in solar energy utilization. *Proc. Natl Acad. Sci. USA* **103**, 15729–15735 (2007).
3. Cook, T. R. *et al.* Solar energy supply and storage for the legacy and nonlegacy worlds. *Chem. Rev.* **110**, 6474–6502 (2010).
4. Walter, M. G. *et al.* Solar water splitting cells. *Chem. Rev.* **110**, 6446–6473 (2010).
5. Jiao, Y., Zheng, Y., Jaroniec, M. & Qiao, S. Z. Design of electrocatalysts for oxygen and hydrogen-involving energy conversion reactions. *Chem. Soc. Rev.* **44**, 2060–2086 (2015).
6. Lee, Y., Suntivich, J., May, K. J., Perry, E. E. & Shao-Horn, Y. Synthesis and activities of rutile IrO_2 and RuO_2 nanoparticles for oxygen evolution in acid and alkaline solutions. *J. Phys. Chem. Lett.* **3**, 399–404 (2012).
7. Yuan, C. Z., Wu, H. B., Xie, Y. & Lou, X. W. Mixed transition metal oxides: design, controllable synthesis and energy-related applications. *Angew. Chem. Int. Ed.* **53**, 1488–1504 (2014).
8. Wang, H. *et al.* Bifunctional non-noble metal oxide nanoparticle electrocatalysts through lithium-induced conversion for overall water splitting. *Nat. Commun.* **6**, 7261 (2015).
9. Hong, W. T. *et al.* Toward the rational design of non-precious transition metal oxides for oxygen electrocatalysis. *Energy Environ. Sci.* **8**, 1404–1427 (2015).
10. Kim, J. M., Yin, X., Tsao, K.-C., Fang, S. H. & Yang, H. $A_2B_2O_5$ as oxygen deficient perovskite electrocatalyst for oxygen evolution reaction. *J. Am. Chem. Soc.* **136**, 14646–14649 (2014).
11. Maiyalagan, T., Jarvis, K. A., Therese, S., Ferreira, P. J. & Manthiram, A. Spinel-type lithium cobalt oxide as a bifunctional electrocatalyst for the oxygen evolution and oxygen reduction reactions. *Nat. Commun.* **5**, 3949 (2014).
12. Lu, Z. *et al.* Electrochemical tuning of layered lithium transition metal oxides for improvement of oxygen evolution reaction. *Nat. Commun.* **5**, 5345 (2014).
13. Huang, J. H. *et al.* CoOOH nanosheets with high mass activity for water oxidation. *Angew. Chem. Int. Ed.* **54**, 8722–8727 (2015).
14. Song, F. & Hu, X. L. Exfoliation of layered double hydroxides for enhanced oxygen evolution catalysis. *Nat. Commun.* **5**, 4477 (2014).
15. Ma, T. Y., Dai, S., Jaroniec, M. & Qiao, S. Z. Metal-organic framework-derived hybrid Co_3O_4-carbon porous nanowire arrays as reversible oxygen evolution electrodes. *J. Am. Chem. Soc.* **136**, 13925–13931 (2014).
16. Hu, H., Guan, B. Y., Xia, B. Y. & Lou, X. W. Designed formation of $Co_3O_4/NiCo_2O_4$ double-shelled nanocages with enhanced pseudocapacitive and electrocatalytic properties. *J. Am. Chem. Soc.* **137**, 5590–5595 (2015).
17. Deng, X. H. & Tüysüz, H. R. Cobalt-oxide-based materials as water oxidation catalyst: recent progress and challenges. *ACS Catal.* **4**, 3701–3714 (2014).
18. Suntivich, J., May, K. J., Gasteiger, H. A., Goodenough, J. B. & Shao-Horn, Y. A perovskite oxide optimized for oxygen evolution catalysis from molecular orbital principles. *Science* **334**, 1383–1385 (2011).
19. Grimaud, A. *et al.* Double perovskites as a family of highly active catalysts for oxygen evolution in alkaline solution. *Nat. Commun.* **4**, 2439 (2013).
20. Kanan, M. W. *et al.* Structure and valency of a cobalt-phosphate water oxidation catalyst determined by *in situ* X-ray spectroscopy. *J. Am. Chem. Soc.* **132**, 13692–13701 (2010).
21. Su, C. *et al.* $SrCo_{0.9}Ti_{0.1}O_{3-\delta}$ as a new electrocatalyst for the oxygen evolution reaction in alkaline electrolyte with stable performance. *ACS Appl. Mater. Interfaces* **7**, 17663–17670 (2015).
22. Risch, M. *et al.* Structural changes of cobalt-based perovskites upon water oxidation investigated by EXAFS. *J. Phys. Chem. C* **117**, 8626–8635 (2013).
23. Zhu, Y. *et al.* $SrNb_{0.1}Co_{0.7}Fe_{0.2}O_{3-\delta}$ perovskite as a next-generation electrocatalyst for oxygen evolution in alkaline solution. *Angew. Chem. Int. Ed.* **54**, 3897–3901 (2015).
24. Raccah, P. M. & Goodenough, J. B. First-order localized-electron ⇆ collective-electron transition in $LaCoO_3$. *Phys. Rev.* **155**, 932 (1967).
25. Señarís-Rodríguez, M. A. & Goodenough, J. B. $LaCoO_3$ revisited. *J. Solid State Chem.* **116**, 224–231 (1995).
26. Korotin, M. A. *et al.* Intermediate-spin state and properties of $LaCoO_3$. *Phys. Rev. B* **54**, 5309 (1996).
27. Zobel, C. *et al.* Evidence for a low-spin to intermediate-spin state transition in $LaCoO_3$. *Phys. Rev. B* **66**, 020402 (2002).
28. Phelan, D. *et al.* Nanomagnetic droplets and implications to orbital ordering in $La_{1-x}Sr_xCoO_3$. *Phys. Rev. Lett.* **96**, 027201 (2006).
29. Hoch, M. J. R. *et al.* Diamagnetic to paramagnetic transition in $LaCoO_3$. *Phys. Rev. B* **79**, 214421 (2009).
30. Křápek, V. *et al.* Spin state transition and covalent bonding in $LaCoO_3$. *Phys. Rev. B* **86**, 195104 (2012).
31. Karolak, M. *et al.* Correlation-driven charge and spin fluctuations in $LaCoO_3$. *Phys. Rev. Lett.* **115**, 046401 (2015).
32. Zhou, S. M. *et al.* Size-dependent structural and magnetic properties of $LaCoO_3$ nanoparticles. *J. Phys. Chem. C* **113**, 13522–13526 (2009).
33. Zhou, S. M. *et al.* Ferromagnetism in $LaCoO_3$ nanoparticles. *Phys. Rev. B* **76**, 172407 (2007).
34. Rossell, M. D. *et al.* Direct evidence of surface reduction in monoclinic $BiVO_4$. *Chem. Mater.* **27**, 3593–3600 (2015).
35. Han, B. H. *et al.* Role of $LiCoO_2$ surface terminations in oxygen reduction and evolution kinetics. *J. Phys. Chem. Lett.* **6**, 1357–1362 (2015).
36. Gazquez, J. *et al.* Atomic-resolution imaging of spin-state superlattices in nanopockets within cobaltite thin films. *Nano Lett.* **11**, 973–976 (2011).
37. Kwon, J. H. *et al.* Nanoscale spin-state ordering in $LaCoO_3$ epitaxial thin films. *Chem. Mater.* **26**, 2496–2501 (2014).
38. Lan, Q. Q. *et al.* Correlation between magnetism and 'dark stripes' in strained $La_{1-x}Sr_xCoO_3$ epitaxial films ($0 \leq x \leq 0.1$). *Appl. Phys. Lett.* **107**, 242404 (2015).
39. Klie, R. F. *et al.* Direct measurement of the low-temperature spin-state transition in $LaCoO_3$. *Phys. Rev. Lett.* **99**, 047203 (2007).
40. Qian, D. *et al.* Electronic spin transition in nanosize stoichiometric lithium cobalt oxide. *J. Am. Chem. Soc.* **134**, 6096–6099 (2012).
41. Zhou, S. M. *et al.* Magnetic phase diagram of nanosized half-doped manganites: role of size reduction. *Daton Trans.* **41**, 7109–7114 (2012).
42. Curiale, J. *et al.* Magnetic dead layer in ferromagnetic manganite nanoparticles. *Appl. Phys. Lett.* **95**, 043106 (2009).
43. Wang, Y. & Fan, H. J. Low-field magnetoresistance effect in core–shell structured $La_{0.7}Sr_{0.3}CoO_3$ nanoparticles. *Small* **8**, 1060–1065 (2012).
44. Vasseur, S. *et al.* Lanthanum manganese perovskite nanoparticles as possible *in vivo* mediators for magnetic hyperthermia. *J. Magn. Magn. Mater.* **302**, 315–320 (2006).
45. Wang, H. Y. *et al.* In operando identification of geometrical-site-dependent water oxidation activity of spinel Co_3O_4. *J. Am. Chem. Soc.* **138**, 36–39 (2016).
46. Malkhandi, S. *et al.* Design insights for tuning the electrocatalytic activity of perovskite oxides for the oxygen evolution reaction. *J. Phys. Chem. C* **119**, 8004–8013 (2015).
47. Bao, J. *et al.* Ultrathin spinel-structured nanosheets rich in oxygen deficiencies for enhanced electrocatalytic water oxidation. *Angew. Chem. Int. Ed.* **127**, 7507–7512 (2015).
48. Kresse, G. & Furthmuller, J. Efficient iterative schemes for *ab* initio total-energy calculations using a plane-wave basis set. *Phys. Rev. B* **54**, 11169 (1996).

Acknowledgements

This project was financially supported by the National Basic Research Programs of China (2012CB927402 and 2014CB932700), the National Science Foundation of China (Grant Nos. U1432134, 21203099, 51371164, and 21573206), the Anhui Provincial Natural Science Foundation (Grant No. 1508085QE109), the Collaborative Innovation Center of Suzhou Nano Science and Technology, Strategic Priority Research Program B of the CAS under Grant No. XDB01020000, Hefei Science Center CAS (2015HSC-UP016), Fundamental Research Funds for the Central Universities, the Doctoral Fund of Ministry of Education of China (20120031120033), the Research Program for Advanced and Applied Technology of Tianjin (13JCYBJC36800), and the Open Research Fund of State Key Laboratory of Luminescent Materials and Devices (2014-skllmd-05). We appreciate the support from the Tianjin Supercomputing Center.

Author contributions

S.Z., Z.H. and J.Ze. designed the studies and wrote the paper. X.M., X.Z., C.M. and J.Zh. performed most of the experiments. Y.Q. and Z.H. carried out DFT calculations. S.Z., C.M., Z.H., L.S. and J.Ze. performed data analysis. All authors discussed the results and commented on the manuscript.

Additional information

23

Transformation of metal-organic frameworks for molecular sieving membranes

Wanbin Li[1], Yufan Zhang[2], Congyang Zhang[1], Qin Meng[3], Zehai Xu[1], Pengcheng Su[1], Qingbiao Li[4], Chong Shen[3], Zheng Fan[1], Lei Qin[1] & Guoliang Zhang[1]

The development of simple, versatile strategies for the synthesis of metal-organic framework (MOF)-derived membranes are of increasing scientific interest, but challenges exist in understanding suitable fabrication mechanisms. Here we report a route for the complete transformation of a series of MOF membranes and particles, based on multivalent cation substitution. Through our approach, the effective pore size can be reduced through the immobilization of metal salt residues in the cavities, and appropriate MOF crystal facets can be exposed, to achieve competitive molecular sieving capabilities. The method can also be used more generally for the synthesis of a variety of MOF membranes and particles. Importantly, we design and synthesize promising MOF membranes candidates that are hard to achieve through conventional methods. For example, our CuBTC/MIL-100 membrane exhibits 89, 171, 241 and 336 times higher H_2 permeance than that of CO_2, O_2, N_2 and CH_4, respectively.

[1] Institute of Oceanic and Environmental Chemical Engineering, State Key Lab Breeding Base of Green Chemical Synthesis Technology and Collaborative Innovation Center of Membrane Separation and Water Treatment of Zhejiang Province, Zhejiang University of Technology, Chaowang Road 18#, Hangzhou 310014, China. [2] Department of Materials Science and Engineering, College of Engineering, University of California, Berkeley, California 94720, USA. [3] Department of Chemical and Biological Engineering, College of Chemical and Biological Engineering, State Key Laboratory of Chemical Engineering, Zhejiang University, Hangzhou 310027, China. [4] Department of Chemical and Biochemical Engineering, College of Chemistry and Chemical Engineering, National Laboratory for Green Chemical Productions of Alcohols, Ethers and Esters, Key Lab for Chemical Biology of Fujian Province, Xiamen University, Xiamen 361005, China. Correspondence and requests for materials should be addressed to G.Z. (email: guoliangz@zjut.edu.cn).

Molecular sieving membranes composed of silica[1], zeolites[2,3], graphene[4,5] or metal-organic frameworks (MOFs)[6,7] have intrinsic advantages, including both larger permeability and better selectivity, compared with those of conventional polymeric membranes[8]. MOFs consisting of metal ions or metal clusters coordinated with organic linkers have great potential as molecular sieves because of their large surface area, diverse structures and tailorable pore sizes. Until now, tens of thousands of MOFs have been reported and investigated[9]. However, compared with the total number of MOFs reports, those that can be assembled into continuous membranes are limited because of the necessary complex fabrication and activation procedures[10]. Substitution reaction chemistry has been extensively employed in various MOF modification procedures, notably to expand pore aperture or enhance adsorption of MOFs, by linker exchange or transmetalation[11–17]. However, MOFs synthesized through substitution protocols usually possess the same topological structure as their precursor, and no such investigation has focused on MOF membranes. Therefore, if we can establish a synthetic connection between different series of MOFs, we may greatly increase their potential in filtration applications.

Here we report a methodology for realizing the connection and complete transformation of different series of MOF membranes and particles based on multivalent cation substitution. This strategy combines three key concepts: (i) facile transformation of unstable and easily fabricated MOF particles to obtain the stable MOFs with completely different topology structure, which are usually fabricated in a relatively harsh synthetic conditions; (ii) *in situ* transformation of one common MOF membrane to another MOF membrane, which is hard to be synthesized in conventional methods at present; (iii) reducing the pore size through immobilizing the metal salt residue in cavities and exposing the appropriate crystal facets of the MOFs to achieve competitive molecular sieving ability by the transformation of MOF membranes. Our strategy can be used more generally to various MOF membranes and particles, but we exhibit our key findings here with two examples, one is the transformation of CuBTC to MIL-100, which takes the advantages of easy preparation and material stability[18–20]. Another is the transformation of CuBTC membrane to CuBTC/MIL-100 membrane, which provides a facile route to design promising candidates of MOF membrane for molecular sieving.

Results

Transformation of CuBTC to MIL-100.
We first conducted the transformation of CuBTC to MIL-100. CuBTC (also known as HKUST-1), a MOF with a cubic Pt_3O_4-type network, is composed of paddle wheel dimeric copper carboxylate units bridged by three-connecting 1,3,5-benzenetricarboxylate (BTC) linkers[18]. As one of the earliest reported and most studied MOF materials, it has been produced in large scale and used to fabricate continuous membranes[10,19] despite its relatively poor chemical stability[20]. The MIL series of MOFs possess the excellent chemical stability, yet their synthetic conditions are much more rigorous compared with those of CuBTC[21]. For example, MIL-100(Fe) (Supplementary Fig. 1), composed of BTC and Fe-based centre with zeotype architecture[22], is typically synthesized at 150 °C for 6 days and the precursor solution contains hazardous hydrofluoric acid, which is harmful to the environment. Although the preparation process of MIL-100(Fe) has been simplified considerably by reflux and microwave-assisted method at about 100 °C recently[23,24] compared with CuBTC, which can be straightforwardly fabricated at low temperature[19,25–27], the synthetic conditions and experimental equipment are still more

rigorous. Moreover, new synthetic methods are desired highly to formation of MOF at room temperature[27]. Thus, we selected CuBTC and MIL-100 as main precursor and product to demonstrate the transformation between the MOFs. Scanning electron microscopy (SEM) image of CuBTC particle and transformation procedure in metal centre and second building unit are shown in Fig. 1a,b. The transformation was implemented by immersing CuBTC into $FeCl_3 \cdot 6H_2O$ solution at room temperature. All water, methanol, ethanol and N,N-dimethylformamide (DMF) were employed as the solvents. From the comparison between the simulated X-ray diffraction (XRD) of MIL-100 and XRD patterns of the prepared materials, we found that only the pattern of MIL-100 transformed in methanol showed all characteristic peaks similar to simulated XRD patterns, which demonstrated that the methanol was a relatively better solvent for achieving pure MIL-100 crystals with excellent crystallinity, but the three other solvents were not the same case (Supplementary Fig. 2). This phenomenon should be attributed to the balance among the diffusion rate of the solvent and Fe^{3+} ions in the pores of crystals, hydrolysis rate of CuBTC and substitution rate. CuBTC is unstable in water, and the Lewis acidic species of Fe^{3+} further accelerate the hydrolysis. Therefore, CuBTC was hydrolysed before the reaction was completed in aqueous solution. Compared with methanol, ethanol and DMF possess larger diameter and smaller polarity, which may conduce to much lower diffusion rate of reagents in CuBTC and substitution rate, and led to little cation substitution. Some monovalent and divalent cations were also employed to transform CuBTC, the result revealed that the MOF crystal structure had not been changed (Supplementary Fig. 3). Moreover, similar molar quantity of Cu^{2+} and BTC as CuBTC cannot be used to synthesize MIL-100 (Supplementary Fig. 4), demonstrating that the importance of the CuBTC framework and the transformation process was based on multivalent cation substitution.

Transformation mechanism.
We characterized the CuBTC/MIL-100 particles with different substitution times to investigate the reaction processes of transformation. After substitution for 1 h, the number of peaks in XRD pattern showed no big change (Fig. 1c). When the duration time was extended to 21 h, peaks for MIL-100 emerged. With the passage of time, the peak intensities of {200} and {220} degraded rapidly, and the peak intensity ratios of {200}/{222} and {220}/{222} also decreased dramatically (Fig. 1e). Moreover, it should be noted that the peak intensity of {333} also displayed a much lower degradation rate compared with other peaks. CuBTC exhibits three window structures, largest pores with diameter of 9 Å, moderate side pores with the diameter of 5 Å and smallest triangular shaped windows with the diameter of 3.5 Å (ref. 28). Correspondingly, the window size of {111} facets is 3.5 Å, which is smaller than the pore size of 5 and 9 Å in {100} facets (Fig. 1d). The octahedron crystal morphology and the much higher peak intensity of {222} compared with {200} and {220} reveal that the CuBTC has mainly exposed {111} facet[29]. Thus, by combining above experimental results and CuBTC structures, we can deduce the following reaction processes: when CuBTC crystals were immersed into the substitution solutions, the exposed {111} facets would greatly hinder the diffusion of the solvent and Fe^{3+} ions, this may be the reason for the little cation substitution of CuBTC in above-mentioned three solvents. After the copper ions were replaced by the Fe^{3+} ions, some MIL-100 nanocrystals formed on the surface of the CuBTC crystals, and subsequently the channel for diffusion of the solvent and Fe^{3+} ion was produced due to the larger window of MIL-100 and the gap between the two kinds of MOFs.

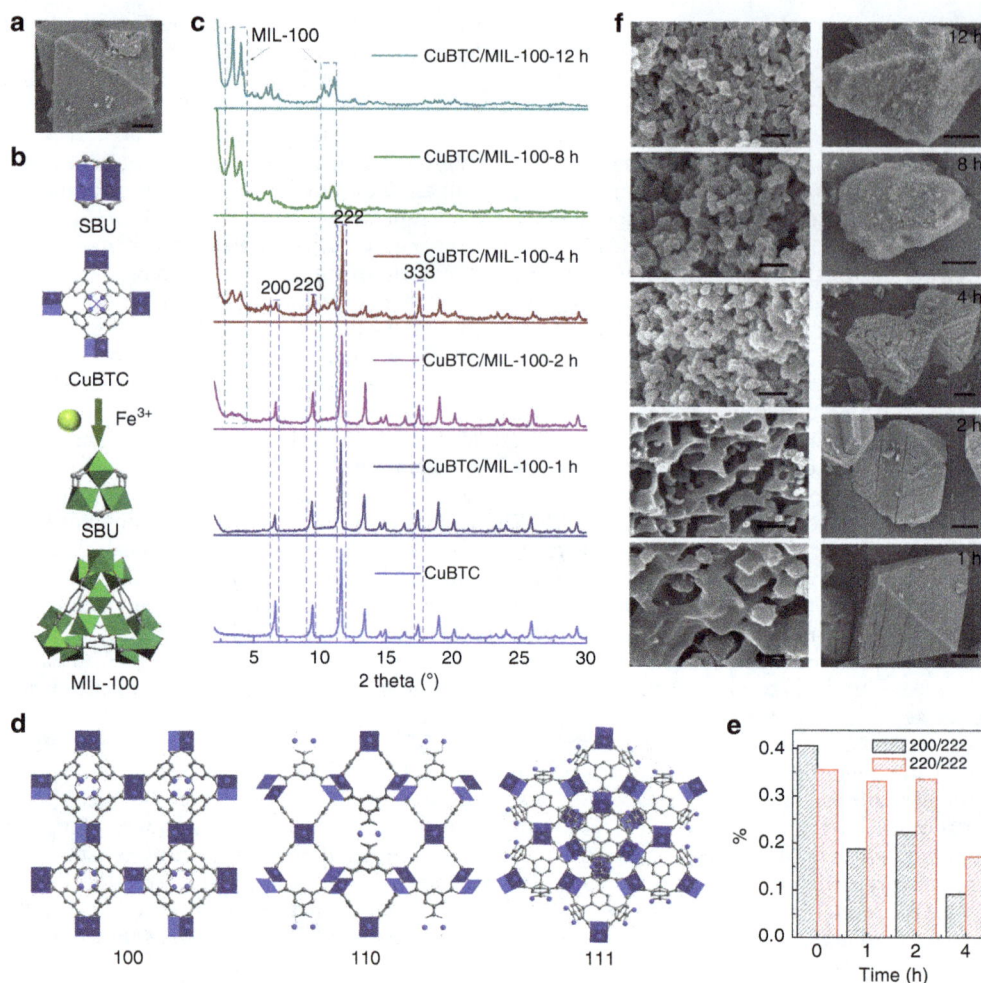

Figure 1 | Transformation of CuBTC to MIL-100. (**a**) SEM image of as-synthesized octahedron CuBTC crystal. Scale bar, 10 μm. (**b**) Transformation procedure in metal centre and second building unit of CuBTC and MIL-100. The Cu and Fe coordination polyhedra are represented in blue and green, respectively, whereas the BTC links are depicted by sticks. (**c**) XRD patterns of CuBTC/MIL-100 with different transformation time. (**d**) CuBTC structure viewed at different facet. (**e**) Peak intensity ratios of {200}/{222} and {220}/{222} with the transformation time increasing. (**f**) SEM images of CuBTC/MIL-100 with different transformation time. The octahedron CuBTC crystal (bottom, 1h) transformed into nanoparticle MIL-100 crystal (top,12h). Scale bar, 200 nm (left); 20 μm (right).

As a result, the solvent and the Fe^{3+} ion would pass through the {100} facets quickly and replace the adjacent Cu^{2+} ions in CuBTC crystals to form MIL-100. Although the window size of {110} facets was the same as that of {111} facets, the {111} facets possessed the large atomic density and accordion-like channel, which greatly decreased the substitution rate and increased the diffusion resistance. This also led to the decrease of the value of {220}/{222}. The morphologies of these CuBTC/MIL-100 crystals are consistent with the XRD results and the deduction (Fig. 1f and Supplementary Figs 5 and 6). After substitution for 1 h, the crystals displayed that CuBTC/MIL-100 had superstructural composite surfaces and three-dimensional CuBTC shelf with interior MIL-100 nanocrystals. When the substitution time increased, the superstructure extended to the middle of the CuBTC crystals. After reaction for 81 h, all CuBTC were transformed into MIL-100. For identification of the inside structure of CuBTC/MIL-100 in transformation, the optical microscopy and energy dispersion spectroscopy (EDS) mapping images of CuBTC/MIL-100 with different transformation time were collected and presented in Fig. 2a–c. We can find that, with the extension of transformation times, the blue part of CuBTC became small but the shape was still octahedron. The optical

microscopy images showed the lamellar structure, which was parallel to the boundary of CuBTC crystal, this may be attributed to the untransformed CuBTC with preferred {111} crystal facet, which was consistent with the results of XRD and showed the important role of {111} facet in transformation. After transformation for 12 h, the whole blue CuBTC was changed to yellow MIL-100. EDS mapping showed the similar phenomenon. When transformation time was extended to 12 h, the Cu/Fe ratio was reduced to 0.4%. These results also revealed that the transformation was conducted from outside to inside. To further demonstrate the importance of the exposed {111} facet for transformation, we synthesized the cubic CuBTC with mainly exposed {100} facet as previous study (Fig. 2d)[30], and used it to obtain the MIL-100. The XRD pattern revealed that this transformation failed to obtain the MIL-100 (Fig. 2f). Compared with the above successful transformation, two differences may lead to the unsuccessful transformation, the difference in exposed facet and the difference in CuBTC particle size. We further prepared the small octahedron CuBTC by simply stirring the precursor at room temperature (Fig. 2e)[26], and found that this material could be transformed to be MIL-100 successfully. This results verified that the difference in exposed

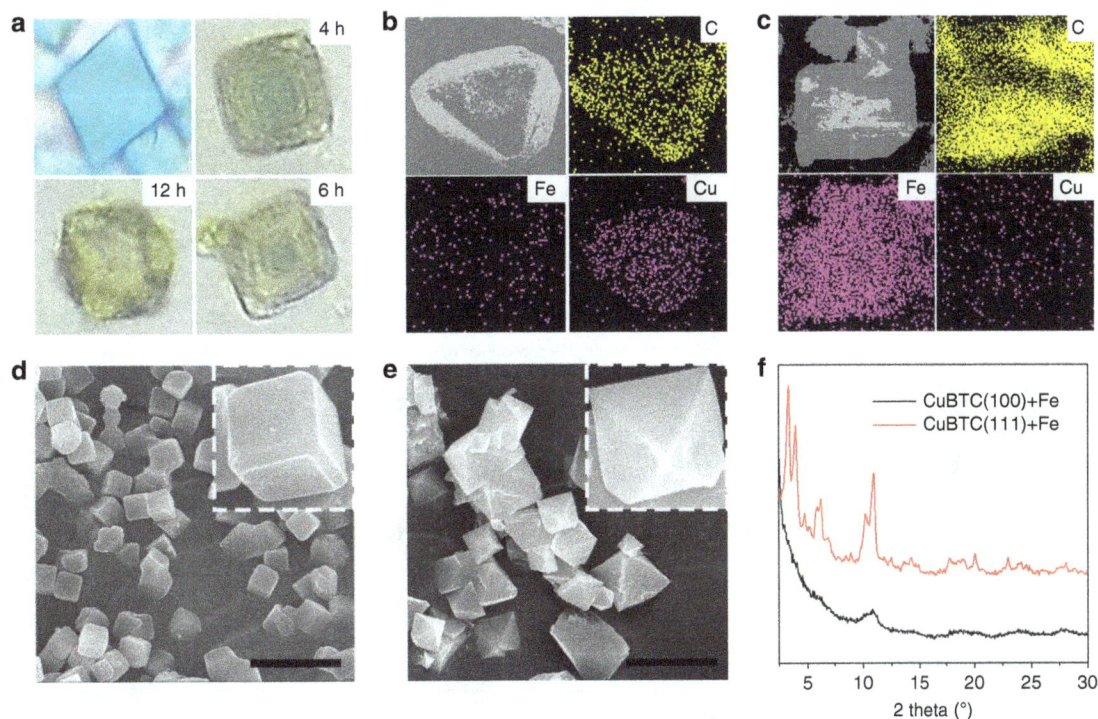

Figure 2 | Transformation process from CuBTC to MIL-100. (**a**) Optical microscopy images of CuBTC with different transformation time, the transformation was carried out under standing. (**b**) EDS mapping of the CuBTC/MIL-100 after transformation for 4 h and purification in methanol, Fe/Cu ratio is 23.5%. (**c**) EDS mapping of the CuBTC/MIL-100 after transformation for 12 h and purification in methanol and hot water, Cu/Fe ratio is 0.4%. (**d**) SEM images of the cubic CuBTC with mainly exposed {100} facet. Scale bar, 5 μm. (**e**) SEM images of the small octahedron CuBTC with mainly exposed {111} facet. (**f**) XRD patterns of cubic CuBTC and small octahedron CuBTC after transformation. Scale bar, 5 μm.

facet was the main reason of the failing transformation of the cubic CuBTC with mainly exposed {100} facet, rather than the difference in CuBTC particle size. The experiment also showed that the MIL-100 can be prepared by transformation of the small octahedron CuBTC in mild route at room temperature without using any special equipment (Supplementary Table 1).

Universality of transformation. To present the versatility of this strategy, we transformed the MOF-5 to CuBDC by Cu^{2+} substitution and transformed the CuBDC to MIL-53 by Fe^{3+} substitution. In this process, MOF-5 was a face-centred cubic material, formed by Zn-based metal cluster and 1,4-ben-zenedicarboxylate (BDC) linker[31], and can be synthesized at room temperature by adding triethylamine[32]; CuBDC consisted of paddle wheel dimeric copper carboxylate units interconnected by BDC[33], and was usually fabricated in DMF at 110 °C or in acetonitrile system[33–36]; MIL-53 was formed by $FeO_4(OH)_2$ clusters with BDC[37], and prepared typically in DMF or in water at high temparature[38–40]. The crystalline structures of these MOFs are presented in Supplementary Fig. 7. The XRD patterns and typical SEM images reveal the successful transformation (Supplementary Figs 7, 8 and 9). Because of the mild and environment-friendly process, the complex synthesis condition of some MOFs can be simplified greatly by transformation, which is of benefit to the scalable production of these MOFs.

Property of transformed MIL-100. We further studied the features of the transformed MIL-100. Nitrogen adsorption experiments reveal that the CuBTC sample has Brunauer–Emmett–Teller (BET)-specific surface area of 960 $m^2 g^{-1}$ and pore volume of 0.48 $cm^3 g^{-1}$ (Fig. 3a). After transformation, BET-specific

surface area and pore volume increase obviously to 1880 $m^2 g^{-1}$ and 1.07 $cm^3 g^{-1}$, respectively, and the hysteresis loop reveals the existence of mesopores in MIL-100 (ref. 41). These values are consistent with those reported in previous studies (Supplementary Table 1), and demonstrate the good porosity of the prepared MIL-100. The DFT pore size distribution curve displays that the prepared MIL-100 possesses the multi-scale pore structure (Fig. 3b), as similar results have been observed in many previous studies[24,42–44]. The small skewing in O 1s X-ray photoelectron spectroscopy demonstrates the successful transformation of the C–O–Cu bond in CuBTC to C–O–Fe bond in MIL-100 (Supplementary Fig. 10). Thermal analysis shows the same good thermal stability of the transformed MIL-100 and normal thermal degradation curve as those in previous reports[22] (Supplementary Fig. 11). From the SEM images (Fig. 1f), we can find that MIL-100 nanocrystals have a relatively small size in the range of 30–60 nm. The integrated MIL-100 particle after treated by ultrasonic treatment demonstrates that MIL-100 nanoparticles grew together tightly, rather than simple accumulation (Supplementary Fig. 12). These features exhibit that MIL-100 was hierarchically porous material. Moreover, we found that some Fe-based materials can be kept in cavities by insufficient purification to reduce the pore size of the obtained MOFs. With the decrease of purification cycle, the adsorption–desorption isotherms change from type-I/IV to type-II. The specific surface area and pore volume reduce to 122 $m^2 g^{-1}$ and 0.18 $cm^3 g^{-1}$ (Fig. 3a), respectively, and the pore size also decreases obviously (Fig. 3b). The relative intensities of XRD patterns of the as-synthesized MIL-100 also show a strong reduction compared with the MIL-100 after purification, especially at low angles (Supplementary Fig. 13), which may be attributed to the impregnation of amorphous $FeCl_3$ (refs 21,42).

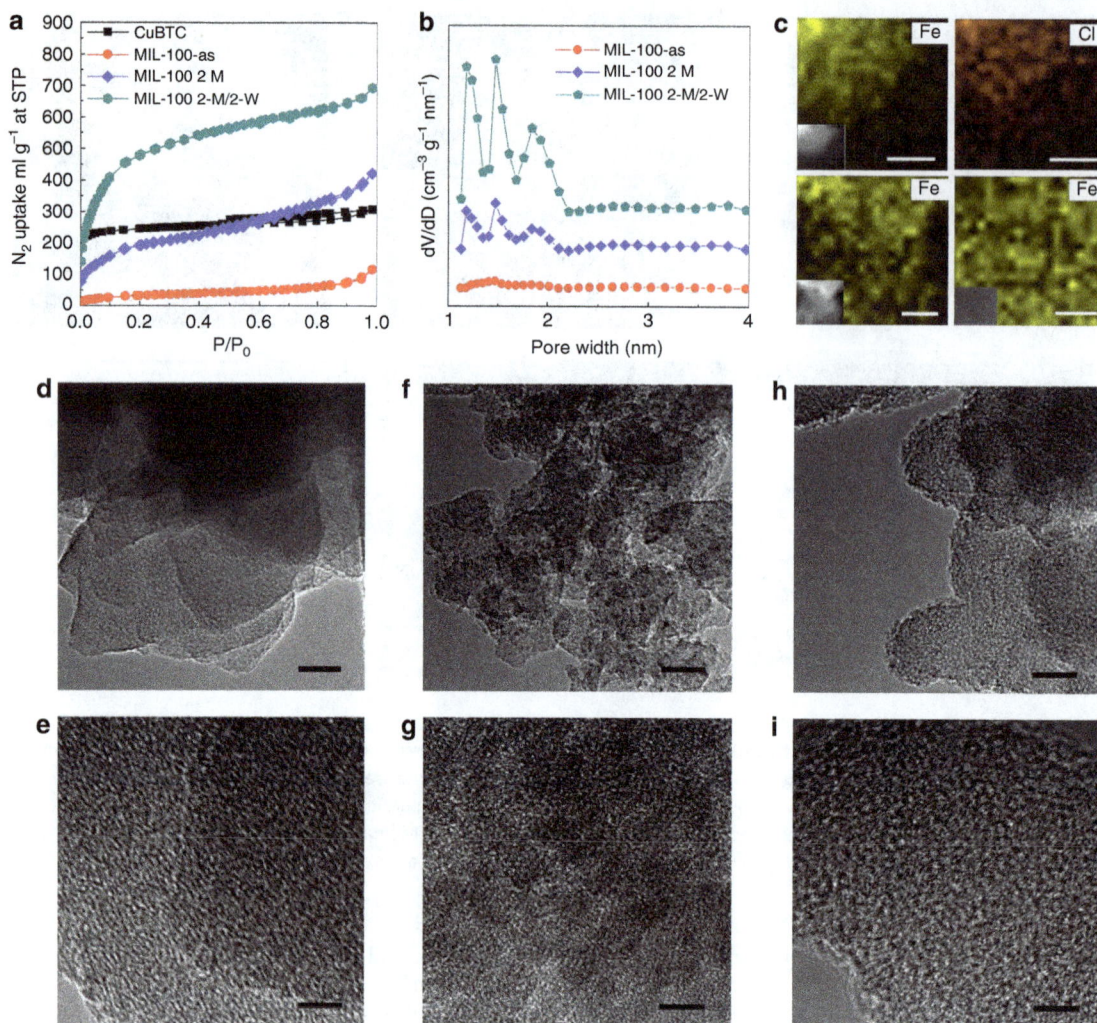

Figure 3 | Characterization of MIL-100. (**a**) N_2 adsorption–desorption isotherms on MIL-100 with various purification cycles, showing the specific surface area reduced with the purification cycle, two cycles in methanol (2-M), two cycles in methanol and then in hot water at 70 °C (2-M/2-W). (**b**) Corresponding pore size distributions calculated by DFT method. (**c**) EDS mapping of the MIL-100 with different post-processing. Top two images: as-synthesized MIL-100 and demonstrating the well-dispersed chlorine, scale bar, 20 nm; bottom left image: as-synthesized MIL-100 after calcination at 200 °C for 4 h, scale bar, 10 nm; bottom right image: MIL-100 after purification; scale bar, 5 nm. (**d,e**) TEM images of as-synthesized MIL-100, showing the amorphous FeCl$_3$ in the cavities. (**f,g**) TEM images of as-synthesized MIL-100 after calcination at 200 °C for 4 h, indicating the presence of nanoclusters in the cavities of the MIL-100. (**h,i**) TEM images of MIL-100 after purification, indicating clear porous structure and pure framework. Scale bar, 20 nm (in **d,f,h**); 5 nm (in **e,g,i**).

Combining with the vast existence of well-dispersed chlorine (10.2%) in the as-synthesized MIL-100 (Supplementary Table 2 and Fig. 3c), and the small amount of chlorine (0.2%) in MIL-100 after purification in methanol and hot water, we come to a conclusion that the cavities of as-synthesized MIL-100 have been occupied by highly dispersed amorphous FeCl$_3$, which can be utilized to artificially control the pores of the transformed MOFs. To further prove the existence of FeCl$_3$ in the cavities of as-synthesized MIL-100, it was calcined at 200 °C for 41 h. The chlorine displays a great reduction (74.9%) and the XRD patterns show some new peaks of Fe$_2$O$_3$ for as-synthesized MIL-100 (Supplementary Fig. 14). All these reveal that the amorphous FeCl$_3$ in the cavities has been converted to Fe$_2$O$_3$ nanoclusters. To further identify the formation of FeCl$_3$ and the structure of the MIL-100, we collected the transmission electron microscopy (TEM) images of as-synthesized MIL-100, as-synthesized MIL-100 after calcination at 200 °C for 4 h and MIL-100 after purification in methanol and hot water (Fig. 3d–i). TEM images clearly indicate that abundant of amorphous materials are left in

the cavities of as-synthesized MIL-100 (Fig. 3d,e), however, the purified MIL-100 has clear porous structure and pure framework (Fig. 3h,i), there are almost no residues left in the cavities of purified MIL-100. For as-synthesized MIL-100 after calcination, the uniformly dispersed dark spots in TEM images demonstrates that the Fe$_2$O$_3$ nanoclusters are immobilized inside the cavities of insufficiently purified MIL-100 (Fig. 3f,g). The nanoclusters in cavities may be beneficial to improve the activity of MOFs. In addition, the reduced pore size will be very important for the performance of molecular sieve.

Transformation of MOF membrane. After demonstrating the transformation of MOFs, our strategy was further employed to transform MOF membranes. CuBTC membrane was synthesized on polymeric hollow fibre by solvothermal method and used as precursor[45] (Supplementary Fig. 15). Figure 4a–c presents the SEM images of CuBTC membrane, as-synthesized CuBTC/MIL-100 membrane and CuBTC/MIL-100 membrane after purification. The CuBTC layer with thickness of ~20 μm is

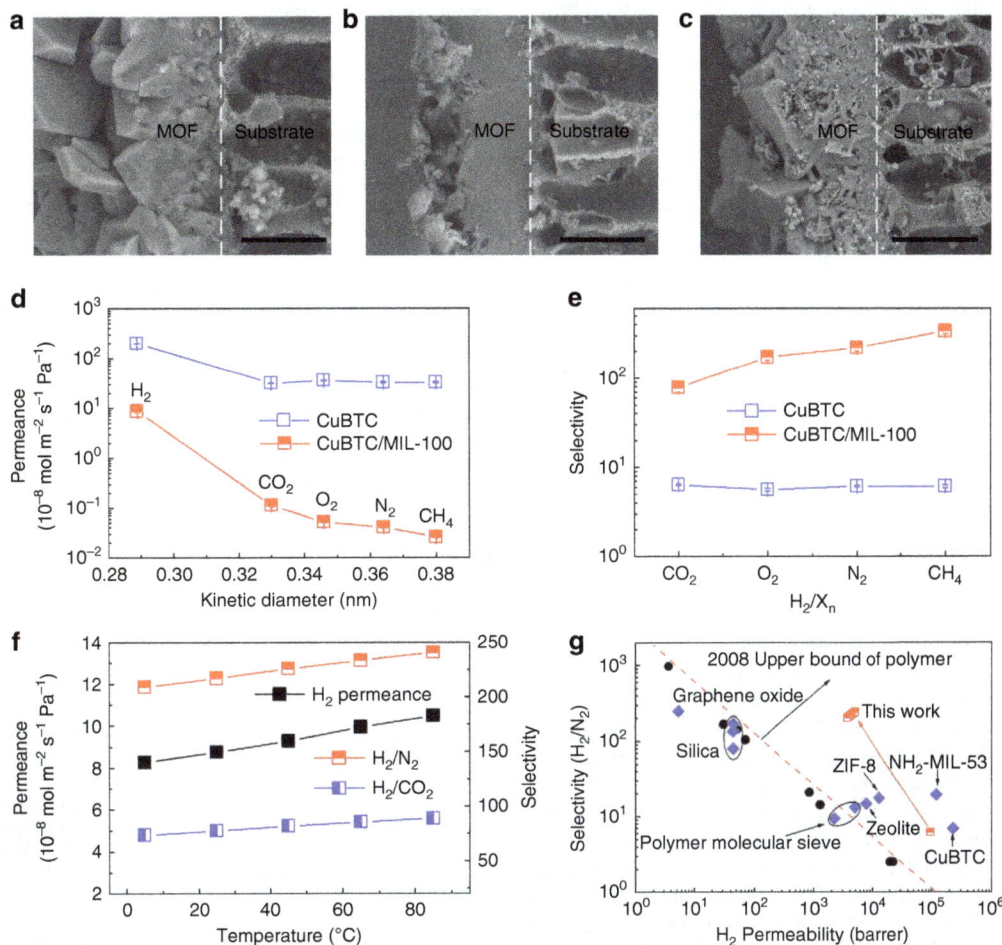

Figure 4 | Transformation of CuBTC membrane and their performance. (a–c) SEM images of original CuBTC membrane, transformed CuBTC/MIL-100 membrane and transformed CuBTC/MIL-100 membrane after purification, respectively. Scale bar, 20 μm. **(d,e)** Gas permeance and selectivities of the CuBTC and CuBTC/MIL-100 membranes. All the average permeation results with standard deviation were calculated from three measurement data. **(f)** Effect of temperature on H_2 permeance and H_2/CO_2 and H_2/N_2 selectivities for CuBTC/MIL-100 membrane. **(g)** Comparison of CuBTC/MIL-100 membrane with polymeric[8,47], silica[1], zeolite[3], other MOF[46,48,49] and graphene oxide membranes[5] for H_2/N_2 system. 1 barrer $= 3.348 \times 10^{-16}$ mol m^{-2} s^{-1} Pa^{-1}, the red dotted line is the Robeson's upper-bound reported in 2008 (ref. 8).

anchored to the substrate continuously. The octahedron morphology of crystal demonstrates that the CuBTC membrane also has mainly exposed {111} facet. After transformation, the as-synthesized CuBTC/MIL-100 membrane seems to become denser due to the FeCl$_3$ residue (Supplementary Figs 16 and 17). However, after the FeCl$_3$ was removed, the massive MIL-100 nanoparticles accompanied by CuBTC flake occurred on the cross-section of purified CuBTC/MIL-100 membrane (Fig. 4c and Supplementary Fig. 16). EDS mapping demonstrates the incomplete transformation (Supplementary Fig. 17). The existence of the peak with 2 theta at 11.6° in XRD pattern also shows the incomplete transformation and the residual CuBTC with exposed {111} facet (Supplementary Fig. 18). This is consistent with the phenomenon observed in incompletely transformed MOF particles, and can be explained by the smaller mass transfer-specific surface area exposed to the solution of membrane. When the duration time was extend to 48 h, all the CuBTC were transformed to MIL-100 (Supplementary Fig. 16).

Performance of transformed membrane. To investigate the performance of the prepared MOF membranes by transformation,

we used the constant-pressure method to measure H_2 (kinetic diameter: 0.289 nm), CO_2 (0.33 nm), O_2 (0.346 nm), N_2 (0.364 nm) and CH_4 (0.38 nm) permeances through the two dense membranes[4]. The CuBTC membrane exhibited H_2 permeance and selectivities of H_2/X_n (X_n: other gases) in the range of 5.5–6.5, similar to those previously reported[46] (Fig. 4d,e). For the transformed CuBTC/MIL-100 membrane, the permeances of all gases were smaller than those of CuBTC membrane, and the largest H_2 permeance was 8.8×10^{-8} mol m^{-2} s^{-1} Pa^{-1}. However, the selectivities of H_2/CO_2, H_2/O_2, H_2/N_2 and H_2/CH_4 displayed great improvement and reached about 77.6, 170.6, 217.0 and 335.7, respectively. Moreover, the selectivities increased with temperature and reached 89.0 and 240.5 for H_2/CO_2 and H_2/N_2 at 85 °C, respectively, as the H_2 permeances grew faster than other gases. Meanwhile, the H_2 permeance also increased to 10.5×10^{-8} mol m^{-2} s^{-1} Pa^{-1} (Fig. 4f). The transformed membrane also presented excellent durability, which maintained good performance with only small fluctuation over a 192-h period (Supplementary Fig. 19). Our transformed molecular sieving membrane had much better performance than polymeric membranes and can easily exceed the Robeson's upper-bound reported in 2008 for all H_2/CO_2, H_2/N_2 and H_2/CH_4 systems[8,47]. With regard to porous zeolites[3] and conventional MOF membranes[46,48,49], the transformed CuBTC/MIL-100 membrane showed superior performance in selectivities (Fig. 4g and Supplementary Fig. 20). Even compared with the silica, graphene oxide and $Zn_2(bim)_4$ nanosheet membranes reported, which exhibited great selectivities for H_2/CO_2 or/and H_2/N_2 systems[1,4,5,7], our membrane also exhibited competitive performance. The transformed CuBTC/MIL-100 membrane possessed excellent H_2 permeability as high as $\sim 4,000$ barrer as well as good stability under high trans-membrane pressure of 0.25 MPa (Supplementary Fig. 21), while the reported membranes were usually operated under very low trans-membrane pressure of 0.1 or even 0 MPa to prevent possible cracks. In binary mixture separation, because of the competitive adsorption, the membrane showed a little smaller H_2 permeance and selectivities compared with the values measured by constant-pressure method, but the membrane still showed the competitive selectivities (Supplementary Table 3). Moreover, the synthetic procedure has good reproducibility, which has been demonstrated by the similar separation performance of two further additional membranes. We speculate that the good molecular sieving performance of transformed CuBTC/MIL-100 membrane is caused by the following three factors (Supplementary Fig. 22): First, the transformed membrane is continuous, which is the basis for good performance. Second, as mentioned above, the reagents first enter into of the {100} facets of CuBTC and then the transformation occurs, the peak at $11.6°$ in XRD pattern displays the exposed {111} facets of the residual CuBTC, so the triangular shaped window with a diameter of 3.5 Å in the residual and the exposed {111} facets is a main channel for gas to pass through. This window size is similar to the kinetic diameter of the CO_2, O_2 and N_2, and is much larger and smaller than that of H_2 and CH_4, respectively. Thus, H_2 molecules can penetrate the transformed membrane quickly. Third, the massive amorphous $FeCl_3$ can fill the gaps between MIL-100 and CuBTC and also occupy the cavities and pores of MOFs. As a result, the void interfaces can be eliminated, the gas channel can be reduced and the gas selectivities of the transformed membrane are increased. Because of the tailorable pore sizes of MOF materials and the diversity in membrane synthesis, we envisage there would be some optimal MOFs and proper preparation conditions, where the membranes with smaller thickness and better separation performance can be achieved by transformation in the future.

Discussion

We have developed a facile and general methodology for realizing the connection and complete transformation of different series of MOF membranes and particles based on multivalent cation substitution. Through this strategy, the unstable and easily fabricated MOF particles can be transformed to obtain the stable MOFs with completely different topology structure, which are usually fabricated in relatively harsh synthetic conditions. The common MOF membranes can also be in situ transformed to another MOF membrane, which is hard to be synthesized in conventional methods so far. The pore size can be controlled through immobilizing the metal salt residue in cavities and the appropriate crystal facets of the MOFs can be exposed to achieve competitive molecular sieving ability by the transformation. The typically transformed CuBTC/MIL-100 membrane with good stability exhibits 89, 171, 241 and 336 times higher H_2 permeance than that of CO_2, O_2, N_2 and CH_4, respectively. The method can be used more generally to various MOF membranes and particles with great potential in wide applications.

Methods

Synthesis of CuBTC. $Cu(NO_3)_2 \bullet 3H_2O$ and BTC were dissolved in water and ethanol, respectively. For crystallization, the two solutions were mixed and

transferred into a Teflon-lined autoclave. Then, the reaction mixture was heated. After cooling to room temperature, the blue particles were isolated by centrifugation. Eventually, CuBTC particles were washed by ethanol and methanol and dried.

Synthesis of MOF-5. $Zn(NO_3)_2 \bullet 6H_2O$ and BDC were dissolved into DMF, respectively. For crystallization, the two solutions were mixed and heated in an autoclave[50]. After cooling to room temperature, the white particles were separated by centrifugation. Eventually, the prepared particles were washed by DMF and methanol and dried.

Synthesis of CuBDC. $Cu(NO_3)_2 \bullet 3H_2O$ and BDC were dissolved into the DMF, respectively. For crystallization, the two solutions were mixed and heated[34]. After cooling, the blue particles were isolated by centrifugation. Eventually, CuBTC particles were washed by DMF and methanol and dried.

Synthesis of CuBTC with mainly exposed {100} facet. $Cu(NO_3)_2 \bullet 3H_2O$ and lauric acid were dissolved into ethanol (10 ml), BTC was also dissolved in ethanol (10 ml). For crystallization, the two solutions were mixed and heated at 150 °C for 24 h (ref. 30). After natural cooling, the resulting blue powders were isolated by centrifugation and washed with ethanol.

Synthesis of small CuBTC with mainly exposed {111} facet. $Cu(NO_3)_2 \bullet 3H_2O$ and BTC were dissolved in water and ethanol, respectively, to obtain the solutions with concentration of 80 mmol l^{-1}. For crystallization, the BTC solution was added into the metal salt solution and stirred at room temperature for 18 h. After reaction, the blue CuBTC crystals were separated by centrifugation and washed by ethanol.

Transformation of CuBTC to MIL-100. $FeCl_3 \bullet 6H_2O$ (2.00 g) was dissolved in different solvents. CuBTC (0.1 g) was dispersed into the above solution. For transformation, the obtained suspension was fixed into a shaker. Reaction was carried out at room temperature. The reaction time was 1, 2, 4, 8 or 12 h. After reaction, the prepared particles were separated by centrifugation (5,000 r.p.m., 5 min) and purified with methanol. The obtained MIL-100 was purified with various purification cycles, two cycles in methanol (2-M), two cycles in methanol and then in hot water at 70 °C (2-M/2- W). To further identify the nanoclusters in MIL-100, the obtained particles were dried and calcined at 200 °C for 4 h. As comparison, water, ethanol and DMF were all employed as the solvent for transformation. To demonstrate the importance of the CuBTC and that the transformation was not a hydrolysis-recrystallization process, the $Cu(NO_3)_2 \bullet 3H_2O$ (0.124 g) and BTC (0.066 g) were applied to displace the CuBTC, which possesed the same molar quantity as the CuBTC. The result revealed that the CuBTC was vital for transformation.

Transformation of MOF-5 to CuBDC. MOF-5 (0.1 g) was dispersed into $Cu(NO_3)_2 \bullet 3H_2O$ methanol solution (0.05 g ml^{-1}). The above suspension was fixed into a shaker. Transformation was carried out at room temperature. The prepared blue particles were collected by centrifugation and purified by methanol and dried at 80 °C.

Transformation of CuBDC to MIL-53. CuBDC (0.1 g) was dispersed into $FeCl_3 \cdot 6H_2O$ methanol solution (0.05 g ml^{-1}). The above suspension was fixed into a shaker. Transformation was carried out at room temperature for 12 h. The prepared particles were collected by centrifugation and purified by methanol and dried at 80 °C.

Synthesis of CuBTC membrane. For dopamine modification[48], the polyvinylidene difluoride (PVDF) was washed using water and dried at room temperature to remove impurities on membrane surface, and levodopa was added in 10 mM Tris-HCl to obtained transparent solution. Then, the PVDF hollow fibre was immersed in the prepared solution for dopamine deposition. After deposition, the membrane was washed using water and dried at atmosphere. For non-activation ZnO array growth, 0.592 g of zinc nitrate hexahydrate, 0.135 g of sodium formate and 0.245 g of 2-methyl-imidazole were dissolved in methanol and transferred into an autoclave. The modified PVDF hollow fibre was immersed vertically into the mixture solution by using a self-made Teflon holder. Then, the autoclave was sealed and heated at 80 °C. After reaction for 12 h, PVDF hollow fibre with non-activation ZnO array was taken out and washed by ultrasound for 60 s to remove the loose powder. Ultimately, the hollow fibre was washed by methanol and dried at atmosphere. For synthesis, $Cu(NO_3)_2 \bullet 3H_2O$ (0.5 g) and BTC (0.25 g) were dissolved in water and ethanol, respectively. These two solutions were mixed to obtain a clear precursor solution. The mixed solution was poured into an autoclave, and the PVDF hollow fibre with non-activation ZnO array was also placed vertically in the autoclave by using a self-made Teflon holder. The crystallization was executed at 85 °C for 48 h. After crystallization, the hollow fibre membrane was taken out, washed with ethanol and dried at room temperature.

Transformation of CuBTC membrane. $FeCl_3 \cdot 6H_2O$ (1.00 g) was dissolved into the methanol. CuBTC hollow fibre membrane was soaked into the prepared solution at room temperature for 12 or 48 h. After reaction, the membrane was taken out and dried directly for obtaining the dense CuBTC/MIL-100 membrane. As comparison, the membrane after transformation was immersed in the pure methanol to remove the excess $FeCl_3$ component.

Gas permeation measurement. The gas permeation properties of the membrane were studied by constant-pressure, variable-volume method[4]. The dense membranes were put in a permeation module and sealed by epoxy glue. The effective area was calculated by the outer surface. The measured gas was used to rinse the permeation module. The feed gas and permeate gas were fed and collected at the shell side and tube side of the membrane, respectively. Upstream pressure and downstream pressure were 2 and 1 bar (atmosphere conditions), respectively, and transmembrane pressure was 1 bar. The experiment was carried out with different kinetic diameters in the following order: H_2 (0.289 nm), CO_2 (0.33 nm), O_2 (0.346 nm), N_2 (0.364 nm), CH_4 (0.38 nm), CH_4, N_2, O_2, CO_2 and H_2. Gas flow rates were measured by a bubble flow-metre. The gas permeation data were calculated by averaging the measured values of two cycles. The data were read and recorded until the system running stably. Gas permeance (P, $mol\, m^{-2}\, s^{-1}\, Pa^{-1}$) and gas permeability ($P_G$, $Barrer = 3.348 \times 10^{-16}\, mol\, m^{-2}\, s^{-1}\, Pa^{-1}$) were calculated by using the following equations:

$$P = \frac{1}{p_u - p_d} \times \frac{1}{A} \times \frac{p_d \times \Delta V}{R \times (273.15 + T) \times \Delta t} \tag{1}$$

$$P_G = \frac{1}{p_u - p_d} \times \frac{1}{A} \times \frac{p_d \times \Delta V}{R \times (273.15 + T) \times \Delta t} \tag{2}$$

where p_u and p_d are the upstream pressure and downstream pressure, respectively, A and l are the membrane effective area and membrane thickness (thickness of the MOF layer), R and T are the gas constant value and temperature (Celsius), ΔV and Δt are the volume of the gas through the membrane and the corresponding time. The permselectivity (α) is defined as the ratio of two kinds of gas permeances.

$$\alpha = \frac{P_1}{P_2} \tag{3}$$

Characterization. XRD patterns were recorded by a PANalytical X' Pert PRO X-ray diffractometer with Cu Kα radiation ($\lambda = 0.154056$ nm) at 40 kV and 40 mA. A field-emission scanning electron microscope (S-4700, Hitachi) was used to observe the morphologies of the membranes. Accelerating voltage was 15 kV. The attached X-ray EDS (GENESIS4000, EDAX) was applied to analyse the element content of prepared MIL-100 particle after tabletting. To keep the cross-section morphology the MOF membranes, it was freeze-fractured in liquid nitrogen. All the prepared samples were coated with an ultrathin layer of platinum using an ion sputter coater to minimize charging effect. X-ray photoelectron spectroscopy experiments were performed on a RBD upgraded PHI-5000C ESCA system (Perkin Elmer) with an incident radiation of monochromatic Mg Kα X-rays (hν = 1253.6 eV) at 250 W. The spectra of all the elements were collected by using RBD 147 interface (RBD Enterprises). A JEM-2100 (JEOL Co.) operated with accelerating voltage of 200 kV was used for obtaining the TEM images. TG measurements were executed on a thermal gravimetric analyser (PERKIN ELMER, Model TGA 7). The samples were heated from 25 or 40 to 700 °C with a heating rate of 20 °C min^{-1} under a flow of synthetic air with a flow rate of 20 ml min^{-1}. N_2 adsorption–desorption isotherms were measured on a Micromeritics-Accelerated Surface Area and Porosimetry system (ASAP 2020M + C, Micromeritics Instrument Co.). Measurements were carried out at 77 K held using a liquid nitrogen bath. The samples were degassed in vacuum at 150 °C for 12 h before the analysis. BET method was used to calculate the specific surface areas in the P/P_0 range of 0.05–0.1. DFT was used to obtaining the pore-size distributions.

References

1. de Vos, R. M. & Verweij, H. High-selectivity, high-flux silica membranes for gas separation. *Science* **279**, 1710–1711 (1998).
2. Lai, Z. P. *et al.* Microstructural optimization of a zeolite membrane for organic vapor separation. *Science* **300**, 456–460 (2003).
3. Guan, G., Tanaka, T., Kusakabe, K., Sotowa, K. & Morooka, S. Characterization of AlPO(4)-type molecular sieving membranes formed on a porous alpha-alumina tube. *J. Membr. Sci.* **214**, 191–198 (2003).
4. Kim, H. W. *et al.* Selective gas transport through few-layered graphene and graphene oxide membranes. *Science* **342**, 91–95 (2013).
5. Li, H. *et al.* Ultrathin, molecular-sieving graphene oxide membranes for selective hydrogen separation. *Science* **342**, 95–98 (2013).
6. Brown, A. J. *et al.* Interfacial microfluidic processing of metal-organic framework hollow fiber membranes. *Science* **345**, 72–75 (2014).
7. Peng, Y. *et al.* Metal-organic framework nanosheets as building blocks for molecular sieving membranes. *Science* **346**, 1356–1359 (2014).
8. Robeson, L. M. The upper bound revisited. *J. Membr. Sci* **320**, 390–400 (2008).
9. Furukawa, H., Cordova, K. E., O'Keeffe, M. & Yaghi, O. M. The chemistry and applications of metal-organic frameworks. *Science* **341**, 1230444 (2013).
10. Qiu, S., Xue, M. & Zhu, G. Metal-organic framework membranes: from synthesis to separation application. *Chem. Soc. Rev.* **43**, 6116–6140 (2014).
11. Han, Y., Li, J. R., Xie, Y. & Guo, G. Substitution reactions in metal-organic frameworks and metal-organic polyhedral. *Chem. Soc. Rev.* **43**, 5952–5981 (2014).
12. Brozek, C. K. & Dincă, M. Cation exchange at the secondary building units of metal-organic frameworks. *Chem. Soc. Rev.* **43**, 5456–5467 (2014).
13. Dincă, M. & Long, J. R. High-enthalpy hydrogen adsorption in cation-exchanged variants of the microporous metal-organic framework $Mn_3[(Mn_4Cl)_3(BTT)_8(CH_3OH)_{10}]_2$. *J. Am. Chem. Soc.* **129**, 11172–11176 (2007).
14. Nouar, F., Eckert, J., Eubank, J. F., Forster, P. & Eddaoudi, M. Zeolite-like metal-organic frameworks (ZMOFs) as hydrogen storage platform: lithium and magnesium ion-exchange and H_2-(rho-ZMOF) interaction studies. *J. Am. Chem. Soc.* **131**, 2864–2870 (2009).
15. Kim, M., Cahill, J. F., Fei, H., Prather, K. A. & Cohen, S. M. Postsynthetic ligand and cation exchange in robust metal-organic frameworks. *J. Am. Chem. Soc.* **134**, 18082–18088 (2012).
16. Li, T., Kozlowski, M. T., Doud, E. A., Blakely, M. N. & Rosi, N. L. Stepwise ligand exchange for the preparation of a family of mesoporous MOFs. *J. Am. Chem. Soc.* **135**, 11688–11691 (2013).
17. Brozek, C. K. & Dincă, M. Ti^{3+}-, $V^{2+/3+}$-, $Cr^{2+/3+}$-, Mn^{2+}-, and Fe^{2+}-Substituted MOF-5 and redox reactivity in Cr- and Fe-MOF-5. *J. Am. Chem. Soc.* **135**, 12886–12891 (2013).
18. Chui, S. S. Y., Lo, S. M. F., Charmant, J. P. H., Guy Orpen, A. & Williams, I. D. A chemically functionalizable nanoporous material $[Cu_3(TMA)_2(H_2O)_3]_n$. *Science* **283**, 1148–1150 (1999).
19. Majano, G. & Pérez-Ramírez, J. Scalable room-temperature conversion of copper(II) hydroxide into HKUST-1 ($Cu_3(btc)_2$). *Adv. Mater.* **25**, 1052–1057 (2013).
20. Burtch, N. C., Jasuja, H. & Walton, K. S. Water stability and adsorption in metal-organic frameworks. *Chem. Rev.* **114**, 10575–10612 (2014).
21. Férey, G. *et al.* A chromium terephthalate-based solid with unusually large pore volumes and surface area. *Science* **309**, 2040–2042 (2005).
22. Horcajada, P. *et al.* Synthesis and catalytic properties of MIL-100(Fe), an iron(III) carboxylate with large pores. *Chem. Commun.* 2820–2822 (2007).
23. Márquez, A. G. *et al.* Green microwave synthesis of MIL-100(Al, Cr, Fe) nanoparticles for thin-film elaboration. *Eur. J. Inorg. Chem.* 5165–5174 (2012).
24. Zhang, F. *et al.* Facile synthesis of MIL-100(Fe) under HF-free conditions and its application in the acetalization of aldehydes with diols. *Chem. Eng. J.* **259**, 183–190 (2015).
25. Mao, Y. *et al.* General incorporation of diverse components inside metal-organic framework thin films at room temperature. *Nat. Commun.* **5**, 5532 (2014).
26. Li, W. *et al.* Assembly of MOF microcapsules with size-selective permeability on cell walls. *Angew. Chem. Int. Ed.* **55**, 955–959 (2016).
27. Zhao, J. *et al.* Facile conversion of hydroxy double salts to metal-organic frameworks using metal oxide particles and atomic layer deposition thin-film templates. *J. Am. Chem. Soc.* **137**, 13756–13759 (2015).
28. Wehring, M. *et al.* Self-diffusion studies in CuBTC by PFG NMR and MD simulations. *J. Phys. Chem. C* **114**, 10527–10534 (2010).
29. Mao, Y. *et al.* Specific oriented metal-organic framework membranes and their facet-tuned separation performance. *ACS Appl. Mater. Interfaces* **6**, 15676–15685 (2014).
30. Umemura, A. *et al.* Morphology design of porous coordination polymer crystals by coordination modulation. *J. Am. Chem. Soc.* **133**, 15506–15513 (2011).
31. Li, H. L., Eddaoudi, M., O'Keeffe, M. & Yaghi, O. M. Design and synthesis of an exceptionally stable and highly porous metal organic framework. *Nature* **402**, 276–279 (1999).
32. Tranchemontagne, D. J., Hunt, J. R. & Yaghi, O. M. Room temperature synthesis of metal-organic frameworks: MOF-5, MOF-74, MOF-177, MOF-199, and IRMOF-0. *Tetrahedron* **64**, 8553–8557 (2008).
33. Mori, W. *et al.* Synthesis of new adsorbent copper(II) terephthalate. *Chem. Lett.* **26**, 1219–1220 (1997).
34. Adams, R., Carson, C., Ward, J., Tannenbaum, R. & Koros, W. Metal organic framework mixed matrix membranes for gas separations. *Microporous Mesoporous Mater.* **131**, 13–20 (2010).
35. Carson, C. G. *et al.* Synthesis and structure characterization of copper terephthalate metal-organic frameworks. *Eur. J. Inorg. Chem.* 2140–2145 (2014).
36. Rodenas, T. *et al.* Metal-organic framework nanosheets in polymer composite materials for gas separation. *Nat. Mater.* **14**, 48–55 (2015).

37. Serre, C. *et al.* Very large breathing effect in the first nanoporous chromium(III)-based solids: MIL-53 or Cr-III(OH){O_2C-C_6H_4-CO_2}{HO_2C-C_6H_4-CO_2H}(x)H_2O_y. *J. Am. Chem. Soc.* **124,** 13519–13526 (2002).

38. Horcajada, P. *et al.* Porous metal-organic-framework nanoscale carriers as a potential platform for drug delivery and imaging. *Nat. Mater.* **9,** 172–178 (2009).

39. Llewellyn, P. L. *et al.* Complex adsorption of short linear alkanes in the flexible metal-organic-framework MIL-53(Fe). *J. Am. Chem. Soc.* **131,** 13002–13008 (2009).

40. Gordon, J., Kazemian, H. & Rohani, S. Rapid and efficient crystallization of MIL-53(Fe) by ultrasound and microwave irradiation. *Microporous Mesoporous Mater.* **162,** 36–43 (2012).

41. Robens, E., Rouquerol, F. & Rouquerol, J. *Adsorption by Powders and Porous Solids* (WILEY-VCH Verlag, 1999).

42. Canioni, R. *et al.* Stable polyoxometalate insertion within the mesoporous metal organic framework MIL-100(Fe). *J. Mater. Chem.* **21,** 1226–1233 (2011).

43. Ahmed, I., Jun, J. W., Jung, B. K. & Jhung, S. H. Adsorptive denitrogenation of model fossil fuels with Lewis acid-loaded metal-organic frameworks (MOFs). *Chem. Eng. J.* **255,** 623–629 (2014).

44. Tan, F. *et al.* Facile synthesis of size-controlled MIL-100(Fe) with excellent adsorption capacity for methylene blue. *Chem. Eng. J.* **281,** 260–367 (2015).

45. Li, W. *et al.* Non-activation ZnO array as a buffering layer to fabricate strongly adhesive metal-organic framework/PVDF hollow fiber membranes. *Chem. Commun.* **50,** 9711–9713 (2014).

46. Guo, H., Zhu, G., Hewitt, I. J. & Qiu, S. "Twin copper source" growth of metal-organic framework membrane: $Cu_3(BTC)_2$ with high permeability and selectivity for recycling H_2. *J. Am. Chem. Soc.* **131,** 1646–1647 (2009).

47. Carta, M. *et al.* An efficient polymer molecular sieve for membrane gas separations. *Science* **339,** 303–307 (2013).

48. Liu, Q., Wang, N., Caro, J. & Huang, A. Bio-inspired polydopamine: a versatile and powerful platform for covalent synthesis of molecular sieve membranes. *J. Am. Chem. Soc.* **135,** 17679–17682 (2013).

49. Zhang, F. *et al.* Hydrogen selective NH_2-MIL-53(Al) MOF membranes with high permeability. *Adv. Funct. Mater.* **22,** 3583–3590 (2012).

50. Yoo, Y. & Jeong, H. K. Heteroepitaxial growth of isoreticular metal-organic frameworks and their hybrid films. *Cryst. Growth Des.* **10,** 1283–1288 (2010).

Acknowledgements

This work was supported by the National Natural Science Foundation of China (Grant Nos. 21236008 and 21476206), Fujian Provincial Department of Ocean (Grant No. 2014-06), Zhejiang Provincial Bureau of Science and Technology (Grant No. 2014C33032), Taishan Scholarship Blue Industry Program from Shandong Provincial Government (Grant No. 2014008) and Huzhou Gemking Biotechnology Co. Ltd. G.Z. thanks Fujian Provincial Government for the Minjiang Scholarship. W.L. also thanks Chinese Government for the first-level prize fellowship.

Author contributions

G.Z. and W.L. conceived the research idea. W.L., Y.Z., C.Z., Z.X. and P.S. synthesized the MOF membranes and particles. W.L., Y.Z., Z.X., P.S., C.S., L.Q. and Z.F. carried out related characterizations and measurements. Q.M., Q.L. and Y.Z. contributed to the general methodology and reviewed the manuscript. G.Z. supervised the project, helped design the experiments and co-drafted the manuscript. All authors contributed to the analysis of the manuscript.

Additional information

Competing financial interests: A patent application related to this work has been filed.

24

Supramolecular macrocycles reversibly assembled by Te···O chalcogen bonding

Peter C. Ho[1], Patrick Szydlowski[1], Jocelyn Sinclair[1,†], Philip J.W. Elder[1,†], Joachim Kübel[1,†], Chris Gendy[1,†], Lucia Myongwon Lee[1], Hilary Jenkins[1], James F. Britten[1], Derek R. Morim[1] & Ignacio Vargas-Baca[1]

Organic molecules with heavy main-group elements frequently form supramolecular links to electron-rich centres. One particular case of such interactions is halogen bonding. Most studies of this phenomenon have been concerned with either dimers or infinitely extended structures (polymers and lattices) but well-defined cyclic structures remain elusive. Here we present oligomeric aggregates of heterocycles that are linked by chalcogen-centered interactions and behave as genuine macrocyclic species. The molecules of 3-methyl-5-phenyl-1,2-tellurazole 2-oxide assemble a variety of supramolecular aggregates that includes cyclic tetramers and hexamers, as well as a helical polymer. In all these aggregates, the building blocks are connected by Te···O–N bridges. Nuclear magnetic resonance spectroscopic experiments demonstrate that the two types of annular aggregates are persistent in solution. These self-assembled structures form coordination complexes with transition-metal ions, act as fullerene receptors and host small molecules in a crystal.

[1]Department of Chemistry and Chemical Biology, McMaster University, 1280 Main Street West, Hamilton, Ontario, Canada L8S 4M1. † Present Addresses: Department of Chemistry, Dalhousie University, 6274 Coburg Road, PO Box 15000, Halifax, Nova Scotia, Canada B3H 4R2 (J.S.); ALS Environmental, 1435 Norjohn Ct #1, Burlington, Ontario, Canada L7L 0E6 (P.J.W.E.); Leibniz Institute of Photonic Technology e.V., Albert-Einstein-Straße 9, Jena 07745, Germany (J.K.); Department of Chemistry, University of Calgary, 2500 University Road, Calgary, Alberta, Canada T2N 1N4 (C.G.). Correspondence and requests for materials should be addressed to I.V.-B. (email: vargas@chemistry.mcmaster.ca).

One of the most remarkable developments in supramolecular chemistry in the last two decades is the evolution of halogen bonding[1,2] from being an intriguing structural feature to becoming a powerful tool in crystal engineering[3–6], which is also applicable to systems and processes as diverse as luminescent[7] and non-linear optical[8] materials, photo-patterning of surfaces[9], the assembly of fractal patterns from molecular building blocks[10], supramolecular gelation[11], organocatalysis[12], macromolecular alignment at macroscopic scale[13], anion recognition[14], transmembrane anion transport[15,16] and mimicking the activity of the deiodinase enzyme[17].

Our current understanding attributes such interactions to the depletion of electron density in a region of space surrounding the nucleus of a halogen atom in a molecule (termed a sigma hole)[11], which consequently is attracted to lone-pairs and π clouds of electrons. This electrostatic factor is supplemented by polarization of the electron density—which in a strong case would be modelled by the mixing of electron donor and acceptor orbitals—and electron correlation manifested in the London dispersion force. Those stabilizing contributions are countered by the Pauli repulsion and the balance results in interaction energies of about 5–70 kJ mol^{-1} and preference for a linear $R\text{-}X\cdots\!:\!B$ geometry (X is a halogen atom and B a Lewis base)[18–20]. The conditions that give rise to each of those stabilizing factors are common to molecules that contain heavy elements from other p-block families[18,21–25]. While it is long-recognized[26–28] that intra- and intermolecular short interatomic contacts are pervasive in structural main-group chemistry, terms such chalcogen and pnictogen bonding have been recently suggested by analogy to the halogen case. There are indeed common traits to all these interactions; for instance, the trends in ratio of interatomic distance to sum of van der Waals radii denote a correlation of interaction strength with the mass of the p-block element and enhancement by electronegative substituents[29–31]. However, one important difference is that atoms of group-16 (refs 32,33) and 15 (refs 34,35) elements can engage in up to two and three concurrent supramolecular interactions, respectively[36,37].

The potential of chalcogen-centered interactions in supramolecular chemistry is illustrated by two well-studied molecular families: the dichalcogena alkynes, which consistently crystalize in tubular structures assembled by chalcogen–chalcogen interactions[38–40], and the 1,2,5-chalcogenadiazoles, in which two pairs of antiparallel chalcogen–nitrogen interactions per molecule tend to build infinite ribbons. The latter structures are of interest because of their charge transport properties[41] and, through moderate steric repulsion, can be distorted to induce non-linear optical properties or chromotropism[42–44]. The auto-association of chalcogenadiazoles is amenable to combination with metal-ion coordination and hydrogen bonding[45,46]. In spite of their charge, N-alkylated chalcogenadiazolium cations do associate in the solid state and, according to electrochemical data, the tellurium derivatives may also be associated in solution[47].

Large assemblies (gels, supramolecular polymers and crystals) provide tangible demonstration of the power of these supramolecular interactions but comparatively less has been investigated at the other end of the size spectrum: small, discrete aggregates of a few molecules held together by supramolecular bonds. In the halogen-bonding case, I$\cdots$N interactions have been employed in the construction of molecular capsules[48] from a pair of complementary molecules; the assembly of such structures in solution was recently demonstrated[49,50].

In the chalcogen case, the crystal of 3-methyl-5-(1,1-dimethylethyl)-1,2-tellurazole 2-oxide (1a)[51] features a tetramer spontaneously assembled by short Te$\cdots$O interactions, its arrangement is specially intriguing because it evokes a macrocycle. However, until now it has been unclear whether such an aggregate would be stable enough to function in the same way as do molecules from the vast category of supramolecular building blocks that encompasses crown ethers, polyazacycloalkanes, tetrapyrroles, phthalocyanines, calixarenes, cyclodextrins, cucurbiturils and cyclophanes. If so, building macrocycles by the spontaneous addition of molecular building blocks would be a uniquely convenient approach because, in general, the synthesis of macrocycles is laborious and low-yielding—even when template methods are used—with notable exceptions such as the recently synthesized macrocyclic cyanostar[52]. We have investigated in detail the stability of the cyclic aggregates of iso-tellurazole N-oxides in solution and probed their chemical behaviour. Here we report that indeed these assemblies are persistent in solution and display properties of actual macrocyclic molecules in their ability to coordinate transition-metal ions, form adducts with fullerenes and host small species in the solid state.

Results

Preparation and structural studies. Improvements from literature procedures[51,53] afford the iso-tellurazole N-oxides 1a and 1b in good yields; these products are remarkably stable in air and tolerant of moisture. Depending on the solvent used for crystallization, 3-methyl-5-phenyl-iso-tellurazole N-oxide (1b) crystallizes in a remarkable variety of polymorphs, all the structures in Fig. 1 were identified by single-crystal X-ray diffraction. In each case, the morphology of the whole sample indicates that only one phase is reproducibly obtained. The unit cells of several polymorphs include solvent molecules and, while the solvent likely influences packing efficiency, the molecules of 1b are all associated to each other only. The aggregates observed in the crystals (Fig. 2a) include an infinite spiral chain (1b$_\infty$) and cyclic tetra- (1b$_4$) and hexamers (1b$_6$). In every case, consistently with the σ-hole/donor–acceptor model, the oxygen atom of one molecule is bound to the tellurium atom of another, always trans to the nitrogen atom and to distances that span 2.171(3) to 2.242(1) Å. These are comparable to the 2.299(2) Å measured in 1a$_4$ (ref. 51) and are slightly longer than the 2.122(1) Å of the axial bonds in β-TeO$_2$ (ref. 54). The Te–N distances (2.197(2) to 2.258(2) Å) are slightly longer than those measured for other single bonds between these elements. In each crystal, the geometry around the chalcogen approximates a T-shape, with N–Te–O†, N–Te–C and O–N–Te average angles of 165.3°, 76.0° and 126.9°, respectively. Individual iso-tellurazole heterocyles are planar, with bond distances consistent with localized single and double bonds. Adjacent iso-tellurazole rings tend to lay perpendicular to each other (Fig. 2b) with inter-planar angles ranging from 60° to 83°.

Crystallization from benzene yields a phase of composition 3(1b)•(C$_6$H$_6$), which is built by infinite spiral chains (1b$_\infty$) coiling in alternating directions along b with a periodicity of 3. However, the P2$_1$2$_1$2$_1$ space group contains no ternary screw axis, thus the unit cell contains three crystallographically distinct molecules. The macrocyclic tetramer 1b$_4$ is formed in non-solvated crystals obtained from CHCl$_3$ or by layering acetonitrile over a CH$_2$Cl$_2$ solution. Its structure resembles that of 1a$_4$ but there are important differences: the structure of the phenyl derivative approaches a chair conformation, has C_i symmetry and is built by two crystallographically independent molecules while the geometry of the t-Bu aggregate corresponds to a boat conformer, belongs to the S_4 point group and the four constituting 1a molecules are all related by symmetry[51]. There are two distinct trans-annular Te–Te distances in 1b$_4$, 5.5895(2) and 5.3043(2) Å. Crystallization from THF and a hexanes/CH$_2$Cl$_2$ mixture produces crystals of compositions 3(1b)•(C$_4$H$_8$O) and

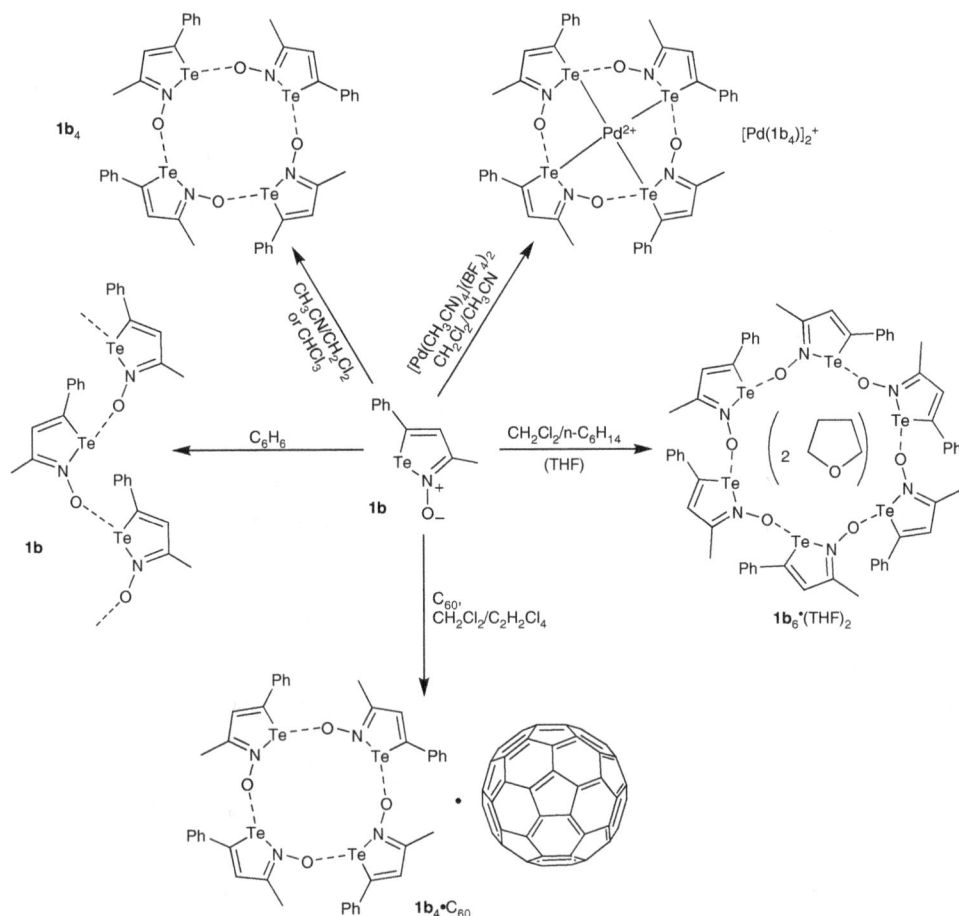

Figure 1 | Summary of supramolecular species derived from 1b. Supramolecular species formed by auto-association of **1b**, alone or in combination with a Pd(II) salt or C$_{60}$.

12(**1b**)•(CH$_2$Cl$_2$), respectively. Both phases contain macrocyclic hexamers. In the latter case, the crystallization solvent occupies voids external to two crystallographically distinct macrocycles, each built from three molecular units that are unique by symmetry. Packing distorts the macrocyle, thus there are three different trans-annular Te–Te distances for each ring: 7.117(1), 7.300(1) and 7.544(1) Å in one case and 7.151(1), 7.227(1) and 7.692(1) Å in the other. As the macrocycles in this crystal stack along b (Fig. 3a) a methyl group of one hexamer extends towards the cavity of the neighbouring macrocycle (Fig. 3b). In contrast, the crystal that contains THF packs in a hexagonal lattice (Fig. 3c); the macrocyclic aggregate is built by six equivalent molecular units of **1b** and the trans-annular Te–Te distances are all 7.638(2) Å. The macrocycles pack in a layer and a second cavity is flanked by the phenyl groups. Vertical stacking of the layers in an ABA sequence alternates the two types of cavity-forming tubular channels (Fig. 3d). THF molecules disordered in three different orientations sit in the macrocycle cavities, above and below the plane defined by the chalcogens. Although these channels are small, the crystals slowly loose solvent and become opaque.

Evidence of persistent auto-association in solution. In spite of the large molecular mass, the electrospray mass spectrum (Supplementary Fig. 1) displays isotopic patterns characteristic of aggregates [**1b**$_n$-H]$^+$ ($n = 1$–7). Given the apparent strength of the Te⋯O supramolecular interactions, it became of interest to probe the aggregation of iso-tellurazole oxides in solution with nuclear magnetic resonance (NMR) spectroscopy. At room temperature, the ^{125}Te NMR spectrum of **1b**, measured in CH$_2$Cl$_2$, displays only a single line. However, a second line appears on cooling and grows in intensity at the expense of the first (Fig. 4a). These changes are fully reverted when room temperature is restored and are paralleled by those in the ^{1}H-NMR spectrum, albeit with some differences. For example, the lines of the methyl resonances coalesce at 230 K in the 500 MHz spectrum. Furthermore, the relative intensities of the methyl lines are dependent on concentration (Fig. 4b). At low temperature, negative nuclear Overhauser effect (NOE) is observed between the ^{1}H nuclei of the methyl and phenyl or t-butyl groups of **1b** (Fig. 4c) and **1a** (Supplementary Fig. 2), respectively. The separation between the ^{1}H nuclei of these pendant groups in the individual molecules of **1** is too large for NOE (>5 Å). However, the crystal structures show that the Te⋯O interactions bring the substituents of neighbouring molecules to shorter distances (<3 Å).

NMR scrambling experiments. Mixtures of **1a** and **1b** were investigated by ^{1}H-NMR to further probe the structure of their aggregates in solution. At room temperature, the spectrum of the 1:1 mixture displays a very broad band (1.9–1.6 p.p.m.) for the methyl resonances, this feature is more clearly discernible at 323 K (Fig. 4d), these observations are indicative of inter-molecular association and exchange. The spectrum with the best

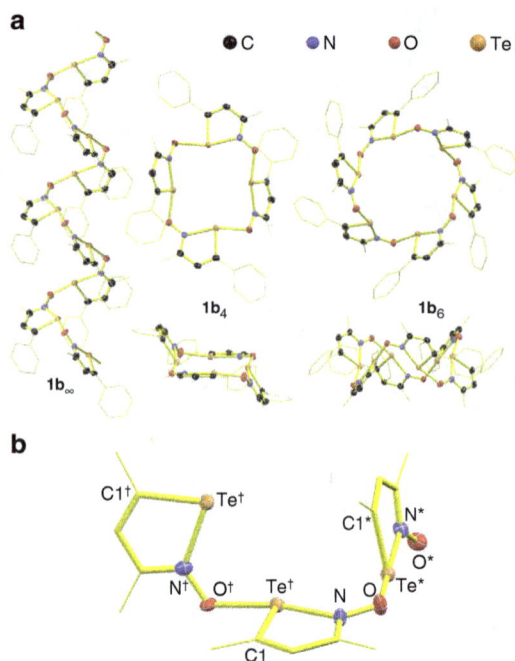

Figure 2 | Crystallographically characterized aggregates of 1b.
(**a**) ORTEPs of the aggregates observed in the crystal structures of
3-methyl-5-phenyl-isotellurazole-N-oxide, **1b**. While **1b**$_\infty$ is contained in
3(**1b**)•(C$_6$H$_6$), the macrocyclic tetramer crystallizes in a solvent-free
polymorph and the hexamer forms the co-crystal 12(**1b**)•(CH$_2$Cl$_2$).
(**b**) Detail of the structure of **1b**$_\infty$ displaying the relative orientations of the
iso-tellurazole planes and the labelling sequence used in the discussion.
Displacement ellipsoids are plotted at 75% probability in all cases. For
clarity, hydrogen atoms are omitted, the phenyl and methyl groups are
portrayed using a wireframe representation and are partially hidden in **b**.

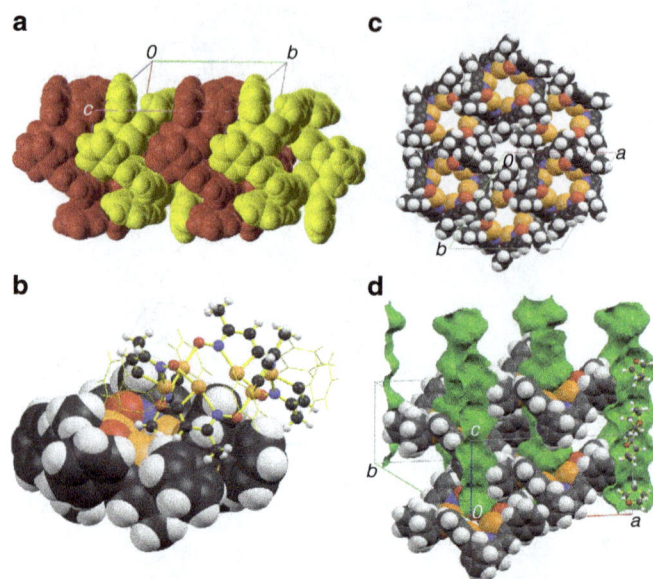

**Figure 3 | Detail of the crystalline structures that feature the hexamer
1b$_6$.** From the crystal of composition 12(**1b**)•(CH$_2$Cl$_2$): (**a**) stacking of the
macrocycles, (**b**) interaction of the methyl group with the cavity of a
neighbouring ring. From the crystal of composition 3(**1b**)•(C$_4$H$_8$O):
(**c**) packing of a layer in the (0,0,1) plane, (**d**) detail of the crystal structure,
highlighting the channels, as well as the location and three orientations of
the THF molecules. Panels **c** and **d** are simplified for clarity; after ruling out
twinning, larger unit cells and less symmetric space groups, the best
approximate model features three distinct orientations for the phenyl
groups but only one orientation for the C$_3$NTe ring. The THF molecules
were treated as rigid groups.

resolution of resonances from this mixture was acquired at 190 K
using a 600 MHz instrument; in these conditions 14 lines are
observed for the methyl and 6 for the *t*-butyl resonances (Fig. 4e).
The methyl lines in the ^{1}H-NMR spectra of mixtures of variable
composition but constant total concentration were classified by
their behaviour as a function of molar fraction (Fig. 4f). It was
possible in this way to identify the patterns characteristic of cyclic
tetramers: resonances that continuously increase or decrease in
intensity following a fourth-degree polynomial trajectory for the
homogeneous aggregates; lines with one maximum each at molar
fractions 0.25 or 0.75 that correspond to the mixed tetramers of
1:3 and 3:1 composition; as well as lines with a maximum at 0.5
that correspond to the 2:2 stoichiometry. Lines with a more
complex behaviour would result from the superposition of more
than one resonance.

As this interpretation attributes the resonance at 1.70 p.p.m. to
the methyl protons of the tetramer **1b$_4$**, it became possible to
explain the effect of concentration on the ^{1}H-NMR spectrum of
1b (Fig. 4b) as the result of further aggregation (equation (1)).
The stoichiometric coefficient (n) and equilibrium constant (K,
equation (2)) of the process were determined from the intensities
(I) of the lines (relative to the resonance of the residual proton on
the solvent) as a function of concentration (equation (3),
Supplementary Fig. 3a). The fitted stoichiometric coefficient
$n = 1.53 \pm 0.02$ implies that the resonance at 1.56 p.p.m. belongs
to the methyl protons of the hexamer. At 190 K, the equilibrium
constant is therefore $K = 0.28 \pm 0.01 \, \text{dm}^{1.5} \, \text{mol}^{-0.5}$; the corresponding van't Hoff analysis (Supplementary Fig. 3b) yielded
$\Delta H = -16 \pm 1 \, \text{kJ} \, \text{mol}^{-1}$ and $\Delta S = -88 \pm 6 \, \text{J} \, \text{mol}^{-1} \text{K}^{-1}$.
However, diffusion-ordered spectroscopy (DOSY) experiments

were unable to provide the hydrodynamic radii and mass of the
aggregates because the magnetization decay during diffusion did
not follow simple Gaussian profiles, which likely is a consequence
of the equilibrium between tetramer and hexamer.

$$n\mathbf{1b_4} \rightleftharpoons \mathbf{1b_{4n}} \qquad (1)$$

$$K = |\mathbf{1b_{4n}}|/|\mathbf{1b_4}|^n \qquad (2)$$

$$\log(I_{\mathbf{1b_4}}) = \log(K) + \log(n) - n\log(I_{\mathbf{1b_{4n}}}) \qquad (3)$$

Computational modelling. The relative stabilities of the oligomeric structures formed by aggregation of iso-tellurazole
N-oxides in solution was assessed with dispersion- and gradient-
corrected relativistic density-functional-theory (DFT) gas-phase
calculations. For computational expediency, the models were
based on the molecular building block with $R = \text{Me}$, **1c**. Calculations for the individual molecule were also used to build a map
of electrostatic potential (Fig. 5a), which demonstrate the
occurrence of two σ holes on the Te atom opposite to the C and
N atoms, the latter hole being the most prominent. The LUMO of
1c has a predominant contribution from the σ*$_{\text{Te-N}}$ orbital
(Fig. 5b). The plot of the electron localization function in the
molecular plane (Fig. 5c) features two dips in the space surrounding the chalcogen. These data confirm that the molecules of
1 are predisposed to associate through contacts between the
electrophilic (O) and nucleophilic (Te) regions and emphasize
that the most favourable position for attachment of a Lewis base
to is opposite to the nitrogen atom.

There is only one published[53] crystallographic determination
of an iso-selenazole N-oxide (3-methyl-phenyl-, **2b**), which

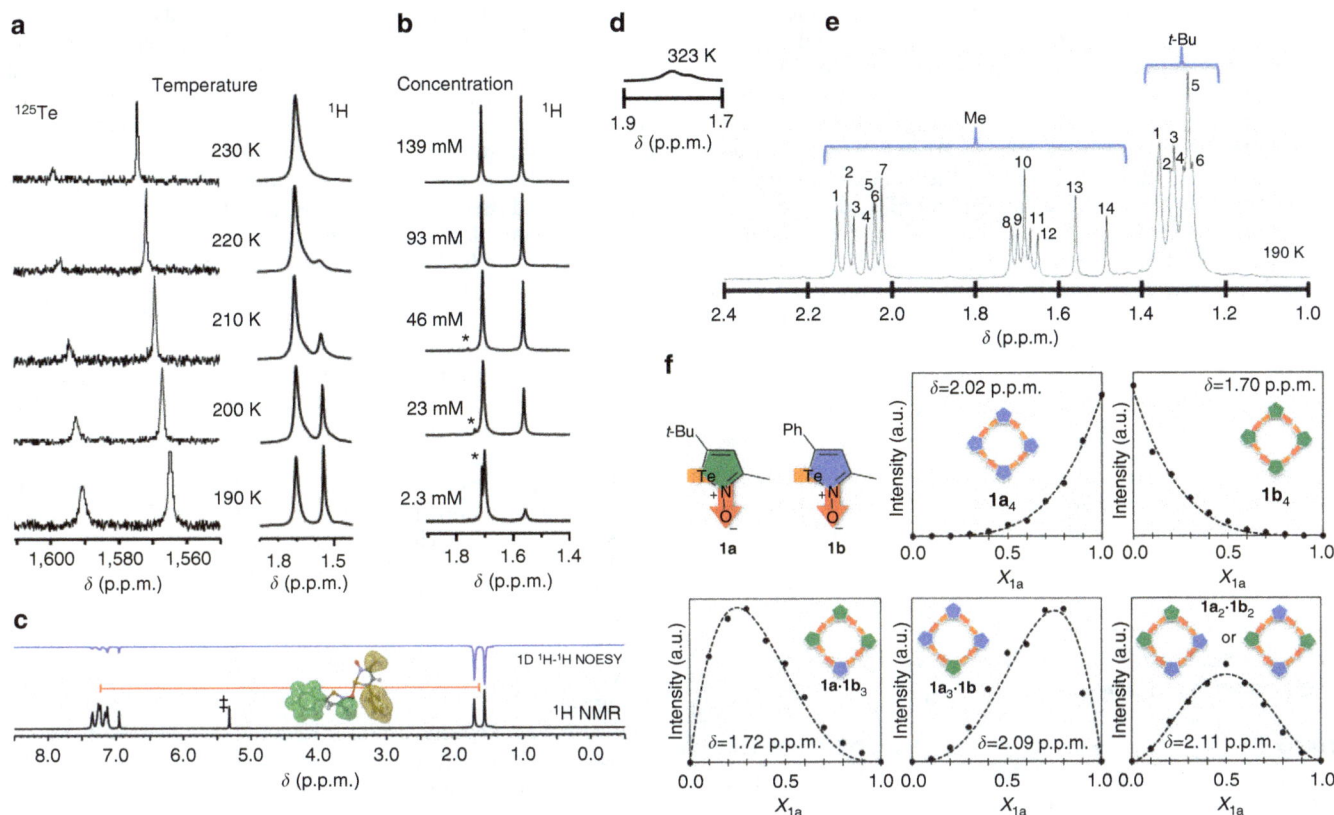

Figure 4 | NMR Investigations of the auto-association of 1 in solution. NMR (500 MHz) spectra of **1b** solutions in CDCl$_3$: (**a**) ^{125}Te and ^{1}H at 74 mmol l^{-1} and variable temperature; (**b**) ^{1}H at 190 K as a function of concentration; *denotes the resonance of a trace amount of H$_2$O. (**c**) ^{1}H-NMR and 1D-NOESY spectra of **1b** in CD$_2$Cl$_2$ at 190 K; the over-imposed structure portrays the van der Waals surfaces of the methyl and phenyl groups of each molecule in a pair modelled with coordinates extracted from the crystal structures of **1b$_4$**; ‡denotes the resonance of residual CHDCl$_2$. ^{1}H-NMR (600 MHz) from the equimolar mixture of **1a** and **1b**: (**d**) methyl resonances above room temperature, (**e**) methyl and *t*-butyl resonances at low temperature. (**f**) evolution of selected methyl ^{1}H-NMR lines as a function of composition (molar fraction of **1a**) of the mixtures, dashed lines correspond to the calculated abundance of each type of aggregate.

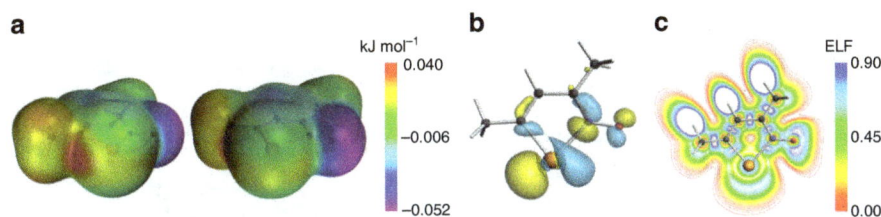

Figure 5 | The aggregation of 1 originates in its electronic properties. (**a**) Two views of the map of electrostatic potential for a molecule of **1c**; (**b**) LUMO of **1c**; (**c**) contour plot of the electron localization function in the iso-tellurazole plane.

features a centro-symmetric dimer formed by a pair of antiparallel Se···O interactions. The analogous structure for **1c$_2$**, could only be optimized by imposing symmetry constraints. In their absence, the geometry converges to a dimer bridged by only one Te···O interaction ($d = 2.40$ Å) in which the two heterocycles define an inter-planar angle of 87.6°, which is consistent with all the observed structures of the aggregates of **1a** and **1b**. The structures of the cyclic **1c$_4$** tetramers (chair and boat conformations) and the hexamer **1c$_6$** were also optimized, in each case vibrational calculations return all real frequencies confirming that all are minima in the potential-energy surface. Structures of hypothetical trimer and pentamer cyclic aggregates could not be satisfactorily optimized, the preferred nearly perpendicular orientation of the iso-tellurazole rings imposes a preference for an even number of molecules in a cyclic aggregate.

Periodic calculations used to optimize the infinite chain **1c$_\infty$**, based on the structure observed in the crystal of 3(**1b**)•(C$_6$H$_6$). The results of these calculations are summarized in Table 1 as thermodynamic parameters for aggregation equilibria.

Coordination of a transition-metal ion. Mixing [Pd(NC-CH$_3$)$_4$](BF$_4$)$_2$ with **1b** dissolved in a CH$_2$Cl$_2$/acetonitrile mixture yields a dark brown mixture, its visible absorption spectrum features a well-defined shoulder at 500 nm. Job's continuous variations method showed that this spectrum is due to a complex of 1:4 stoichiometry (Fig. 1), the composition of which was confirmed by a structural determination from crystals grown by slow diffusion of an acetonitrile solution of the metal salt into a CH$_2$Cl$_2$ solution of **1b**. Figure 6a,b displays two views of the structure of the coordination complex in the crystal of $[Pd(\mathbf{1b_4})](BF_4)_2 \bullet 2(CH_2Cl_2)_2$. The crystal structure features the tetrameric aggregate of **1b** in the boat conformation while the metal centre displays a square planar coordination geometry with Pd–Te distances (2.5804(4)Å) that are comparable to those measured in complexes of anionic tellurium ligands[55]; the Te–Pd–Te trans bond angles of 172.38(2)° denote a slight pseudo D_{2d} distortion. This crystal structure features metal depletion due to partial occupation of the coordination sites. CH_2Cl_2 molecules replaced the tetrafluoroborate anions in proportion to the missing metal ions. After refinement, the final ratios of occupancies from two crystals grown in separate batches, 0.863(7) and 0.797(5), are different as expected for independently prepared samples.

C_{60} adduct. Mixing **1b** with C_{60} in chloroform immediately yields a solid that is not soluble enough for spectroscopic investigations. However, slow diffusion of C_{60} into a solution of **1b** produces crystals of composition $4(\mathbf{1b}) \bullet C_{60}$ (Fig. 1). Along the b-axis, the

crystal structure features stacks of alternating fullerene molecules and distorted boat conformers of the $\mathbf{1b_4}$ aggregate (Fig. 6c). Compared with the boat $\mathbf{1a_4}$, in the adduct crystal the iso-tell-urazole heterocycles are tilted towards the meridional plane of the macrocycle to maximize their contact with the fullerene. The stack is not symmetrical, there are two distinct distances between the centroids defined by the 4 tellurium atoms of each $\mathbf{1b_4}$ aggregate and the fullerene molecule, the closest macrocycle engaged in two short Te⋯C contacts (3.457(4) Å, cf., the sum of van der Waals radii 3.76 Å) with the fullerene. The C_{60} molecule is slightly distorted, it features three crystallographically distinct diameter values (6.952(5), 6.9393(5) and 6.9223(5) Å), which evokes a Jahn–Teller distortion that would result from electron transfer into the t_{1u} LUMO[56]. However, there are no significant changes in the bond distances and angles of $\mathbf{1b_4}$ in this structure and the material is diamagnetic. As compared with the other crystalline phases in this report, the main differences are in the torsion angles only. Along c, the C_{60} molecules are organized in a columnar arrangement with even C⋯C spacing of 3.496 Å (Fig. 6d). The distance between C_{60} centroids is 10.533(2) Å, longer than the 10.008 Å observed in the crystal of pure C_{60} (ref. 57), and may be determined by the size of the macrocycle.

Discussion

The variety of supramolecular structures obtained from **1b** and the ease with which the crystallization conditions select the aggregate and polymorph clearly indicate that these assemblies undergo reversible dissociation in solution. On the other hand, the short Te⋯O distances in the crystals structures and the results of mass spectrometry, NMR spectroscopy and DFT calculations show that the interaction between the tellurium and oxygen atoms is very strong.

Table 1 | Calculated (PBE-D3) thermodynamic parameters of aggregation in gas phase.

Equilibrium	ΔH per Te⋯O interaction (kJ mol^{-1})	ΔS per Te⋯O interaction (J mol^{-1}K^{-1})
$2\ \mathbf{1c} \rightleftharpoons \mathbf{1c_2}$	− 68.7	− 185
$4\ \mathbf{1c} \rightleftharpoons \mathbf{1c_4}$ (chair)	− 68.6	− 158
$4\ \mathbf{1c} \rightleftharpoons \mathbf{1c_4}$ (boat)	− 69.3	− 140
$6\ \mathbf{1c} \rightleftharpoons \mathbf{1c_6}$	− 75.7	− 171
$\infty\ \mathbf{1c} \rightleftharpoons \mathbf{1c_\infty}$	− 82.1	− 291

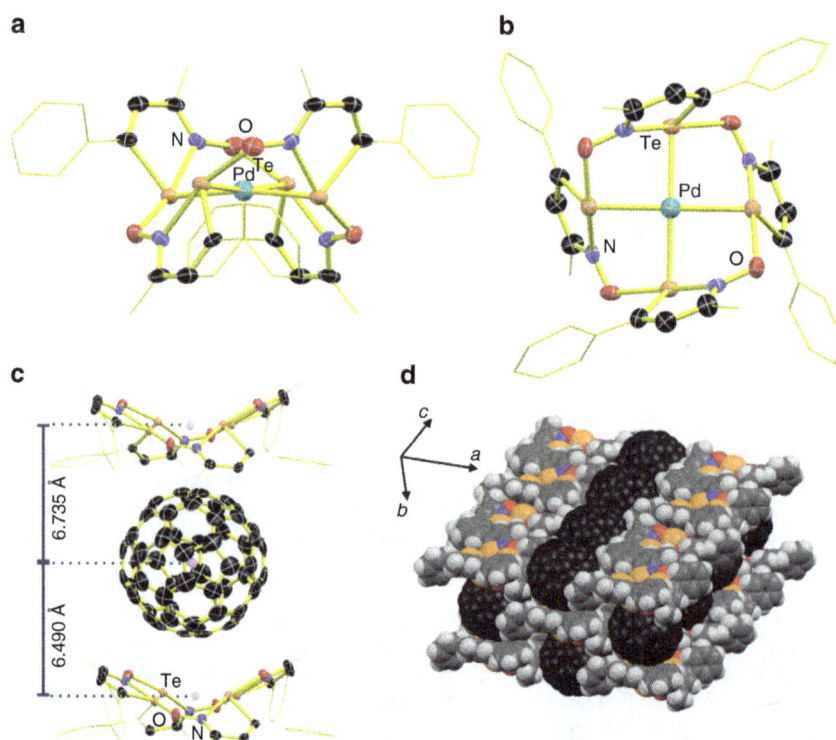

Figure 6 | Detail of the crystal structures of the derivatives of the macrocyclic tetramer 1b₄. (**a,b**) ORTEP perspectives of the $[Pd(\mathbf{1b_4})]^{2+}$ complex along (0,1,0) and along (2,1,0), respectively. (**c**) ORTEP of the crystal structure of $\mathbf{1b_4} \bullet C_{60}$. All displacement ellipsoids are shown at 75% probability. (**d**) Space-filling depiction of molecular packing the same crystal.

The observation of aggregates in the electrospray mass spectrum **1b** is remarkable; supramolecular dimers assembled by Te···N interactions have been observed in mass spectra acquired from the laser-ablation plume of benzotelluradiazoles[58] but the detection of oligomers with aggregation numbers 3–7 is unprecedented for organo-tellurium molecules.

Multinuclear NMR spectroscopic experiments demonstrate that the annular tetra- and hexamers are persistent and exist in equilibrium in solution. Direct observation of σ-hole interactions in solution is usually difficult[59,60] but encapsulation of halogen-bonded adducts in cavitands[61,62] has been helpful and in some cases can be monitored by spectroscopic methods[63]. Earlier observations of broadening in the ^{1}H-NMR spectrum of **1a** at low temperature hinted at the existence of a dynamic process but in that instance the ^{125}Te resonance could not be located and the nature of the process could not be conclusively established. The observation of NOE is one of the strongest evidences of the association of iso-tellurazole N-oxide molecules in solution at low temperature. Such spin cross-relaxation is only observable when the distance between the interacting nuclei is <5 Å, which is not possible within individual molecules of **1a** or **1b**. Moreover, that the NOE is negative indicates the zero quantum path is dominant in these systems, such situation is characteristic of restricted mobility due to large molecular weights, high viscosity and—arguably—cyclic structures. Provided there is no significant difference in the association energies of these molecules, the combination of **1a** and **1b** in solution would result in an even distribution of mixed structures that could be identified by their NMR spectra. For instance, an equimolar mixture that only forms centro-symmetric dimers would yield three different structures and display four lines from the methyl protons and two from the t-butyl groups; more complex patterns would arise from the dimers (Me: 8, t-Bu: 4), the tetramers (Me: 16, t-Bu: 8) and hexamers (Me: 41, t-Bu: 26). Of course, those are the maximum number of lines that would arise in each case; whether each of those lines could actually be observed would depend on the actual separation of their resonance frequencies, as well as on the dispersion and resolution provided by the instrument. The experimental result from the 1:1 mixture (Me: 14, t-Bu: 6; Fig. 4e) points to the tetramers, the size of these aggregates is confirmed by the observation of the mixed 1:3 and 2:2 macrocycles in the continuous variations experiments (Fig. 4f). Furthermore, the study of the concentration dependency of the ^{1}H-NMR of **1a** at low temperature is consistent with the equilibrium between tetramers and hexamers in solution (Supplementary Fig. 3).

The thermodynamic parameters calculated with DFT-D3 for the model compound **1c** are indeed favourable for supramolecular association; however, their magnitudes are taken with caution as it has been argued that this method overestimates the binding energies of this type of interaction due to delocalization error. Also, solvation is likely to have an important role but these calculations do not account for it nor the effect of packing in a crystalline lattice. The calculations indicate that the binding energies of the Te···O interactions are nearly additive, there is little strain in the annular structures and only a small energetic difference between the two tetramer conformations. By enthalpy alone the hexamer would be the most stable cyclic structure, although entropy favours the smaller aggregates and individual molecules. Even more enthalpically favourable would be the infinite polymer chain but its formation naturally imposes the highest entropic cost.

The annular aggregates of iso-tellurazole N-oxides not only are persistent in solution but also display properties of actual macrocycles. The crystal structures of the hexamers already showcase their ability to host small molecules and suggest the construction of rotaxanes and inclusion compounds. Here we

further demonstrate that they form coordination complexes and act as fullerene receptors.

The cyclic arrangement of chalcogen atoms and the trans-annular Te–Te distances of 5.0–5.6 Å suggest that the tetramers would be suitable to host transition-metal ions, this is indeed the case with Pd(II). The formation of the macrocyclic complex $[Pd(1b_4)]^{2+}$ is particularly significant; while the reversibility of the Te···O interactions favours the discreet oligomeric aggregates as the predominant species in solution, the lack of kinetic stabilization could compromise the structural integrity of the macrocycle. Iso-tellurazole oxides are potentially ambidentate ligands, coordination by oxygen would likely compete with the Te···O interactions. Such complication is possible even with soft metal ions; pyridine oxides, for example, easily coordinate palladium(II)[64]. As a macrocyclic ligand, the tetramer will enable the study of metal ions in a uniquely soft coordination sphere, which is difficult to achieve using more traditional approaches. Indeed, metal complexes of telluracrown ethers are difficult to obtain because their Te–C bonds are very reactive[65].

Fullerenes form adducts with a variety of macrocyclic and polycyclic molecules in solution and are amenable to structural characterization by X-ray diffraction[66–68]. In the case of **1b**, poor solubility restricted the study of the product of reaction with C_{60} to the crystallographic determination but the ability of the tetramer $1b_4$ to bind the fullerene receptor is well-demonstrated. The fullerene adduct is an intriguing material in its own right; its columnar arrangement of C_{60} molecules could facilitate charge transport, which calls for further investigations of applications in photovoltaics and molecular semiconductors.

As shown here, iso-tellurazole N-oxides have an unparalleled ability to spontaneously assemble functional macrocycles and thus hold great promise as supramolecular building blocks.

Methods

Experimental. The manipulation of air-sensitive materials was carried out in a glove box or using standards Schlenk techniques under an atmosphere of UHP argon (Praxair). Photosensitive materials were handled under a red LED illumination source. Elemental tellurium (CERAC), DMF (EMD), sodium hydroxide (EMD), acetic anhydride (Sigma-Aldrich), boron trifluoride diethyl etherate (Sigma-Aldrich), Boron trifluoride diethyl etherate (Sigma-Aldrich), hydroxylamine-O-sulfonic acid (Sigma-Aldrich), sodium borohydride (Sigma-Aldrich), tetrakis(acetonitrile)palladium(II) tetrafluoroborate (Sigma-Aldrich), Fullerene C_{60} (Sigma-Aldrich), dimethylcarbamoyl chloride (Alfa Aesar), phenylacetylene (Alfa Aesar), t-butylacetylene (Alfa Aesar), Chloroform (Caledon), dichloromethane (Caledon), diethyl ether (Caledon), ethyl acetate (Caledon), methanol (Caledon), Sodium sulphate (Caledon), toluene (Caledon), n-butyllithium (Acros Organics), Silica gel 60 (VWR) and sodium carbonate (VWR) were used as received from the commercial suppliers without further purification. Solvents for used synthesis were dehydrated within an Innovative Technologies solvent purification system (THF, acetonitrile) or by reflux with an appropriate dehydrating agent (Methanol over magnesium). 4-Phenylbut-3-yn-2-one and 5,5-dimethyl-hex-3-yn-2-one were prepared by literature methods. All NMR spectra were acquired in solution with a deuterated solvent. Spectra were obtained using Bruker AVANCE 500 MHz (Bruker 5-mm Broad Band Inverse probe) or Bruker AVANCE 600 MHz (Bruker 5-mm BROAD BAND OBSERVE probe) Spectrometers at 287.5 K unless otherwise indicated. Variable temperature spectra were acquired using either a cold or ambient temperature gas flow with a BV-T 2000 variable temperature controller. The sample temperature in the spectrometers was calibrated with a chemical-shift thermometer consisting of a 4% solution of methanol in methanol-d$_4$. The ^{1}H, ^{13}C and ^{125}Te spectra were processed using Bruker TopSpin 2.1 or 3.2 software packages. The ^{1}H and ^{13}C spectra were referenced to tetramethyl silane using the deuterated solvent signal as a secondary reference. The ^{125}Te chemical shifts are reported with respect to the room-temperature resonance of TeMe$_2$ ($\delta = 0.00$ p.p.m.) but were measured using a secondary reference of diphenyl ditelluride in CD$_2$Cl$_2$ ($\delta = 420.36$ p.p.m.). Electrospray ionization mass spectra were acquired in positive ion mode on a Waters/Micromass Quattro Ultima Global ToF mass spectrometer operating in W Mode. Pure samples were dissolved in dichloromethane followed by dilution with methanol. High resolution Mass spectra were obtained in a Waters Global and Ultima (ES Q-TOF) Mass Spectrometer (capillary = 3.20 V, cone = 100 V, source temp = 80 °C and resolving power = 10,000). Infrared vibrational spectra were acquired in a Bio-Rad FTS-40 FT-IR spectrometer or a Thermo Scientific Nicolet 6700 FT-IR spectrometer. Melting points were determined with Uni-Melt Thomas

Hoover capillary melting point apparatus and are reported uncorrected. Combustion elemental analyses were carried out by Guelph Analytical Laboratories (Guelph, Ontario, Canada).

Synthesis overview. The iso-tellurazole N-oxides were prepared by a method (Supplementary Fig. 4) that includes the trans addition of an *in situ*-generated tellurocarbamic acid to an ynone. The resulting enone undergoes condensation with hydoxylamine-O-sulfonic acid to introduce the nitrogen atom and the heterocycle is closed by hydrolysis of the intermediate product. The process yields DMF and sulfuric acid as by-products that are separated in an aqueous workup.

Bis-(N,N-dimethylcarbamoyl)-ditelluride. Sodium hydrogen telluride was prepared *in situ* from elemental tellurium (1.58 g), anhydrous sodium borohydride (2.33 g, 5 eq.) and anhydrous deoxygenated DMF (70 ml) at 95 °C under argon in a single-piece glass vessel. Shortly after heating started, the tellurium began to dissolve into a dark red–purple solution. After about 1 h, all the tellurium was consumed and the mixture became a light yellow suspension. The NaTeH dispersion was cooled to room temperature with a water bath and dimethylcarbamoyl chloride (3.97 g, 3 eq.) was added by cannula under argon; the reaction mixture was then stirred at 95 °C. The light yellow slurry was removed from the heat and cooled to room temperature in a water bath. Argon was removed with vacuum, and oxygen was introduced into the apparatus at 1 atm and the slurry became dark brown after 1 h. About 700 ml of distilled water was added into the mixture with stirring, the brown slurry turned black and was extracted repeatedly with 70 ml of diethyl ether until the aqueous solution was no longer yellow. The yellow organic solution was washed with aqueous sodium carbonate followed by distilled water, then dehydrated with Na_2SO_4. The organic fraction was concentrated under vacuum until a dark yellow solid began to precipitate at room temperature. The mixture was placed in a freezer (-20 °C) to promote crystallization of the pure ditelluride. Yield: 68%; mp: 105–110 °C (decomposed); ^{1}H-NMR (500 MHz, CDCl$_3$): δ 3.11 (s, 3H), 3.08 (s, 3H) (cf. ref. 66 3.11 (s, 3H), 3.08 (s, 3H)); ^{13}C NMR (500 MHz, CDCl$_3$): δ 145.4, 40.6, 36.1; IR (KBr): 1,660, 1,353, 1,249, 1,076, 872, 665 cm^{-1}.

(Z)-4-[(dimethylamino)carbonyltelluro]-4-phenyl-3-buten-2-one. Based on the procedure from the study by Shimada *et al.*[53], bis(N,N-dimethylcarbamoyl)-ditelluride (1.54 g, 3.86 mmol) was dissolved in 35 ml of anhydrous DMF under argon; the solution was dark yellow. Anhydrous NaBH$_4$ (0.321 g, 8.49 mmol) was dissolved in anhydrous methanol (18 ml) then added dropwise into the ditelluride solution while maintaining the temperature between -50 and -78 °C. The mixture was stirred at 0 °C for 30 min, the evolution of gas was observed and the colour of the mixture became dark red. The 4-phenylbut-3-yn-2-one (1.95 g, 13.51 mmol) was added to the reaction at 0 °C dropwise. The solution turned light yellow and stirring continued for 3 h. The reaction was quenched with 5 ml of distilled water followed by extraction with toluene in 50 ml portions from a 500 ml brine solution. The organic solution was washed with distilled water, dried with Na_2SO_4 and evaporated under high vacuum at 35 °C. The organic residue was purified by silica gel column chromatography with CH$_2$Cl$_2$:ethyl acetate (95:5 v/v). The solvent from eluate was evaporated and the product was a yellow solid. Yield: 72%; mp: 70–71 °C; ^{1}H-NMR (500 MHz, CD$_2$Cl$_2$): δ 7.35–7.65 (m, 5H), 2.77 (s, 3H), 2.60 (s, 3H), 2.34 (s, 3H) (cf. ref. 53 7.33–7.41 (m, 5H), 2.81 (s, 3H), 2.60 (s, 3H), 2.56 (s, 3H)); ^{13}C-DEPTq NMR (500 MHz, CD$_2$Cl$_2$): δ 197.2, 160.7, 156.6, 143.9, 131, 129.17, 128.9, 128.2, 41.7, 33.9, 30.1.

(Z)-4-[(dimethylamino)carbonyltelluro]-4-t-butyl-3-buten-2-one. This derivative was synthesized in a similar way and obtained as a yellow solid. Yield: 37%; ^{1}H-NMR (500 MHz, CDCl$_3$): δ 6.81 (s, 1H), 3.00 (s, 6H), 2.28 (s, 3H), 1.25 (s, 9H); ^{13}C-DEPTq NMR (500 MHz, CD$_2$Cl$_2$): δ 199.6, 156.9, 151.6, 133.4, 41.4, 35.9, 31.2, 30.7.

3-methyl-5-phenyl-1,2-tellurazole N-oxide, R = Ph (1b). This compound was synthesized following the method described by Kübel[51] with some modifications. The tellurocarbamate (1.42 g, 4.13 mmol) was refluxed with hydroxylamine-O-sulfonic acid (2.05 g, 18.16 mmol) for 1 h in anhydrous methanol (90 ml). The product was extracted with chloroform, washed with distilled water, dehydrated and dried under vacuum. The crude product was dissolved again in methylene chloride and was deposited on a layer of silica (2 cm). Impurities were eluted with methylene chloride through the silica. The pure product was then eluted with a CH$_2$Cl$_2$/methanol solution (50:50 v/v) and the solvent was removed under vacuum. The product was obtained as a pale yellow solid. Yield: >95%; mp: 207–211 °C (decomposed); ^{1}H-NMR (500 MHz, CD$_2$Cl$_2$): δ 7.26–7.42 (m, 5H), 7.10 (s, 1H), 1.77 (s, 3H); ^{13}C NMR (500 MHz, CD$_2$Cl$_2$): δ 157.9, 152.6, 140.8, 129.9, 128.2, 128.0, 127.7, 15.9; ^{125}Te NMR (500 MHz, CD$_2$Cl$_2$): δ 1,595.2; IR (KBr): 3,050, 3,022, 2,918, 1,571, 1,493, 1,468, 1,443, 1,373, 1,343, 1,222, 1,109, 1,028, 927, 908, 869, 832, 759, 713, 696, 617, 584, 534 cm^{-1}; HRMS (m/z): [M − H]$^+$ calcd. for C$_{10}$H$_{10}$NOTe, 289.7961; found, 289.9831.

3-methyl-5-t-butyl-1,2-tellurazole N-oxide, R = t-Bu (1a). This compound was synthesized in a similar way. Yield: 80%; mp: 180–185 °C (decomposed); ^{1}H-NMR (500 MHz, CDCl$_3$): δ 6.96 (s, 1H), 2.17 (s, 3H), 1.42 (s, 9H). ^{13}C-DEPTq NMR (125.8 MHz, CD$_2$Cl$_2$): δ 168.9, 156.4, 122.7, 41.5, 32.1, 16.0; IR (KBr): 2,953, 2,912, 2,865, 1,565, 1,466, 1,424, 1,389, 1,370, 1,361, 1,337, 1,243, 1,231, 1,202, 1,125, 1,030, 1,001, 967, 896, 842, 828, 794, 760, 756, 697 cm^{-1}; HRMS (m/z): [M-H]$^+$ calcd. for C$_8$H$_{14}$ON^{129}Te, 270.0138; found, 270.0122.

[Pd(1b$_4$)](BF$_4$)$_2$. A solution of isotellurazole-N-oxide in anhydrous dichloromethane (0.031 g, 0.108 mmol) was added dropwise to a solution of tetrakis(acetonitrile)palladium(II) tetrafluoroborate in anhydrous acetonitrile (0.012 g, 0.027 mmol). The solution turned from light yellow to deep red and a reddish–brown solid precipitated. The mixture was stirred under nitrogen for a day and the solid was filtered off, washed with dichloromethane and dried under vacuum. Yield: 98%; mp: 190–191 °C; ^{1}H-NMR (500 MHz, CD$_3$CN): δ 7.49-7.42 (m, 6H), 2.11 (s, 3H); ^{13}C NMR (500 MHz, DMSO-d$_6$): δ 138.4, 131.9, 129.2, 129.0, 128.7, 128.4, 128.1, 15.7; Not soluble enough for ^{125}Te NMR; IR (KBr): 1,615, 1,575, 1,490, 1,442, 1,384, 1,219, 1,113, 1,084, 1,062, 928, 866, 760, 696, 614, 573, 533 cm^{-1}; analysis (calcd., found for C$_{40}$H$_{36}$N$_4$O$_4$B$_2$F$_8$Te$_4$Pd): C (33.66, 33.48), H (2.54, 2.28), N (3.93, 4.08). Slow diffusion in long tube yielded instead single crystals of idealized composition [Pd(1b$_4$)](BF$_4$)$_2$.(CH$_2$Cl$_2$)$_2$ in the mixing zone and crystals of pure 1b$_4$ at the bottom. The former loose the crystallization solvent under vacuum.

[1b$_4$]C$_{60}$. A concentrated solution of 1b (0.048 g, 0.167 mmol) was dissolved in chloroform. The layer of 1b solution was allowed to diffuse with a layer concentrated solution of fullerene (0.030 g, 0.041 mmol) in tetrachloroethane that was filtered through an activated neutral alumina. Crystals suitable for X-ray diffraction were obtained by slow diffusion with the two layers of solution until the growth of crystals reached equilibrium. Yield: 76.9%; the material did not melt or appear to decompose up to 280 °C; [1b$_4$]C$_{60}$ is not soluble enough to acquire meaningful ^{1}H, ^{13}C and ^{125}Te NMR spectra. IR (KBr): 1,572, 1,491, 1,468, 1,428, 1,372, 1,221, 1,182, 1,107, 1,028, 927, 908, 869, 833, 755, 715, 694, 616, 577, 527 cm^{-1}; analysis (calcd., found for C$_{100}$H$_{36}$N$_4$O$_4$Te$_4$): C (64.30, 64.16), H (1.94, 1.77), N (3.00, 2.75).

Scrambling experiment. The NMR samples were prepared by mixing 1b with 1a in 10:0, 9:1, 8:2, 7:3, 6:4, 5:5, 4:6, 3:7, 2:8, 1:9, 0:10 molar ratios while maintaining a total amount of 8.717 × 10^{-3} mmol. Each sample was dissolved in 0.7 ml of deuterated methylene chloride, yielding a total concentration 12.6 mmol l^{-1}. ^{1}H-NMR spectra were acquired at both 179.9 and 287.5 K using a Bruker Avance 600 MHz spectrometer and are provided as Supplementary Fig. 5

Single-crystal X-ray diffraction. Single crystals were grown under the following conditions: 3(1b)•(C$_6$H$_6$): slow evaporation from a benzene solution. 1b: slow evaporation from a CH$_2$Cl$_2$ solution. 12(1b)•(CH$_2$Cl$_2$): slow evaporation from a concentrated solution in a mixture of or CH$_2$CH$_2$/pentane (90:10% v/v). 3(1b)•(C$_4$H$_8$O): slow evaporation from a THF solution. [Pd(1b$_4$)](BF$_4$)$_2$(CH$_2$Cl$_2$)$_2$: diffusion of a solution of 1b in CH$_2$Cl$_2$ into [Pd(CH$_3$CN)$_4$)](BF$_4$)$_2$ in acetonitrile. (1b$_4$)•C$_{60}$: diffusion of a 1b solution in CHCl$_3$ into saturated C$_{60}$ in tetrachloroethane. All crystals were mounted on a MiTeGen Micromounts with Paratone-n oil. Crystals were mounted on nylon loops (Hampton, CA) or MiTeGen Micromount (Ithica, NY) with Paratone-n oil. A Bruker APEX2 diffractometer was used to collect data at 100 K with Mo-Kα radiation ($\lambda = 0.71073$ Å). A CCD area detector was used and equipped with a low-temperature accessory Oxford cryostream. Solution and refinement procedures are presented in the Supplementary Methods and specific details are compiled in Supplementary Table 1. Selected distances and angles are provided in Supplementary Table 2.

Computational. All DFT calculations were performed using the ADF/BAND DFT package (versions 2013 and 2014). Details of the method are provided in the Supplementary Methods. Coordinates of all optimized structures are provided in Supplementary Tables 3–8.

References

1. Desiraju, G. R. *et al.* Definition of the halogen bond (IUPAC Recommendations 2013). *Pure Appl. Chem.* **85**, 1711–1713 (2013).
2. Pedireddi, V. R. *et al.* The nature of halogen halogen interactions and the crystal structure of 1,3,5,7-tetraiodoadamantane. *J. Chem. Soc., Perkin Trans. 2*, 2353–2360 (1994).
3. Rissanen, K. Halogen bonded supramolecular complexes and networks. *CrystEngComm.* **10**, 1107–1113 (2008).
4. Metrangolo, P., Meyer, F., Pilati, T., Resnati, G. & Terraneo, G. Halogen bonding in supramolecular chemistry. *Angew. Chem. Int. Ed.* **47**, 6114–6127 (2008).

5. Metrangolo, P. & Resnati, G. Chemistry: halogen versus hydrogen. *Science* **321**, 918–919 (2008).

6. Raatikainen, K. & Rissanen, K. Breathing molecular crystals: halogen- and hydrogen-bonded porous molecular crystals with solvent induced adaptation of the nanosized channels. *Chem. Sci.* **3**, 1235–1239 (2012).

7. Bolton, O., Lee, K., Kim, H.-J., Lin, K. Y. & Kim, J. Activating efficient phosphorescence from purely organic materials by crystal design. *Nat. Chem.* **3**, 205–210 (2011).

8. Virkki, M. et al. Halogen bonding enhances nonlinear optical response in poled supramolecular polymers. *J. Mater. Chem. C* **3**, 3003–3006 (2015).

9. Saccone, M. et al. Supramolecular hierarchy among halogen and hydrogen bond donors in light-induced surface patterning. *J. Mater. Chem. C* **3**, 759–768 (2015).

10. Shang, J. et al. Assembling molecular Sierpiński triangle fractals. *Nat. Chem.* **7**, 389–393 (2015).

11. Meazza, L. et al. Halogen-bonding-triggered supramolecular gel formation. *Nat. Chem.* **5**, 42–47 (2013).

12. Kniep, F. et al. Organocatalysis by neutral multidentate halogen-bond donors. *Angew. Chem. Int. Ed.* **52**, 7028–7032 (2013).

13. Houbenov, N. et al. Halogen-bonded mesogens direct polymer self-assemblies up to millimetre length scale. *Nat. Commun.* **5**, 4043 (2014).

14. Langton, M. J., Robinson, S. W., Marques, I., Félix, V. & Beer, P. D. Halogen bonding in water results in enhanced anion recognition in acyclic and rotaxane hosts. *Nat. Chem.* **6**, 1039–1043 (2014).

15. Jentzsch, A. V. et al. Transmembrane anion transport mediated by halogen-bond donors. *Nat. Commun.* **3**, 905 (2012).

16. Jentzsch, A. V. & Matile, S. Transmembrane halogen-bonding cascades. *J. Am. Chem. Soc.* **135**, 5302–5303 (2013).

17. Manna, D. & Mugesh, G. Regioselective deiodination of thyroxine by iodothyronine deiodinase mimics: an unusual mechanistic pathway involving cooperative chalcogen and halogen bonding. *J. Am. Chem. Soc.* **134**, 4269–4279 (2012).

18. Politzer, P., Murray, J. S. & Clark, T. Halogen bonding and other σ-hole interactions: a perspective. *Phys. Chem. Chem. Phys.* **15**, 11178–11189 (2013).

19. Rosokha, S. V., Stern, C. L. & Ritzert, J. T. Experimental and computational probes of the nature of halogen bonding: complexes of bromine-containing molecules with bromide anions. *Chem. Eur. J.* **19**, 8774–8788 (2013).

20. Ramasubbu, N., Parthasarathy, R. & Murray-Rust, P. Angular preferences of intermolecular forces around halogen centers: preferred directions of approach of electrophiles and nucleophiles around carbon-halogen bond. *J. Am. Chem. Soc.* **108**, 4308–4314 (2002).

21. Brezgunova, M. E. et al. Chalcogen bonding: experimental and theoretical determinations from electron density analysis. geometrical preferences driven by electrophilic–nucleophilic interactions. *Crystal Growth Des.* **13**, 3283–3289 (2013).

22. Sarkar, S., Pavan, M. S. & Row, T. N. G. Experimental validation of 'pnicogen bonding' in nitrogen by charge density analysis. *Phys. Chem. Chem. Phys.* **17**, 2330–2334 (2014).

23. Scheiner, S. Sensitivity of noncovalent bonds to intermolecular separation: hydrogen, halogen, chalcogen, and pnicogen bonds. *CrystEngComm* **15**, 3119–3124 (2013).

24. Bauza, A., Quiñonero, D., Deyà, P. M. & Frontera, A. Halogen bonding versus chalcogen and pnicogen bonding: a combined Cambridge structural database and theoretical study. *CrystEngComm* **15**, 3137–3144 (2013).

25. Zahn, S., Frank, R., Hey Hawkins, E. & Kirchner, B. Pnicogen bonds: a new molecular linker? *Chem. Eur. J.* **17**, 6034–6038 (2011).

26. Alcock, N. W. Secondary bonding to nonmetallic elements. *Adv. Inorg. Chem. Radiochem.* **15**, 1–58 (1972).

27. Landrum, G. A. & Hoffmann, R. Secondary bonding between chalcogens or pnicogens and halogens. *Angew. Chem. Int. Ed.* **37**, 1887–1890 (1998).

28. Rosenfield, R. E., Parthasarathy, R. & Dunitz, J. D. Directional preferences of nonbonded atomic contacts with divalent sulfur. 1. Electrophiles and nucleophiles. *J. Am. Chem. Soc.* **99**, 4860–4862 (1977).

29. de Paul, N, Nziko, V. & Scheiner, S. Intramolecular S...O chalcogen bond as stabilizing factor in geometry of substituted phenyl-SF₃ molecules. *J. Org. Chem.* **80**, 2356–2363 (2015).

30. Chaudhary, P. et al. Lewis acid behavior of SF₄: synthesis, characterization and computational study of adducts of SF₄ with pyridine and pyridine derivatives. *Chem. Eur. J.* **21**, 6247–6256 (2015).

31. Alikhani, E., Fuster, F., Madebene, B. & Grabowski, S. J. Topological reaction sites—very strong chalcogen bonds. *Phys. Chem. Chem. Phys.* **16**, 2430–2442 (2014).

32. Chivers, T. & Laitinen, R. S. Tellurium: a maverick among the chalcogens. *Chem. Soc. Rev.* **44**, 1725–1739 (2015).

33. Tiekink, E. R. T. & Zukerman-Schpector, J. Stereochemical activity of lone pairs of electrons and supramolecular aggregation patterns based on secondary interactions involving tellurium in its 1,1-dithiolate structures. *Coord. Chem. Rev.* **254**, 46–76 (2010).

34. Politzer, P., Murray, J., Janjić, G. & Zarić, S. σ-hole interactions of covalently-bonded nitrogen, phosphorus and arsenic: a survey of crystal structures. *Crystals* **4**, 12–31 (2014).

35. Setiawan, D., Kraka, E. & Cremer, D. Strength of the pnicogen bond in complexes involving group Va elements N, P and As. *J. Phys. Chem. A* **119**, 1642–1656 (2014).

36. Cozzolino, A. F., Elder, P. J. W. & Vargas-Baca, I. A survey of tellurium-centered secondary-bonding supramolecular synthons. *Coord. Chem. Rev.* **255**, 1426–1438 (2011).

37. Politzer, P., Murray, J. S. & Clark, T. Halogen bonding: an electrostatically-driven highly directional noncovalent interaction. *Phys. Chem. Chem. Phys.* **12**, 7748–7757 (2010).

38. Werz, D. B. et al. Self-organization of chalcogen-containing cyclic alkynes and alkenes to yield columnar structures. *Org. Lett.* **4**, 339–342 (2002).

39. Gleiter, R. & Werz, D. B. Elastic cycles as flexible hosts: how tubes built by cyclic chalcogenaalkynes individually host their guests. *Chem. Lett.* **34**, 126–131 (2005).

40. Lari, A., Gleiter, R. & Rominger, F. Supramolecular organization based on van der Waals forces: syntheses and solid state structures of isomeric [6.6]cyclophanes with 2,5-diselenahex-3-yne bridges. *Eur. J. Org. Chem.* **2009**, 2267–2274 (2009).

41. Pushkarevsky, N. A. et al. First charge-transfer complexes between tetrathiafulvalene and 1,2,5-chalcogenadiazole derivatives: Design, synthesis, crystal structures, electronic and electrical properties. *Synth. Metals* **162**, 2267–2274 (2012).

42. Cozzolino, A. F., Vargas-Baca, I., Mansour, S. & Mahmoudkhani, A. H. The nature of the supramolecular association of 1,2,5-chalcogenadiazoles. *J. Am. Chem. Soc.* **127**, 3184–3190 (2005).

43. Cozzolino, A. F., Britten, J. F. & Vargas-Baca, I. The effect of steric hindrance on the association of telluradiazoles through Te − N secondary bonding interactions. *Crystal Growth Des.* **6**, 181–186 (2005).

44. Cozzolino, A. F., Whitfield, P. S. & Vargas-Baca, I. Supramolecular chromotropism of the crystalline phases of 4,5,6,7-tetrafluorobenzo-2,1,3-telluradiazole. *J. Am. Chem. Soc.* **132**, 17265–17270 (2010).

45. Zhou, A.-J., Zheng, S.-L., Fang, Y. & Tong, M.-L. Molecular tectonics: self-complementary supramolecular Se...N synthons directing assembly of 1D silver chains into 3D porous molecular architectures. *Inorg. Chem.* **44**, 4457–4459 (2013).

46. Lee, L. M. et al. The size of the metal ion controls the structures of the coordination polymers of benzo-2,1,3-selenadiazole. *CrystEngComm* **15**, 7434–7437 (2013).

47. Berionni, G., Pégot, B., Marrot, J. & Goumont, R. Supramolecular association of 1,2,5-chalcogenadiazoles: an unexpected self-assembled dissymetric [Se⋯N]₂ four-membered ring. *CrystEngComm* **11**, 986–988 (2009).

48. Aakeröy, C. B. et al. The quest for a molecular capsule assembled via halogen bonds. *CrystEngComm* **14**, 6366–6368 (2012).

49. Dumele, O., Trapp, N. & Diederich, F. Halogen bonding molecular capsules. *Angew. Chem. Int. Ed.* **54**, 12339–12344 (2015).

50. Beyeh, N. K., Pan, F. & Rissanen, K. A halogen-bonded dimeric resorcinarene capsule. *Angew. Chem. Int. Ed.* **54**, 7303–7307 (2015).

51. Kübel, J., Elder, P. J. W., Jenkins, H. A. & Vargas-Baca, I. Structure and formation of the first (−O−Te−N−)₄ ring. *Dalton Trans.* **39**, 11126–11128 (2010).

52. Lee, S., Chen, C.-H. & Flood, A. H. A pentagonal cyanostar macrocycle with cyanostilbene CH donors binds anions and forms dialkylphosphate [3]rotaxanes. *Nat. Chem.* **5**, 704–710 (2013).

53. Shimada, K. et al. Synthesis of isochalcogenazole rings by treating β-(N,N-dimethylcarbamoylchalcogenenyl)alkenyl ketones with hydroxylamine-O-sulfonic acid. *Bull Chem. Soc. Jpn* **80**, 567–577 (2007).

54. Thomas, P. A. The crystal structure and absolute optical chirality of paratellurite, α-TeO₂. *J. Phys. C: Solid State Phys.* **21**, 4611–4627 (1988).

55. Robertson, S. D., Ritch, J. S. & Chivers, T. Palladium and platinum complexes of tellurium-containing imidodiphosphinate ligands: nucleophilic attack of Li[(PiPr₂)(TePiPr₂)N] on coordinated 1,5-cyclooctadiene. *Dalton Trans.* 8582–8592 (2009).

56. Reed, C. A. & Bolskar, R. D. Discrete fulleride anions and fullerenium cations. *Chem. Rev.* **100**, 1075–1120 (2000).

57. Krätschmer, W., Lamb, L. D., Fostiropoulos, K. & Huffman, D. R. Solid C₆₀: a new form of carbon. *Pure Appl. Chem.* **347**, 354–358 (1990).

58. Cozzolino, A. F., Dimopoulos-Italiano, G., Lee, L. M. & Vargas-Baca, I. Chalcogen–nitrogen secondary bonding interactions in the gas phase—spectrometric detection of ionized benzo-2,1,3-telluradiazole dimers. *Eur. J. Inorg. Chem.* **2013**, 2751–2756 (2013).

59. Beale, T. M., Chudzinski, M. G., Sarwar, M. G. & Taylor, M. S. Halogen bonding in solution: thermodynamics and applications. *Chem. Soc. Rev.* **42**, 1667–1680 (2013).

60. Sarwar, M. G., Dragisic, B., Salsberg, L. J., Gouliaras, C. & Taylor, M. S. Thermodynamics of halogen bonding in solution: substituent, structural and solvent effects. *J. Am. Chem. Soc.* **132**, 1646–1653 (2010).

61. El-Sheshtawy, H. S., Bassil, B. S., Assaf, K. I., Kortz, U. & Nau, W. M. Halogen bonding inside a molecular container. *J. Am. Chem. Soc.* **134,** 19935–19941 (2012).

62. Sarwar, M. G., Ajami, D., Theodorakopoulos, G., Petsalakis, I. D. & Rebek, Jr J. Amplified halogen bonding in a small space. *J. Am. Chem. Soc.* **135,** 13672–13675 (2013).

63. Garrett, G. E., Gibson, G. L., Straus, R. N., Seferos, D. S. & Taylor, M. S. Chalcogen bonding in solution: interactions of benzotelluradiazoles with anionic and uncharged Lewis bases. *J. Am. Chem. Soc.* **137,** 4126–4133 (2015).

64. Cho, S. H., Hwang, S. J. & Chang, S. Palladium-catalyzed C – H functionalization of pyridine N-oxides: highly selective alkenylation and direct arylation with unactivated arenes. *J. Am. Chem. Soc.* **130,** 9254–9256 (2008).

65. Levason, W., Reid, G. & Zhang, W. The chemistry of the p-block elements with thioether, selenoether and telluroether ligands. *Dalton Trans.* **40,** 8491–8506 (2011).

66. Zingaro, R. A., Herrera, C. & Meyers, E. A. Isolation and crystal structure of an unusual ditelluride: bis(N,N-dimethylaminoformyl) ditelluride. *J. Organomet. Chem.* **306,** C36–C40 (1986).

Acknowledgements

The support of the Natural Sciences and Engineering Research Council Canada (I.V.-B.—DG, L.M.L.—PSD, J.S.—USRA) and McMaster University (P.C.H.—Summer Work Program) is gratefully acknowledged. We thank Professor Goward and Drs Berno and Krachkovskiy (McMaster) for their valuable advice on the NMR experiments. Portions of this work were made possible by the facilities of the Shared Hierarchical Academic Research Computing Network (SHARCNET: www.sharcnet.ca) and Compute/ Calcul Canada.

Author contributions

P.C.H. optimized the synthetic methods, isolated 3(**1b**)•(C_6H_6), 12(**1b**)•(CH_2Cl_2), [Pd(**1b**$_4$)](BF_4)$_2$•2(CH_2Cl_2)$_2$ and 4(**1b**)C_{60}, performed the NOESY and scrambling NMR experiments, analysed the data and assisted writing the manuscript. J.S. isolated 3(**1b**)•(C_4H_8O), analysed the structural data and assisted with the scrambling and other experiments. J.K. and P.J.W.E. performed the synthesis and spectroscopic characterization of **1a** and DFT calculations. P.S. and C.G. performed the synthesis and preliminary spectroscopic characterization of **1b**. L.M.L., H.J. and J.F.B. screened samples for X-ray crystallography, acquired diffraction data, solved and refined the structures. D.R.M. performed synthetic experiments and some DFT calculations. I.V.-B. designed the experiments, performed some DFT calculations and wrote the manuscript.

Additional information

Accession codes: The X-ray crystallographic coordinates for all structures reported in this article have been deposited at the Cambridge Crystallographic Data Centre, under deposition numbers CCDC 1414076-1414081 and 1415229. These data can be obtained free of charge from The Cambridge Crystallographic Data Centre (www.ccdc.cam.ac.uk/ data_request/cif).

25

Ambiphilic boron in 1,4,2,5-diazadiborinine

Baolin Wang[1], Yongxin Li[2], Rakesh Ganguly[2], Hajime Hirao[1] & Rei Kinjo[1]

Boranes have long been known as the archetypal Lewis acids owing to an empty p-orbital on the boron centre. Meanwhile, Lewis basic tricoordinate boranes have been developed in recent years. Here we report the synthesis of an annulated 1,4,2,5-diazadiborinine derivative featuring boron atoms that exhibit both Lewis acidic and basic properties. Experimental and computational studies confirmed that two boron atoms in this molecule are spectroscopically equivalent. Nevertheless, this molecule cleaves C–O, B–H, Si–H and P–H bonds heterolytically, and readily undergoes [4 + 2] cycloaddition reaction with non-activated unsaturated bonds such as C=O, C=C, C≡C and C≡N bonds. The result, thus, indicates that the indistinguishable boron atoms in 1,4,2,5-diazadiborinine act as both nucleophilic and electrophilic centres, demonstrating ambiphilic nature.

[1] Division of Chemistry and Biological Chemistry, School of Physical and Mathematical Sciences, Nanyang Technological University, 21 Nanyang Link, Singapore 637371, Singapore. [2] NTU-SPMS-CBC Crystallography Facility, Nanyang Technological University, 21 Nanyang Link, Singapore 637371, Singapore. Correspondence and requests for materials should be addressed to H.H. (email: hirao@ntu.edu.sg) or R.K. (email: rkinjo@ntu.edu.sg).

As classical trivalent boranes inherently possess an unoccupied p-orbital on the boron centre, they have been widely utilized as electron-pair acceptors or Lewis acids in synthetic chemistry[1]. Apart from this line of research, isolable nucleophilic and low-valent boron species have also attracted great attention in recent years since the seminal work by Nozaki, Yamashita and co-workers[2-11]. Among them, by installing two carbene ligands, the Bertrand group and our group developed neutral tricoordinate organoboron species isoelectronic with amines[12-15]. Braunschweig and co-workers reported relevant species, including the first complex featuring two carbon monoxide ligands coordinating to a boron centre[16,17]. In these twofold base adducts of monovalent boron, electrons in the filled p-orbital of the boron centre are delocalized into the formally empty p-orbitals or the π^*-orbitals of the ligands. Thus, the bonding situation therein is reminiscent of the σ donation and π back donation between a metal and a ligand in conventional transition metal complexes[18,19], and there seems to exist similarity between this type of boron and transition metals, although there is controversy regarding how to interpret such bonding in main group compounds[20-22].

Recently, it has been demonstrated that various nonmetallic systems based on p-block elements featuring both strong electron-donor and electron-acceptor sites or small highest occupied molecular orbital–lowest unoccupied molecular orbital (HOMO–LUMO) gap, may be utilized for small-molecule activation[23-28]. Depending on the number of sites that participate in the activation, these nonmetallic systems can be classified into two major types. The first type has a nucleophilic centre and an electrophilic centre as independent active sites. Representative examples include phosphine-borane-containing species, pioneered by Stephan et $al.$[29-31], in which the phosphorus centre acts as a Lewis base, whereas the boron centre acts as a Lewis acid. The second type possesses a single active site with an ambiphilic property. Bertrand and co-workers reported the first successful, facile splitting of dihydrogen and ammonia by (alkyl)(amino)carbenes bearing an ambiphilic divalent carbon centre with a lone pair of electrons and a vacant orbital[32]. In the boron series, only one boron derivative isoelectronic with singlet carbenes has been crystallographically characterized, which exhibits a considerable electrophilic property[33]. Construction of organoboron compounds containing ambiphilic elemental centres still remains extremely challenging[34]. To date, a system with two ambiphilic sites for small-molecule activation, having both of the features mentioned above, has never been described.

Very recently, we have synthesized a 1,3,2,5-diazadiborinine bearing two boron atoms that are spectroscopically inequivalent. We showed that these two boron centres in an aromatic ring cooperatively activate small molecules in which one of them behaves as a Lewis acid centre, whereas the other serves as a Lewis base centre[35,36]. The result demonstrated that incorporation of two boron atoms into an aromatic skeleton allows for an effective remote interaction between them through the π-system[37]. Herein, we report the synthesis of 1,4,2,5-diazadiborinine in which two equivalent boron atoms act as both nucleophilic and electrophilic centres; thus, the compound features ambiphilic nature.

Results

Synthesis and characterization of 2.
Reaction of compound **1** with excess amounts of potassium graphite (KC$_8$) in benzene slowly proceeded under ambient condition, and after work-up, 1,4,2,5-diazadiborinine derivative **2** was obtained as an orange powder in 52% yield (Fig. 1a). The ^{11}B NMR spectrum of **2** shows a singlet at $\delta = 18.3$ p.p.m., which is shifted downfield with respect to that $(-1.9$ p.p.m.) of **1**. In the ^{1}H NMR spectrum, a peak for the

methyl groups on two nitrogen atoms was observed at $\delta = 2.90$ p.p.m., and two signals for the CH protons of the imidazole ring moieties appeared at $\delta = 6.07$ and 7.40 p.p.m. These data indicate the highly symmetric nature of the product **2**, which was further confirmed by an X-ray diffraction study (Fig. 1b). The annulated B$_2$C$_2$N$_2$ six-membered ring is essentially planar, and the boron atoms display trigonal–planar geometry with the N1–B1–C1$'$ bond angle of $112.50(15)°$ (the sum of the bond angles: B1 $= 359.75°$). Both the B1–C1$'$ $(1.491(3)$ Å) and B1–N1 $(1.458(2)$ Å) distances are significantly shorter than those $(1.610(2)$ and $1.551(2)$ Å) in **1**. The N1–C1 distance of $1.403(2)$ Å is only slightly longer than the C1–N2 bond $(1.397(2)$ Å). The C2–C3 bond $(1.346(2)$ Å) lies in the range of typical double-bond distances of carbon–carbon bonds. These structural properties suggest the delocalization of electrons over the π-system including the six-membered B$_2$C$_2$N$_2$ ring, which can be represented by the average of the several canonical forms, including **2a–g** (Fig. 1c). In the electronic paramagnetic resonance spectrum of a benzene solution of **2**, no signal was detected at room temperature.

To investigate the electronic property of **2**, we carried out quantum chemical density functional theory calculations. Natural bond orbital analysis gave a Wiberg bond index value slightly < 1.0 for the B1–N1 bond (0.94). Meanwhile, Wiberg bond index values were > 1.0 for the B1–C1$'$ bond (1.19), the C1–N1 bond (1.13) and the C1–N2 bond (1.10), thus suggesting the partial double-bond character of these bonds. The HOMO of **2** exhibits a π-system over the six-membered B$_2$C$_2$N$_2$ ring with a node found between two CBN π-units and also has large amplitudes on the N atoms and the C $=$ C moieties of the annulated imidazole rings (Fig. 1d). The LUMO is a π-type orbital that contains a mixture of C–B π-bonding interactions and the π-orbital of the phenyl ring on the B atoms. Natural population analysis shows that the two boron atoms possess the same charge of $+0.55$.

Figure 2 summarizes the nucleus-independent chemical-shift (NICS) values for **2**, a model compound **2$'$** and other related compounds for comparison. The NICS values for **2** and **2$'$** are less negative than that of annulated indole, but comparable to that of benzene, and more negative than those of other heterocycles, suggesting the considerable aromatic property of **2**. The resonance stabilization energy (RSE) value of parent 1,4,2,5-diazadiborinine **2$''$** estimated at the B3LYP/6-311 + G(d,p) level is 37.9 kcal mol^{-1} smaller than that of benzene (Supplementary Table 4)[38,39]. Since the RSE value of benzene is estimated to be about 34 kcal mol^{-1}, although the reported values vary depending on the method used[38,39], it can be inferred that the aromatic character of parent 1,4,2,5-diazadiborinine **2$''$** is significantly weak, in line with the less negative NICS values for **2$''$** than that of benzene (Fig. 2).

Reactivity. Compound **2** is thermally stable both in the solid state and in solutions at ambient temperature, but decomposes rapidly on exposure to air. To examine the reactivity, we first treated **2** with methyl trifluoromethanesulfonate (MeOTf). A stoichiometric amount of MeOTf was added to a benzene solution of **2** at ambient temperature. After removing the solvent under vacuum, **3** was obtained as a mixture of diastereomers (1:1) in 87% yield (Fig. 3). An X-ray diffraction study revealed the solid-state structure of one of the diastereomers. The methyl group is attached to one of the boron atoms, while an oxygen atom of the triflate forms a bond with the other boron atom (Fig. 4), suggesting that two spectroscopically indistinguishable boron atoms in **2** may act as both nucleophilic and electrophilic centres. Because two boron atoms interact with each other through the π-system, it is reasonable to predict that when one

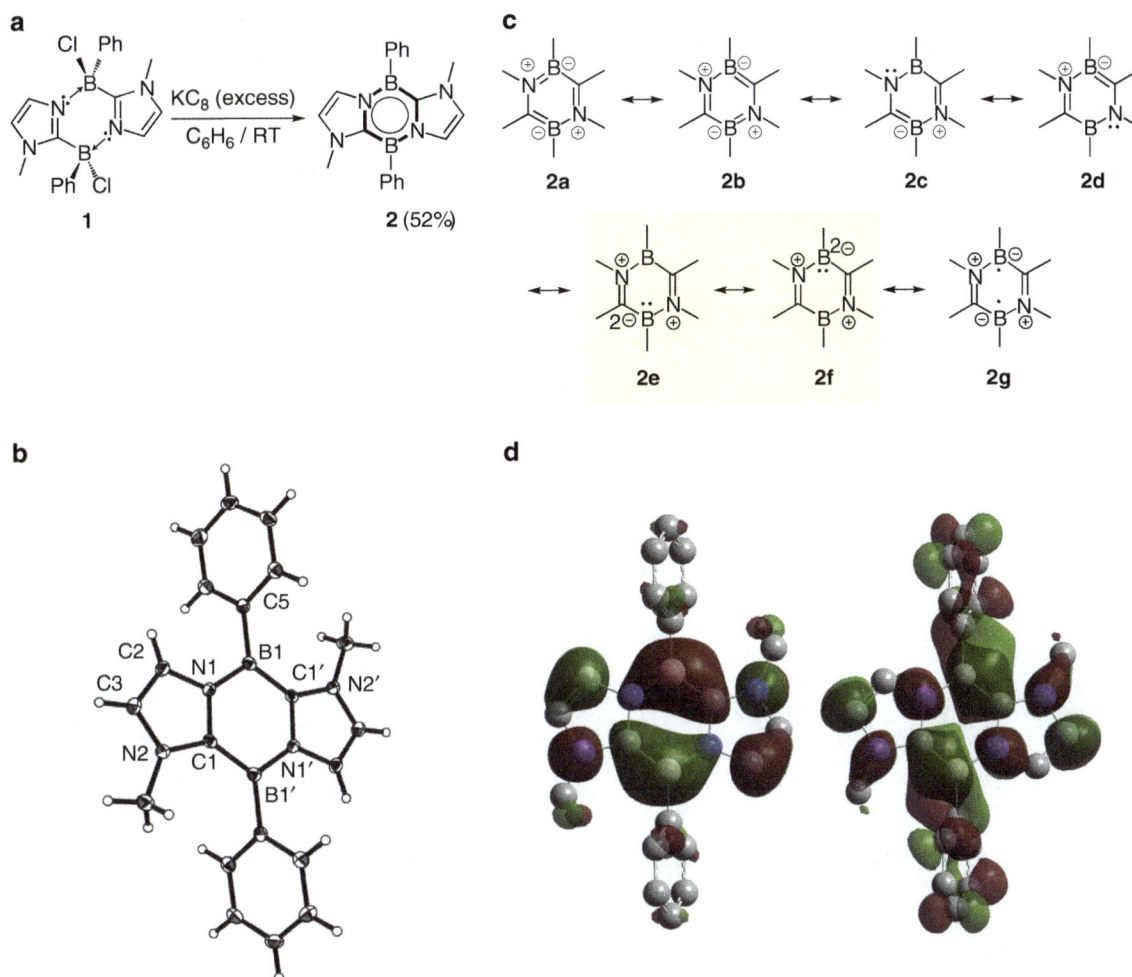

Figure 1 | Characterization of annulated 1,4,2,5-diazadiborinine derivative 2. (a) Preparation of **2** (Ph = phenyl). (b) Solid-state structure of **2** (thermal ellipsoids are set at the 50% probability level). (c) Schematic representations of selected canonical forms regarding the central $C_2B_2N_2$ ring of **2**. (d) Plots of the HOMO (left) and the LUMO (right) of **2**. Calculated at the B3LYP/6-311 + G(d,p) level of theory. Hydrogen atoms are omitted for clarity.

	2	**2′**		**2″**			
NICS(0)	−8.2	−8.5	−11.7	−4.8	−4.3	−8.0	−1.6
NICS(1)	−8.0	−8.8	−11.9	−6.9	−6.5	−10.2	−2.7

Figure 2 | Theoretical evaluation of aromaticity. Calculated NICS(0) and NICS(1) values for **2**, **2′**, annulated indole derivative, parent 1,4,2,5-diazadiborinine **2″**, parent 1,3,2,5-diazadiborinine, benzene and borazine. Calculated at the B3LYP/6-311 + G(d,p) level of theory.

boron centre in **2** behaves as a Lewis base, the other boron centre takes on Lewis acidic character, as presented by the resonance forms **2e–f** (Fig. 1c). Having observed the ambiphilic nature of the boron atoms in **2**, we considered it likely that **2** would act as a frustrated Lewis pair (FLP)[30,31]. This hypothesis was borne out by further examination of the reactions between **2** and pinacolborane (HBpin) as well as arylsilane derivatives (PhSiH$_3$ and Ph$_2$SiH$_2$) because the reactions proceeded cleanly, and after work-up, adducts **4** and **5** were isolated in good yields (**4**, 83%; **5a**, 87%; **5b**, 85%). In contrast to the case of **3**, both products **4** and **5** were obtained as single diastereomers. **2** could readily activate the P–H bond of diarylphosphine

[(p-FC$_6$H$_4$)$_2$PH] as well to afford product **6** in 85% yield. Products **3**, **4**, **5a** and **6** were fully characterized by standard spectroscopic methodologies and X-ray diffractometry (Fig. 4). In the solid-state structures of **4** and **5a**, the Bpin or Ph$_2$HSi group and the H atom on the B atom are attached on the same side of the six-membered B$_2$C$_2$N$_2$ ring. By contrast, the (FH$_4$C$_6$)$_2$P group and the H atom on the B atom in **6** point in opposite directions.

Next, we employed molecules involving non-activated unsaturated bonds as reactants. When CO$_2$ gas was introduced into a benzene solution of **2** at 1 bar, a white precipitate was instantaneously formed at room temperature. After work-up,

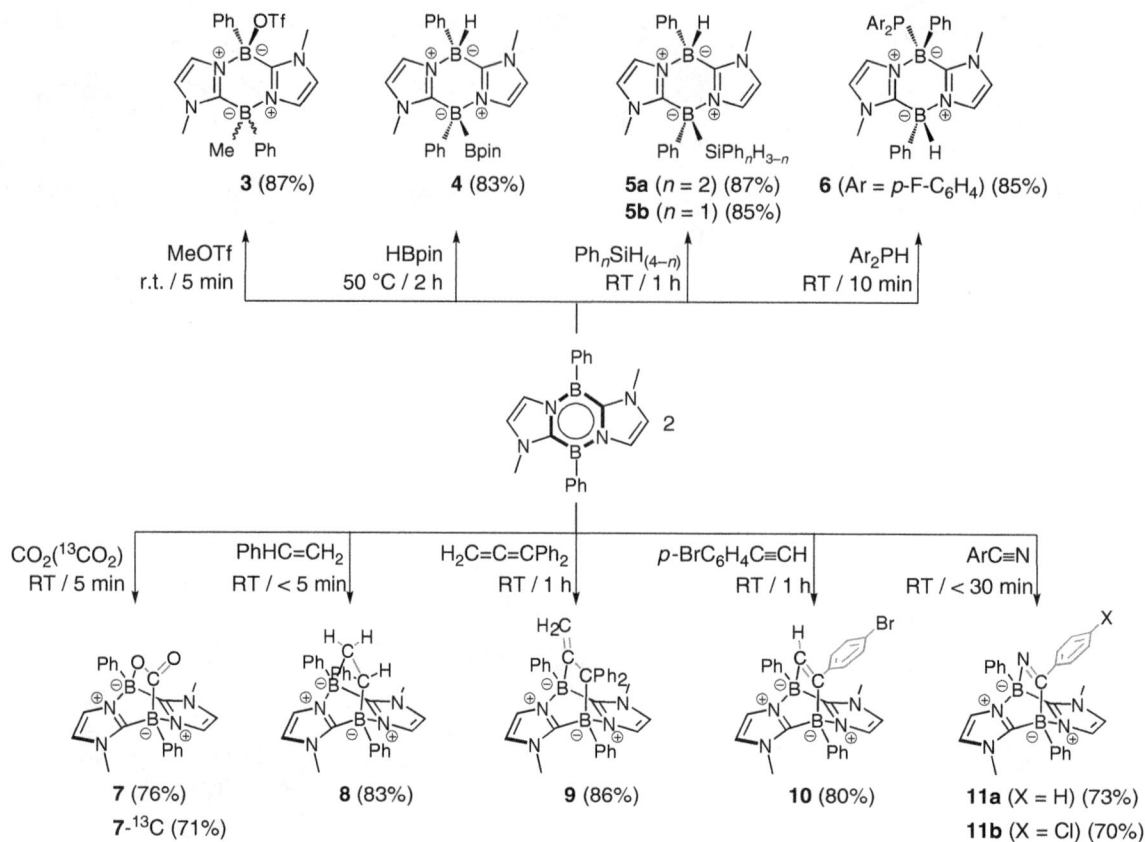

Figure 3 | Reactivity of 2. Reactions of **2** with MeOTf, HBpin, $Ph_nSiH_{(4-n)}$ ($n = 1, 2$), Ar_2PH (Ar = p-FC_6H_5), CO_2 ($^{13}CO_2$), PhHC=CH_2, H_2C=C=CPh_2, p-BrC_6H_4C≡CH, and ArC≡N (Ar = Ph, p-ClC_6H_4).

compound **7** was isolated in 76% yield. We also performed a ^{13}C-labelling study employing $^{13}CO_2$, which afforded **7-^{13}C** quantitatively. The ^{13}C NMR spectrum of **7-^{13}C** showed a broad singlet at 195.6 p.p.m. In the ^{11}B NMR spectrum, a set of new broad peaks was detected at $\delta = -1.0$ and -10.8 p.p.m. An X-ray diffraction study revealed the bicyclo[2.2.2] structure involving two boron atoms at the bridgehead, which was formed via [4 + 2] cycloaddition between **2** and one of the two C=O double bonds of CO_2. **2** was also able to activate styrene in a similar manner, to afford the corresponding bicyclo[2.2.2] derivative **8** stereoselectively as a single diastereomer in 83% yield. Analogously, the reaction of **2** with 1,1-diphenylpropa-1,2-diene (H_2C=C=CPh_2) proceeded smoothly, and clean formation of **9** through regioselective [4 + 2] cycloaddition was confirmed (86% yield), demonstrating a rare example of allene activation with a FLP[40]. Moreover, both alkyne and nitrile derivatives (p-BrC_6H_4C≡CH, PhC≡N and p-ClC_6H_4C≡N) instantly reacted with **2**, to provide **10** (80% yield), **11a** (73% yield) and **11b** (70% yield), respectively. All of the cycloadducts **8–10** and **11b** were thoroughly characterized by various spectrometric methods and X-ray crystallography (Fig. 4).

Proposed mechanism based on DFT calculations. To gain insight into the reaction mechanism, pathways for the diastereo-selective formation of **4**, **5a**, **8** and **9** were explored computationally using Gaussian 09 at the B3LYP-D3(BJ) (SCRF)/6-311 + G(d,p)//B3LYP/6-311 + G (d,p) level of theory (Fig. 5)[41–53]. The default SCRF method was used to describe the solvent effect of benzene. For the formation of **4** and **5a**, both concerted and stepwise pathways were examined. However, only concerted pathways could be determined (Fig. 5a,b), suggesting

that these reactions proceed in a concerted mechanism. This explains why the H atom on the B atom and the Bpin group in **4** or the Ph_2HSi group in **5** are attached on the same side of the six-membered $B_2C_2N_2$ ring. The relatively high free energy barrier (22.0 kcal mol^{-1}) for the formation of **4** is consistent with the fact that elevated temperature (50 °C) was required to accelerate the reaction. We compared the energies of **4** and **5** with their respective diastereomers **4*** and **5*** that possess the H atom on the B atom and Bpin group (**4***) or Ph_2HSi group (**5***) on the opposite side of the $B_2C_2N_2$ ring. Both **4** and **5** are only slightly more stable than **4*** ($+0.4$ kcal mol^{-1}) and **5*** ($+1.8$ kcal mol^{-1}), respectively, indicating that the reactions may not be thermodynamically controlled.

For the reaction of **2** with styrene, we could determine two plausible concerted pathways, which revealed that the formation of **8** is favoured both thermodynamically and kinetically (Fig. 5c). Thus, product **8** is 0.6 kcal mol^{-1} more stable than its diastereomer **8***, and the activation barrier for the formation of **8** is 1.9 kcal mol^{-1} lower than that of **8***. Importantly, when the dispersion effect was not included, the difference in barrier height between the two pathways was only 0.4 kcal mol^{-1} (Supplementary Table 6), indicating the important role played by attractive dispersion interactions, especially between the phenyl group of styrene and the methyl group of **2**, in determining the diastereoselectivity. As shown in Fig. 1d, both the HOMO and LUMO of **2** have significant amplitude over the B–C moieties rather than over the B–N moieties of the $B_2C_2N_2$ six-membered ring. At the transition state, hence, to maximize the interaction between the frontier orbitals of **2** and the π or π* orbital of styrene, the C=C bond of styrene does not lie completely parallel to the line connecting the two B atoms of **2**, but the two carbon atoms are directed slightly towards the

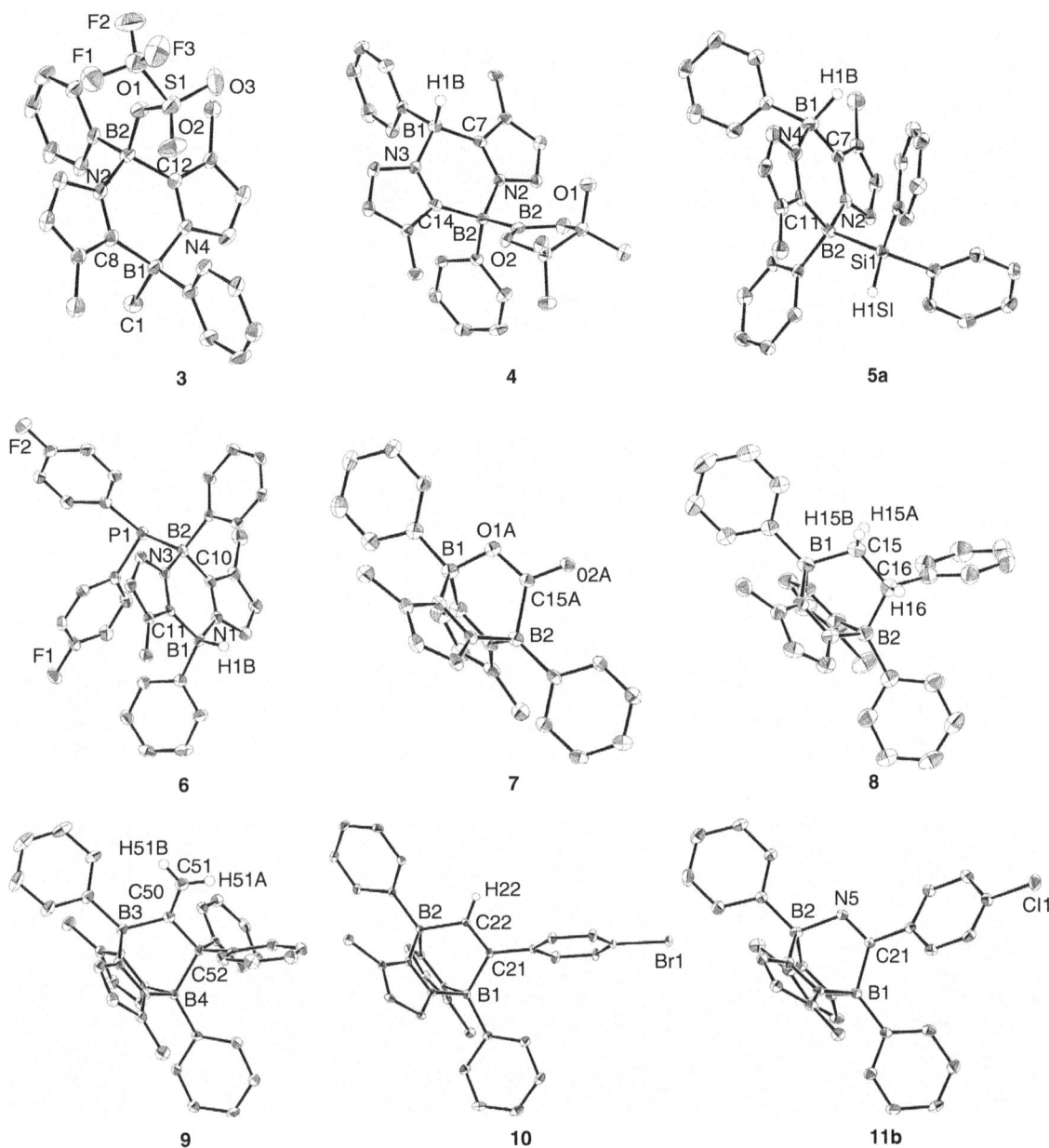

Figure 4 | Structural characterization of products. Solid-state structures of **3**, **4**, **5a**, **6-10** and **11b**.

midpoints of the two B–C bonds of **2**. In such pathways, directing the phenyl group of styrene towards the B atom of **2** would cause significant steric repulsion between the Ph rings of styrene and **2**. The transition state for the favoured pathway looks less sterically encumbered, which could also contribute to the diastereo-selectivity of the cycloaddition. Figure 5d shows the pathways obtained for the concerted cycloaddition between **2** and H₂C=C=CPh₂. The lower-energy pathway for the formation of **9** has an energy barrier of 15.1 kcal mol⁻¹, which is 1.0 kcal mol⁻¹ lower than that for the stereoisomer **9***. Interest-ingly, the pathway to **9*** has a lower barrier when dispersion is not taken into account, but the relative energy of the transition states is inverted on inclusion of the dispersion effect (Supplementary Table 6). This result again highlights the important role of dispersion in determining the regioselectivity. Compound **9** is 7.9 kcal mol⁻¹ less stable than **9***, and thus formation of **9** is kinetically preferred.

The formation of compound **3** as a mixture of two diastereomers suggests that a stepwise reaction mechanism may

be involved, in addition to a concerted pathway (Fig. 6). However, our calculations could determine only a concerted pathway that gives **3B** (Supplementary Fig. 59). This might be attributed to the limited accuracy of density functional theory (DFT). We infer that the initial step would be methylation of **2** by MeOTf to afford an ionic intermediate **INT-3** (Fig. 6a). In the second step, triflate (⁻OTf) could attack the tricoordinate boron centre of **INT-3** from either the same side or the opposite side of the B₂C₂N₂ ring, to form a B–O bond. The former attack would afford **3A**, whereas the latter would give **3B**. Compound **3A** is estimated to be 0.2 kcal mol⁻¹ more stable than **3B**. Similarly, compound **6** may be formed through a stepwise mechanism because the (FH₄C₆)₂P group and the H atom on the B atom in **6** point in opposite directions (Fig. 6b). However, DFT calculations could determine only a concerted pathway to afford **6*** (Supplementary Fig. 59), and thus the reaction mechanism for the formation of **6** remains unclear, although again, a stepwise mechanism might operate here. The possible stepwise pathways for the formation of **3** and **6** are in contrast to the concerted one in the formation of **4** and **5**.

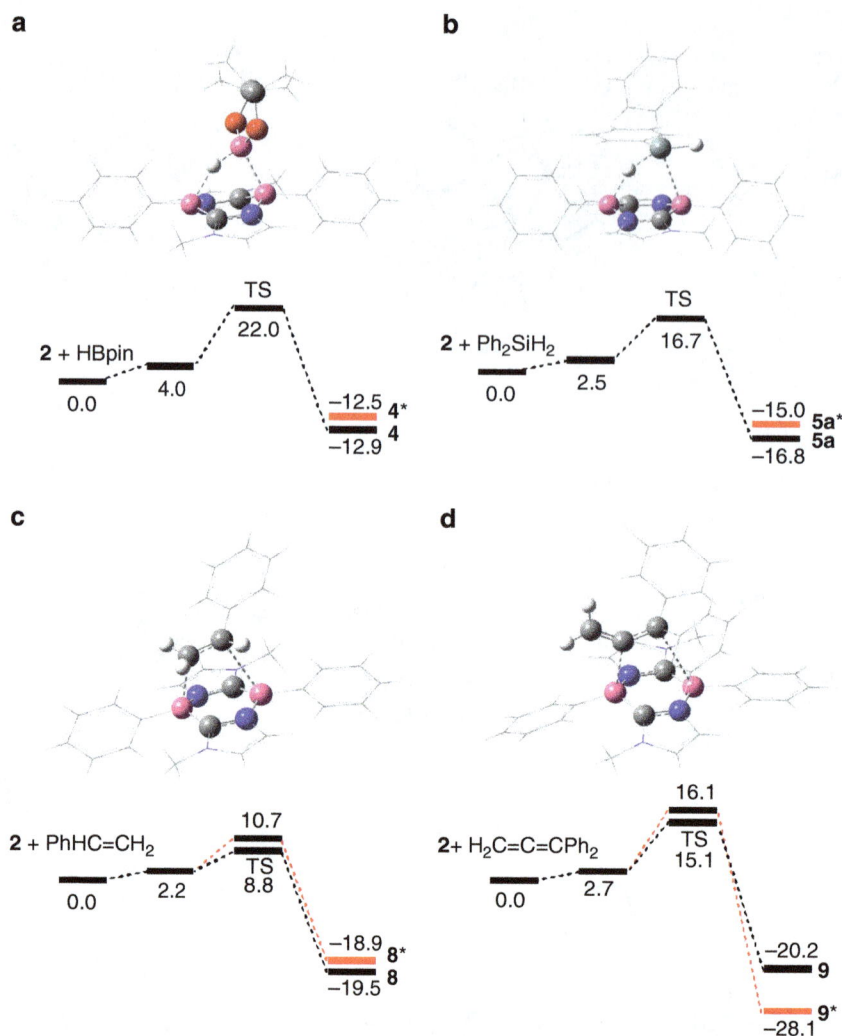

Figure 5 | DFT-calculated free energy profiles of plausible concerted mechanism. Energy profiles of possible mechanism for the stereo- and region-selective formation of **4**, **5a**, **8** and **9** from **2** with relative Gibbs free energies in kcal mol^{-1} obtained at the B3LYP-D3(BJ)(SCRF)/6-311 + G(d,p)//B3LYP/6-311 + G(d,p) level, and dispersion force, all the compound numbers are in conjunction with Fig. 3). (**a**) Pathway for the formation of **4**. (**b**) Pathway for the formation of **5a**. (**c**) Pathways for the formation of **8** and the other diastereomer **8***. (**d**) Pathways for the formation of **9** and the other diastereomer **9***.

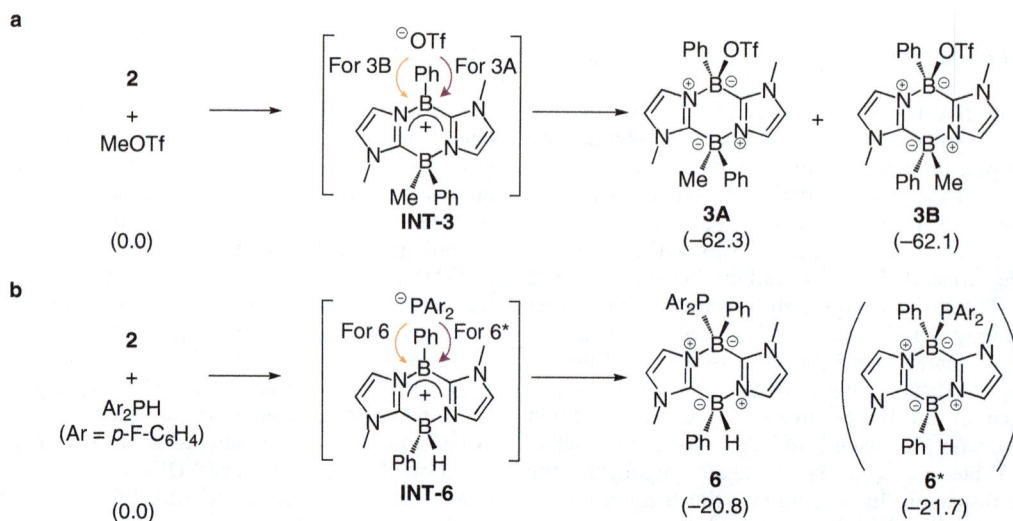

Figure 6 | Proposed stepwise mechanism for the formation of 3 and 6. Relative Gibbs free energies in kcal mol^{-1} (estimated by optimization obtained at the B3LYP-D3(BJ)(SCRF)/6-311 + G(d,p)//B3LYP/6-311 + G(d,p) level. (**a**) Stepwise mechanism for the formation of **3A** and **3B** via **INT-3**. (**b**) Stepwise mechanism for the formation of **6** via **INT-6**.

Because the H atoms in H-Bpin and Ph$_2$SiH$_2$ are hydridic, the formation of the corresponding ionic intermediates such as [**2**-H]$^-$[E]$^+$ (E = Bpin or SiPh$_2$H) would be disfavoured due to the instability of the boryl or silyl cation fragment, which could be, at least in part, the origin of different stereo-selectivity observed in the formation of these products **4**, **5** and **6**. Isolation of ionic species as **INT-3** and **INT-6** is a subject of our ongoing study.

Discussion

Stephan and co-workers reported that an aromatic triphosphabenzene featuring a C$_3$P$_3$ six-membered ring activates a hydrogen molecule under relatively mild conditions to give a [3.1.0]bicyclo reduction product[54]. It has also been computationally confirmed that the initial transition state involves the deformation of the flexible C$_3$P$_3$ ring to a boat configuration. Similar boat character is predicted in the transition structure for the uncatalysed 1,4-hydrogenation of benzene[55]. Likewise, at the transition states for the formation of **4**, **5a**, **8** and **9** (Fig. 5), we observed distortion of the B$_2$C$_2$N$_2$ six-membered ring moiety in **2**, which exhibits some boat-like character. The less significant deformation is probably due to the lack of innate flexibility of the B$_2$C$_2$N$_2$ ring skeleton. It is inferred that such deformation may polarize frontier orbitals of **2** and enhance zwitterionic nature at two boron atoms, which would allow them to play cooperatively as a nucleophilic and an elctrophilic centres.

Previously, we reported reversible [4 + 2] cycloaddition of 1,3,2,5-diazadiborinine with CO$_2$ as well as alkenes including ethylene[35,36]. A significant relevant example showed that 1-bora-4-tellurocyclohexa-2,5-diene undergoes subsequent [4 + 2] cycloaddition/alkyne-elimination via a Te/B FLP-type mechanism, reported by Stephan et al.[56]. These results prompted us to examine the thermal stability of the [4 + 2] cycloadducts **7** and **8**. A C$_6$D$_6$ solution of **7** or **8** in a degassed J-young NMR tube was heated and monitored by NMR spectroscopy. However, even at 150 °C, neither retro-[4 + 2] cycloaddition nor a pronounced decomposition was observed, demonstrating considerable stability of these cycloadducts. This is in good agreement with the greater computed thermodynamic stability of **8**.

Collectively, we have developed a synthetic protocol for aromatic 1,4,2,5-diazadiborinine **2**. According to the NICS values of **2**, **2′**, **2″** and benzene as well as the gap of the RSE values between **2″** and benzene, the annulation of the B$_2$C$_2$N$_2$ ring in 1,4,2,5-diazadiborinine may increase the aromatic nature of **2′**, which is comparable to that of benzene. Nevertheless, **2** exhibits reactivity that is peculiar and much higher than that of benzene. Experimental and computational studies on the reactivity demonstrate a manifesting dual character of the boron atoms in **2**, which is capable of activating a series of small molecules under mild reaction conditions. Such chemical behaviour has never been observed for the previously reported B$_2$C$_2$N$_2$ heterocycles[57–59]. As the substituents on the boron and nitrogen atoms in 1,4,2,5-diazadiborinine can be readily modified, the steric and electronic properties as well as the reactivity could be finely tunable. The isolation of this molecule provides a new strategy for the development of myriad B/N-containing π-systems, which involve two equivalent boron centres displaying ambiphilic nature. This approach could be of particular interest to the future design of various main group compounds featuring a FLP-type property.

Methods
Materials. For details of spectroscopic analyses of compounds in this manuscript, see Supplementary Figs 1–55. For details of X-ray analysis, see Supplementary Tables 1–3, Supplementary Methods and Supplementary Data 1–11. For details of

density functional theory calculations, see Supplementary Figs 56–59, Supplementary Tables 4–8 and Supplementary Methods.

General synthetic procedures. All reactions were performed under an atmosphere of argon or nitrogen using standard Schlenk or dry box techniques; solvents were dried over Na metal, K metal or CaH$_2$. Reagents were of analytical grade, obtained from commercial suppliers and used without further purification. ^{1}H, ^{11}B, ^{13}C and ^{19}F NMR spectra were obtained with a Bruker AVIII 400 MHz BBFO1 spectrometer at 298 K unless otherwise stated. NMR multiplicities are abbreviated as follows: s = singlet, d = doublet, t = triplet, m = multiplet and br = broad signal. Coupling constants J are given in Hz. Most of signals for the quaternary carbon atoms could not be detected, presumably due to coupling with the B atom. Electrospray ionization (ESI) mass spectra were obtained at the Mass Spectrometry Laboratory at the Division of Chemistry and Biological Chemistry, Nanyang Technological University. Melting points were measured with an OpticMelt Stanford Research System. Infrared spectra were measured with the Bruker Alpha-FT-IR Spectrometer with an ECO-ATR module. Continuous wave X-band electron paramagnetic resonance (EPR) spectrum was checked using a Bruker ELEXSYS E500 EPR spectrometer.

Synthesis of 1. To a tetrahydrofuran (THF; 20 ml) solution of 1-methyl-1H-imidazole (0.55 g, 6.7 mmol), n-butyl lithium (1.6 M in hexane; 4.6 ml, 7.4 mmol) was added dropwise at − 40 °C. The reaction mixture was allowed to warm to room temperature over 60 min, and then transferred to a THF (40 ml) solution of dimethyl phenylboronate (1.0 g, 6.7 mmol) at − 78 °C. The mixture was warmed to room temperature and stirred overnight. Then, trimethylsilane chloride (1.09 g, 10 mmol) was added to the mixture at − 78 °C, and the solution was slowly warmed to room temperature and stirred for 5 h. After removal of all the volatiles, the residual solid was re-dissolved in dichloromethane (DCM; 40 ml). To the solution, boron trichloride (1 M in hexane; 6.7 ml, 6.7 mmol) was added at 0 °C, and the reaction mixture was warmed to room temperature, and stirred for 1 h. After filtration, all volatiles were removed under vacuum, and then the residue was washed with hexane and dried under vacuum to afford compound **1** as a pale yellow powder (1.02 g, 75%). Melting point: 235 °C (dec). ^{1}H NMR (400 MHz, C$_6$D$_6$): δ = 7.95 (m, 4H, Ar-H), 7.30 (t, 4H, J = 7.4 Hz, Ar-H), 7.18 (t, 2H, J = 7.4 Hz, Ar-H), 6.76 (d, 2H, J = 1.8 Hz, CH), 5.52 (d, 2H, J = 1.8 Hz, CH) and 2.84 (s, 6H, N-CH$_3$); ^{13}C{^{1}H} NMR (100.56 MHz, C$_6$D$_6$): δ = 133.0 (Ar-CH), 128.3 (Ar-CH), 127.5 (Ar-CH), 123.0 (CH), 122.2 (CH) and 34.8 (N-CH$_3$); and ^{11}B{^{1}H} NMR (128.3 MHz, C$_6$D$_6$): δ = − 1.9 (s). High resolution mass spectrometry (ESI): m/z calculated for C$_{20}$H$_{21}$B$_2$Cl$_2$N$_4$: 409.1329 [(M + H)]$^+$; found: 409.1338.

Synthesis of 2. Potassium graphite (3.76 g, 27.8 mmol) was added into a benzene (80 ml) solution of **1** (0.71 g, 1.74 mmol) at room temperature, and the reaction mixture was stirred for 5 days. After removal of graphite and inorganic salt by filtration, the solvent was removed under vacuum to afford **2** as an orange solid (305 mg, 52%). **2** decomposes at 155 °C without melting. ^{1}H NMR (400 MHz, C$_6$D$_6$): δ = 7.79 (m, 4H, Ar-H), 7.40 (m, 6H, Ar-H × 4 and CH × 2), 7.32 (t, 2H, J = 7.4 Hz, Ar-H), 6.07 (d, 2H, J = 1.9 Hz, CH) and 2.90 (s, 6H, N-CH$_3$); ^{13}C{^{1}H} NMR (100.56 MHz, C$_6$D$_6$): δ = 135.4 (Ar-CH), 127.9 (Ar-CH), 126.9 (Ar-CH), 124.1 (CH), 113.8 (CH) and 36.1 (N-CH$_3$); and ^{11}B{^{1}H} NMR (128.3 MHz, C$_6$D$_6$): δ = 18.3 (s). Ultraviolet–visible (ε, in hexane): λ = 496 nm (7,600), 468 nm (4,280), 354 nm (2,140) and 273 nm (3,410). HRMS (ESI): m/z calculated for C$_{20}$H$_{21}$B$_2$N$_4$: 339.1952 [(M + H)]$^+$; found: 339.1962.

Synthesis of 3. MeOTf (0.060 ml, 0.53 mmol) was added into a benzene (2.0 ml) solution of **2** (180 mg, 0.53 mmol) at room temperature, and the reaction mixture was stirred for 5 min. After removal of the solvent, the solid residue was washed with hexane and dried under vacuum to afford a yellow solid of **3** as a 1:1 mixture of diastereomers (216 mg, 87%). Recrystallization of **3** from a mixture of dichloromethane and hexane solution afforded single crystals of one of the diastereomers, which was confirmed by an X-ray diffraction analysis. When the single crystals were re-dissolved in C$_6$D$_6$ and checked by NMR spectroscopy, two diastereomers were observed in 1:1 ratio, suggesting that these two diastereomers are in equilibrium in solution at room temperature (For the estimated energy difference between two diastereomers, see Supplementary Figure 57.) *Mixture of two diastereomers of **3**: ^{1}H NMR (400 MHz, C$_6$D$_6$): δ = 7.80 (d, 2H, J = 6.8 Hz, Ar-H), 7.64 (d, 2H, J = 6.8 Hz, Ar-H), 7.54 (d, 2H, J = 6.8 Hz, Ar-H), 7.48 (d, 2H, J = 6.8 Hz, Ar-H), 7.37 (d, 2H, J = 7.5 Hz, Ar-H), 7.31–7.18 (m, 10H, Ar-H), 6.73 (d, 1H, J = 1.6 Hz, CH), 6.64 (d, 1H, J = 1.6 Hz, CH), 6.50 (d, 1H, J = 1.6 Hz, CH), 6.42 (d, 1H, J = 1.6 Hz, CH), 5.75 (dd, 2H, J = 3.3, 1.6 Hz, CH), 5.68 (t, 2H, J = 1.6 Hz, CH), 3.00 (s, 3H, N-CH$_3$), 2.90 (s, 3H, N-CH$_3$), 2.59 (s, 3H, N-CH$_3$), 2.57 (s, 3H, N-CH$_3$), 0.80 (s, 3H, B-CH$_3$) and 0.61 (s, 3H, B-CH$_3$); ^{13}C{^{1}H} NMR (100.56 MHz, C$_6$D$_6$): δ = 134.0 (Ar-CH), 133.2 (Ar-CH), 131.7 (Ar-CH), 131.4 (Ar-CH), 128.6 (Ar-CH), 128.5 (Ar-CH), 128.3 (Ar-CH), 128.2 (Ar-CH), 128.0 (Ar-CH), 127.9 (Ar-CH), 126.8 (Ar-CH), 126.5 (Ar-CH), 123.0 (CH), 122.9 (CH), 122.8 (CH), 122.7 (CH), 122.5 (CH), 122.4 (CH), 121.4 (CH), 121.0 (CH), 34.7 (N-CH$_3$), 34.5 (N-CH$_3$), 34.4 (N-CH$_3$) and 34.3 (N-CH$_3$); ^{11}B{^{1}H} NMR (128.3 MHz, C$_6$D$_6$): δ = 1.0 (br) and -7.7 (br); and ^{19}F{^{1}H} NMR (376 MHz,

C_6D_6): $\delta = -77.7$ and -77.9. HRMS (ESI): *m/z* calculated for $C_{22}H_{24}B_2F_3N_4O_3S$: 503.1707 [$(M+H)]^+$; found: 503.1715.

Synthesis of 4. Pinacolborane (0.087 ml, 0.59 mmol) was added to a benzene (2.0 ml) solution of **2** (200 mg, 0.59 mmol) at room temperature, and the reaction mixture was stirred for 2 h at 50 °C. After removal of the solvent, the solid residue was washed with hexane and dried under vacuum to afford **4** as a yellow powder (229 mg, 83%). Melting point: 159 °C (dec). ^1H NMR (400 MHz, C_6D_6): $\delta = 7.86$ (d, 2H, $J = 7.2$ Hz, Ar-*H*), 7.70 (d, 2H, $J = 7.2$ Hz, Ar-*H*), 7.39 (t, 2H, $J = 7.2$ Hz, Ar-*H*), 7.33 (t, 2H, $J = 7.2$ Hz, Ar-*H*), 7.26–7.20 (m, 2H, Ar-*H*), 7.12 (s, 1H, C*H*), 6.75 (s, 1H, C*H*), 5.92 (s, 1H, C*H*), 5.87 (s, 1H, C*H*), 3.20 (s, 3H, N-CH_3), 2.68 (s, 3H, N-CH_3) and 1.09 (s, 12H, CH_3); $^{13}C\{^1H\}$ NMR (100.56 MHz, C_6D_6): $\delta = 135.3$ (Ar-CH), 135.2 (Ar-CH), 128.2 (Ar-CH), 128.1 (Ar-CH), 126.2 (Ar-CH), 126.1 (Ar-CH), 123.8 (CH), 123.2 (CH), 121.5 (CH), 120.5 (CH), 82.0 ($C(CH_3)_2$), 35.2 (N-CH_3), 34.1 (N-CH_3), 25.4 (CH_3) and 25.3 (CH_3); and $^{11}B\{^1H\}$ NMR (128.3 MHz, C_6D_6): $\delta = 22.7$ (s), -10.1 (s) and -11.1 (s). HRMS (ESI): *m/z* calculated for $C_{26}H_{34}B_3N_4O_2$: 467.2961 [$(M+H)]^+$; found: 467.2937.

Synthesis of 5a. Diphenylsilane (0.110 ml, 0.59 mmol) was added to a benzene (3.0 ml) solution of **2** (200 mg, 0.59 mmol), and the reaction mixture stirred for 1 h at room temperature. After removal of the solvent, the residue was washed with hexane and dried under vacuum to afford **5a** as a white powder (269 mg, 87%). Melting point: 182 °C (dec). ^1H NMR (400 MHz, C_6D_6): $\delta = 7.95$ (d, 2H, $J = 7.2$ Hz, Ar-*H*), 7.66 (m, 2H, Ar-*H*), 7.62 (d, 2H, $J = 7.2$ Hz, Ar-*H*), 7.55–7.53 (m, 2H, Ar-*H*), 7.38–7.30 (m, 4H, Ar-*H*), 7.24–7.16 (m, 2H, Ar-*H*), $\delta = 7.14$–7.11 (m, 6H, Ar-*H*), 6.65 (s, 1H, C*H*), 6.56 (s, 1H, C*H*), 5.67 (s, 1H, C*H*), 5.59 (s, 1H, C*H*), 5.57 (s, 1H, Si*H*), 2.64 (s, 1H, N-CH_3) and 2.56 (s, 1H, N-CH_3); $^{13}C\{^1H\}$ NMR (100.56 MHz, C_6D_6): $\delta = 139.3$ (Si^{q}C), 139.0 (Si^{q}C), 136.3 (Ar-CH), 136.0 (Ar-CH), 135.6 Ar-CH), 135.1 (Ar-CH), 128.4 (Ar-CH), 128.31 (Ar-CH), 128.25 (Ar-CH), 128.22 (Ar-CH), 127.9 (Ar-CH), 127.8 (Ar-CH), 126.6 (Ar-CH), 126.4 (Ar-CH), 123.5 (CH), 123.1 (CH), 121.8 (CH), 120.9 (CH), 35.2(N-CH_3) and 34.2 (N-CH_3); $^{11}B\{^1H\}$ NMR (128.3 MHz, C_6D_6): $\delta = -9.2$ (br, two signals are overlapped). HRMS (ESI): *m/z* calculated for $C_{32}H_{33}B_2N_4Si$: 523.2661 [$(M+H)]^+$; found: 523.2671.

Synthesis of 5b. Phenylsilane (0.091 ml, 0.74 mmol) was added to a benzene (3.0 ml) solution of **2** (250 mg, 0.74 mmol), and the reaction mixture stirred for 1 h at room temperature. After removal of the solvent, the residue was washed with hexane and dried under vacuum to afford **5b** as a white powder (280 mg, 85%). Melting point: 163 °C (dec). ^1H NMR (400 MHz, C_6D_6): $\delta = 7.82$ (d, 2H, $J = 6.8$ Hz, Ar-*H*), 7.60 (d, 2H, $J = 6.8$ Hz, Ar-*H*), 7.35–7.29 (m, 6H, Ar-*H*), 7.23–7.19 (m, 2H, Ar-*H*), 7.11–7.09 (m, 3H, Ar-*H*), 6.71 (d, 1H, $J = 1.7$ Hz, C*H*), 6.63 (d, 1H, $J = 1.7$ Hz, C*H*), 5.75 (d, 1H, $J = 1.7$ Hz, C*H*), 5.73 (s, 1H, $J = 1.7$ Hz, C*H*), 4.82 (s, 1H, Si*H*), 4.77 (s, 1H, Si*H*), 2.75 (s, 3H, N-CH_3) and 2.63 (s, 3H, N-CH_3); $^{13}C\{^1H\}$ NMR (100.56 MHz, C_6D_6): $\delta = 136.2$ (Si^{q}C), 136.0 (Ar-CH), 135.2 (Ar-CH), 135.0 (Ar-CH), 128.5 (Ar-CH), 128.3 (Ar-CH), 128.2 (Ar-CH), 127.8 (Ar-CH), 126.7 (Ar-CH), 126.3 (Ar-CH), 123.5 (CH), 122.6 (CH), 121.7 (CH), 121.4 (CH), 35.0 (N-CH_3) and 34.2 (N-CH_3); and $^{11}B\{^1H\}$ NMR (128.3 MHz, C_6D_6): $\delta = -9.7$ (br, two signals are overlapped). HRMS (ESI): *m/z* calculated for $C_{26}H_{29}B_2N_4Si$: 447.2348 [$(M+H)]^+$; found: 447.2368.

Synthesis of 6. Bis(4-fluorophenyl)phosphine (164 mg, 0.74 mmol) was added to a benzene (3.0 ml) and THF (0.5 ml) solution of **2** (250 mg, 0.74 mmol), and the reaction mixture stirred for 10 min at room temperature. After removal of the solvent, the residue was washed with hexane and dried under vacuum to afford **6** as a white powder (353 mg, 85%). Melting point: 210 °C. ^1H NMR (400 MHz, C_6D_6): $\delta = 7.83$ (d, 2H, $J = 7.7$ Hz, Ar-*H*), 7.47–7.43 (m, 2H, Ar-*H*), 7.35 (t, 2H, $J = 7.3$ Hz Ar-*H*), 7.29–7.23 (m, 3H, Ar-*H*), 7.19–7.09 (m, 3H, Ar-*H*), 7.04 (d, 2H, $J = 7.3$ Hz, Ar-*H*), 6.76 (d, 1H, $J = 1.5$ Hz, C*H*), 6.67 (dd, 4H, $J = 14.9$, 8.2 Hz, Ar-*H*), 6.60 (d, 1H, $J = 1.5$ Hz, C*H*), 5.72 (d, 1H, $J = 1.5$ Hz, C*H*), 5.69 (s, 1H, $J = 1.5$ Hz, C*H*), 3.25 (s, 3H, N-CH_3) and 2.63 (s, 3H, N-CH_3); $^{13}C\{^1H\}$ NMR (100.56 MHz, C_6D_6): $\delta = 136.6$ (d, $J = 7.2$ Hz, Ar-CH), 136.4 (d, $J = 7.2$ Hz, Ar-CH), 135.5 (Ar-CH), 134.8 (Ar-CH), 134.6 (Ar-CH), 128.1 (Ar-CH), 126.8 (Ar-CH), 126.6 (Ar-CH), 123.5 (CH), 123.1 (CH), 122.4 (CH), 120.8 (CH), 115.5 (d, $J = 6.3$ Hz, Ar-CH), 115.3 (d, $J = 6.3$ Hz, Ar-CH), 36.3 (d, $J = 18.6$ Hz, N-CH_3) and 34.7 (N-CH_3); $^{11}B\{^1H\}$ NMR (128.3 MHz, C_6D_6): $\delta = -5.1$ (s) and -10.2 (s); $^{19}F\{^1H\}$ NMR (376 MHz, C_6D_6): $\delta = -114.95$ (d, $J = 3.8$ Hz) and -115.98 (d, $J = 4.9$ Hz); and $^{31}P\{^1H\}$ NMR (162 MHz, C_6D_6): $\delta = -34.1$. HRMS (ESI): *m/z* calculated for $C_{32}H_{30}B_2N_4F_2P$: 561.2362 [$(M+H)]^+$; found: 561.2370.

Synthesis of 7. A benzene (2.0 ml) solution of **2** (200 mg, 0.59 mmol) was degassed using a freeze-pump-thaw method, and then CO_2 (1 bar) was introduced into the schlenk tube. After stirring for 5 min at room temperature, a white precipitate was collected by filtration and dried under vacuum to afford **7** as a white solid (172 mg, 76%). Melting point: 223 °C (dec). ^1H NMR (400 MHz, $CDCl_3$): $\delta = 8.23$ (d, 2H, $J = 6.9$ Hz, Ar-*H*), 7.90 (d, 2H, $J = 6.9$ Hz, Ar-*H*), 7.47–7.35 (m, 6H, Ar-*H*), 6.99 (d, 1H, $J = 0.8$ Hz, C*H*), 6.95 (d, 1H, $J = 0.8$ Hz, C*H*), 6.57 (d, 2H, $J = 0.8$ Hz, C*H*), 3.27 (s, 3H, N-CH_3) and 3.13 (s, 3H, N-CH_3); $^{13}C\{^1H\}$ NMR (100.56 MHz, $CDCl_3$): $\delta = 135.1$ (Ar-CH), 133.1 (Ar-CH), 128.1 (Ar-CH), 128.0 (Ar-CH), 127.8 (Ar-CH), 127.3 (Ar-CH), 121.9 (CH), 121.4 (CH), 121.2 (CH), 120.5 (CH), 35.51 (N-CH_3) and 35.45 (N-CH_3); and $^{11}B\{^1H\}$ NMR (128.3 MHz, $CDCl_3$): $\delta = -1.0$ (s) and -10.8 (s). Infrared ν cm^{-1} (solid): 1667 (s). HRMS (ESI): *m/z* calculated for $C_{21}H_{21}B_2N_4O$: 383.1851 [$(M+H)]^+$; found: 383.1858.

Synthesis of 7-^{13}C. By following the same procedure utilized for the synthesis of 7, the reaction employing $^{13}CO_2$ afforded **7-^{13}C** (161 mg, 71%). Melting point: 223 °C (dec). $^{13}C\{^1H\}$ NMR (100.56 MHz, $CDCl_3$): $\delta = 195.6$ ($C{=}O$), 135.0 (Ar-CH), 133.0 (Ar-CH), 128.0 (Ar-CH), 127.9 (Ar-CH), 127.7 (Ar-CH), 127.2 (Ar-CH), 121.8 (CH), 121.3 (CH), 121.2 (CH), 120.6 (CH), 35.43 (N-CH_3) and 35.36 (N-CH_3); and $^{11}B\{^1H\}$ NMR (128.3 MHz, $CDCl_3$): $\delta = -0.8$ (s) and -10.8 (br). HRMS (ESI): *m/z* calculated for $C_{20}{}^{13}CH_{21}B_2N_4O$: 384.1884 [$(M+H)]^+$; found: 384.1903.

Synthesis of 8. Styrene (0.085 ml, 0.74 mmol) was added to a benzene (3.0 ml) solution of **2** (250 mg, 0.74 mmol), and the reaction mixture stirred for 2 min at room temperature. After removal of the solvent, the residue was washed with hexane and dried under vacuum to afford **8** as a yellow powder (271 mg, 83%). Melting point: 238 °C. ^1H NMR (400 MHz, C_6D_6): $\delta = 7.83$–7.82 (m, 4H, Ar-*H*), 7.44 (t, 2H, $J = 7.3$ Hz, Ar-*H*), 7.38–7.32 (m, 3H, Ar-*H*), 7.26 (t, 1H, $J = 7.3$ Hz, Ar-*H*), 7.13 (d, 2H, $J = 7.3$ Hz, Ar-*H*), 7.02 (t, 1H, $J = 7.3$ Hz, Ar-*H*), 6.84 (d, 1H, $J = 1.5$ Hz, C*H*), 6.56 (d, 2H, $J = 7.3$ Hz, Ar-*H*), 6.36 (d, 1H, $J = 1.5$ Hz, C*H*), 5.88 (d, 1H, $J = 1.5$ Hz, C*H*), 5.67 (s, 1H, $J = 1.5$ Hz, C*H*), 2.73 (dd, 1H, $J = 10.1$, 4.4 Hz, C*H*), 2.61 (s, 3H, N-CH_3), 2.55 (s, 3H, N-CH_3), 1.93 (dd, 1H, $J = 13.5$, 10.1 Hz, C*H*) and 1.04 (dd, 1H, $J = 13.5$, 4.4 Hz, C*H*); $^{13}C\{^1H\}$ NMR (100.56 MHz, C_6D_6): $\delta = 155.6$ (Ar-^{q}C), 136.4 (Ar-CH), 135.2 (Ar-CH), 128.4 (Ar-CH), 128.1 (Ar-CH), 128.0 (Ar-CH), 127.8 (Ar-CH), 126.8 (Ar-CH), 126.4 (Ar-CH), 123.2 (Ar-CH), 121.1 (CH), 120.9 (CH), 120.1 (CH), 119.3 (CH), 35.3 (N-CH_3) and 34.7 (N-CH_3); and $^{11}B\{^1H\}$ NMR (128.3 MHz, C_6D_6): $\delta = -5.8$ (s) and -6.7 (s). HRMS (ESI): *m/z* calculated for $C_{28}H_{29}B_2N_4$: 443.2578 [$(M+H)]^+$; found: 443.2581.

Synthesis of 9. 1,1-Diphenylpropa-1,2-diene (114 mg, 0.59 mmol) was added to a benzene (2.0 ml) solution of **2** (200 mg, 0.59 mmol), and the reaction mixture stirred for 1 h at room temperature. After removal of the solvent, the residue was washed with hexane and dried under vacuum to afford **9** as a white powder (270 mg, 86%). Melting point: 245 °C. ^1H NMR (400 MHz, C_6D_6): $\delta = 8.12$ (d, 2H, $J = 6.8$ Hz, Ar-*H*), 7.56 (d, 4H, $J = 7.4$ Hz, Ar-*H*), 7.46 (t, 2H, $J = 7.4$ Hz, Ar-*H*), 7.36 (t, 1H, $J = 7.4$ Hz, Ar-*H*), 7.21–7.17 (m, 5H, Ar-*H*), 7.12 (t, 2H, $J = 7.4$ Hz, Ar-*H*), 7.06–6.99 (m, 3H, Ar-*H*), 6.92 (t, 1H, $J = 7.4$ Hz, Ar-*H*), 6.50 (d, 1H, $J = 1.5$ Hz, C*H*), 6.26 (d, 1H, $J = 1.5$ Hz, C*H*), 6.16 (d, 1H, $J = 3.2$ Hz, CH_2), 5.68 (d, 1H, $J = 1.5$ Hz, C*H*), 5.45 (d, 1H, $J = 1.5$ Hz, C*H*), 5.14 (d, 1H, $J = 3.2$ Hz, CH_2), 2.70 (s, 3H, N-CH_3) and 2.43 (s, 3H, N-CH_3); $^{13}C\{^1H\}$ NMR (100.56 MHz, C_6D_6): $\delta = 153.6$ (Ar-^{q}C), 152.0 (Ar-^{q}C), 136.6 (Ar-CH), 130.9 (Ar-CH), 130.2 (Ar-CH), 128.2 (Ar-CH), 127.4 (Ar-CH), 127.3 (Ar-CH), 126.9 (Ar-CH), 126.6 (Ar-CH), 124.1 (Ar-CH), 123.9 (CH), 122.8 (CH), 121.1 (CH), 119.8 (CH_2), 119.7 (CH) and 35.2 (overlap, N-CH_3); $^{11}B\{^1H\}$ NMR (128.3 MHz, C_6D_6): $\delta = -3.5$ (s) and -6.2 (s). HRMS (ESI): *m/z* calculated for $C_{35}H_{33}B_2N_4$: 531.2891 [$(M+H)]^+$; found: 531.2896.

Synthesis of 10. 1-Bromo-4-ethynylbenzene (80 mg, 0.44 mmol) was added to a benzene (2.0 ml) solution of **2** (150 mg, 0.44 mmol), and the reaction mixture was stirred for 1 h at room temperature. After removal of the solvent, the residue was washed with hexane and dried under vacuum to afford **10** as a yellow solid (183 mg, 80%). Melting point: 238 °C (dec). ^1H NMR (400 MHz, C_6D_6): $\delta = 8.02$ (d, 2H, $J = 7.1$ Hz, Ar-*H*), 7.86 (s, 1H, C=C*H*), 7.67 (d, 2H, $J = 7.1$ Hz, Ar-*H*), 7.48 (t, 2H, $J = 7.4$ Hz, Ar-*H*), 7.38 (d, 1H, $J = 7.4$ Hz, Ar-*H*), 7.34 (d, 2H, J = 8.3 Hz, Ar-*H*), 7.28–7.23(m, 3H, Ar-*H*), 7.13 (d, 2H, $J = 8.3$ Hz, Ar-*H*), 6.83 (s, 1H, C*H*), 6.68 (s, 1H, C*H*), 5.54 (s, 1H, C*H*), 5.50 (s, 1H, C*H*), 2.53 (s, 3H, N-CH_3) and 2.49 (s, 3H, N-CH_3); $^{13}C\{^1H\}$ NMR (100.56 MHz, C_6D_6): $\delta = 149.1$ (Ar-^{q}C), 136.7 (Ar-CH), 135.3 (Ar-CH), 130.7 (Ar-CH), 129.4 (Ar-CH), 128.4 (Ar-CH), 127.9 (Ar-CH), 127.1 (Ar-CH), 126.7 (Ar-CH), 121.5 (CH), 121.3 (CH), 119.2 (CH), 118.7 (CH), 118.4 (Ar-^{q}C), 34.8 (N-CH_3) and 34.7 (N-CH_3); and $^{11}B\{^1H\}$ NMR (128.3 MHz, $CDCl_3$): $\delta = -6.4$ (br, two signals are overlapped). HRMS (ESI): *m/z* calculated for $C_{28}CH_{26}B_2BrN_4$: 519.1527 [$(M+H)]^+$; found: 519.1549.

Synthesis of 11a. Benzonitrile (0.061 ml, 0.59 mmol) was added into a benzene (2.0 ml) solution of **2** (200 mg, 0.59 mmol), and the reaction mixture was stirred for 5 min at room temperature. A white precipitate was collected by filtration and dried under vacuum to afford **11a** as a white solid (191 mg, 73%). Melting point: 156 °C (dec). ^1H NMR (400 MHz, $CDCl_3$): $\delta = 8.33$ (d, 2H, $J = 7.3$ Hz, Ar-*H*), 7.61 (m, 2H, Ar-*H*), 7.50 (t, 2H, $J = 7.3$ Hz, Ar-*H*), 7.35–7.22 (m, 9H, Ar-*H*), 6.97 (s, 1H, C*H*), 6.95 (s, 1H, C*H*), 6.43 (s, 2H, C*H*), 3.27 (s, 3H, N-CH_3) and 3.08 (s, 3H, N-CH_3); $^{13}C\{^1H\}$ NMR (100.56 MHz, $CDCl_3$): $\delta = 136.1$ (Ar-CH), 135.2 (Ar-CH), 128.5 (Ar-^{q}C), 127.8 (Ar-CH), 127.6 (Ar-CH), 127.5 (Ar-CH), 127.4 (Ar-CH), 127.2 (Ar-CH), 127.0 (Ar-CH), 126.5 (Ar-CH), 121.2 (CH), 120.9 (CH), 120.0 (CH),

119.6 (CH), 35.6 (N-CH_3) and 35.2 (N-CH_3); and ^{11}B{^{1}H} NMR (128.3 MHz, $CDCl_3$): $\delta = -3.4$ (s) and -8.2 (s). HRMS (ESI): m/z calculated for $C_{27}H_{26}B_2N_5$:442.2374 $[(M+H)]^+$; found: 442.2369.

Synthesis of 11b. 4-Chlorobenzonitrile (81 mg, 0.59 mmol) was added into a benzene (2.0 ml) solution of **2** (200 mg, 0.59 mmol), and the reaction mixture was stirred for 30 min at room temperature. A white precipitate was collected by filtration and dried under vacuum to afford **11b** as a white solid (197 mg, 70%). Melting point: 160 °C (dec). ^{1}H NMR (400 MHz, $CDCl_3$): $\delta = 8.31$ (d, 2H, $J = 6.7$ Hz, Ar-H), 7.61 (m, 2H, Ar-H), 7.51 (t, 2H, $J = 7.5$ Hz, Ar-H), 7.33–7.29 (m, 6H, Ar-H), 7.19 (m, 2H, Ar-H), 7.00 (d, 1H, $J = 1.6$ Hz, CH), 6.98 (d, 1H, $J = 1.6$ Hz, CH), 6.44 (dd, 2H, $J = 2.6$ Hz and 1.7 Hz, CH), 3.27 (s, 3H, N-CH_3) and 3.05 (s, 3H, N-CH_3); ^{13}C{^{1}H} NMR (100.56 MHz, $CDCl_3$): $\delta = 135.9$ (Ar-CH), 135.1 (Ar-CH), 129.1 (Ar-CH), 128.5 (Ar-^qC), 127.9 (Ar-CH), 127.8 (Ar-CH), 127.5 (Ar-CH), 127.1(Ar-CH), 126.7 (Ar-CH), 121.2 (CH), 120.9 (CH), 120.2 (CH), 119.7 (CH), 35.6 (N-CH_3) and 35.2 (N-CH_3); and ^{11}B{^{1}H} NMR (128.3 MHz, $CDCl_3$): $\delta = -3.3$ (s) and -8.5 (s). HRMS (ESI): m/z calculated for $C_{27}H_{25}B_2ClN_5$:476.1985 $[(M+H)]^+$; found: 476.1984.

References

1. Suzuki, A. Organoboranes in organic syntheses including Suzuki coupling reaction. *Heterocycles* **80**, 15–43 (2010).
2. Segawa, Y., Yamashita, M. & Nozaki, K. Boryllithium: isolation, characterization, and reactivity as a boryl anion. *Science* **314**, 113–115 (2006).
3. Yamashita, M. & Nozaki, K. Boryl anion: syntheses and properties of novel borylmetals. *J. Syn. Org. Chem. Jpn* **68**, 359–369 (2010).
4. Segawa, Y., Suzuki, Y., Yamashita, M. & Nozaki, K. Chemistry of boryllithium: synthesis, structure, and reactivity. *J. Am. Chem. Soc.* **130**, 16069–16079 (2008).
5. Nozaki, K. Not just any old anion. *Nature* **464**, 1136–1137 (2010).
6. Braunschweig, H., Chiu, C.-W., Radacki, K. & Kupfer, T. Synthesis and structure of a carbene-stabilized π-boryl anion. *Angew. Chem. Int. Ed.* **49**, 2041–2044 (2010).
7. Braunschweig, H., Chiu, C.-W., Kupfer, T. & Radacki, K. NHC-stabilized 1-hydro-1H-borole and its nondegenerate sigmatropic isomers. *Inorg. Chem.* **50**, 4247–4249 (2011).
8. Bertermann, R. *et al.* Evidence for extensive single-electron-transfer chemistry in boryl anions: isolation and reactivity of a neutral borole radical. *Angew. Chem. Int. Ed.* **53**, 5453–5457 (2014).
9. Braunschweig, H., Burzler, M., Dewhurst, R. D. & Radacki, K. A linear, anionic dimetalloborylene complex. *Angew. Chem. Int. Ed.* **47**, 5650–5653 (2008).
10. Dewhurst, R. D., Neeve, E. C., Braunschweig, H. & Marder, T. B. sp²-sp³ diboranes: astounding structural variability and mild sources of nucleophilic boron for organic synthesis. *Chem. Commun.* **51**, 9594–9607 (2015).
11. Cid, J., Gulyás, H., Carbó, J. J. & Fernández, E. Trivalent boron nucleophile as a new tool in organic synthesis: reactivity and asymmetric induction. *Chem. Soc. Rev.* **41**, 3558–3570 (2012).
12. Kinjo, R., Donnadieu, B., Celik, M. A., Frenking, G. & Bertrand, G. Synthesis and characterization of a neutral tricoordinate organoboron isoelectronic with amines. *Science* **333**, 610–613 (2011).
13. Ruiz, D. A., Melaimi, M. & Bertrand, G. An efficient synthetic route to stable bis(carbene)borylenes [(L₁)(L₂)BH]. *Chem. Commun.* **50**, 7837–7839 (2014).
14. Kong, L., Li, Y., Ganguly, R., Vidovic, D. & Kinjo, R. Isolation of a bis(oxazol-2-ylidene)–phenylborylene adduct and its reactivity as a boron-centered nucleophile. *Angew. Chem., Int. Ed.* **53**, 9280–9283 (2014).
15. Kong, L., Ganguly, R., Li, Y. & Kinjo, R. Diverse reactivity of a tricoordinate organoboron L₂PhB: (L = oxazol-2-ylidene) towards alkali metal, group 9 metal, and coinage metal precursors. *Chem. Sci.* **6**, 2893–2902 (2015).
16. Braunschweig, H. *et al.* Multiple complexation of CO and related ligands to a main-group element. *Nature* **522**, 327–330 (2015).
17. Braunschweig, H. *et al.* Dative bonding between group 13 elements using a boron-centered Lewis base. *Angew. Chem. Int. Ed.* **55**, 436–440 (2016).
18. Frenking, G. & Shaik, S. *The Chemical Bond 2* (Wiley-VCH, Weinheim, 2014).
19. Celik, M. A. *et al.* Borylene complexes (BH)L₂ and nitrogen cation complexes (N⁺)L₂: isoelectronic homologues of carbones CL₂. *Chem. A Eur. J.* **18**, 5676–5692 (2012).
20. Himmel, D., Krossing, I. & Schnepf, A. Dative bonds in main-group compounds: a case for fewer arrows! *Angew. Chem. Int. Ed.* **53**, 370–374 (2014).
21. Frenking, G. Dative bonds in main-group compounds: a case for more arrows! *Angew. Chem. Int. Ed.* **53**, 6040–6046 (2014).
22. Himmel, D., Krossing, I. & Schnepf, A. Dative or not dative? *Angew. Chem. Int. Ed.* **53**, 6047–6048 (2014).
23. Power, P. P. Main-group elements as transition metals. *Nature* **463**, 171–177 (2010).

24. Power, P. P. Interaction of multiple bonded and unsaturated heavier main group compounds with hydrogen, ammonia, olefins, and related molecules. *Acc. Chem. Res.* **44**, 627–637 (2011).
25. Mandal, S. K. & Roesky, H. W. Group 14 hydrides with low valent elements for activation of small molecules. *Acc. Chem. Res.* **45**, 298–307 (2012).
26. Martin, D., Soleilhavoup, M. & Bertrand, G. Stable singlet carbenes as mimics for transition metal centers. *Chem. Sci.* **2**, 389–399 (2011).
27. Yao, S., Xiong, Y. & Driess, M. Zwitterionic and donor-stabilized N-heterocyclic silylenes (NHSis) for metal-free activation of small molecules. *Organometallics* **30**, 1748–1767 (2011).
28. Braunschweig, H. *et al.* Metal-free boron binding and coupling of carbon monoxide at a boron-boron triple bond. *Nat. Chem.* **5**, 1025–1028 (2013).
29. Welch, G. C., San Juan, R. R., Masuda, J. D. & Stephan, D. W. Reversible, metal-free hydrogen activation. *Science* **314**, 1124–1126 (2006).
30. Stephan, D. W. & Erker, G. Frustrated lewis pair chemistry: development and perspectives. *Angew. Chem. Int. Ed.* **54**, 6400–6441 (2015).
31. Stephan, D. W. Frustrated Lewis pairs: from concept to catalysis. *Acc. Chem. Res.* **48**, 306–316 (2015).
32. Frey, G. D., Lavallo, V., Donnadieu, B., Schoeller, W. W. & Bertrand, G. Facile splitting of hydrogen and ammonia by nucleophilic activation at a single carbon center. *Science* **316**, 439–441 (2007).
33. Dahcheh, F., Martin, D., Stephan, D. W. & Bertrand, G. Synthesis and reactivity of a CAAC–aminoborylene adduct: a hetero-allene or an organoboron isoelectronic with singlet carbenes. *Angew. Chem. Int. Ed.* **53**, 13159–13163 (2014).
34. Braunschweig, H. *et al.* Ditopic amphiphilicity of an anionic dimetalloborylene complex. *J. Am. Chem. Soc.* **135**, 2313–2320 (2013).
35. Wu, D., Kong, L., Li, Y., Ganguly, R. & Kinjo, R. 1,3,2,5-Diazadiborinine featuring nucleophilic and electrophilic boron centres. *Nat. Commun.* **6**, 7340 (2015).
36. Wu, D., Ganguly, R., Li, Y., Hirao, H. & Kinjo, R. Reversible [4 + 2] cycloaddition reaction of 1,3,2,5-diazadiborinine with ethylene. *Chem. Sci.* **6**, 7150–7155 (2015).
37. Campbell, P. G., Marwitz, A. J. V. & Liu, S.-Y. Recent advances in azaborine chemistry. *Angew. Chem. Int. Ed.* **51**, 6074–6092 (2012).
38. Mo, Y. The resonance energy of benzene: a revisit. *J. Phys. Chem. A* **113**, 5163–5169 (2009).
39. Cyrański, M. K. Energetic aspects of cyclic pi-electron delocalization: evaluation of the methods of estimating aromatic stabilization energies. *Chem. Rev.* **105**, 3773–3811 (2005).
40. Melen, R. L. *et al.* Diverging pathways in the activation of allenes with Lewis acids and bases: addition, 1,2-carboboration, and cyclization. *Organometallics* **34**, 4127–4137 (2015).
41. Frisch, M. J. *et al. Gaussian 09, Revision D.01* (Gaussian, Inc., 2009).
42. Becke, A. D. Density-functional thermochemistry. III. The role of exact exchange. *J. Chem. Phys.* **98**, 5648–5652 (1993).
43. Lee, C., Yang, W. & Parr, R. G. Development of the Colle-Salvetti correlation-energy formula into a functional of the electron density. *Phys. Rev. B* **37**, 785–789 (1988).
44. Vosko, S. H., Wilk, L. & Nusair, M. Accurate spin-dependent electron liquid correlation energies for local spin density calculations: a critical analysis. *Can. J. Phys.* **58**, 1200–1211 (1980).
45. Grimme, S., Ehrlich, S. & Goerigk, L. Effect of the damping function in dispersion corrected density functional theory. *J. Comput. Chem.* **32**, 1456–1465 (2011).
46. Grimme, S., Antony, J., Ehrlich, S. & Krieg, H. A consistent and accurate *ab initio* parametrization of density functional dispersion correction (DFT-D) for the 94 elements H-Pu. *J. Chem. Phys.* **132**, 154104 (2010).
47. Johnson, E. R. & Becke, A. D. A post-Hartree–Fock model of intermolecular interactions: Inclusion of higher-order corrections. *J. Chem. Phys.* **124**, 174104 (2006).
48. Becke, A. D. & Johnson, E. R. A density-functional model of the dispersion interaction. *J. Chem. Phys.* **123**, 154101 (2005).
49. Johnson, E. R. & Becke, A. D. A post-Hartree–Fock model of intermolecular interactions. *J. Chem. Phys.* **123**, 024101 (2005).
50. Clark, T., Chandrasekhar, J., Spitznagel, G. W. & Schleyer, P. v. R. Efficient diffuse function-augmented basis sets for anion calculations. III. The 3-21 + G basis set for first-row elements, Li–F. *J. Comput. Chem.* **4**, 294–301 (1983).
51. Krishnan, R., Binkley, J. S., Seeger, R. & Pople, J. A. Self-consistent molecular orbital methods. XX. A basis set for correlated wave functions. *J. Chem. Phys.* **72**, 650 (1980).
52. Hariharan, P. C. & Pople, J. A. The influence of polarization functions on molecular orbital hydrogenation energies. *Theor. Chim. Acta* **28**, 213–222 (1973).
53. Tomasi, J., Mennucci, B. & Cammi, R. Quantum mechanical continuum solvation models. *Chem. Rev.* **105**, 2999–3094 (2005).
54. Longobardi, L. E. *et al.* Hydrogen activation by an aromatic triphosphabenzene. *J. Am. Chem. Soc.* **136**, 13453–13457 (2014).
55. Zhong, G., Chan, B. & Radom, L. Hydrogenation of simple aromatic molecules: a computational study of the mechanism. *J. Am. Chem. Soc.* **129**, 924–933 (2007).

56. Tsao, F. A., Cao, L., Grimme, S. & Stephan, D. W. Double FLP-alkyne exchange reactions: a facile route to Te/B heterocycles. *J. Am. Chem. Soc.* **137**, 13264–13267 (2015).

57. Forster, T. D. *et al.* A σ-donor with a planar six-π-electron $B_2N_2C_2$ framework: anionic N-heterocyclic carbene or heterocyclic terphenyl anion? *Angew. Chem. Int. Ed.* **45**, 6356–6359 (2006).

58. Krahulic, K. E., Tuononen, H. M., Parvez, M. & Roesler, R. Isolation of free phenylide-like carbanions with N-heterocyclic carbene frameworks. *J. Am. Chem. Soc.* **131**, 5858–5865 (2009).

59. Jaska, C. A. *et al.* Triphenylene analogues with $B_2N_2C_2$ cores: synthesis, structure, redox behavior, and photophysical properties. *J. Am. Chem. Soc.* **128**, 10885–10896 (2006).

Acknowledgements

We gratefully acknowledge financial support from Nanyang Technological University and Singapore Ministry of Education (MOE2013-T2-1-005).

Author contributions

B.W. performed the synthetic experiments Y.L. and R.G. performed the X-ray crystallographic measurements. H.H. conducted theoretical studies. R.K. conceived and supervised the study, and drafted the manuscript. All authors contributed to discussions.

Additional information

Accession codes. CCDC 1442910–1442920 contain the supplementary crystallographic data for this paper. These data can be obtained free of charge from The Cambridge Crystallographic Data Centre via www.ccdc.cam.ac.uk/data_request/cif.

Competing financial interests: The authors declare no competing financial interests.

26

Cotton-textile-enabled flexible self-sustaining power packs via roll-to-roll fabrication

Zan Gao[1], Clifton Bumgardner[1], Ningning Song[1], Yunya Zhang[1], Jingjing Li[2] & Xiaodong Li[1]

With rising energy concerns, efficient energy conversion and storage devices are required to provide a sustainable, green energy supply. Solar cells hold promise as energy conversion devices due to their utilization of readily accessible solar energy; however, the output of solar cells can be non-continuous and unstable. Therefore, it is necessary to combine solar cells with compatible energy storage devices to realize a stable power supply. To this end, supercapacitors, highly efficient energy storage devices, can be integrated with solar cells to mitigate the power fluctuations. Here, we report on the development of a solar cell-supercapacitor hybrid device as a solution to this energy requirement. A high-performance, cotton-textile-enabled asymmetric supercapacitor is integrated with a flexible solar cell via a scalable roll-to-roll manufacturing approach to fabricate a self-sustaining power pack, demonstrating its potential to continuously power future electronic devices.

[1] Department of Mechanical and Aerospace Engineering, University of Virginia, 122 Engineer's Way, Charlottesville, Virginia 22904-4746, USA. [2] Department of Mechanical Engineering, University of Hawaii at Manoa, 2540 Dole Street, Honolulu, Hawaii 96822, USA. Correspondence and requests for materials should be addressed to X.L. (email: xl3p@virginia.edu).

With ever-increasing global energy consumption and the depletion of fossil fuels, finding a sustainable and clean energy supply has become one of the most important scientific and technological challenges facing humanity today[1]. Fossil fuels such as coal, oil and natural gas are limited and nonrenewable. With growing energy demand in many urbanizing nations, particularly in India and China, the threat of depleting the planet's fossil fuel reserves is of increasing concern. Relief may be as simple as utilizing the most accessible solar energy. Only recently, thanks to the development of less expensive and more efficient photovoltaic cells, solar energy has been harnessed and widely used as an alternative energy source. However, due to the fluctuation of light intensity and the diurnal cycle, the output of a solar cell can be non-continuous and unstable. Therefore, there is a clear need to combine solar cells with energy storage devices (serving as a buffer) to mitigate the power fluctuations and to allow operation of the cell and storage device as a reliable source of energy. Among various emerging energy storage technologies, supercapacitors have been regarded as a very promising energy source for next-generation electronics and electric vehicles because of their excellent properties, such as high power density, long lifespan, environment benignancy, safety and low maintenance cost[2].

Compared with lithium-ion batteries, the relatively low energy density of supercapacitors limits their wide application. Developing new electrode materials with well-defined nano-architecture and high active surface area is key to improving the electrochemical performance of supercapacitors. To date, according to the different mechanisms of charge storage, typical supercapacitor electrode materials can be categorized into electric double-layer capacitive materials including various carbon materials[3,4] and pseudo-capacitive materials including transition metal oxides[5-8], and conductive polymers[9-11]. However, different materials have different advantages and disadvantages for supercapacitor applications. For example, carbon materials usually possess higher power density and longer lifespans but lower energy density. Transition metal oxides/hydroxides and conductive polymers possess higher energy density but poorer cyclic life and lower power density. One intelligent strategy to strengthen the electrochemical performance of supercapacitors is to develop a hybrid material of carbon and metal oxides/hydroxides, which is anticipated to jointly improve the overall performance of supercapacitors in terms of their energy density, power density and cyclic stability[12].

Alternatively, a more attractive approach for balancing the contradiction between energy density and power density of supercapacitors is to assemble an asymmetric supercapacitor, which consists of a battery-type Faradic electrode as energy source (usually pseudo-capacitive materials) and a capacitor-type electrode as power source (usually carbon materials)[13,14]. In the last few decades, metallic layered double hydroxides (LDHs) with the general chemical formula $[M^{II}_{1-x}M^{III}_x(OH)_2]^{x+}[A^{n-}]_{x/n} \cdot mH_2O$, where M^{II} and M^{III} are divalent and trivalent metal cations and A^{n-} are the charge-balancing anions[15], have aroused great interest in catalysis[16], biotechnology[17], separation[18] and electrochemistry[19,20]. Moreover, LDHs have been proven to be a promising class of electrode materials for supercapacitors because of their relatively low cost, high redox activity and environmentally friendliness[21,22]. For example, Huang et al. reported that nickel-aluminium LDH (Ni-Al LDH) deposited on nickel foam exhibits an ultrahigh specific capacitance of 2,123 F g^{-1} at 1 A g^{-1} (ref. 23). Chen et al. also demonstrated that nickel-cobalt LDH (Ni-Co LDH) film had a significantly enhanced specific capacitance of 2,682 F g^{-1} at 3 A g^{-1} (ref. 24). In our previous work, Ni-Al LDH was coupled with graphene nanosheets to prepare a hybrid electrode, which also demonstrated excellent electrochemical performance[25].

Therefore, high-energy LDHs are ideal candidates for asymmetric supercapacitors. On the other hand, two-dimensional (2D) monolayered graphene has been intensively explored for energy storage application because of its superior electrical conductivity, high specific surface area, outstanding mechanical properties and relative wide operation windows[26-29]. Graphene-based films and papers have been shown to be ideal 'power sources' for flexible, asymmetric supercapacitors[30-33].

Consumer demand for portable/wearable electronics has triggered a technological race to drive innovation in flexible mobile phones, bendable displays, electronic skin and distributed sensors[34-36]. Many attempts have been devoted to developing safe, lightweight and flexible power sources to meet the urgent demand for flexible and wearable electronics[37-40]. In addition, 'self-powered nanotechnology' has been proposed to enable future electronics to operate independently and sustainably without batteries, or with a battery possessing extended lifespan[41]. Self-powered nanosystems with multi-functionalities will be a prominent driving force for future world economies by employing transformative nanomaterials and nanofabrication technologies[42]. To date, many attempts and achievements have been made in developing self-powered nanosystems by combining nanogenerators or solar cells with lithium-ion batteries or supercapacitors[43-45]. However, until now, a streamlined manufacturing process for integrating a flexible energy harvesting unit with a flexible energy storage unit has not been achieved due to the lack of effective packaging technology.

Cotton textile, a source of flexible, 'green', renewable, breathable clothing, has been shown to be an excellent wearable platform for constructing flexible energy storage devices as activated cotton textiles (ACTs) exhibit eminent flexibility and excellent conductivity[46,47]. In this work, flower-like cobalt-aluminium LDH (Co-Al LDH) nanoarrays consisting of interconnected nano-petals and pompon-like nano-stamens are anchored *in situ* on flexible ACTs to realize ACT/Co-Al LDH hybrid by a facile hydrothermal method. A highly conductive graphene coating is wrapped around ACT fibres to form ACT/graphene composite by a simple dipping, drying and reducing process. A flexible, all-solid-state asymmetric supercapacitor (ACT/Co-Al LDH//ACT/graphene) is assembled using the nanostructured ACT/Co-Al LDH as the positive electrode, ACT/graphene as the negative electrode, and PVA-KOH gel as both the solid-state electrolyte and separator. The asymmetric cell works synergically to achieve excellent electrochemical performance in terms of its working potential (1.6 V), energy density (55.04 Wh kg^{-1}), power density (5.44 kW kg^{-1}) and cycling stability (capacitance retention rate of 87.54% after 2,000 cycles). Moreover, we demonstrate a practical roll-to-roll manufacturing approach to combine the flexible asymmetric supercapacitor with a flexible solar cell to fabricate an integrated self-sustaining power pack, which is scalable for industrial manufacturing. Importantly, the assembled power pack can power a commercial light-emitting-diode (LED) continuously, with or without sunlight, demonstrating its potential for flexible, self-powered energy devices.

Results

Positive electrode materials. Figure 1a illustrates the two-step design and synthesis procedures of Co-Al LDH nanoarrays on ACT fibres. In a typical experiment, highly conductive and flexible ACTs were first prepared by direct conversion of cotton T-shirt textiles through a dipping, drying and annealing process (step i). Then, flower-like Co-Al LDH nanoarrays were radially grown *in situ* on ACT fibres using a facile hydrothermal method (step ii), which resulted in a pink coating on the surface of ACT

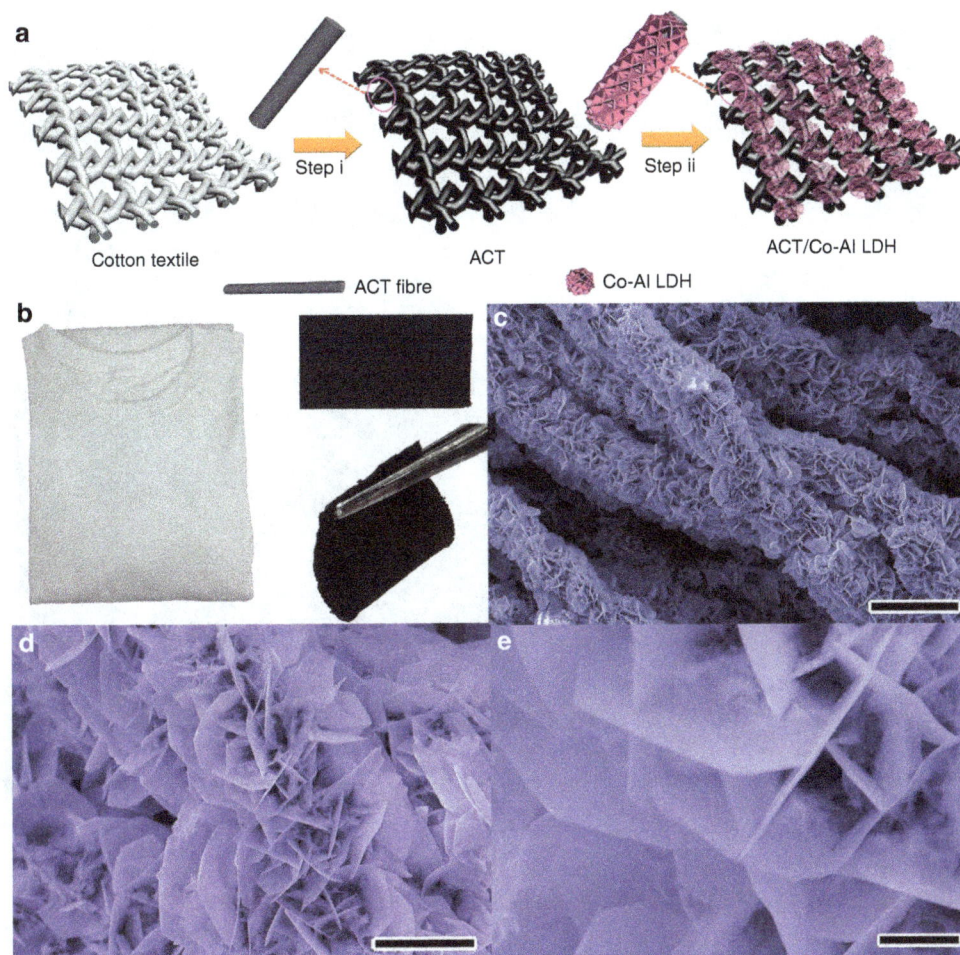

Figure 1 | Synthesis and characterization of ACT/Co-Al LDH composite. (a) Schematic illustration of the formation processes of Co-Al LDH nanostructure on the ACT. **(b)** Photographs of a cotton T-shirt, a piece of ACT and a piece of ACT under folded state. **(c–e)** SEM images of Co-Al LDH nanosheets on ACT fibres at different magnifications. Scale bars, 20 µm for **c**, 5 µm for **d** and 1 µm for **e**.

fibres. Figure 1b shows the digital photographs of the cotton T-shirt and the converted ACT under normal and folded states, demonstrating its mechanical flexibility. Figure 1c–e show the scanning electron microscopy (SEM) images of the as-synthesized Co-Al LDH nanoarrays on ACT fibres. High-density hexagonal Co-Al LDH nanosheets, with the length of about 2–5 µm and thickness of around 40–70 nm, were radially aligned on individual ACT fibres, forming flower-like nanoarrays that consist of interconnected LDH nano-petals and pompon-like LDH nano-stamens. Such three-dimensional (3D) hierarchical, porous nanostructure *in situ* anchored on ACT fibres not only enhances the contact between the active material and ACT substrate but also serves as a reservoir for electrolyte ions, shortening ion diffusion path and facilitating charge transfer, jointly enhancing the electrochemical performance.

The structure and morphology of the Co-Al LDH nanosheets were further investigated by transmission electron microscopy (TEM) and X-ray diffraction (XRD). TEM inspection (Fig. 2a) reveals that the LDH nanosheets are of hexagonal shape with 2- to 5-nm-sized nanopores on the surface of individual LDH nanosheets (Fig. 2b), which is expected to have a higher specific area. The ACT/Co-Al LDH composite achieved a high Brunauer–Emmett–Teller (BET) surface area of $814.4 \, m^2 \, g^{-1}$. A lattice spacing of 0.26 nm (Fig. 2c) can be ascribed to the (012) crystal plane of Co-Al LDH, which is in consistent with the XRD results (Fig. 2f). The corresponding selected area electron

diffraction (SAED) pattern (Fig. 2d) shows hexagonally arranged bright spots, indicating the single-crystal nature of LDH nanosheets. The XRD pattern (Fig. 2f) shows a series of diffraction peaks at 11.59°, 23.26°, 34.56°, 38.78° and 46.7°, which can be indexed to the (003), (006), (012), (015) and (018) planes of the layered LDH phase (JCPDS #51-0045), respectively. The small peak at ~19° probably resulted from cobalt hydroxide impurity in the ACT/Co-Al LDH composite.

Negative electrode material. Highly conductive graphene was coated on flexible ACTs to fabricate negative electrode material for constructing asymmetric supercapacitors. Conductive ACTs (sheet resistance: $\sim 10\text{–}20 \, \Omega \, sq.^{-1}$) were first prepared by direct conversion of a cotton T-shirt. As shown in Fig. 3a, the ACT fibres have diameters ranging from 5 to 10 µm, which inherit the cellulose fibre structure of the cotton textile. After activation, a piece of ACT was dipped into graphene oxide solution for a few minutes to coat graphene oxide nanosheets. Figure 3b shows the typical SEM images of the corrugated and scrolled graphene oxide sheets, resembling crumpled silk veil waves. The thickness of the as-synthesized graphene oxide sheets was measured to be 1.1 nm (inset of Fig. 3b), which is close to the theoretical value of 0.78 nm for single-layer graphene oxide, indicating that the as-prepared graphene oxide sheets were predominantly in monolayered manner[48]. Usually, the graphene oxide sheets tend to agglomerate

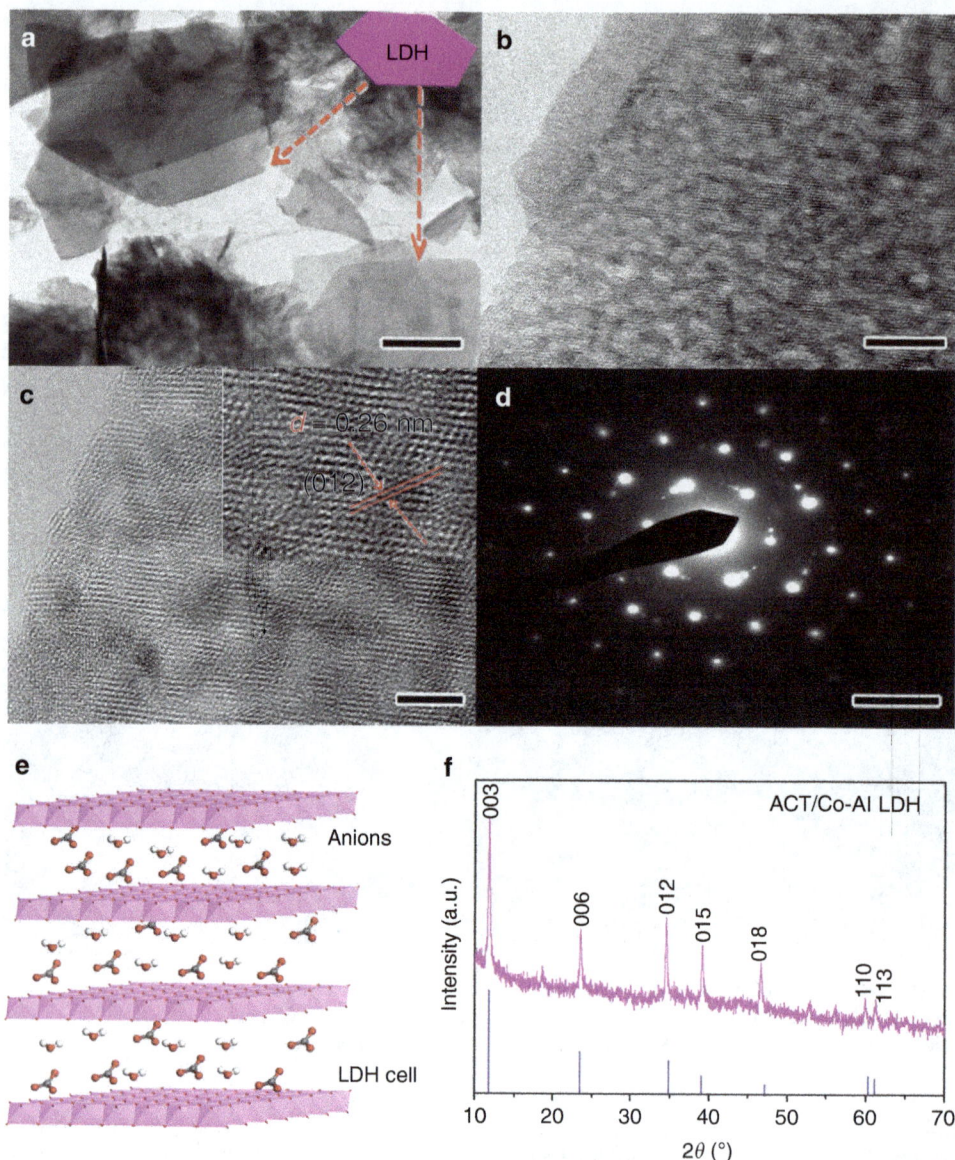

Figure 2 | Characterization of the as-obtained ACT/Co-Al LDH composite. (**a**) TEM image of Co-Al LDH (scale bar, 1 μm), inset is the illustration of LDH. (**b-d**) HRTEM images of Co-Al LDH and the corresponding SAED pattern. Scale bars, 10 nm for **b**, 5 nm for **c** and 5 nm for **d**. (**e**) Schematic illustration of the molecular structure of Co-Al LDH. (**f**) XRD patterns of ACT/Co-Al LDH and the standard peaks of Co-Al LDH (JCPDS # 510045).

due to the electrostatic interactions of the oxygen-containing functional groups (epoxide, hydroxyl, carbonyl and carboxyl groups) on the surfaces and the edges of the sheets (inset of Fig. 3b). After dipping, individual ACT fibres were wrapped with curled and entangled graphene oxide sheets (a mixture of multilayered/monolayered sheets). During the thermal reduction, the conductivity of the graphene oxide sheets was enhanced due to partial overlapping or coalescing via π–π stacking or hydrogen bonding[49,50], resulting in an interconnected 3D conductive graphene network on the ACT fibres (Fig. 3c–e). The insets of Fig. 3d are the high-resolution transmission electron microscopy (HRTEM) image and corresponding SAED pattern of the graphene nanosheets. The deformed crystal fringes with a *d*-spacing of ∼0.37 nm in the HRTEM image suggest the existence of defects in the graphene nanosheets. The crystalline nature of the graphene nanosheets was validated by the well-defined diffraction spots in the SAED pattern. More importantly, the ACT fibres showed a porous structure as revealed by the TEM image (inset of Fig. 3e). The entangled graphene sheets wrapped

around the porous ACT fibres, forming a 3D porous conductive net. The BET surface area of the ACT/graphene composite was measured to be ∼450 m^2 g^{-1}. The 3D graphene/fibre conductive net is expected to facilitate electrolyte ion diffusion and electron transport in the charge/discharge processes, improving the overall power density of the asymmetric cell.

Figure 3f shows the XRD spectra of cotton, ACT, ACT/ graphene oxide and ACT/graphene composites, respectively. The diffraction peaks at 14.8°, 22.7° and 34.3° can be indexed, respectively, to the (101), (002) and (040) peaks of the cellulose polymorphs from Cellulose I ingredients in pure cotton[14,51]. After activation, the cotton peaks disappeared, alternately, a broad diffraction peak appeared at ∼21° for ACT, indicating the breakage of cellulose polymorphs and the formation of amorphous carbon during the activation process. The peak at 8.8° for ACT/graphene oxide corresponds to the (001) lattice plane of graphene oxide with an interplanar spacing of 0.78 nm, indicating the complete exfoliation of graphite to graphene oxide[52]. After the thermal reduction process, the peak at 8.8°

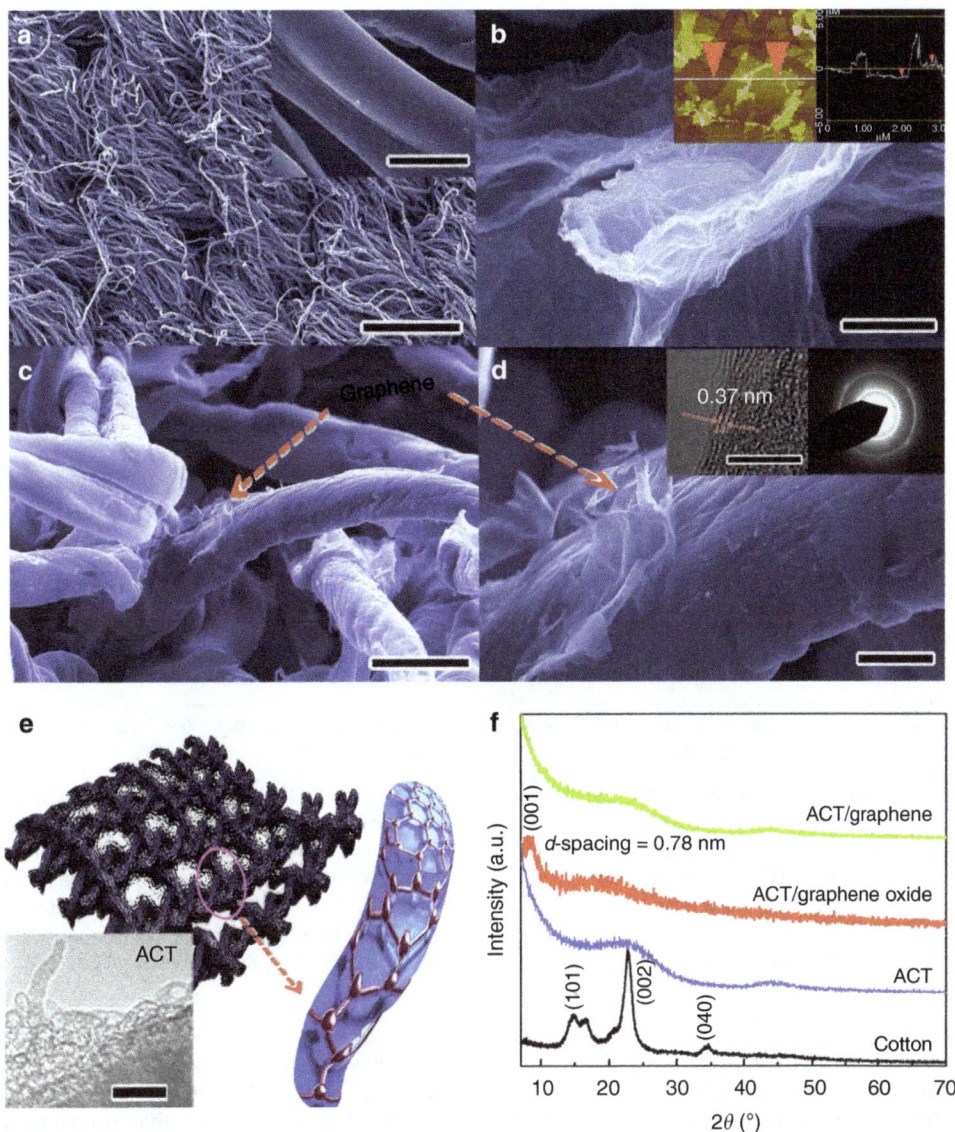

Figure 3 | Characterization of the as-obtained ACT/graphene composite. (**a**) SEM image of ACT (scale bar, 250 μm), inset is the amplified SEM image (scale bar, 5 μm). (**b**) SEM image of graphene oxide, inset is the atomic force microscopy image of graphene oxide nanosheets (with a height profile showing a step of 1.1 nm marked by the red arrows). (**c**,**d**) SEM images of ACT/graphene composite at different magnifications (scale bars, 10 μm for **c** and 2 μm for **d**), insets of **d** are the HRTEM image (scale bar, 5 nm) and corresponding SAED pattern of graphene nanosheets. (**e**) Schematic illustration of the ACT fibre with graphene coating, inset is the TEM image of ACT fibre (scale bar, 25 nm). (**f**) XRD patterns of pure cotton textile, ACT, ACT/graphene oxide and ACT/graphene composite.

for the ACT/graphene oxide disappeared, suggesting the oxygen-containing groups on the graphene oxide had been removed, and the conductivity of graphene oxide was restored[53,54], ensuring improved electrochemical performance of the assembled asymmetric cell. Raman spectroscopy (Supplementary Fig. 1) was used to further characterize the ACT/graphene composite (reduced from ACT/graphene oxide). Generally, graphene exhibits two characteristic bands in its Raman spectrum: the G band at $\sim$1,575 cm^{-1} resulting from the first-order scattering of the E$_{1g}$ phonon of sp^2 C atoms and the D band at $\sim$1,350 cm^{-1} resulting from a breathing mode of point photons of A$_{1g}$ symmetry[55]. The ACT/graphene showed an increased I_D/I_G ratio ($\sim$1.21) compared with the ACT/graphite oxide with the I_D/I_G ratio of $\sim$1.03. The increased I_D/I_G ratio is ascribed to the re-establishment of graphene network during the reduction process, indicating smaller sizes, more defects and disordered structures of the graphene[14]. The XRD and Raman

spectroscopy results jointly demonstrate successful removal of the oxygen-containing groups on the surfaces and edges of graphene oxide sheets during the reduction process.

Cell performance. The as-prepared ACT/Co-Al LDH was used as the positive electrode, ACT/graphene as the negative electrode and the PVA-KOH gel as both the solid electrolyte and separator to construct a flexible asymmetric supercapacitor. The energy storage performances of the assembled cells and individual electrodes were studied by cyclic voltammetry (CV), galvanostatic (GV) charge/discharge and electrochemical impedance spectroscopy (EIS) in both two-electrode and three-electrode testing systems.

The three-electrode testing system was used to study the electrochemical reaction mechanisms of the as-prepared ACT, ACT/graphene and ACT/Co-Al LDH electrodes. The CV curves

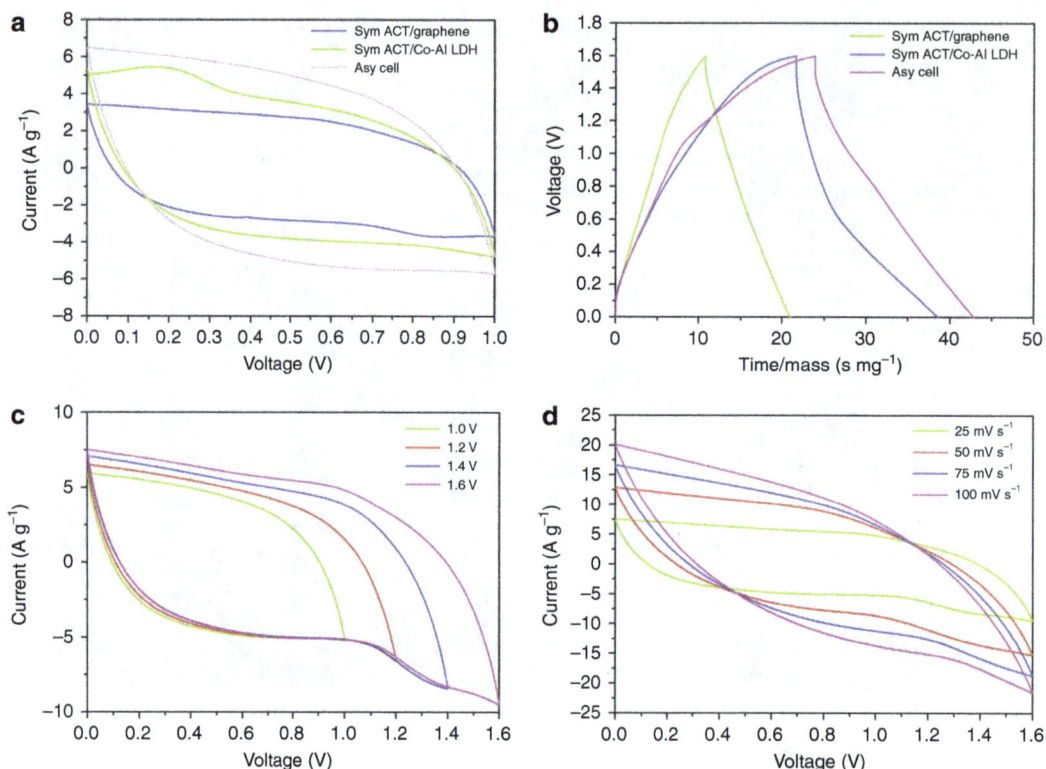

Figure 4 | Electrochemical performance of symmetric and asymmetric cells. (a) CV curves of ACT/graphene//ACT/graphene and ACT/Co-Al LDH//ACT/Co-Al LDH symmetric cells as well as ACT/Co-Al LDH//ACT/graphene asymmetric cell with PVA/KOH polymer gel electrolyte in the voltage window of 1.0 V at a scan rate of 25 mV s^{-1}. **(b)** Charge/discharge curves of ACT/graphene//ACT/graphene and ACT/Co-Al LDH//ACT/Co-Al LDH symmetric cells as well as ACT/Co-Al LDH//ACT/graphene asymmetric cell with PVA/KOH polymer gel electrolyte in the voltage window of 1.6 V at a current density of 12.5 mA cm^{-2} (x axis has been transformed to time/mass, based on the mass of the electrode). **(c)** CV curves of ACT/Co-Al LDH//ACT/graphene asymmetric cell with PVA/KOH polymer gel electrolyte under the voltage windows of 1, 1.2, 1.4 and 1.6 V at the scan rate of 25 mV s^{-1}, respectively. **(d)** CV curves of ACT/Co-Al LDH//ACT/graphene asymmetric cell with PVA/KOH polymer gel electrolyte in the voltage windows of 1.6 V at different scan rates.

of the aforementioned electrodes at the scan rate of 25 mV s^{-1} with the potential windows ranging from -0.2 to 0.6 V are shown in Supplementary Fig. 2a. Both ACT and ACT/graphene electrodes showed quasi-rectangular-shaped CV curves, indicating ideal electrical double layer capacitive behaviour. The small peak at ~ 0.25 V in the CV curve of ACT/graphene electrode is ascribed to the pseudo-capacitive reaction from the residual oxygen-containing groups on graphene sheets, resulting in a higher capacitance. The CV curve of the ACT/Co-Al LDH hybrid electrode showed a more complex shape with two pairs of redox peaks, which resulted from the typical Faradaic redox reactions of Co hydroxides, as described below[56]:

$$Co(OH)_2 + OH^- \leftrightarrow CoOOH + H_2O + e^- \tag{1}$$

$$CoOOH + OH^- \leftrightarrow CoO_2 + H_2O + e^- \tag{2}$$

Supplementary Fig. 2b shows the GV charge/discharge performances of the ACT, ACT/graphene and ACT/Co-Al LDH electrodes at the current density of 2.5 A g^{-1}. Compared with the triangular-shaped charge/discharge curves of the ACT and ACT/graphene electrodes, the ACT/Co-Al LDH electrode exhibited two charge/discharge voltage plateaus, which are the pseudocapacitive plateaus from the Co-Al LDH. The specific capacitance values of the ACT, ACT/graphene and ACT/Co-Al LDH electrodes were measured to be 90.1, 136 and 977.3 F g^{-1}, respectively. The CV curves of the ACT/Co-Al LDH electrode at the scan rates of 25, 50, 75 and 100 mV s^{-1} in 6 M KOH electrolyte solution are shown in Supplementary Fig. 2c. All of

them exhibited similar redox peaks, even at increased scan rates, indicating the quasi-reversible and continuous faradic redox reactions from Co-Al LDH that contributed remarkably to the pseudocapacitance. The GV charge/discharge curves of the ACT/Co-Al LDH electrode at different current densities are shown in Supplementary Fig. 2d. The specific capacitance values obtained from the ACT/Co-Al LDH at the current densities of 2.5, 5, 10 and 20 A g^{-1} are 977.3, 863, 727 and 545 F g^{-1}, respectively, with a capacitance retention of $\sim 55.8\%$ even when the current density increased from 2.5 to 20 A g^{-1}.

Figure 4a shows the CV curves of ACT/graphene//ACT/graphene and ACT/Co-Al LDH//ACT/Co-Al LDH symmetric cells as well as ACT/Co-Al LDH//ACT/graphene asymmetric cell with PVA/KOH polymer gel electrolyte at the scan rate of 25 mV s^{-1}. The rectangular-shaped CV curves of the ACT/graphene symmetric cell suggest that the ACT/graphene mainly worked as an electrochemical double-layered capacitor. The broad peak at 0.25 V of the ACT/Co-Al LDH symmetric cell indicates the Faradic reaction from the battery-type Co-Al LDH. The CV area of the ACT/Co-Al LDH//ACT/graphene asymmetric cell was much larger than that of the ACT/graphene//ACT/graphene and ACT/Co-Al LDH//ACT/Co-Al LDH symmetric cells, suggesting the larger specific capacitance of ACT/Co-Al LDH//ACT/graphene asymmetric cell resulting from the high accessibility of electrolyte ions. Figure 4b shows the charge/discharge curves of the ACT/graphene//ACT/graphene and ACT/Co-Al LDH//ACT/Co-Al LDH symmetric cells as well as ACT/Co-Al LDH//ACT/graphene asymmetric cell. The

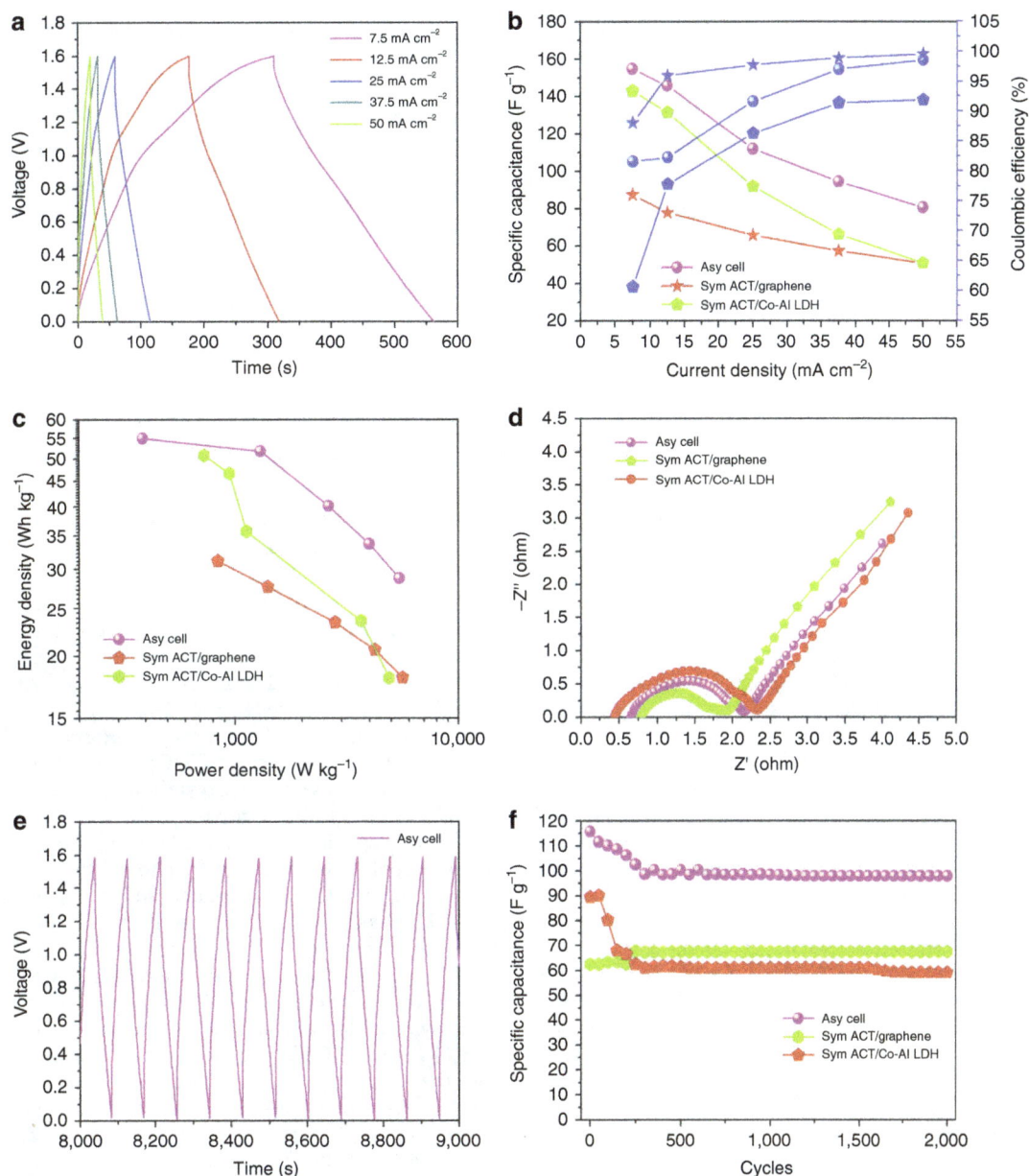

Figure 5 | Electrochemical performance of the asymmetric cell. (**a**) Charge/discharge curves of ACT/Co-Al LDH//ACT/graphene asymmetric cell with PVA/KOH polymer gel electrolyte in the voltage window of 1.6 V at different current densities. (**b**) Specific capacitances and Coulombic efficiency of the ACT/graphene//ACT/graphene and ACT/Co-Al LDH//ACT/Co-Al LDH symmetric cells as well as ACT/Co-Al LDH//ACT/graphene asymmetric cell with PVA/KOH polymer gel electrolyte in the voltage window of 1.6 V. (**c**) Ragone plots of the ACT/graphene//ACT/graphene and ACT/Co-Al LDH//ACT/Co-Al LDH symmetric cells as well as ACT/Co-Al LDH//ACT/graphene asymmetric cell with PVA/KOH polymer gel electrolyte in the voltage window of 1.6 V. (**d**) Nyquist plots of the ACT/graphene//ACT/graphene and ACT/Co-Al LDH//ACT/Co-Al LDH symmetric cells as well as ACT/Co-Al LDH//ACT/graphene asymmetric cell with PVA/KOH polymer gel electrolyte in the voltage window of 1.6 V. (**e**) Representative charge/discharge curve of ACT/Co-Al LDH//ACT/graphene asymmetric cell with PVA/KOH polymer gel electrolyte in the voltage window of 1.6 V. (**f**) Cyclic performance of the ACT/graphene//ACT/graphene and ACT/Co-Al LDH//ACT/Co-Al LDH symmetric cells as well as ACT/Co-Al LDH//ACT/graphene asymmetric cell with PVA/KOH polymer gel electrolyte.

ACT/graphene symmetric cell exhibited a triangular-shaped charge/discharge curve, implying an ideal capacitor character. Note that the ACT/Co-Al LDH symmetric cell exhibited a distorted charge/discharge profile, indicating the Faradic reaction during the charging/discharging process. The discharging capacities were measured to be 145.8, 131.75 and 77.9 F g^{-1} for the ACT/Co-Al LDH//ACT/graphene asymmetric cell, ACT/Co-Al LDH symmetric cell and ACT/graphene symmetric cell at the current density of 12.5 mA cm^{-2}, respectively. Figure 4c shows the CV curves of the ACT/Co-Al LDH//ACT/

graphene asymmetric cell with PVA/KOH polymer gel electrolyte in the respective voltage windows of 1.0, 1.2, 1.4 and 1.6 V at the scan rate of 25 mV s^{-1}. The working voltage of the ACT/Co-Al LDH//ACT/graphene asymmetric cell can be extended to 1.6 V, which is essential for practical application. Figure 4d shows the CV curves of the ACT/Co-Al LDH//ACT/graphene asymmetric cell at different scan rates ranging from 25 to 100 mV s^{-1} at an operation window of 1.6 V. The current density increases with increasing scan rate. All the CV curves at different scan rates in Fig. 4e have similar shape, indicating stable reversibility and good

Table 1 | Comparison of the as-prepared Co-Al LDH asymmetric supercapacitor with previously published results.

Reference	Electrode materials	Specific capacitance $(F g^{-1})$	Working potential (V)	Maximum energy density $(Wh kg^{-1})$	Maximum Power density $(kW kg^{-1})$	Cyclic performance (retention)
24	Ni-Co LDH//rGO	550	1.6	188	7.32	82% after 5,000 cycles
56	GSP/Co-Al LDH//SGC	—	1.6	41.2	9.3	84% after 2,000 cycles
61	Ni-Al LDH@CNPs//AC	138	1.6	47.7	51	88.9% after 2,000 cycles
62	Ni-Co LDH//AC	125.2	1.2	23.7	5.82	92.7% after 5,000 cycles
63	Co-Al LDHs-CNTs//AC	80.6	1.6	28	~6	88.9% after 1,000 cycles
64	Ni-Co LDH//AC	—	1.5	25.3	10.5	91.2% after 10,000 cycles
65	NiCo-LDHs@CNT/NF//APDC/NF	210.9	1.8	89.7	8.7	78% after 1,200 cycles
Our work	Co-Al LDH/ACT//ACT/graphene	145.8	1.6	55.04	5.4	87.54% after 2,000 cycles

AC, activated carbon; CNPs, carbon nanoparticles; GSP, integrated porous Co-Al hydroxide nanosheets; LDH, layered double hydroxide; NF, nickel foam; rGO, reduced graphene oxide; SGC, sandwiched graphene/porous carbon.

rate performance of the ACT/Co-Al LDH//ACT/graphene asymmetric cell.

Rate performance and coulombic efficiency are important factors for the real power application of supercapacitors. Figure 5a shows the charge/discharge curves of ACT/Co-Al LDH//ACT/graphene asymmetric supercapacitor at different current densities. A high capacitance retention of $80.78 F g^{-1}$ was achieved even when the current density increased from 7.5 to $50 mA cm^{-2}$, indicating the excellent rate performance of the ACT/Co-Al LDH//ACT/graphene asymmetric cell. The rate performance and corresponding coulombic efficiency of ACT/graphene//ACT/graphene and ACT/Co-Al LDH//ACT/Co-Al LDH symmetric cells as well as ACT/Co-Al LDH//ACT/graphene asymmetric cell are comprehensively compared in Fig. 5b. The assembled ACT/graphene symmetric cell exhibited an excellent electrochemical stability (58% capacitance retention) and high coulombic efficiency (99.5% at $50 mA cm^{-2}$) in a wide range of current densities ($7.5-50 mA cm^{-2}$). Whereas the ACT/Co-Al LDH symmetric cell showed poor rate performance (35% capacitance retention) and low coulombic efficiency (91.8% at $50 mA cm^{-2}$). Compared with the ACT/Co-Al LDH symmetric cell, the ACT/Co-Al LDH//ACT/graphene asymmetric cell exhibited not only improved capacitance retention (52.3% capacitance retention) but also enhanced coulombic efficiency (98.5% at $50 mA cm^{-2}$). At high charge/discharge rates the ions on electrode decreased rapidly with increasing rate, and the ions in the electrolyte diffused too slowly to satisfy the need of ions near the solid–liquid interface, leading to the decrease of capacitance[57]. The improvements in both rate capability and coulombic efficiency for the asymmetric cell result from the synergetic effects between the two distinct electrodes where the high conductive ACT/graphene with high rate capacity balanced the poor rate capacity of ACT/Co-Al LDH.

Energy density and power density are important factors for evaluating the practical application of supercapacitors. Figure 5c shows the Ragone plots of ACT/graphene//ACT/graphene and ACT/Co-Al LDH//ACT/Co-Al LDH symmetric cells as well as ACT/Co-Al LDH//ACT/graphene asymmetric cell. Compared with the symmetric cells, the ACT/Co-Al LDH//ACT/graphene asymmetric cell exhibited a higher energy density of $55.04 Wh kg^{-1}$ at the power density of $387.9 W kg^{-1}$ and maintained $28.72 Wh kg^{-1}$ at the power density of $5.44 kW kg^{-1}$. For an in-depth understanding of the electrochemical behaviour of the assembled symmetric and asymmetric cells, the EIS tests were carried out on the aforementioned cells (Fig. 5d). A similar shape can be found for all the impedance spectra, with a straight line at the

low-frequency regime and an arc at the high-frequency region. The high-frequency arc is ascribed to the double-layer capacitance (C_{dl}) and the charge transfer resistance (R_{ct}) at the electrode and electrolyte interface, corresponding to the charge transfer-limiting process[58]. R_{ct} was directly measured from the diameter of the semicircle arc in the Niquist plot. Clearly, compared with the ACT/graphene (1.1Ω) and ACT/Co-Al LDH (1.89Ω) symmetric cells, the relatively smaller R_{ct} of the asymmetric cell resulted from the synergistic effect of the ACT/graphene and hierarchical ACT/Co-Al LDH architectures that facilitated the access of electrolyte ions to the active surface and shortened the ion diffusion path.

Cycling capability is another crucial requirement for the practical application of supercapacitors. The cyclic performances of ACT/graphene//ACT/graphene and ACT/Co-Al LDH//ACT/Co-Al LDH symmetric cells as well as ACT/Co-Al LDH//ACT/graphene asymmetric cell were evaluated at the current density of $12.5 mA cm^{-2}$ using a GV charge–discharge technique. An obvious capacity decay (~35%) was observed for the ACT/Co-Al LDH symmetric cell, whereas there was a capacity increase (~8%) for the ACT/graphene symmetric cell after 2,000 cycles. For the ACT/Co-Al LDH//ACT/graphene asymmetric cell, ~16% decay of its original capacitance was noted after 2,000 cycles. The asymmetric cell exhibited linear and symmetrical characteristics in its charge/discharge curves, and no obvious 'IR drop' was observed, indicating its low internal resistance (Fig. 5e). Compared with previously reported LDH-based asymmetric supercapacitors (see Table 1 for details), the electrochemical properties of the ACT/Co-Al LDH//ACT/graphene asymmetric cell, benefiting from the synergistic effects of ACT/graphene and flower-like ACT/Co-Al LDH, are highly competitive in terms of specific capacity, maximum energy and power densities, and cyclability. The assembled flexible asymmetric cell also worked well under a 180° folded state (Supplementary Fig. 3), a clear indication of the cell's excellent coupled mechanical and electrochemical robustness.

Supercapacitors have been proven to be efficient and powerful energy storage devices to drive various electronic components. Furthermore, if a renewable energy source can be used to sustain an energy charge, the combined supercapacitor/energy source cell will provide continuous power for consumer electronics, forming a self-sustaining system without need for large, heavy batteries[59]. As the most sustainable and cleanest source of energy in the world, solar energy is usually limited by access to sunlight as restricted by time of day, location and weather. Combining solar cells with energy storage devices provides a promising solution to extend the practical applications of solar energy beyond the

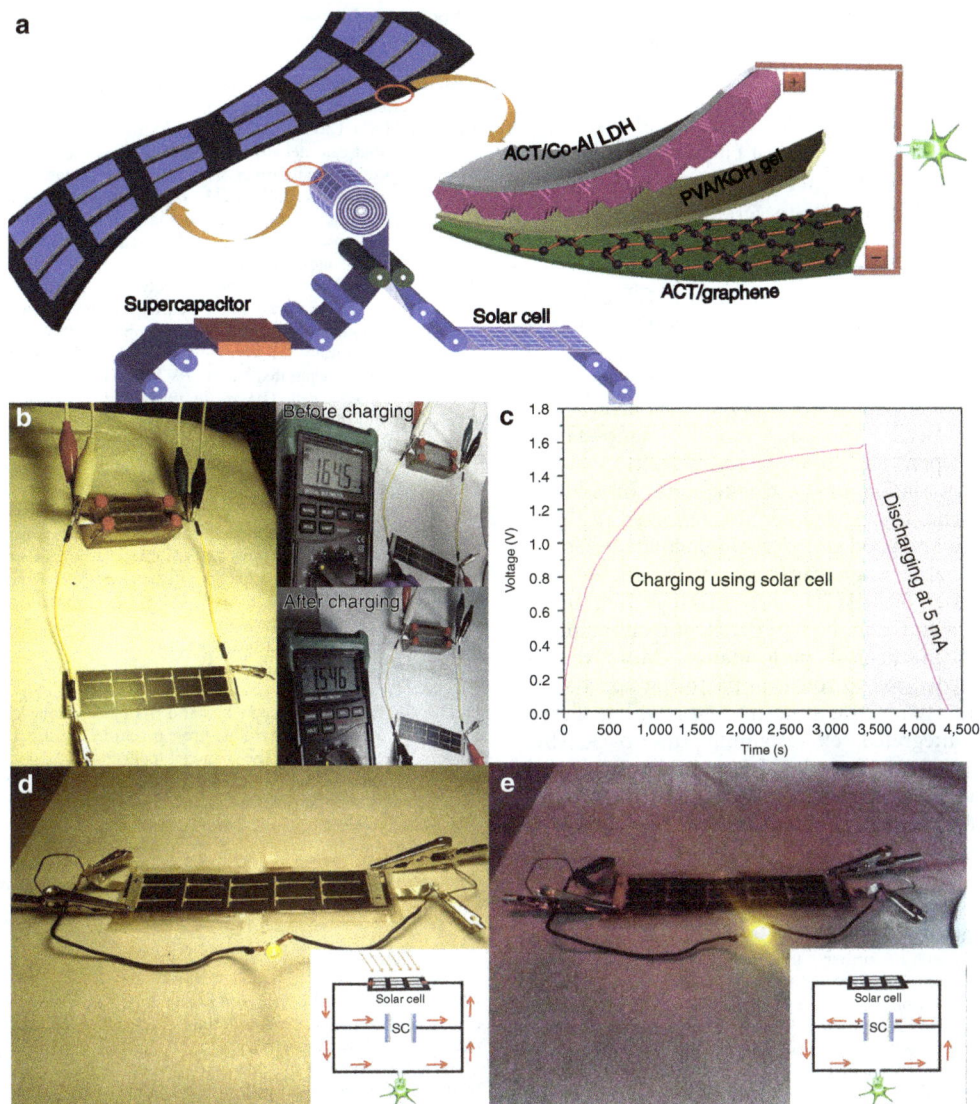

Figure 6 | The charging/discharging processes of the assembled self-sustaining power pack. (a) Schematic illustration of the roll-to-roll manufacturing process for integrating a flexible solar cell with the supercapacitor into a self-sustaining power pack. (b) Digital photograph of the assembled solar cell/supercapacitor hybrid energy conversion and storage system, insets are the photographs of the open circuit potential of the ACT/Co-Al LDH//ACT/graphene asymmetric cell before charging and after charging with solar cell. (c) Charging curve of the ACT/Co-Al LDH//ACT/graphene asymmetric cell with solar cell, and discharging curve of the ACT/Co-Al LDH//ACT/graphene asymmetric cell at a current density of $5\,mA\,cm^{-2}$. (d) Digital photograph of the assembled solar cell/supercapacitor hybrid power pack worked under the light, inset shows schematically the working circuit connection. (e) Digital photograph of the assembled solar cell/supercapacitor hybrid power pack worked without the light, inset shows schematically the working circuit connection.

imposed restrictions of sunlight availability. As schematically illustrated in Fig. 6a, a flexible thin-film solar cell with an open circuit potential of 3 V under light was integrated with the flexible asymmetric cell to realize a combined energy conversion and storage system in a single device by a simple roll-to-roll process. Figure 6b illustrates the operation mechanism of the assembled solar cell/supercapacitor hybrid energy conversion and storage system. Insets of Fig. 6b are the photographs of the open circuit potential of the ACT/Co-Al LDH//ACT/graphene asymmetric supercapacitor before and after charging with the solar cell. Before charging by the solar cell, the open circuit potential of the asymmetric supercapacitor was 0.1645 V. Under the light source, the solar cell (with an open circuit potential of 3 V) serving as the power source charged the supercapacitor. After charging, the open circuit potential increased to 1.546 V. The corresponding charging (under light) and discharging curves (no light) of the asymmetric supercapacitor at the current density of $5\,mA\,cm^{-2}$

were recorded by the electrochemical workstation (Fig. 6c), demonstrating the self-powered function of such hybrid energy conversion and storage system. Encouragingly, a high energy transfer efficiency of ~43% was achieved from the solar cell to the capacitor. To further demonstrate the practical application of such integrated flexible energy system, a flexible solar cell was integrated with two flexible asymmetric cells connected in series by a facile roll-to-roll process, which holds a great promise for large-scale manufacturing of such hybrid cells (Fig. 6d,e). Insets of Fig. 6d,e illustrate the schematics of working circuit connection. Under light, the solar cell served as the power source to provide energy for the supercapacitor and LED. When the light was turned off, the stored energy in the supercapacitor in turn served as the power source for the LED, which enabled this hybrid energy system to work continuously for ~10 min, overcoming the limitation of solar discontinuity. Such flexible, self-sustaining energy systems hold great potential for future portable/wearable electronics where they can reliably power devices as consumers move in and out of sunlight during their normal daily activities.

Discussion

Two-dimensional battery-type Co-Al LDH nanosheets were anchored *in situ* on ACTs by a simple hydrothermal process. Separately, ACT fibres were wrapped with a highly conductive graphene coating by a simple dipping, drying and reducing process (ACT/graphene). A flexible, all-solid-state asymmetric supercapacitor (ACT/Co-Al LDH//ACT/graphene) was assembled using the nanostructured ACT/Co-Al LDH as the positive electrode, ACT/graphene as the negative electrode and PVA-KOH gel as both the solid-state electrolyte and separator. The hierarchical flower-like Co-Al LDH nanoarrays with interconnected nano-petals and pompon-like nano-stamens provided a highly open and porous scaffold-like structure, facilitating the transportation of electrolyte ions. In addition, the wrapped graphene on porous ACTs rendered a highly accessible surface area and good electrical conductivity, endowing the flexible asymmetric supercapacitor with higher rate performance. The assembled ACT/Co-Al LDH//ACT/graphene asymmetric supercapacitor exhibited an excellent combination of electrochemical and mechanical performance. Moreover, we demonstrated a low-cost, roll-to-roll manufacturing approach to combine the flexible asymmetric supercapacitor with a flexible solar cell to build an integrated, self-sustaining power pack, which is scalable for industrial manufacturing. Importantly, such hybrid energy storage devices could continuously power a commercial LED, demonstrating a great potential for the future of self-powered nanotechnology.

Methods

Preparation of flexible ACT/Co-Al LDH positive electrode. All of the chemicals were used after purchasing without further purification. A piece of commercial cotton T-shirt was first cleaned using distilled water in an ultrasonic bath before activation. Activation of the cotton T-shirt was performed as described in detail in our previous report[47]. First, a piece of cotton T-shirt was dipped into 1 M NaF solution and soaked for 1 h. The wet textile was then dried at 120 °C for 3 h. Second, the NaF-treated cotton textile was transferred into a horizontal tube furnace and kept at 1,000 °C for 1 h with a continuous argon gas flow (300 s.c.c.m.). After cooling down to room temperature, the as-obtained ACTs were washed with distilled water to remove the residual NaF and then dried at 80 °C for 6 h. Co-Al LDH nanoarrays were grown *in situ* on ACT fibres via a simple hydrothermal process. Briefly, 0.582 g of $Co(NO_3)_2 \cdot 6H_2O$, 0.518 g of $Al(NO_3)_3 \cdot 6H_2O$, 0.296 g of NH_4F and 0.6 g of urea were dissolved in 36 ml distilled water. The resulting solution was transferred into a 50-ml Teflon-lined stainless autoclave with a piece of vertically suspended ACT ($1 \times 2\ cm^2$) in the solution. Then, the autoclave was placed in an electric oven at 100 °C for 24 h. Finally, the as-prepared products were washed thoroughly with ethanol and distilled water and dried at 80 °C overnight to produce ACT/Co-Al LDH composite.

Preparation of flexible ACT/graphene-negative electrode. Graphite oxide was synthesized by exfoliating natural graphite flakes using a modified Hummers method[60]. The obtained graphite oxide solution was further exfoliated by ultrasonication in an ultrasonic bath for 1 h to prepare graphene oxide. Then, the above graphene oxide solution was centrifuged at 3,000 r.p.m. for 5 min to remove residual aggregates, forming a brown aqueous colloid with a concentration of ~4 mg ml^{-1}. A piece of ACT ($1 \times 2\ cm^2$) was then soaked with the graphene oxide aqueous colloid. After drying at 80 °C for 6 h, the as-prepared ACT with graphene oxide coating was heated at 450 °C for 1 h with argon/hydrogen mixture gas (v/v, 90/10) to reduce flexible ACT/graphene oxide to ACT/graphene-negative electrode.

Characterization methods. The morphology and microstructure of the as-prepared products were characterized by SEM (FEI Quanta 650), TEM (JEOL 2000FX), HRTEM (FEI Titan), and atomic force microscopy (Nanoscope IIIa). The crystallographic structure of the synthesized materials was determined by a PANalytical X'Pert Pro Multi-Purpose Diffractometer equipped with Cu K_α radiation ($\lambda = 0.15406$ nm). The Raman spectra of the ACT/graphene were recorded by a Renishaw InVia Raman microscope at 785 nm. Surface areas of the active materials were measured by physical adsorption of N_2 at 77 K (Quantachrome Autosorb iQ surface area and pore size analyser) and calculated by the BET method.

Fabrication and characterization of integrated flexible self-sustaining power pack. The electrochemical performances of the prepared ACT, ACT/graphene and ACT/Co-Al LDH electrodes were characterized by both three-electrode and two-electrode systems. The three-electrode tests were carried out in 6 M KOH aqueous electrolyte at room temperature. The prepared ACT, ACT/graphene and ACT/Co-Al LDH were used as the working electrodes, platinum foil ($1 \times 1\ cm^2$) and a saturated calomel electrode were used as the counter and reference electrodes, respectively. The mass of the ACT, ACT/graphene and ACT/Co-Al LDH was measured to be 2, 3.45 and 4.1 mg cm^{-2}, respectively. The asymmetric supercapacitor used in this paper was assembled with two pieces of flexible, face-to-face electrodes (ACT/Co-Al LDH-positive and ACT/graphene-negative) separated by the solid-state, polymer PVA/KOH gel electrolyte (ACT/Co-Al LDH//ACT/graphene). The polymer gel electrolyte was prepared by mixing 3 g KOH and 6 g PVA in 60 ml deionized water by magnetically stirring the solution at 80 °C until the solution became clear. Then, the above solution was transferred into a flat Petri-dish to naturally solidify to PVA/KOH gel film electrolyte. Two pieces of flexible electrodes separated by the PVA/KOH gel film were used to assemble the flexible asymmetric cell. The as-obtained solid-state, polymer PVA/KOH gel film severed as both the separator and electrolyte. The electrochemical properties of the assembled flexible asymmetric cell were measured using a CHI 660E electrochemical workstation. CV, GV charge/discharge curves and EIS in the frequency range from 100 kHz to 0.05 Hz with an AC perturbation of 5 mV were used to evaluate the electrochemical performance of the flexible solid-state asymmetric supercapacitors. The integrated flexible energy conversion and storage unit was assembled by combining a piece of commercial flexible solar panel (with an open circuit potential of 3 V and energy efficiency of 22%, from PowerFilm) with an energy storage unit consisting of two pieces of asymmetric cell connected in series using an electrical rolling machine (MSK-HRP-MR100A, MTI). Two pieces of assembled asymmetric flexible supercapacitors were connected in series and then coated with double-side adhesive tapes. At the same time, the bottom surface of the flexible solar cell was also coated with double-side adhesive tapes. As illustrated in Fig. 6a, the flexible supercapacitor and the flexible solar cell were integrated together by the roll-to-roll rolling press machine. During the roll-to-roll fabrication process, the top surface of the solar cell (collecting sunlight) was protected by using the soften cloth to obtain a flexible, solar cell/supercapacitor, self-sustaining power pack. The open circuit potential and charge/discharge processes were monitored by the CHI 660E electrochemical workstation.

Calculation. Specific capacitances derived from GV tests were calculated from equation 1 as follows:

$$C_{sp} = \frac{It}{(\Delta V)m} \tag{1}$$

where C_{sp} (F/g), I (A), t (s), m (g) and ΔV (V) are the specific capacitance, the discharge current, the discharge time, the mass of the active material and the total potential window, respectively.

Energy density (E) and power (P) density derived from GV tests were calculated from the following equations:

$$E = \frac{1}{2}CV^2 \tag{2}$$

$$P = \frac{E}{t} \tag{3}$$

where E (Wh/kg), C (F/g), V (V), P (W/kg) and t (s) are the energy density, specific capacitance, potential window, power density and discharge time, respectively.

References

1. Lewis, N. S. & Nocera, D. G. Powering the planet: chemical challenges in solar energy utilization. *Proc. Natl Acad. Sci. USA* **103**, 15729–15735 (2007).
2. Burke, A. Ultracapacitors: Why, how, and where is the technology. *J. Power Sources* **91**, 37–50 (2000).
3. Frackowiak, E. Carbon materials for supercapacitor application. *Phys. Chem. Chem. Phys.* **9**, 1774–1785 (2007).
4. Yang, Z. *et al.* Recent advancement of nanostructured carbon for energy applications. *Chem. Rev.* **115**, 5159–5223 (2015).
5. Hu, C. C., Chang, K. H., Lin, M. C. & Wu, Y. T. Design and tailoring of the nanotubular arrayed architecture of hydrous RuO_2 for next generation supercapacitors. *Nano Lett.* **6**, 2690–2695 (2006).
6. Jiang, J. *et al.* Recent advances in metal oxide-based electrode architecture design for electrochemical energy storage. *Adv. Mater.* **24**, 5166–5180 (2012).
7. Lang, X., Hirata, A., Fujita, T. & Chen, M. Nanoporous metal/oxide hybrid electrodes for electrochemical supercapacitors. *Nat. Nanotechnol.* **6**, 232–236 (2011).
8. Yang, W. *et al.* Hierarchical $NiCo_2O_4$@NiO core–shell hetero-structured nanowire arrays on carbon cloth for a high-performance flexible all-solid-state electrochemical capacitor. *J. Mater. Chem. A* **2**, 1448–1457 (2014).
9. Fan, L. Z. & Maier, J. High-performance polypyrrole electrode materials for redox supercapacitors. *Electrochem. Commun.* **8**, 937–940 (2006).

10. Yang, W. *et al.* Synthesis of hollow polyaniline nano-capsules and their supercapacitor application. *J. Power Sources* **272**, 915–921 (2014).

11. Snook, G. A., Kao, P. & Best, A. S. Conducting-polymer-based supercapacitor devices and electrodes. *J. Power Sources* **196**, 1–12 (2011).

12. Yu, G., Xie, X., Pan, L., Bao, Z. & Cui, Y. Hybrid nanostructured materials for high-performance electrochemical capacitors. *Nano Energy* **2**, 213–234 (2013).

13. Tang, Z., Tang, C. & Gong, H. A high energy density asymmetric supercapacitor from nano-architectured Ni(OH)$_2$/carbon nanotube electrodes. *Adv. Funct. Mater.* **22**, 1272–1278 (2012).

14. Gao, Z., Song, N. & Li, X. Microstructural design of hybrid CoO@NiO and graphene nano-architectures for flexible high performance supercapacitors. *J. Mater. Chem. A* **3**, 14833–14844 (2015).

15. Wang, Q. & O'Hare, D. Recent advances in the synthesis and application of layered double hydroxide (LDH) nanosheets. *Chem. Rev.* **112**, 4124–4155 (2012).

16. Yang, R., Gao, Y., Wang, J. & Wang, Q. Layered double hydroxide (LDH) derived catalysts for simultaneous catalytic removal of soot and NO(x). *Dalton Trans.* **43**, 10317–10327 (2014).

17. Ladewig, K., Xu, Z. P. & Lu, G. Q. M. Layered double hydroxide nanoparticles in gene and drug delivery. *Expert Opin. Drug Deliv.* **6**, 907–922 (2009).

18. Liu, X., Ge, L., Li, W., Wang, X. & Li, F. Layered double hydroxide functionalized textile for effective oil/water separation and selective oil adsorption. *ACS Appl. Mater. Interfaces* **7**, 791–800 (2015).

19. Su, L. H., Zhang, X. G., Mi, C. H. & Liu, Y. Insights into the electrochemistry of layered double hydroxide containing cobalt and aluminum elements in lithium hydroxide aqueous solution. *J. Power Sources* **179**, 388–394 (2008).

20. Yang, W. *et al.* Solvothermal one-step synthesis of Ni-Al layered hydroxide/carbon nanotube/reduced graphene oxide sheet ternary nanocomposite with ultrahigh capacitance for supercapacitors. *ACS Appl. Mater. Interfaces* **5**, 5443–5454 (2013).

21. Lu, Z. *et al.* High pseudocapacitive cobalt carbonate hydroxide films derived from CoAl layered double hydroxides. *Nanoscale* **4**, 3640–3643 (2012).

22. Vialat, P. *et al.* High-performing monometallic cobalt layered double hydroxide supercapacitor with defined local structure. *Adv. Funct. Mater.* **24**, 4831–4842 (2014).

23. Huang, J. *et al.* Effect of Al-doped β-Ni(OH)$_2$ nanosheets on electrochemical behaviors for high performance supercapacitor application. *J. Power Sources* **232**, 370–375 (2013).

24. Chen, H., Hu, L., Chen, M., Yan, Y. & Wu, L. Nickel-cobalt layered double hydroxide nanosheets for high-performance supercapacitor electrode materials. *Adv. Funct. Mater.* **24**, 934–942 (2014).

25. Gao, Z. *et al.* Graphene nanosheet/Ni^{2+}/Al^{3+} layered double-hydroxide composite as a novel electrode for a supercapacitor. *Chem. Mater.* **23**, 3509–3516 (2011).

26. Novoselov, K. S. *et al.* Two-dimensional gas of massless Dirac fermions in graphene. *Nature* **438**, 197–200 (2005).

27. Chen, D., Tang, L. & Li, J. Graphene-based materials in electrochemistry. *Chem. Soc. Rev.* **39**, 3157–3180 (2010).

28. Stoller, M. D., Park, S., Zhu, Y., An, J. & Ruoff, R. S. Graphene-based ultracapacitors. *Nano Lett.* **8**, 6–10 (2008).

29. Zhu, Y. *et al.* Carbon-based supercapacitors produced by activation of graphene. *Science* **332**, 1537–1542 (2011).

30. Wu, Z. S. *et al.* High-energy MnO$_2$ nanowire/graphene and graphene asymmetric electrochemical capacitors. *ACS Nano* **4**, 5835–5842 (2010).

31. Films, G. H. *et al.* Flexible solid-state supercapacitors based on three-dimensional. *ACS Nano* **7**, 4042–4049 (2013).

32. Ji, J. *et al.* Nanoporous Ni (OH)$_2$ thin film on 3D ultrathin-graphite foam for asymmetric supercapacitor. *ACS Nano* **7**, 6237–6243 (2013).

33. Gao, Z., Yang, W., Wang, J., Song, N. & Li, X. Flexible all-solid-state hierarchical NiCo$_2$O$_4$/porous graphene paper asymmetric supercapacitors with an exceptional combination of electrochemical properties. *Nano Energy* **13**, 306–317 (2015).

34. Liu, Z., Xu, J., Chen, D. & Shen, G. Flexible electronics based on inorganic nanowires. *Chem. Soc. Rev.* **44**, 161–192 (2015).

35. Jung, Y. H. *et al.* High-performance green flexible electronics based on biodegradable cellulose nanofibril paper. *Nat. Commun.* **6**, 7170 (2015).

36. Gao, Z., Song, N., Zhang, Y. & Li, X. Cotton textile enabled, all-solid-state flexible supercapacitors. *RSC Adv.* **5**, 15438–15447 (2015).

37. Huang, Y. *et al.* From industrially weavable and knittable highly conductive yarns to large wearable energy storage textiles. *ACS Nano* **9**, 4766–4775 (2015).

38. Gao, P. X., Song, J., Liu, J. & Wang, Z. L. Nanowire piezoelectric nanogenerators on plastic substrates as flexible power sources for nanodevices. *Adv. Mater.* **19**, 67–72 (2007).

39. Jost, K. *et al.* Natural fiber welded electrode yarns for knittable textile supercapacitors. *Adv. Energy Mater* **5**, 1401286 (2015).

40. Xiao, X. *et al.* High-strain sensors based on ZnO nanowire/polystyrene hybridized flexible films. *Adv. Mater.* **23**, 5440–5444 (2011).

41. Xu, S. *et al.* Self-powered nanowire devices. *Nat. Nanotechnol* **5**, 366–373 (2010).

42. Wang, X. *et al.* Fiber-based all-solid-state flexible supercapacitors for self-powered systems. *Adv Mater* **3**, 6790–6797 (2014).

43. Yang, P. *et al.* Hydrogenated ZnO core-shell nanocables for flexible supercapacitors and self-powered systems. *ACS Nano* **7**, 2617–2626 (2013).

44. Xu, X. *et al.* A power pack based on organometallic perovskite solar cell and supercapacitor. *ACS Nano* **9**, 1782–1787 (2015).

45. Wang, Z., Chen, J. & Lin, L. Progress in triboelectric nanogenerators as new energy technology and self-powered sensors. *Energy Environ. Sci.* **8**, 2250–2282 (2015).

46. Liu, Y., Wang, X., Qi, K. & Xin, J. H. Functionalization of cotton with carbon nanotubes. *J. Mater. Chem.* **18**, 3454 (2008).

47. Bao, L. & Li, X. Towards textile energy storage from cotton T-shirts. *Adv. Mater.* **24**, 3246–3252 (2012).

48. Dreyer, D. R., Park, S., Bielawski, C. W. & Ruoff, R. S. The chemistry of graphene oxide. *Chem. Soc. Rev.* **39**, 228–240 (2010).

49. Bai, H., Li, C., Wang, X. & Shi, G. On the gelation of graphene oxide. *J. Phys. Chem. C* **115**, 5545–5551 (2011).

50. Chen, W. & Yan, L. In situ self-assembly of mild chemical reduction graphene for three-dimensional architectures. *Nanoscale* **3**, 3132–3137 (2011).

51. Ford, E. N. J., Mendon, S. K., Thames, S. F. & Rawlins, J. W. X-ray diffraction of cotton treated with neutralized vegetable oil-based macromolecular crosslinkers. *J. Eng. Fiber Fabr* **5**, 10–20 (2010).

52. Liu, Z. H., Wang, Z. M., Yang, X. & Ooi, K. Intercalation of organic ammonium ions into layered graphite oxide. *Langmuir* **18**, 4926–4932 (2002).

53. Hassan, H. M. A. *et al.* Microwave synthesis of graphene sheets supporting metal nanocrystals in aqueous and organic media. *J. Mater. Chem.* **19**, 3832–3837 (2009).

54. Mcallister, M. J. *et al.* Single sheet functionalized graphene by oxidation and thermal expansion of graphite. *Chem. Mater.* **19**, 4396–4404 (2007).

55. Wang, Y. *et al.* Magnetic graphene oxide nanocomposites: nanoparticles growth mechanism and property analysis. *J. Mater. Chem. C* **2**, 9478–9488 (2014).

56. Wu, X., Jiang, L., Long, C., Wei, T. & Fan, Z. Dual support system ensuring porous Co-Al hydroxide nanosheets with ultrahigh rate performance and high energy density for supercapacitors. *Adv. Funct. Mater.* **25**, 1648–1655 (2015).

57. Di Fabio, A., Giorgi, A., Mastragostino, M. & Soavi, F. Carbon-Poly (3-methylthiophene) Hybrid Supercapacitors. *J. Electrochem. Soc.* **148**, A845–A850 (2001).

58. Wang, J. *et al.* Green synthesis of graphene nanosheets/ZnO composites and electrochemical properties. *J. Solid State Chem.* **184**, 1421–1427 (2011).

59. Wang, Z. L. Toward self-powered sensor networks. *Nano Today* **5**, 512–514 (2010).

60. William, S., Hummers, J. & Offeman, R. E. Preparation of graphitic oxide. *J. Am. Chem. Soc.* **80**, 1339 (1958).

61. Liu, X. *et al.* A Ni-Al layered double hydroxide@carbon nanoparticles hybrid electrode for high-performance asymmetric supercapacitors. *J. Mater. Chem. A* **2**, 1682–1685 (2014).

62. Wang, X., Sumboja, A., Lin, M., Yan, J. & Lee, P. S. Enhancing electrochemical reaction sites in nickel–cobalt layered double hydroxides on zinc tin oxide nanowires: a hybrid material for an asymmetric supercapacitor device. *Nanoscale* **4**, 7266 (2012).

63. Yu, L. *et al.* Facile synthesis of exfoliated Co-Al LDH–carbon nanotube composites with high performance as supercapacitor electrodes. *Phys. Chem. Chem. Phys.* **16**, 17936 (2014).

64. Jing, M. *et al.* Alternating voltage introduced NiCo double hydroxide layered nanoflakes for an asymmetric supercapacitor. *ACS Appl. Mater. Interfaces* **7**, 22741–22744 (2015).

65. Li, X. *et al.* An asymmetric supercapacitor with super-high energy density based on 3D core-shell structured NiCo-layered double hydroxide@carbon nanotube and activated polyaniline-derived carbon electrodes with commercial level mass loading. *J. Mater. Chem. A* 13244–13253 (2015).

Acknowledgements

Financial support for this study was provided by the US National Science Foundation (CMMI-1418696 and CMMI-1358673) and the i6 Virginia Innovation Partnership. We thank the staff members at the University of Virginia NMCF for electron microscopy technical support.

Author contributions

Z.G. and X.L. conceived the idea. Z.G. carried out the materials synthesis, cell assembly and electrochemical tests. Z.G., C.B., N.S., and Y.Z. performed the materials characterization. Z.G. and X.L. co-wrote the paper. J.L. together with other authors discussed and commented on the paper.

Additional information

Unfolding the physics of URu_2Si_2 through silicon to phosphorus substitution

A. Gallagher[1], K.-W. Chen[1], C.M. Moir[1], S.K. Cary[2], F. Kametani[3], N. Kikugawa[1,4], D. Graf[1], T.E. Albrecht-Schmitt[2], S.C. Riggs[1], A. Shekhter[1] & R.E. Baumbach[1]

The heavy fermion intermetallic compound URu_2Si_2 exhibits a hidden-order phase below the temperature of 17.5 K, which supports both anomalous metallic behavior and unconventional superconductivity. While these individual phenomena have been investigated in detail, it remains unclear how they are related to each other and to what extent uranium f-electron valence fluctuations influence each one. Here we use ligand site substituted $URu_2Si_{2-x}P_x$ to establish their evolution under electronic tuning. We find that while hidden order is monotonically suppressed and destroyed for $x \leq 0.035$, the superconducting strength evolves non-monotonically with a maximum near $x \approx 0.01$ and that superconductivity is destroyed near $x \approx 0.028$. This behavior reveals that hidden order depends strongly on tuning outside of the U f-electron shells. It also suggests that while hidden order provides an environment for superconductivity and anomalous metallic behavior, it's fluctuations may not be solely responsible for their progression.

[1] National High Magnetic Field Laboratory, Florida State University, Tallahassee, Florida 32310, USA. [2] Department of Chemistry and Biochemistry, Florida State University, Tallahassee, Florida 32306, USA. [3] Applied Superconductivity Center, Florida State University, Tallahassee, Florida 32310, USA. [4] National Institute for Materials Science 3-13 Sakura, Tsukuba 305-0003, Japan. Correspondence and requests for materials should be addressed to R.E.B. (email: baumbach@magnet.fsu.edu).

Materials that defy straightforward description in terms of either localized or itinerant electron behavior are a longstanding challenge to understanding novel electronic matter[1–4]. The intermetallic URu_2Si_2 is a classic example, where the itinerant electrons exhibit a giant magnetic anisotropy that is normally a characteristic of localized electrons[5,6]. URu_2Si_2 further displays an unknown broken symmetry state ('hidden-order') and unconventional superconductivity for temperatures below $T_0 = 17.5$ K and $T_c = 1.4$ K, respectively[7–10]. The development of new experimental techniques and access to ultra-high-purity single crystal specimens has recently advanced our understanding of the ordered states in this compound. For instance, excitations in the A_{2g} channel have been identified by electronic Raman spectroscopy as a signature of the hidden order[11], while elastoresistance[12], resonant ultrasound[13] (Ramshaw, B.J., Private communication.) and spectroscopic measurements[14] suggest the presence of fluctuations in the B_{2g}, B_{1g} and E_g channels, respectively. These studies follow high resolution X-ray diffraction[15], torque magnetometry[16] and polar Kerr effect measurements[17], which provide further insight into hidden order. It is also noteworthy that URu_2Si_2 differs from most other unconventional superconductors, which are typically found near the zero temperature termination point of a line of phase transitions, a 'quantum critical point,' where strong fluctuations are believed to be favourable for the superconducting pairing[18–21]. In contrast, the superconductivity in URu_2Si_2 is fully contained inside the ordered phase, as revealed by numerous tuning studies[22–33].

To explore the mechanisms of hidden order, anomalous metallic behavior and superconductivity in this material we synthesized high purity single crystal specimens of chemically substituted $URu_2Si_{2-x}P_x$, where ligand site substitution is a 'gentle' way to tune the electronic state. Recent advances developing a molten metal flux growth technique to produce high quality single crystal specimens of URu_2Si_2 enabled these experiments, which were previously inaccessible due to metallurgical challenges associated with the high-vapor pressure of phosphorus[34]. We report electrical transport and thermodynamic measurements, which reveal that while hidden order is destroyed for $x \leq 0.035$, the superconducting strength evolves non-monotonically with a maximum near $x \approx 0.01$ and that superconductivity disappears near $x \approx 0.028$. While the rapid suppression of hidden order indicates the importance of itinerant electrons, the unexpected maximum in superconducting strength may suggest the presence of a critical point that is defined by the termination of a phase boundary other than that of the hidden order. Owing to the relatively small chemical difference between silicon and phosphorus, we propose that in this series the dominant tuning effect is simply to change the chemical potential. Extrinsic factors such as disorder, as well as some intrinsic factors including changes in the unit cell volume or bond angles, spin orbit coupling and ligand hybridization strength play a minor role. Phosphorus substitution further has the advantage that it does not directly affect either the local electron count or the balance of the spin–orbit and Coulomb interactions of the d- or f-electrons on the Ru and U sites, making it an ideal and long desired tool for unraveling the physics of URu_2Si_2.

Results

Electrical transport and heat capacity. In Fig. 1 we show normalized electrical resistance R and heat capacity C versus temperature T for several phosphorus concentrations x (see Supplementary Fig. 1 for magnetic susceptibility $\chi(T)$). The Kondo lattice behavior of the electrical resistance (that is, non-monotonic temperature dependence below room temperature

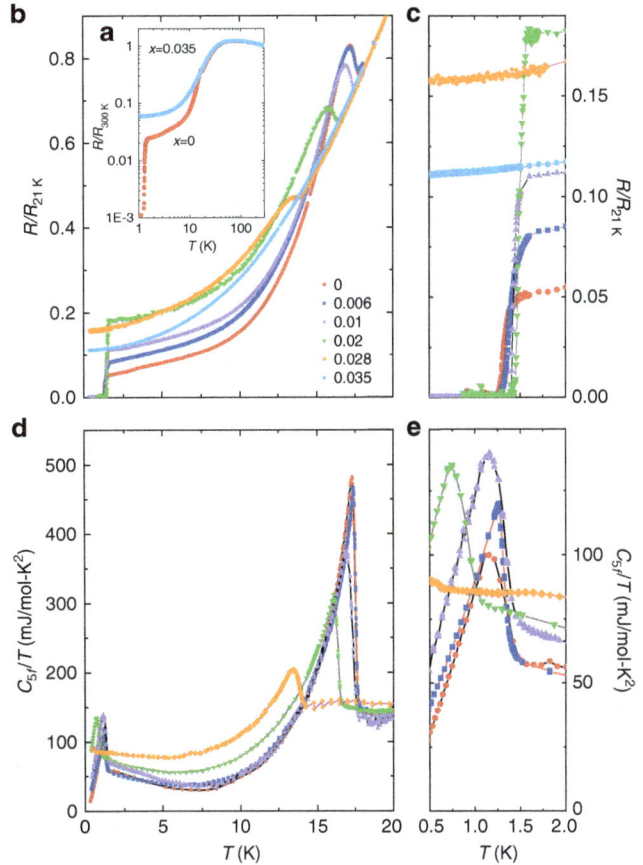

Figure 1 | Electrical transport and heat capacity for $URu_2Si_{2-x}P_x$. (**a**) Electrical resistance normalized to the value at 300 K $R/R_{300 K}$ versus temperature T for phosphorus concentrations $x = 0$ and 0.035. (**b**) Electrical resistance normalized to the value at 21 K $R/R_{21 K}$ versus T for $0 \leq x \leq 0.035$. (**c**) $R/R_{21 K}$ versus T in the low temperature region, emphasizing the superconducting transitions. (**d**) The $5f$ contribution to the heat capacity C_{5f} divided by T versus T for $0 \leq x \leq 0.028$. (**e**) C_{5f}/T versus T in the low T region, showing the bulk superconducting transitions. See Supplementary section for a description of the phonon background subtraction.

with a resistive peak near 80 K) for $T \geq T_0$ is unaffected by phosphorus substitution for $x \leq 0.035$ (Fig. 1a), suggesting that the strength of the hybridization between the f- and conduction electron states does not change much in the range $0 < x < 0.035$. At lower temperatures, the hidden order transition temperature T_0 and the size of the anomalies in R and C associated with it are monotonically suppressed with increasing x (Fig. 1b,d). There is an excellent agreement between the values of T_0 as extracted from R, C/T, and χ, indicating that disorder effects are negligible. We find no evidence for hidden order in $x = 0.035$ for $T > 20$ mK, showing that there is a quantum phase transition from hidden order to a paramagnetic correlated electron metal between $0.028 < x < 0.035$. However, the data does not preclude the possibility of a broad fluctuation regime around the quantum phase transition. The paramagnetic region subsequently extends up to $x \approx 0.25$ (nearly a factor of 10 larger x) where correlated electron antiferromagnetism appears (See Supplementary Fig. 2) (Results for the larger x $URu_2Si_{2-x}P_x$ substitution series will be reported separately). Unlike for other tuning strategies[22–33], magnetism is distant from hidden order in this phase diagram.

The resistive superconducting transition temperature $T_{c,\rho}$ (Fig. 1c) initially weakly increases with x and subsequently vanishes, with no evidence for bulk superconductivity above 20 mK for $x > 0.02$. While the value of $T_{c,C}$ extracted from heat capacity is in close agreement with $T_{c,\rho}$ for $x \leq 0.01$, these values separate for $x = 0.02$, where $T_{c,\rho} > T_{c,C}$. We note that a similar discrepancy between $T_{c,\rho}$ and $T_{c,C}$ is seen for high quality single crystal specimens of the correlated electron superconductor CeIrIn$_5$ and may be an intrinsic feature of the unconventional superconducting state[35]. From both ρ and C/T, we find that for $x = 0.028$ there is a transition into the hidden order state near 13.5 K, but no bulk superconductivity down to 20 mK. There is no evidence for superconductivity in $x = 0.035$ for $T > 20$ mK.

Silicon to phosphorus phase diagram. These results are summarized in Fig. 2a, where the superconducting region is enclosed by hidden order in the $T - x$ phase space. Over this concentration range, the ground state is mainly tuned by electronic variation, as indicated by the comparably small changes in other intrinsic and extrinsic factors. The lowest residual resistivity ratio ($RRR \approx \rho_{300\,K}/\rho_0$) for the specimens reported here is $RRR = 10$ (see Supplementary Fig. 3), which is comparable with the typical values for parent URu$_2$Si$_2$, where T_0 and T_c depend weakly on RRR in the range 10–500 (ref. 34). The high crystal–chemical quality of these specimens is further highlighted by the observation of quantum oscillations in electrical transport measurements (see Supplementary Fig. 4), indicating that disorder effects are negligible. The unit cell volume and bond angles are also unchanged by phosphorus substitution (Fig. 2e), in contrast to some previous studies[26,27].

Having established the $T - x$ phase diagram (Fig. 2a), we now discuss the region beneath the hidden order phase boundary, where unexpectedly rich behavior occurs. As evidenced by the jump size in C_{5f}/T at T_c ($\Delta C_{5f}/T_c$) (and the transition width), which evolve though a maximum (and a minimum) between 0.006 and 0.01, respectively (Figs 1e and 2b), phosphorus substitution non-monotonically enhances the thermodynamic signature of the superconductivity. Here, C_{5f} refers to the heat capacity following subtraction of the nonmagnetic ThRu$_2$Si$_2$ lattice term, as described in the Supplementary section (Supplementary Fig. 5). The non-monotonic behavior is reflected in the behavior of $S_{5f,T_c}(x)$ (Fig. 2c), which goes through a maximum near $x = 0.01$. In contrast, for the hidden order $\Delta C/T_0$ and $S_{5f,T0}$ are monotonically suppressed with increasing x (see Supplementary Fig. 6). Further evidence for non-monotonic evolution of the superconductivity is provided by the doping evolution of the ratio $\zeta = \Delta C_{5f}/\gamma T_c(x)$, (Fig. 2b) which, for conventional superconductors, is a numeric constant $\zeta_{BCS} = 1.43$. By using C_{5f}/T at T_c for the value of the normal state γ we find that $\zeta(x)$ evolves non-monotonically through a maximum value of 1.2 at $x = 0.01$ (Fig. 2b), where the $x = 0$ value is near 0.7 as previously reported[18,19]. This suggests that the superconducting coupling strength may evolve through a maximum. We note that there is no evidence that all electrons from phosphorus substitution contribute directly to the conduction bands that are important for superconductivity.

Electrical transport in magnetic fields. Magnetoresistance data (Fig. 3) show quantum oscillations, emphasizing the high quality of these specimens (see Supplementary Fig. 4). Similar to the parent compound, the upper critical field H_{c2} (Fig. 3d) is highly aniso-tropic at all x and follows $H_{c2}(\theta) \propto 1/\sqrt{g_c^2\cos^2\theta + g_a^2\sin^2\theta}$ dependence (where θ is measured from the c-axis), suggesting that the upper critical field is Pauli limited[5]. While there is little x dependence in $g_c(x)$, the a-axis g-factor $g_a(x)$ significantly decreases before the superconductivity is destroyed near $x \approx 0.028$ (Fig. 3d). We note that the actual value of the g-factor in the a-direction may differ significantly from the fitted $g_a(x)$ because of the increased importance of diamagnetic effects as the field rotates into the ab-

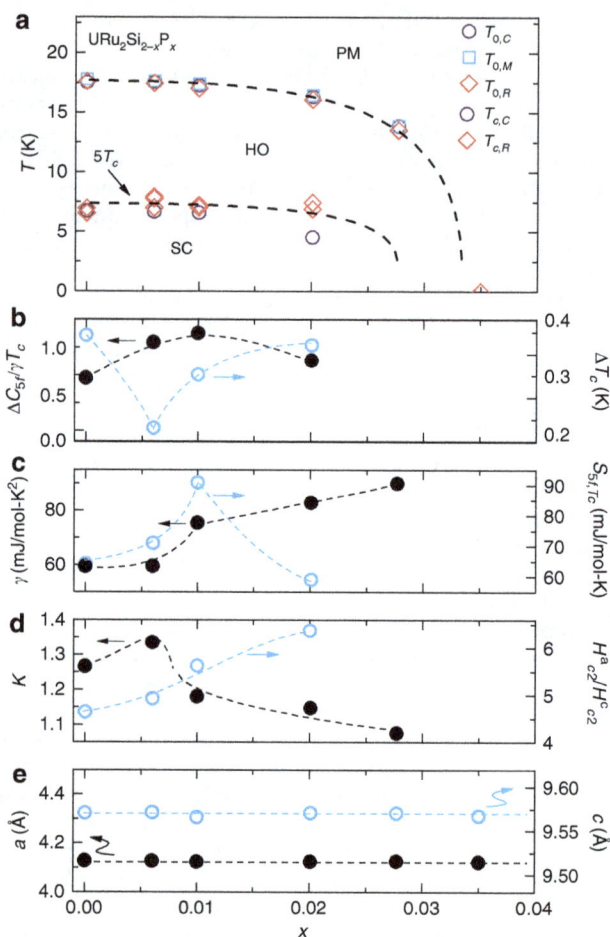

Figure 2 | Summary of physical quantities for URu$_2$Si$_{2-x}$P$_x$.
(**a**) Temperature T versus phosphorus concentration x phase diagram for URu$_2$Si$_{2-x}$P$_x$ constructed from heat capacity (circles), magnetic susceptibility (squares) and electrical resistance (diamonds). The $T - x$ phase boundary $T_0(x)$ separates the paramagnetic heavy electron liquid phase from the hidden order phase. $T_c(x)$ separates the hidden order and superconducting phases. The dotted lines are guides to the eye. (**b**) Left axis: The size of the discontinuity in the heat capacity divided by the superconducting transition temperature T_c and the electronic coefficient of the heat capacity γ, $\Delta C_{5f}/\gamma T_c$ versus x. Right axis: The width of the superconducting phase transition ΔT_c versus x. (**c**) Left axis: The electronic coefficient of the heat capacity γ versus x. Right axis: The $5f$ contribution to the entropy S_{5f} at T_c versus x. (**d**) Left axis: The value of the Kohler scaled curve at $H/R_0 = 50$ versus x. Right axis: The anisotropy of the upper critical field curves H_{c2}^a/H_{c2}^c versus x. (**e**) The lattice constants, $a(x)$ (left axis) and $c(x)$ (right axis), obtained from single crystal X-ray diffraction measurements.

plane. It remains to be seen whether these trends are consistent with recent theoretical proposals such as ref. 6.

Magnetoresistance measurements further highlight the non-monotonic evolution with x of the superconductivity and the underlying metallic state. Figure 3e demonstrates Kohler scaling for $H < 9$ T applied parallel to the c-axis at all dopings[36], suggesting that the magnetotransport is controlled by the same (temperature dependent) relaxation time as the zero field resistivity. At each composition in the range $0 < x < 0.028$ the normalized magnetoresistance is described by a distinct $f_x(h)$ (where $h = H/\rho(0, T)$), which itself evolves with doping. Notably, the function $f_x(h)$ evolves non-monotonically with x with a maximum near $x = 0.006$ (Fig. 2d). The maximum in the value of

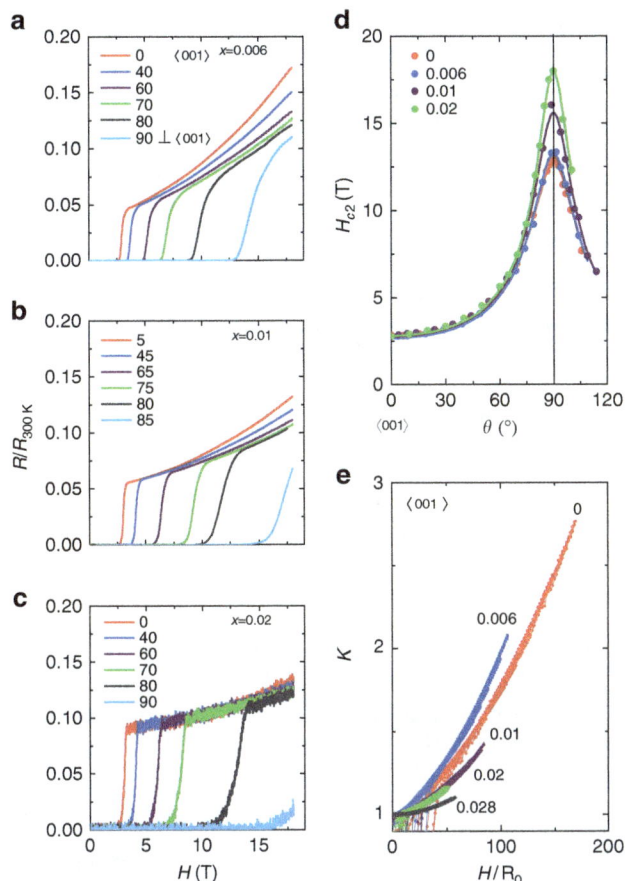

Figure 3 | Electrical transport in magnetic fields for $URu_2Si_{2-x}P_x$. (a) The $x = 0.006$ electrical resistance normalized to the room temperature value $R/R_{300\,K}$ versus magnetic field H for several different angles θ. The data were collected at the temperature $T = 20$ mK. The electrical current was applied in the ab-plane and $\theta = 0$ is the configuration where, H is parallel (||) the crystallographic c-axis. **(b)** $R/R_{300\,K}$ versus H for $x = 0.01$ at $T = 20$ mK for select θ. **(c)** $R/R_{300\,K}$ versus H at $T = 20$ mK for $x = 0.02$ for select θ. **(d)** The upper critical field H_{c2}, defined as the extrapolated zero resistance intercept, for $T = 20$ mK for $0 \leq x \leq 0.02$. Data for $x = 0$ is from ref. 38. **(e)** The Kohler scaled electrical resistivity $K = \frac{\rho(H,T)}{\rho(0,T)}$ versus the reduced field H/R_0 for $0 \leq x \leq 0.028$.

$f_x(h)$ nearly coincides with the maximum in the thermodynamic signatures of the superconductivity inside the hidden order phase.

Discussion

Although much of the recent excitement surrounding URu_2Si_2 has focused on the uranium electronic structure and the symmetry of the hidden-order phase, a more fundamental question is the degree to which the f-electrons can be treated as being localized and the role of valence fluctuations. The continuity of experimental information extracted from well-developed applied pressure (P) and chemical substitution (x) series has proven essential to disentangle such effects in other correlated systems including high temperature superconducting cuprates, pnictides and heavy fermion compounds. To some extent, URu_2Si_2 has also benefited from such studies. For example, pressure drives a first order phase transition from hidden order into antiferromagnetism near $P_c = 0.5$ GPa, with a simultaneous evolution of the Fermi surface[22-25], but the resulting insight is limited by the small number of pressure-cell compatible experimental probes. Ruthenium site substitution

with Fe and Os produces $T - x$ phase diagrams that closely resemble the $T - P$ phase diagram[26-28], but the information gained from these series is constrained by strong disorder. Moreover, ruthenium site substitution is particularly disruptive, as evidenced by the rather different phase diagrams resulting from Rh and Re substitution studies where the hidden order and superconductivity are rapidly destroyed[29-33]. To understand the complex interplay between different phenomena in this compound a more 'gentle' tuning scheme has long been desired, which could provide access to the physics of URu_2Si_2 in clean single crystals at ambient pressure. In this context, ligand site substitution in URu_2Si_2 is an obvious target for investigation.

While in many theoretical scenarios for hidden order in URu_2Si_2 the U-5f electrons are treated as having mostly fixed valence in a particular atomic crystal field state[6,11], it is now believed that they have a dual character: that is, the dynamic nature of the U-5f valence electrons allows for fluctuations between different configurations. However, measurements of the pure compound so far give no insight into the role of these fluctuations in favouring hidden order and superconductivity. The rapid changes in the hidden order and superconductivity in our measurements confirm the importance of the itinerant electron states. This is further supported by the weak evolution of Kondo lattice physics[37] (which tracks the hybridization strength between f- and conduction electrons) and strong evolution in the g-factor anisotropy (which is a marker for local moment character). Together these results point towards this series as a platform for unraveling the relationship between local and itinerant behavior in URu_2Si_2.

The stark contrast in the evolution of the hidden order and superconductivity in $URu_2Si_{2-x}P_x$ (monotonic versus non-monotonic) further suggests that hidden order, although necessary, may not be directly responsible for the superconducting pairing. Instead, the observation of superconductivity completely contained inside the hidden order region and non-monotonic evolution of thermodynamic and electrical transport properties may indicate the presence of an independent collapsing phase boundary within the hidden order state, as is ubiquitous in other unconventional superconductors. This scenario is reinforced by the observed non-monotonic evolution of the normal state electrical transport, which is also common in correlated electron systems[18-21], where the strongest deviation from Fermi liquid behavior is seen near the critical point. Alternatively, the independent evolution of hidden order and superconductivity may suggest several competing order parameters in the hidden order phase, as evidenced by electronic Raman (A_{2g})[11], elastoresistance (B_{2g})[12], resonant ultrasound (B_{1g})[13] (Ramshaw, B.J., Private communication.) and spectroscopic measurements (E_g)[14]. These studies should be extended into ligand site substituted URu_2Si_2.

Finally, the existing theoretical landscape focuses on f-electron physics with no guidance regarding the specificity of the transition metal ion. It is especially puzzling that hidden order and superconductivity are only observed in the U–Ru duo. Examination of silicon site substituted transition metal analogues ($UT_2Si_{2-x}P_x$, $T = $ transition metal), which can now be synthesized using molten metal flux growth (We have already successfully synthesized several transition metal analogues UT_2Si_2 ($T = $ Mn, Fe, Ir and Pt) using the same flux growth recipe.), may be particularly illuminating in addressing the universality of hidden order and superconductivity in this fascinating uranium compound.

Methods

Single crystal synthesis using molten indium flux. Single crystals of $URu_2Si_{2-x}P_x$ were grown from elements with purities > 99.9% in a molten in flux, as previously reported[34]. The reaction ampoules were prepared by loading the elements into a 5 cm^3 tantalum crucible in the ratio 1(U):2(Ru):2(Si):22(In). The crucible was then loaded into an alumina tube spanning the bore of a high-temperature horizontal tube furnace. Argon gas was passed through the tube and a zirconium getter was placed in a pot before the tantalum crucible in order to purify the argon at high temperatures. The crucible was heated to 500 °C at 50 °C/h, dwelled for 5 h, heated to 600 °C at 50 °C/h, dwelled for 5 h and heated to 1,450 °C at 70 °C/h. The dwells at intermediate temperature are intended to allow the phosphorus to completely dissolve into the indium flux without producing a dangerous high-vapor pressure. The crucible was then cycled between 1,450 and 1,400 °C at 100 °C/h 10 times. Finally, the furnace was turned off and quickly cooled to room temperature. The indium flux was subsequently removed using hydrochloric acid, to which the $URu_2Si_{2-x}P_x$ crystals are insensitive. This technique produced single-crystal platelets similar to the ones previously reported.

Bulk thermodynamic and electrical transport measurements. Heat capacity measurements were performed for mosaics of single crystals using the He3 option in a Quantum Design Physical Properties Measurement System for temperatures 400 mK $< T <$ 20 K. Magnetization $M(T, H)$ measurements were carried out for mosaics of single crystals for temperatures $T = 1.8$–350 K under an applied magnetic field of $H = 5$ kOe applied parallel to the c-axis using a Quantum Design Magnetic Property Measurement System. Magnetic susceptibility χ is defined as the ratio M/H. Zero magnetic field electrical resistance R was measured using the He3 option in Quantum Design Physical Properties Measurement System for temperatures 400 mK $< T <$ 300 K. Several individual crystals were measured for each concentration, which revealed a high degree of batch uniformity. The angular dependence of the superconducting upper critical field was measured using the superconducting magnet (SCM-1) dilution refrigerator system at the National High Magnetic Field Laboratory for $H <$ 18 T and $T = 20$ mK. Additional magnetoresistance measurements were performed at the National High Magnetic Field Laboratory, Tallahassee, up to magnetic fields of 35 T and at $T = 50$ mK.

References

1. Herring, C. The state of d electrons in transition metals. *J. Appl. Phys.* **31,** S3 (1960).
2. Mott, N. in *Progress in Metal Physics* 3 (ed. Chalmers, B.) (London Pergamon Press, 1952).
3. Keimer, B., Kivelson, S. A., Norman, M. R., Uchida, S. & Zaanen, J. From quantum matter to high-temperature superconductivity in copper oxides. *Nature* **518,** 179 (2015).
4. Moore, K. T. & van der Laan, G. Nature of the 5f states in actinide metals. *Rev. Mod. Phys.* **81,** 235 (2009).
5. Altarawneh, M. M. *et al.* Superconducting pairs with extreme uniaxial anisotropy in URu_2Si_2. *Phys. Rev. Lett.* **108,** 066407 (2012).
6. Chandra, P., Coleman, P. & Flint, R. Hastatic order in the heavy-fermion compound URu_2Si_2. *Nature* **493,** 621 (2013).
7. Palstra, T. T. M. *et al.* Superconducting and magnetic transitions in the heavy-fermion system URu_2Si_2. *Phys. Rev. Lett.* **55,** 2727 (1985).
8. Schlabitz, W. *et al.* Superconductivity and magnetic order in a strongly interacting fermi-system: URu_2Si_2. *Z. Phys. B* **62,** 171 (1986).
9. Maple, M. B. *et al.* Partially gapped Fermi surface in the heavy-electron superconductor URu_2Si_2. *Phys. Rev. Lett.* **56,** 185 (1986).
10. Mydosh, J. A. & Oppeneer, P. M. Colloquium: Hidden order, superconductivity, and magnetism: The unsolved case of URu_2Si_2. *Rev. Mod. Phys.* **83,** 1301 (2011).
11. Kung, H.-H. *et al.* Chirality density wave of the "hidden order" phase in URu_2Si_2. *Science* **347,** 1339 (2015).
12. Riggs, S. C. *et al.* Evidence for a nematic component to the hidden-order parameter in URu_2Si_2 from differential elastoresistance measurements. *Nat. Commun.* **6,** 6425 (2015).
13. Yanagisawa, T. *et al.* Hybridization-driven orthorhombic lattice instability in URu_2Si_2. *Phys. Rev. B* **88,** 195150 (2013).
14. Wray, L. A. *et al.* Spectroscopic determination of the atomic f-electron symmetry underlying hidden order in URu_2Si_2. *Phys. Rev. Lett.* **114,** 236401 (2015).
15. Tonegawa, S. *et al.* Direct observation of lattice symmetry breaking at the hidden-order transition in URu_2Si_2. *Nat. Commun.* **5,** 4188 (2014).
16. Okazaki, R. *et al.* Rotational symmetry breaking in the hidden-order phase of URu_2Si_2. *Science* **331,** 439 (2011).
17. Schemm, E. R. *et al.* Evidence for broken time-reversal symmetry in the superconducting phase of URu_2Si_2. *Phys. Rev. B* **91,** 140506 (2015).
18. Löhneysen, H. v., Rosch, A., Vojta, M. & Wölfle, P. Fermi-liquid instabilities at magnetic quantum phase transitions. *Rev. Mod. Phys.* **79,** 1015 (2007).
19. Pfleiderer, C. Superconducting phases of f-electron compounds. *Rev. Mod. Phys.* **81,** 1551 (2009).
20. Tallon, J. L. & Loram, J. W. The doping dependence of T*—what is the real high-Tc phase diagram? *Physica C* **349,** 53 (2001).
21. Walmsley, P. *et al.* Quasiparticle mass enhancement close to the quantum critical point in $BaFe_2As_{(1-x}P_x)_2$. *Phys. Rev. Lett.* **110,** 257002 (2013).
22. McElfresh, M. W. *et al.* Effect of pressure on competing electronic correlations in the heavy-electron system URu_2Si_2. *Phys. Rev. B* **35,** 43 (1987).
23. Amitsuka, H. *et al.* Effect of pressure on tiny antiferromagnetic moment in the heavy-electron compound URu_2Si_2. *Phys. Rev. Lett.* **83,** 5114 (1999).
24. Jeffries, J. R., Butch, N. P., Yukich, B. T. & Maple, M. B. Competing ordered phases in URu_2Si_2: Hydrostatic pressure and rhenium substitution. *J. Phys. Condens. Matt.* **20,** 095225 (2008).
25. Hassinger, E. *et al.* Similarity of the fermi surface in the hidden order state and in the antiferromagnetic state of URu_2Si_2. *Phys. Rev. Lett.* **105,** 216409 (2010).
26. Kanchanavatee, N. *et al.* Twofold enhancement of the hidden-order/large-moment antiferromagnetic phase boundary in the $URu_{2-x}Fe_xSi_2$ system. *Phys. Rev. B* **84,** 245122 (2011).
27. Kanchanavatee, N., White, B. D., Burnett, V. W. & Maple, M. B. Enhancement of the hidden order/large moment antiferromagnetic transition temperature in the $URu_{2-x}Os_xSi_2$ system. *Phil. Mag.* **94,** 3681 (2014).
28. Das, P. *et al.* Chemical pressure tuning of URu_2Si_2 via isoelectronic substitution of Ru with Fe. *Phys. Rev. B* **91,** 085122 (2015).
29. Amitsuka, H., Hyomi, K., Nishioka, T., Mitakato, Y. & Suzuki, T. Specific heat and susceptibility of $U(Ru_{1-x}Rh_x)_2Si_2$. *J. Mag. Magn. Mat.* **76,** 168 (1988).
30. Dalichaouch, Y. *et al.* Effect of transition-metal substitutions on competing electronic transitions in the heavy-electron compound URu_2Si_2. *Phys. Rev. B* **41,** 1829 (1990).
31. Dalichaouch, Y., Maple, M. B., Torikachvili, M. S. & Giorgi, A. L. Ferromagnetic instability in the heavy-electron compound URu_2Si_2 doped with Re or Tc. *Phys. Rev. B* **39,** 2423 (1989).
32. Bauer, E. D. *et al.* Non-Fermi-liquid behavior within the ferromagnetic phase in $URu_{2-x}Re_xSi_2$. *Phys. Rev. Lett.* **94,** 046401 (2005).
33. Butch, N. P. & Maple, M. B. The suppression of hidden order and the onset of ferromagnetism in URu_2Si_2 via Re substitution. *J. Phys. Condens. Matt.* **22,** 164204 (2010).
34. Baumbach, R. E. *et al.* High purity specimens of URu_2Si_2 produced by a molten metal flux technique. *Phil. Mag.* **94,** 3663 (2014).
35. Petrovic, C. *et al.* A new heavy-fermion superconductor $CeIrIn_5$: A relative of the cuprates? *Europhys. Lett.* **53,** 354 (2001).
36. Kohler, V. M. Zur magnetischen Widerstandsänderung reiner Metalle. *Ann. Phys.* **32,** 211 (1938).
37. Hewson, A. C. *The Kondo Problem to Heavy Fermions* (Cambridge University Press, 1993).
38. Ohkuni, H. *et al.* Fermi surface properties and de Haas-van Alphen oscillations in both the normal and superconducting mixed states of URu_2Si_2. *Phil. Mag. B* **79,** 1045 (1999).

Acknowledgements

This work was performed at the National High Magnetic Field Laboratory (NHMFL), which is supported by National Science Foundation Cooperative agreement number DMR-1157490, the State of Florida and the DOE. A portion of this work was supported by the NHMFL User Collaboration Grant Program (UCGP). TAS and SC acknowledge support from the US Department of Energy, Office of Science, Office of Basic Energy Sciences, Heavy Elements Chemistry Program, under award number DE-FG02-13ER16414.

Author contributions

R.E.B., A.S. and S.C.R. conceived and designed the experiments and supervised measurements. A.G. synthesized single crystals. A.G., K.-W.C., C.M.M., N.K. and D.G. performed electrical transport and heat capacity measurements. F.K. performed chemical analysis measurements. S.K.C. and T.E.A.-S. performed single crystal x-ray diffraction measurements.

Additional information

Strain-relief by single dislocation loops in calcite crystals grown on self-assembled monolayers

Johannes Ihli[1,*,†], Jesse N. Clark[2,3,*], Alexander S. Côté[4,†], Yi-Yeoun Kim[1], Anna S. Schenk[1], Alexander N. Kulak[1], Timothy P. Comyn[5], Oliver Chammas[6], Ross J. Harder[7], Dorothy M. Duffy[4], Ian K. Robinson[8] & Fiona C. Meldrum[1]

Most of our knowledge of dislocation-mediated stress relaxation during epitaxial crystal growth comes from the study of inorganic heterostructures. Here we use Bragg coherent diffraction imaging to investigate a contrasting system, the epitaxial growth of calcite ($CaCO_3$) crystals on organic self-assembled monolayers, where these are widely used as a model for biomineralization processes. The calcite crystals are imaged to simultaneously visualize the crystal morphology and internal strain fields. Our data reveal that each crystal possesses a single dislocation loop that occupies a common position in every crystal. The loops exhibit entirely different geometries to misfit dislocations generated in conventional epitaxial thin films and are suggested to form in response to the stress field, arising from interfacial defects and the nanoscale roughness of the substrate. This work provides unique insight into how self-assembled monolayers control the growth of inorganic crystals and demonstrates important differences as compared with inorganic substrates.

[1] School of Chemistry, University of Leeds, Woodhouse Lane, Leeds LS2 9JT, UK. [2] Stanford PULSE Institute, SLAC National Accelerator Laboratory, 2575 Sand Hill Road, Menlo Park, California 94025, USA. [3] Center for Free-Electron Laser Science (CFEL), Deutsches Elektronensynchrotron (DESY) Notkestrasse 85, 22607 Hamburg, Germany. [4] Department of Physics and Astronomy, University College London, Gower Street, London WC1E 6BT, UK. [5] Institute for Materials Research, University of Leeds, Leeds LS2 9JT, UK. [6] School of Physics and Astronomy, University of Leeds, Leeds LS2 9JT, UK. [7] Advanced Photon Source, Argonne, Illinois 60439, USA. [8] London Centre for Nanotechnology, University College London, 17–19 Gordon Street, London WC1H 0AH, UK. * These authors contributed equally to this work. † Present addresses: Paul Scherrer Institute, 5232 Villigen PSI, Switzerland. (J.I.); London Centre for Nanotechnology, University College London, 17–19 Gordon Street, London WC1H 0AH, UK (A.S.C.). Correspondence and requests for materials should be addressed to J.I. (email: johannes.ihli@psi.ch) or to I.K.R. (email: i.robinson@ucl.ac.uk) or to F.C.M. (email: F.Meldrum@leeds.ac.uk).

The control of crystal growth at interfaces is fundamental to a wide range of processes of scientific, environmental and technological importance. The epitaxial growth of crystals, where a single crystal of one compound grows with a unique orientation on the surface of a second[1], attracts particular interest. This phenomenon has received considerable attention for the fabrication of thin film and nanoparticulate semiconductor, ferroelectric and superconducting devices, where the properties of these heterostructures can be tuned according to the degree of strain introduced into the supported thin film[2–4]. The strain, in turn, arises from the lattice mismatch present between the substrate and crystalline thin film, where this increases in value until a critical film thickness is reached[5]. Beyond this point, the strain energy can be relieved through the introduction of misfit dislocations, where these must have edge character parallel to the substrate/crystal interface to reduce strain[1,6,7]. Due to their technological importance and suitability for study with techniques such as transmission electron microscopy (TEM), these systems have provided the vast majority of our current knowledge about stress relaxation and dislocation formation during epitaxial crystal growth.

In the work described here, we profit from recent advances in imaging methods[8] to investigate strain and associated dislocation formation in a quite different example of epitaxial crystal growth—the growth of calcite ($CaCO_3$) crystals on organothiol self-assembled monolayers (SAMs). The ability of organized organic matrices such as Langmuir monolayers[9], Langmuir–Shaeffer films[10] and SAMs to direct the orientation, and sometimes even the polymorph, of inorganic crystals has received significant attention, where these systems provide excellent models for biomineralization processes such as the formation of mollusc shell nacre. Of these studies, the precipitation of calcite on organothiol SAMs on coinage metals is the best-characterized, where the nucleation plane can be selected according to the SAM chain length[11], packing geometry/tilt[12], the terminal group[13] and its degree of ionization[14,15], and the type of metal substrate[13,16]. That the precipitated calcite crystals are co-aligned within monocrystalline Au {111} domains demonstrates an epitaxial relationship between the SAM and the crystal lattice[17], and selection of the nucleation face has been proposed to arise from a stereochemical match between the orientation of the SAM headgroups and the ions in the crystal nucleation face[13,18]. Subsequent studies have refined this view and have suggested that templating of inorganic crystals by the organic interface is a cooperative process[12,19,20], in which structural feedback between the crystal and monolayer ensures selection of the most favourable combination of the orientation of the SAM and crystal[21–25].

However, many questions remain regarding the mechanisms by which organic matrices can control crystallization. Indeed, while Langmuir–Shaeffer films[10] and SAMs[13,19] have been observed to change conformation when directing the growth of oriented calcite, little is known about whether the organic matrix can cause parallel changes in the crystal lattice. The work described in this article uses Bragg coherent diffraction imaging (BCDI)[8] to address this question. BCDI provides a unique method for simultaneously visualizing both the morphology of a crystal, and the strain fields within it at a spatial resolution of ~100 nm, without the requirement for sample preparation methods. Our data demonstrate how a 'soft' organic matrix can inform the structure of a 'hard' inorganic crystal and show that heterogeneous nucleation of a calcite single crystal on a SAM results in deformation of the calcite crystal lattice. As a key finding, we show that the stresses originating at the crystal/SAM interface give rise to the formation of a single dislocation loop within each crystal, where the geometry of this dislocation is entirely different to that of the misfit dislocation loops, characteristically seen in purely inorganic epitaxial heterostructures. The roughness of the substrate—where this is intrinsic to all SAMs prepared on evaporated metal films—is considered to lie at the heart of this effect. These results, therefore, provide a unique insight into the mechanisms by which SAMs control both the growth and defect structures of crystals, where the latter is intimately linked to the mechanical properties.

Results

Morphological development of calcite crystals on SAMs. Oriented calcite crystals were precipitated on 11-mercaptoundecanoic acid SAMs on Au (111)/Si (001) using a hanging drop set-up, in which $200\,\mu l$ drops of $5\,mM$ $CaCl_2$ solution were suspended from a SAM (Fig. 1)[26]. These substrates were then placed in a closed desiccator containing solid ammonium carbonate, and crystallization was allowed to proceed for up to 30 min. Characterization of the crystal orientations using acquired pole measurements showed that the majority were oriented with the {012} or the {113} plane parallel to the SAM (Supplementary Fig. 1). Analysis of the morphological development of the crystals demonstrated that the initial form was roughly pyramidal and that it appears to comprise an aggregate of smaller particles (Fig. 1). This is indicative of crystallization via an amorphous calcium carbonate precursor phase, as is expected under these reaction conditions[12,27]. These particles then convert to irregular tetrahedra with three, well-defined {104} faces directed into the solution, while further growth leads to the truncation of the vertex, where the longest sides meet. This generates an additional {104} face. Finally, as growth normal to the substrate begins to dominate over growth adjacent to the substrate, the crystals undergo a morphological transition to full rhombohedra.

This sequence of morphologies can be readily explained in terms of surface free energy minimization, where the surface-to-volume ratio of the calcite crystals and the relative contributions of the crystal/solution and the crystal/SAM interfaces change during growth. Winterbottom constructions, which provide a phenomenological prediction of the equilibrium shapes of crystals located on solid substrates, were performed on an interface for a fixed crystal orientation and volume, under variation of the SAM/crystal interfacial energy. A comparable morphological transition to that seen experimentally was observed under the increasing interfacial energy, as reflects the decreasing influence of the substrate/crystal interface on the crystal morphology during growth (Fig. 1). Significantly, a larger variation between the experimental and predicted morphologies was observed at later growth stages, where the Winterbottom constructions terminate with a regular rhombohedron, as compared with the elongated rhombohedron seen experimentally[28].

Bragg coherent diffraction imaging. Calcite crystals of sizes $1–4\,\mu m$ were characterized at different stages of the morphological development using BCDI, where this provides a simultaneous visualization of the crystal morphology and its internal strain. BCDI is achieved by illumination of a crystal with a coherent X-ray beam, whose coherence volume is larger than that of the crystal. A series of two-dimensional (2D) diffraction patterns is collected around a selected Bragg reflection for different points on the rocking curve, and is used to generate a coherent, three-dimensional (3D) X-ray diffraction pattern, which is formed by scattering from all parts of the crystal. In this lens-free form of microscopy, analysis of the coherent, 3D X-ray diffraction pattern through application of iterative phase-retrieval algorithms[29] generates a complex-valued reconstructed crystal

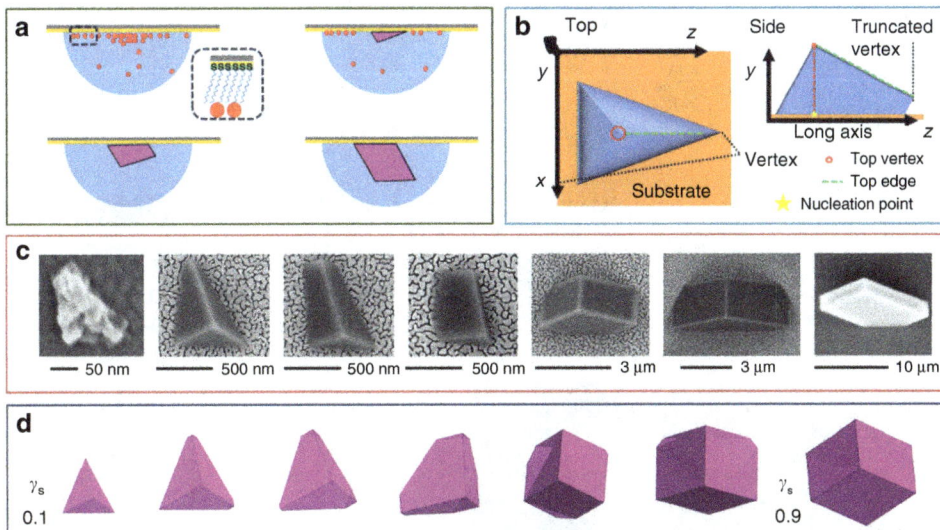

Figure 1 | Morphological progression of calcite nucleated on COOH terminated SAMs. (a) A schematic of the experimental set-up, where CO_2 and NH_3 diffuse into hanging droplets of $CaCl_{2(aq)}$, causing amorphous calcium carbonate formation (red dot). An oriented tetrahedron of calcite bounded by planar {104} faces then forms, whose growth leads to the development of an additional facet as a truncation of the long axis. Further growth then results in a transformation to rhombohedral calcite. **(b)** Schematic images of the tetrahedral growth form, showing the location of the truncation face. **(c)** Scanning electron microscope images showing the morphological development of the calcite crystals. **(d)** Winterbottom reconstructions predicting the morphological development of the calcite crystals, where these are of identical volume with stepwise increasing relative interfacial energy of the crystal/SAM (γ_s; 0.1–0.9). Crystal/ water interfacial energy values used were taken from Duffy and colleagues[36].

$\rho(r)$. The reconstructed amplitude provides a 3D representation of the specimen's electron density distribution $|\rho(r)|$, where this is sensitive to the crystallinity of the specimen. The phase shifts in the reconstructed complex amplitude, $\arg[\rho(r)] = \phi(r)$, are sensitive to small variations in lattice deformation and are proportional to a projection of the vector displacement field $u(r)$ of the atoms (from the ideal lattice points) and the scattering vector Q via $\phi(r) = u(r) Q$ (ref. 30). Here, a series of 2D diffraction patterns were collected from oriented calcite crystals at an angle corresponding to the off-specular {104} reflection. Stacks of 2D diffraction patterns were then inverted using an approach based on guided phase retrieval[31,32]. Detailed descriptions of the BCDI experiments and image reconstructions are provided in the Method section and the Supplementary Methods[8].

BCDI and TEM analysis of calcite crystals on SAMs. Representative BCDI reconstructions from two randomly selected calcite crystals are shown in Fig. 2. The images shown are top-down and bottom-up projections of the iso-surface renderings of (Fig. 2a) the reconstructed electron densities (amplitudes), where this provides a visualization of the crystal morphologies and (Fig. 2b) the projected displacements (phase), which correspond to lattice strains. The displacements are represented by a cyclic colour map projected onto the recorded electron density. A colour shift towards red ($+ d/2$) corresponds to a lattice contraction, while a shift towards blue ($- d/2$) equates to lattice dilation, where d is equal to the spacing between adjacent lattice planes. A colour change across the whole scale corresponds to a displacement of one unit cell in a particular direction. The crystals labelled (i) and (ii) correspond to different stages of growth. Crystal (i) is 1.4 μm in size, approximately tetrahedral in shape, and shows the beginning of a new truncation face. At 2 μm in size, crystal (ii) appears to be at a later stage of development, where it shows smoother faces and a well-defined {104} truncation. In addition, the 'top edge' of this crystal is now almost parallel to the substrate (as is also seen at later stages in the Winterbottom constructions).

The BCDI reconstructions also provide a unique opportunity to examine the influence of the SAM on the crystal structure. In both crystals, the face adjacent to the SAM exhibits a degree of roughness that is consistent with atomic force microscopy measurements of the SAM functionalized gold substrate (Supplementary Fig. 2)[33]. That the nucleation face of the calcite crystals are themselves roughened suggests that the crystal grows, so as to preserve interfacial contact. Interestingly, examination of the nucleation faces of crystals (i) and (ii) also reveals the presence of two adjacent surface cusps of sizes ~ 70–100 nm on each face. These intriguing features can be seen more clearly in cross-sections of the crystals (Fig. 2c), which show that they lie in a plane approximately parallel to the truncated vertex. Confirmation that these cusps correspond to physical features in the crystals was obtained by TEM of thin sections prepared by focused ion beam (FIB) milling. Figure 3 shows electron micrographs of a prepared section, and its location with respect to the original crystal. These reveal a linear feature of length 85 nm and width 10–20 nm (arrowed), whose location is commensurate with the surface cusps observed using BCDI (Fig. 3c).

The strain present within the calcite crystals (as inferred from the projected displacements; Fig. 2b,d) shows that lattice deformation is concentrated at the edges and corners of each of the crystals. This strain becomes more localized around the corners as the crystal grows in size. These distributions of lattice displacements can be attributed to their elastic anisotropy, as supported by finite element (FE) modelling (Supplementary Fig. 3). Placing a {012} oriented, tetrahedral calcite crystal under a uniform surface stress of $1.5\,N\,m^{-1}$ resulted in a comparable displacement profile within the crystal to that observed experimentally.

Returning to the surface cusps visualized within crystals (i) and (ii), each of these are associated with localized strain fields that radiate from the substrate into the crystal. Importantly, these regions possess both a hollow core and a spiral phase/displacement (turquoise circle in Fig. 2d), where this combination of features identifies them as dislocations[8,32].

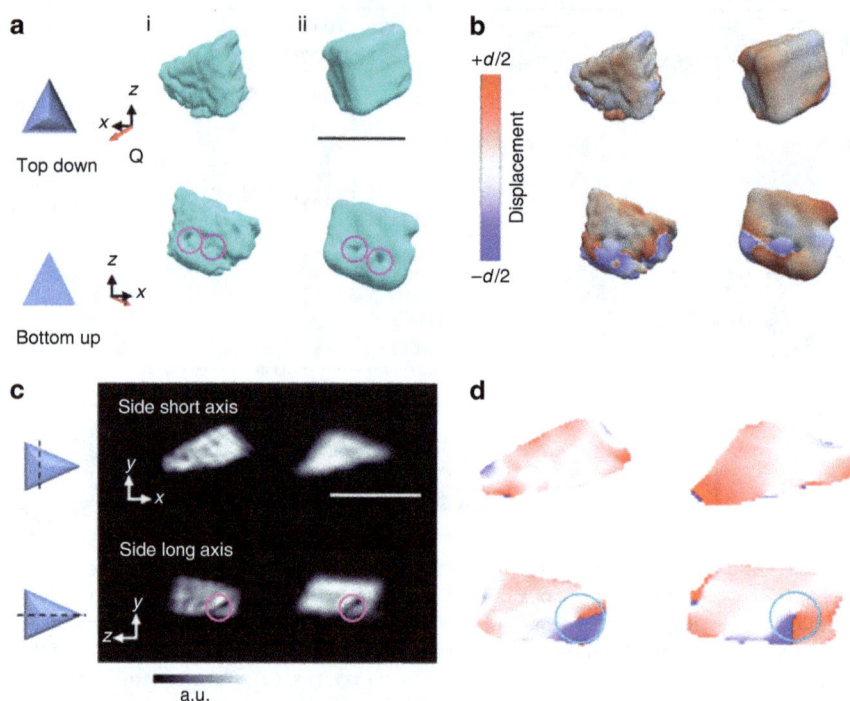

Figure 2 | Bragg coherent diffraction imaging (BCDI) reconstructions. Reconstructions of two different calcite crystals (i) and (ii), which were nucleated on carboxylate-terminated SAMs, are shown. The reconstructed crystal shapes from BCDI amplitude measurements are shown in **a**, where these are viewed from the directions indicated. The surface cusps, which appear on the bases of both crystals, are circled. (**b**) The projected displacements ($-d/2$ blue lattice dilation and $+d/2$ red lattice contraction) of the crystals. (**c**) Sections through the electron density maps and (**d**) sections cut through the displacement maps, where these made through the centres of the crystals, normal to the substrate. The surface cusps (magenta circle) and areas of spiral displacement (cyan circle) are highlighted. The beam direction is along the z axis, with the y axis vertical, while the sample/substrate is located at a set scattering angle towards the beam direction (z) with **Q** the scattering vector. Scale bar, 1.8 μm.

Further examination of the strain fields associated with these cusps then demonstrated that each pair of surface cusps actually form part/are the surface expressions of a single dislocation loop (Fig. 4i,ii; Supplementary Movies 1 and 2). As shown in Fig. 2d, the planes of these dislocation loops are approximately parallel to the new 'truncation' faces, which suggests that they are not randomly located. Notably, with the exception of these single dislocation loops, crystals (i) and (ii) appear mostly dislocation free.

Discussion

Crystallization at interfaces has been well studied for the epitaxial growth of crystalline thin films on solid substrates, and misfit dislocations often form in the growing layer to relieve the elastic strain associated with lattice mismatch with the substrate[1,5,6]. This strain energy increases, as the volume of the crystal increases, and when the stored elastic energy exceeds the energy cost of making a dislocation, dislocation loops may be nucleated to relieve the stress. In the absence of existing dislocations, such loops will generally nucleate at the free surface of the growing crystal, where the nucleation barriers are lower than in the bulk (Fig. 5a). The loops then grow under the influence of the misfit stress until they reach the interface[6]. The final loop has edge character parallel to the interface, to relieve misfit strain, and some screw character on the sections perpendicular to the interface.

Our BCDI study of calcite crystals grown on SAMs shows that, while dislocations also form in this system, they are entirely different from conventional misfit dislocations in that the ends of the loop lie at the crystal/substrate interface rather than the free surface of the crystal. The observed dislocation loops must either

have been created during crystal growth or have nucleated and grown in response to stress. Although we are not currently in a position to conclusively identify their origins, their location and orientation is strongly indicative of a mechanism in which the dislocation loop nucleates at the crystal/SAM interface and then expands on a slip plane into the interior of the crystal during crystal growth. That they terminate at the crystal/SAM interface, as opposed to a growth surface, appears to rule out a mechanism, where they form as a result of a growth defect. For that to be the case, two independent dislocations would have to grow in from the substrate, before curving round to form a continuous loop at a later stage of the growth. While it is possible that two dislocations that were sufficiently close could combine due to the interaction of their stress fields, the end of the ones formed here are ≈1 μm apart. Further, the two separated dislocations arms could only unite if their ends have the same Burgers vectors. A far more credible explanation is that the dislocations nucleate and grow under the internal stress field of the crystal to generate a loop. Further, their configuration is consistent with their nucleation at the base of the (012) oriented calcite tetrahedron, where the stress—which reduces the barrier to dislocation nucleation—is concentrated (Fig. 5b).

Both modelling and experimental studies have shown that the innate flexibility of SAMs is fundamental to their ability to support the epitaxial growth of calcite crystals. There is a very large lattice mismatch between an idealized 11-mercaptoundecanoic acid SAM on Au and the (012) calcite face, which can be compensated by a high density of surface vacancies or surface steps in the calcite[34]. If it is assumed that the SAM is commensurate with the gold substrate, then the lattice mismatch is 0.2% in the calcite <001> direction and 26% in the perpendicular <12$\bar{1}$> direction[14]. Unlike SAMs, which are

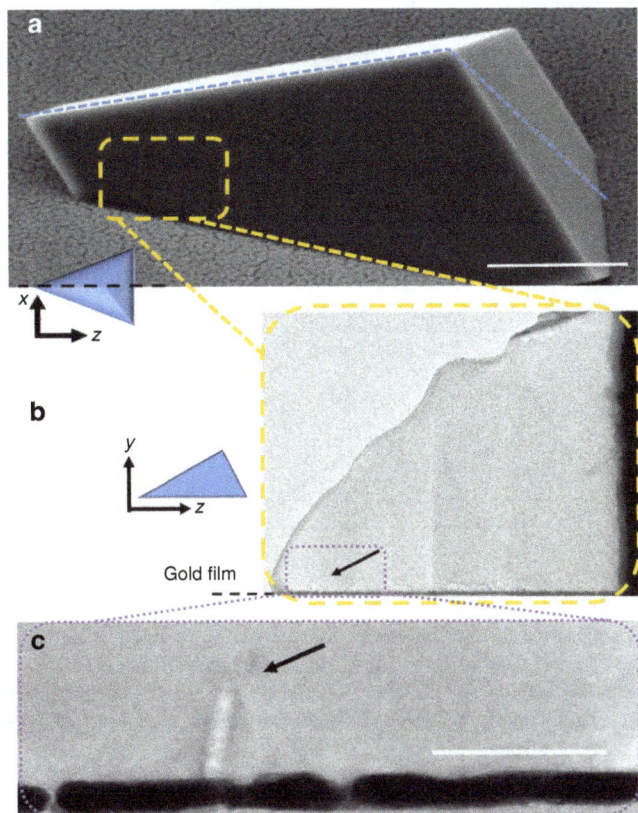

Figure 3 | Electron micrographs of oriented sections of precipitated calcite crystals. (**a**) Scanning electron microscope image of the sample crystal, where the dotted blue line shows the direction of the cut. Scale bar, 1 μm. (**b**) The selectively thinned tip of the prepared lamella, whose location with respect to the original crystal is indicated in the yellow box shown in **a**. (**c**) Shows a higher-magnification image of the front end of the tip (arrowed in **b**). Arrowed in **c** is a feature of size ∼85 × 15 nm, which corresponds to the 'surface cusps' imaged using BCDI. Scale bar, 100 nm.

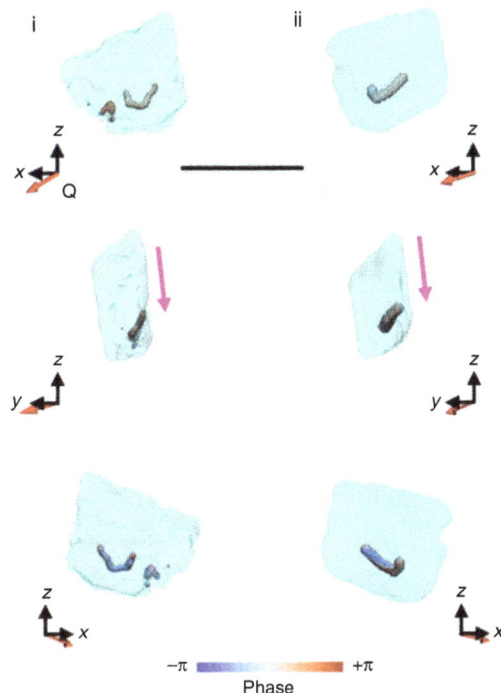

Figure 4 | Iso-surface rendering of the defects present within the oriented calcite crystals. The defects shown have a low electron density core that is surrounded by a spiral deformation field/phase, and correspond to dislocation loops. In both crystals (i and ii) the plane of the loop is directed towards the truncation facet (pink arrow). Defects are given in respect to their location within the crystal, as given by semi-transparent projections of the electron density. **Q** is the scattering vector. Scale bar, 1.8 μm.

easily deformable, calcite has a high elastic modulus and so it will not readily deform to match the SAM. A cycle of mutual control, in which both the organic and mineral components induce complementary local order across the interface then leads to the formation of a critical crystalline region, which defines the nucleation face.

Although stress due to interfacial defects undoubtedly contributes to the nucleation of the dislocation loops viewed here, the magnitude of the stress is unlikely to be high enough to overcome the barrier for dislocation growth. An additional source of internal stress is the roughness of the Au/SAM substrate, which has surface features of ≈50 × 10 nm. If the calcite crystal grows so as to maintain contact with the curved substrate, as suggested by our data, this will result in bending of the lattice planes close to the substrate, which will induce a complex stress field in the crystal. This, in turn, could be partially relieved by the formation of a dislocation. The total stored elastic energy in a crystal due to bending can be calculated using FE modelling by applying a stress to the base of the tetrahedron (Fig. 5c), while the elastic energy of a dislocation loop can be estimated ($E_{\text{dislocation}} \approx Gb^2 l/2$). Given typical values for calcite of the Burgers vector $b \approx 0.8$ nm, shear modulus $G = 35$ GPa, and the dislocation length, $l = 1$ μm, the dislocation energy can be estimated as ∼10^{-14} J. Therefore, it can be seen from Fig. 5c that for an interfacial stress of 6 N m^{-1} on the base of the tetrahedron, a crystal of volume 0.2 μm^3 (comparable to the crystals analysed here) will have more stored

elastic energy than it costs to create a 1-μm dislocation loop. This lends support to the argument that the dislocation relieves interface induced stress.

After nucleation, growth of the dislocation loop is governed by the preferential calcite slip system of {104} and {012} planes, and $<\bar{4}\,\bar{2}\,1>$ slip directions[35], and the growing dislocation loop is constrained to lie in one of these planes. As the plane of the observed dislocation loops is approximately parallel to the {104} truncation face, we can conclude that the {104} planes are the preferred slip planes for stress relief on SAMs. A schematic representation of the dislocation configuration is illustrated in Fig. 5d. This dislocation is consistent with the configurations of both of the observed dislocations. It is noted that the {104} $<\bar{4}\,\bar{2}\,1>$ slip system generally requires high temperature and/or pressure for the activation[35]. However, the dislocation loops are unambiguously observed in our crystals, and nucleation and growth appears to be the only plausible explanation for their origin. Rough surfaces have been shown to have very high local stresses (up to 4 GPa in TiO$_2$ nanoparticles[36]) that would make the nucleation of dislocation loops entirely feasible. It is also possible that dislocation motion is easier in micron-sized crystals than in their bulk counterparts, which may also favour their growth in our small crystals.

It remains striking, however, that each calcite crystal imaged here possesses a single dislocation loop, where each are of similar size and in an identical location with respect to the final crystal morphology. This suggests that the control over calcite nucleation and growth exerted by the monolayer is highly reproducible, and that the developments of the dislocation loop and crystal morphology are intimately linked. Finally, it is interesting to

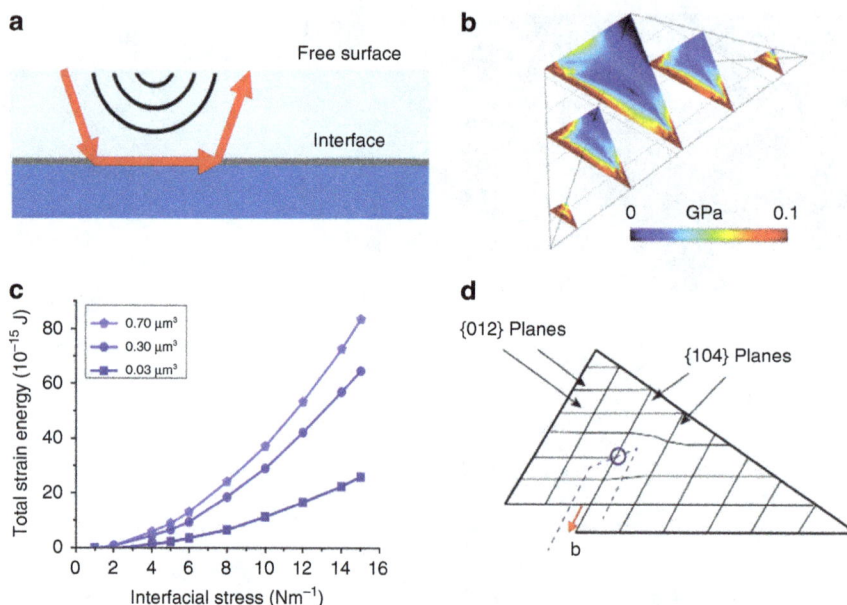

Figure 5 | Dislocation geometry. (**a**) Schematic of a classical 'epitaxy' misfit dislocations (red arrow), between two materials with a close lattice match. (**b**) Calculated stress distribution inside a calcite tetrahedron caused by an interfacial stress. The von Mises stress distribution is shown in vertical slices through the crystal. The interfacial stress was set to 10 N m^{-1}. (**c**) Total elastic energy, calculated using FE, due to interfacial stress for three different sized crystals. (**d**) Lattice planes on a cross-section of a (012) tetrahedron, showing the configuration of a dislocation on a (104) slip plane, with a $[4\,\bar{2}\,1]$ Burgers vector. The blue circle shows where the dislocation line cuts the plane and the dashed line shows the dislocation line, which has screw character on the vertical segments and edge character on the horizontal segment (perpendicular to the plane). The red arrow shows the Burgers vector (**b**). This configuration is consistent with the experimental observations.

note that the tetrahedral calcite crystals grown at the gas/liquid interface frequently reach diameters of ≈20 μm (refs 37,38), as compared with the maximal value of ≈2 μm observed on SAMs here. Again, the energy cost associated with the interfacial stress and dislocation loops would be expected to give rise to this reduced footprint on the substrate.

In summary, we have used X-ray BCDI to gain unique insight into the epitaxial growth of calcite crystals on organic SAMs, where this topic has received enormous interest in the literature, thanks to its relevance to biomineralization processes. High-resolution images of the internal strain present within micron-scale calcite crystals precipitated at SAMs reveal the presence of single dislocation loops within each crystal, where their configurations suggest that they play a role in the morphological development of the crystal. This provides an alternative explanation to the elastic strain model[24], which proposes that elastic deformation of both the SAM and calcite lattice gives rise to anisotropic growth. Importantly, the observed dislocation loops also exhibit entirely different geometries to the misfit dislocations generated during conventional epitaxial thin film formation. Our data strongly suggest that the nanoscale roughness present in SAM/evaporated metal film systems is fundamental to the formation of these defects, where this provides new insight into the factors that govern crystallization on these substrates. The results presented, therefore, provide an important contribution to our knowledge about stress relaxation and dislocation formation during epitaxial crystal growth, and demonstrate that the roughness of the substrate—be it organic or inorganic in nature—should be considered when controlling interfacial strain and defects during epitaxial crystal growth.

Methods

Materials. Analytical grade $(NH_4)_2CO_3$ and $CaCl_2 \cdot 2H_2O$, were used as received. Solutions were prepared using Milli-Q standard 18.2 MΩcm and experiments were

performed at a temperature of 21 °C. Glassware was soaked overnight in 10% w/v NaOH, followed by rinsing with dilute HCl and washing with Milli-Q water.

Substrate preparation. Functionalized SAMs were prepared on freshly deposited noble metal films. Thin films were deposited either on silicon wafer or cleaned glass slides using a Mantis Qprep 250 deposition system at a base pressure $<10^{-6}$ mbar. A 2-nm Cr was initially deposited to promote substrate adhesion, followed by the evaporation of 25–50 nm of Au at ≤0.1 nm s^{-1}. Monolayer formation on metal substrate was initiated by immersion in a 1 mM solution of 11-mercaptoundecanoic acid in ethanol. The prepared SAMs were then thoroughly rinsed with ethanol and Milli-Q water, and were subsequently dried under nitrogen.

Mineral deposition. Preferentially oriented calcite was obtained by diffusion methods[26]. A measure of 200-μl droplets of 5 mM $CaCl_2$ solution were hung from an inverted substrate, which was in turn placed in a sealed container (2 l) in the presence of $(NH_4)_2CO_{3(s)}$ (2 g). This geometry prevented 'homogenously' formed calcite from settling onto the surface. $(NH_4)_2CO_{3(s)}$ decomposition into $CO_{2(g)}$ and $NH_{3(g)}$ created the required supersaturation for $CaCO_3$ precipitation, and provided a gradual increase in supersaturation that ensured a sufficient number density of single crystals. Samples were removed after 30 min of incubation, giving rise to crystals 1–4 μm in diameter, which are suitable for coherent imaging. They were washed in ethanol and left to dry.

CDI set-up. Bragg CDI experiments were performed at beamline 34-ID-C of the Advanced Photon Source, Argonne National Laboratory, USA. An undulator produced X-rays that were monochromatized using a silicon (111) double-crystal monochromator to an energy of 9 keV. Calcite crystals on a substrate were placed on a diffractometer that had its rotation centre aligned with the X-ray beam. Slits were used to aperture the X-rays to reduce the illuminated area. An X-ray sensitive charge-coupled device (Princeton instruments) with 1300 × 1300 square pixels of side length 22.5 μm was positioned at the desired diffraction angle for an off-specular (104) reflection 2.5 m from the sample. To measure its full 3D diffraction patterns, the crystal was rotated by 0.3 degree with 0.003 degree step size. At each rotation angle, a 2D slice of the 3D far-field diffraction pattern was recorded. By stacking all 2D diffraction frames together, a complete 3D diffraction pattern is obtained, from which real-space images can be reconstructed. Due to the small size of the crystals (1–4 μm), the illumination can be considered almost completely coherent.

Reconstruction algorithm. Images were obtained by performing iterative phase retrieval[29] on 3D coherent diffraction patterns. Complete knowledge (both amplitude and phase) of the diffracted wavefield allows an image to be obtained via an inverse Fourier transform. Provided the diffraction data is oversampled, that is the sample has its Fourier transform sampled at least twice the Nyquist frequency (or alternatively its auto-correlation is sampled at least at the Nyquist frequency) and the crystal is isolated, phase retrieval can be performed. The basic phase-retrieval process begins with a guess for the diffracted phase before applying an inverse Fourier transform to yield a first estimate of the crystal. After enforcing the constraint that the crystal is isolated, this new crystal iterate is Fourier transformed to yield an estimate for the 3D-diffracted wavefield. Consistency with the measured intensity is enforced, while retaining the current estimate of the phase. This process is repeated until a self-consistent solution is reached using combinations of current and previous estimates for the crystal. For this work an approach was used that combined guided phase retrieval[31], with low- to high-resolution (or multi-resolution) reconstructions[32]. A detailed description is provided in the Supplementary Methods.

Finite element calculation. The calculations were performed with COMSOL Multiphysics 4.2a. Each element of the elasticity tensor of calcite was rotated to the crystallographic orientation of a crystal with a (012) nucleation plane. Surface stress was modelled by applying a 1-nm thick membrane, which was then uniformly contracted on each crystal facet (Supplementary Fig. 3).

Characterization. Scanning electron micrographs of uncoated specimen were obtained using an FEI Nova NanoSEM 650. Crystal growth was followed using an inverted Olympus IX-70 confocal microscope. Crystal orientation was determined using a Bruker D8 Advanced diffractometer equipped with a CuKα_1 X-ray source in pole configuration, using a step size 1.5° at 2.5 s (Psi 0–90, Phi 0–360). Substrates were characterized using atomic force microscopy (Bruker dimensions 3100 AFM) in tapping mode (Brucker Tespa; resonance frequency 345–385 kHz, K 20–80 Nm^{-1}) at a scan rate of 1.98 Hz with pixel dimension of 512 × 512. Images of the internal structure of the oriented calcite crystals were obtained using high-resolution TEM imaging of thin sections prepared by FIB milling. Sample preparation was performed using an FEI Nova200 Dual Beam FIB/scanning electron microscopy. The ion beam was operated at 30 kV and at beam currents between 0.1 and 5 nA. Lift-out was performed *in situ* using a Kleindiek micromanipulator. The samples were then analysed with a FEI Tecnai F20 200 kV field emission gun–TEM fitted with an Oxford Instruments INCA 350 EDX system/80 mm X-Max SDD detector and a Gatan Orius SC600A Charge Coupled Device (CCD) camera.

References

1. Van der Merwe, J. H. Misfit dislocation generation in epitaxial layers. *Crit. Rev. Solid State Mater. Sci.* **17**, 187–209 (1991).
2. Shiraki, Y. & Sakai, A. Fabrication technology of SiGe hetero-structures and their properties. *Surf. Sci. Rep.* **59**, 153–207 (2005).
3. O'Reilly, E. P. Valence band engineering in strained-layer structures. *Semicond. Sci. Technol.* **4**, 121–137 (1989).
4. Oh, S. H. & Park, C. G. Misfit strain relaxation by dislocations in SrRuO₃/SrTiO₃ (001) heteroepitaxy. *J. Appl. Phys.* **95**, 4691–4704 (2004).
5. Jain, S. C., Harker, A. H. & Cowley, R. A. Misfit strain and misfit dislocations in lattice mismatched epitaxial layers and other systems. *Phil. Mag. A* **75**, 1461–1515 (1997).
6. Hull, R. & Bean, J. C. Misfit dislocations in lattice-mismatched epitaxial films. *Crit. Rev. Solid State Mater. Sci.* **17**, 507–546 (1992).
7. Chen, Y. & Washburn, J. Structural transition in large-lattice-mismatch heteroepitaxy. *Phys. Rev. Letts.* **77**, 4046–4049 (1996).
8. Clark, J. N. *et al.* Three-dimensional imaging of dislocation propagation during crystal growth and dissolution. *Nat. Mater.* **14**, 780–784 (2015).
9. Heywood, B. R. & Mann, S. Template-directed nuclatiosn and growth of inorganic materials. *Adv. Mater.* **6**, 9–20 (1994).
10. Berman, A. *et al.* Total alignment of calcite at acidic polydiacetylene films: cooperativity at the organic-inorganic interface. *Science* **269**, 515–518 (1995).
11. DiMasi, E. *et al.* Complementary control by additives of the kinetics of amorphous CaCO₃ mineralization at an organic interface: in-situ synchrotron X-ray observations. *Phys. Rev. Letts.* **97**, 045503 (2006).
12. Lee, J. R. I. *et al.* Structural development of mercaptophenol self-assembled monolayers and the overlying mineral phase during templated CaCO₃ crystallization from a transient amorphous film. *J. Am. Chem. Soc.* **129**, 10370–10381 (2007).
13. Aizenberg, J., Black, A. J. & Whitesides, G. M. Control of crystal nucleation by patterned self-assembled monolayers. *Nature* **398**, 495–498 (1999).
14. Travaille, A. M. *et al.* Highly oriented self-assembled monolayers as templates for epitaxial calcite growth. *J. Am. Chem. Soc.* **125**, 11571–11577 (2003).
15. Darkins, R., Côté, A. S., Freeman, C. L. & Duffy, D. M. Crystallization rates of calcite from an amorphous precursor in confinement. *J. Cryst. Growth* **367**, 110–114 (2013).
16. Nielsen, M. H. & Lee, J. R.I in *Methods in Enzymology* Vol. 532 (ed. De Yoreo, J. J.) 209–224 (Academic Press, 2013).
17. Travaille, A. M. *et al.* Aligned growth of calcite crystals on a self-assembled monolayer. *Adv. Mater.* **14**, 492–495 (2002).
18. Aizenberg, J., Black, A. J. & Whitesides, G. M. Oriented growth of calcite controlled by self-assembled monolayers of functionalized alkanethiols supported on gold and silver. *J. Am. Chem. Soc.* **121**, 4500–4509 (1999).
19. Ahn, D. J., Berman, A. & Charych, D. Probing the dynamics of template-directed calcite crystallization with *in situ* FTIR. *J. Phys. Chem.* **100**, 12455–12461 (1996).
20. Lee, J. R. I. *et al.* Cooperative reorganization of mineral and template during directed nucleation of calcium carbonate. *J. Phys. Chem. C* **117**, 11076–11085 (2013).
21. Quigley, D., Rodger, P. M., Freeman, C. L., Harding, J. H. & Duffy, D. M. Metadynamics simulations of calcite crystallization on self-assembled monolayers. *J. Chem. Phys.* **131**, 094703 (2009).
22. Volkmer, D., Fricke, M., Vollhardt, D. & Siegel, S. Crystallization of (012) oriented calcite single crystals underneath monolayers of tetra(carboxymethoxy)calix 4 arenes. *Dalton. Trans.* **24**, 4547–4554 (2002).
23. Freeman, C. L., Harding, J. H. & Duffy, D. M. Simulations of calcite crystallization on self-assembled monolayers. *Langmuir* **24**, 9607–9615 (2008).
24. Pokroy, B. & Aizenberg, J. Calcite shape modulation through the lattice mismatch between the self-assembled monolayer template and the nucleated crystal face. *CrystEngComm.* **9**, 1219–1225 (2007).
25. Harding, J. H., Freeman, C. L. & Duffy, D. M. Oriented crystal growth on organic monolayers. *CrystEngComm.* **16**, 1430–1438 (2014).
26. Ihli, J., Bots, P., Kulak, A., Benning, L. G. & Meldrum, F. C. Elucidating mechanisms of diffusion-based calcium carbonate synthesis leads to controlled mesocrystal formation. *Adv. Funct. Mater.* **23**, 1965–1973 (2013).
27. Stephens, C. J., Kim, Y.-Y., Evans, S. D., Meldrum, F. C. & Christenson, H. K. Early Stages of crystallization of calcium carbonate revealed in picoliter droplets. *J. Am. Chem. Soc.* **133**, 5210–5213 (2011).
28. Zucker, R., Chatain, D., Dahmen, U., Hagège, S. & Carter, W. C. New software tools for the calculation and display of isolated and attached interfacial-energy minimizing particle shapes. *J. Mater. Sci.* **47**, 8290–8302 (2012).
29. Fienup, J. R. Phase retrieval algorithms: a comparison. *Appl. Opt.* **21**, 2758–2769 (1982).
30. Clark, J. N. *et al.* Ultrafast three-dimensional imaging of lattice dynamics in individual gold nanocrystals. *Science* **341**, 56–59 (2013).
31. Chen, C.-C., Miao, J., Wang, C. W. & Lee, T. K. Application of optimization technique to noncrystalline x-ray diffraction microscopy: guided hybrid input-output method. *Phys. Rev. B* **76**, 064113 (2007).
32. McCallum, B. C. & Bates, R. H. T. Towards a strategy for automatic phase retrieval from noisy fourier intensities. *J. Mod. Opt.* **36**, 619–648 (1989).
33. Borukhin, S. & Pokroy, B. Formation and elimination of surface nanodefects on ultraflat metal surfaces produced by template stripping. *Langmuir* **27**, 13415–13419 (2011).
34. Côté, A. S., Darkins, R. & Duffy, D. M. Modeling calcite crystallization on self-assembled carboxylate-terminated alkanethiols. *J. Phys. Chem. C* **118**, 19188–19193 (2014).
35. DeBresser, J. H. P. & Spiers, C. J. Strength characteristics of the r, f, and c slip systems in calcite. *Tectonophysics* **272**, 1–23 (1997).
36. Darkins, R., Sushko, M. L., Liu, J. & Duffy, D. M. Stress in titania nanoparticles: an atomistic study. *Phys. Chem. Chem. Phys.* **16**, 9441–9447 (2014).
37. Hashmi, S. M., Wickman, H. H. & Weitz, D. A. Tetrahedral calcite crystals facilitate self-assembly at the air-water interface. *Phys. Rev. E* **72**, 041605 (2005).
38. Loste, E., Díaz-Martí, E., Zarbakhsh, A. & Meldrum, F. C. Study of calcium carbonate precipitation under a series of fatty acid langmuir monolayers using brewster angle microscopy. *Langmuir* **19**, 2830–2837 (2003).
39. Ihli, J. *et al.* Dataset for BCDI study of Calcite Crystals Grown on Self-Assembled Monolayers, Research Data Leeds Repository. Available at http://doi.org/10.5518/53.

Acknowledgements

This work was supported by FP7-advanced grant from the European Research Council (J.N.C. and I.K.R.), by an Engineering and Physical Sciences Research Council Leadership Fellowship (F.C.M., J.I. and Y.Y.K.; EP/H005374/1) and EPSRC grant EP/K006304/1 (A.N.K.). F.C.M., A.S.S., D.M.D. and A.S.C. are also supported by an EPSRC Programme Grant (grant EP/I001514/1) that funds the Materials Interface with Biology (MIB) consortium. Some of the experimental work was carried out at Advanced Photon Source Beamline 34-ID-C, built with funds from the US National Science Foundation under Grant DMR-9724294 and operated by the US Department of Energy, Office of Science, Office of Basic Energy Sciences, under contract DE-AC02-06CH11357.

Author contributions

J.I. and J.N.C. designed the project; J.I. prepared the samples; J.N.C., J.I. and R.J.H., performed the BCDI experiments; J.N.K. performed the image reconstructions; J.I. and T.P.C. acquired the pole figures; A.N.K., Y.Y.K. and J.I. acquired the FIB images; O.C. and J.I. acquired the AFM data; J.N.C., J.I. and A.S.S. analysed the data, D.M.D. and A.S.C. performed the FE studies; and J.I., J.N.C., F.C.M., D.M.D. and I.K.R. wrote the paper. All the authors read and commented on the manuscript.

Additional information

Competing financial interests: The authors declare no competing financial interests.

Permissions

List of Contributors

Li Peng, Zhen Xu, Zheng Liu, Yangyang Wei, Haiyan Sun, Zheng Li, Xiaoli Zhao and Chao Gao
MOE Key Laboratory of Macromolecular Synthesis and Functionalization, Department of Polymer Science and Engineering, Zhejiang University, Polymer Building, 38 Zheda Road, Hangzhou 310027, P.R. China

Sebastian Grundner, Monica A.C. Markovits, Andreas Jentys and Maricruz Sanchez-Sanchez
Department of Chemistry and Catalysis Research Center, Technische Universität München, Lichtenbergstrasse 4, Garching 85748, Germany

Johannes A. Lercher
Department of Chemistry and Catalysis Research Center, Technische Universität München, Lichtenbergstrasse 4, Garching 85748, Germany
Institute for Integrated Catalysis, Pacific Northwest National Laboratory, PO Box 999, Richland, Washington 99352, USA

Guanna Li and Emiel J.M. Hensen
Schuit Institute of Catalysis, Inorganic Materials Chemistry Group, Department of Chemical Engineering and Chemistry, Eindhoven University of Technology, PO Box 513, Eindhoven 5600 MB, The Netherlands.

Evgeny A. Pidko
Schuit Institute of Catalysis, Inorganic Materials Chemistry Group, Department of Chemical Engineering and Chemistry, Eindhoven University of Technology, PO Box 513, Eindhoven 5600 MB, The Netherlands
Institute for Complex Molecular Systems, Eindhoven University of Technology, PO Box 513, Eindhoven 5600 MB, The Netherlands

Moniek Tromp
Van't Hoff Institute for Molecular Sciences, University of Amsterdam, PO Box 94215, Amsterdam 1090GE, The Netherlands

Dragos Neagu, David N. Miller, Syed M. Bukhari, Stephen R. Gamble and John T.S. Irvine
School of Chemistry, University of St Andrews, St Andrews, KY16 9ST Scotland, UK

Tae-Sik Oh, Raymond J. Gorte and John M. Vohs
Department of Chemical and Biomolecular Engineering, University of Pennsylvania, Philadelphia, Pennsylvania 19104, USA

Hervé Ménard
Sasol Technology (UK) Ltd., St Andrews, KY16 9ST Scotland, UK. These authors contributed equally to this work

Tong-Liang Hu
Department of Chemistry, TKL of Metal- and Molecule-Based Material Chemistry, Collaborative Innovation Center of Chemical Science and Engineering (Tianjin), Nankai University, Tianjin 300071, China
Department of Chemistry, University of Texas at San Antonio, One UTSA Circle, San Antonio, Texas 78249-0698, USA

Hailong Wang, Bin Li and Banglin Chen
Department of Chemistry, University of Texas at San Antonio, One UTSA Circle, San Antonio, Texas 78249-0698, USA

Rajamani Krishna
Van 't Hoff Institute for Molecular Sciences, University of Amsterdam, Science Park 904, Amsterdam 1098 XH, The Netherlands

Hui Wu and Wei Zhou
NIST Center for Neutron Research, Gaithersburg, Maryland 20899-6102, USA

Yunfeng Zhao and Yu Han
Advanced Membranes and Porous Materials Center, Physical Sciences and Engineering Division, King Abdullah University of Science and Technology, Thuwal 23955-6900, Saudi Arabia

Xue Wang and Weidong Zhu
Key Laboratory of the Ministry of Education for Advanced Catalysis Materials, Institute of Physical Chemistry, Zhejiang Normal University, Jinhua 321004, China

Zizhu Yao and Shengchang Xiang
College of Chemistry and Chemical Engineering, Fujian Provincial Key Laboratory of Polymer Materials, Fujian Normal University, Fuzhou 350007, China

Yoann Roux, Rémy Ricoux, Frédéric Avenier and Jean-Pierre Mahy
Laboratoire de Chimie Bioorganique et Bioinorganique, Institut de Chimie Moléculaire et des Matériaux d'Orsay (UMR 8182), Univ Paris Sud, Université Paris Saclay, rue du doyen Georges Poitou, 91405 Orsay, France

Yi Li1, Xu Li, Jiancong Liu, Fangzheng Duan and Jihong Yu
State Key Laboratory of Inorganic Synthesis and Preparative Chemistry, Jilin University, Qianjin Street 2699, Changchun 130012, China

Shunsuke Yagi and Akihiro Seno
Nanoscience and Nanotechnology Research Centre, Osaka Prefecture University, Osaka 599-8570, Japan

Ikuya Yamada
Nanoscience and Nanotechnology Research Centre, Osaka Prefecture University, Osaka 599-8570, Japan
Precursory Research for Embryonic Science and Technology, Japan Science and Technology Agency, Tokyo 102-0075, Japan

Hirofumi Tsukasaki, Makoto Murakami, Hiroshi Fujii and Shigeo Mori
Department of Materials Science and Engineering, Osaka Prefecture University,

Osaka 599-8531, Japan
Hungru Chen
National Institute for Materials Science, Tsukuba 305-0044, Japan

Naoto Umezawa and Hideki Abe
Precursory Research for Embryonic Science and Technology, Japan Science and Technology Agency, Tokyo 102-0075, Japan
National Institute for Materials Science, Tsukuba 305-0044, Japan

Norimasa Nishiyama
Precursory Research for Embryonic Science and Technology, Japan Science and Technology Agency, Tokyo 102-0075, Japan
Deutsches Elektronen Synchrotron, Hamburg 22607, Germany

Zhi Li, Yusheng Shi, Diyang Zhang and Dingsheng Wang
Department of Chemistry and Collaborative Innovation Center for Nanomaterial Science and Engineering, Tsinghua University, Beijing 100084, China

Rong Yu, Jinglu Huang and Xiaoyan Zhong
Beijing National Center for Electron Microscopy, School of Materials Science and Engineering, Tsinghua University, Beijing 100084, China

Yuen Wu and Yadong Li
Department of Chemistry and Collaborative Innovation Center for Nanomaterial Science and Engineering, Tsinghua University, Beijing 100084, China

Center of Advanced Nanocatalysis (CAN-USTC), University of Science and Technology of China, Hefei, Anhui 230026, China

Ivan A. Popov and Alexander I. Boldyrev
Department of Chemistry and Biochemistry, Utah State University, Logan, Utah 84322, USA

Tian Jian, Gary V. Lopez and Lai-Sheng Wang
Department of Chemistry, Brown University, Providence, Rhode Island 02912, USA. These authors contributed equally to this work

Pengfei Zhang, Li Zhang and Zili Wu
Chemical Sciences Division, Oak Ridge National Laboratory, Oak Ridge, Tennessee 37831, USA

Hanfeng Lu, Ying Zhou, Qiulian Zhu and Yinfei Chen
Institute of Catalytic Reaction Engineering, College of Chemical Engineering, Zhejiang University of Technology, Hangzhou 310014, China

Shize Yang and Hongliang Shi
Materials Science and Technology Division, Oak Ridge National Laboratory, Oak Ridge, Tennessee 37831, USA

Sheng Dai
Chemical Sciences Division, Oak Ridge National Laboratory, Oak Ridge, Tennessee 37831, USA
Department of Chemistry, University of Tennessee, Knoxville, Tennessee 37996, USA

Jung-Hoon Lee, Kyle J. Gibson, Gang Chen and Yossi Weizmann
Department of Chemistry, The University of Chicago, 929 East 57th Street, Chicago, Illinois 60637, USA. These authors contributed equally to this work

Keke Wang, Tongtong Han, Minman Tong, Liangsha Li, Qingyuan Yang, Dahuan Liu and Chongli Zhong
State Key Laboratory of Organic-Inorganic Composites, Beijing University of Chemical Technology, Beijing 100029, China

Hongliang Huang
State Key Laboratory of Organic-Inorganic Composites, Beijing University of Chemical Technology, Beijing 100029, China
Beijing Key Laboratory for Green Catalysis and Separation, Department of Chemistry and Chemical Engineering, College of Environmental and Energy Engineering, Beijing University of Technology, Pingleyuan 100, Chaoyang, Beijing 100124, China

Jian-Rong Li and Yabo Xie
Beijing Key Laboratory for Green Catalysis and Separation, Department of Chemistry and Chemical Engineering, College of Environmental and Energy Engineering, Beijing University of Technology, Pingleyuan 100, Chaoyang, Beijing 100124, China

Sheng-Qun Su, Takashi Kamachi, Zi-Shuo Yao, You-Gui Huang, Yoshihito Shiota, Kazunari Yoshizawa, Soonchul Kang, Shinji Kanegawa and Osamu Sato
Institute for Materials Chemistry and Engineering, Kyushu University, 744 Motooka, Nishi-ku, Fukuoka 819-0395, Japan
Nobuaki Azuma, Yuji Miyazaki and Motohiro Nakano
Research Center for Structural Thermodynamics, Graduate School of Science, Osaka University, Toyonaka, Osaka 560-0043, Japan

Goro Maruta and Sadamu Takeda
Department of Chemistry, Faculty of Science, Hokkaido University, Sapporo 060-0810, Japan

Shubo Tian, Man-Bo Li and Zhikun Wu
Key Laboratory of Materials Physics, Anhui Key Laboratory of Nanomaterials and Nanotechnology, Institute of Solid State Physics, Chinese Academy of Sciences, Hefei, Anhui 230031, China

Yi-Zhi Li
State Key Laboratory of Coordination Chemistry, School of Chemistry and Chemical Engineering, Nanjing University, Nanjing 210093, China

Jinyun Yuan and Jinlong Yang
Hefei National Laboratory for Physical Sciences at the Microscale and Synergetic Innovation Center of Quantum Information and Quantum Physics, University of Science and Technology of China, Hefei, Anhui 230026, China

Rongchao Jin
Department of Chemistry, Carnegie Mellon University, Pittsburgh, Pennsylvania 15213, United States

Qiangfeng Xiao, Fang Dai, Li Yang, Zhongyi Liu, Xingcheng Xiao and Mei Cai
Gereral Motors Research and Development Center, 30500 Mound Road, Warren, Michigan 48090, USA

Meng Gu and Chongmin Wang
Environmental Molecular Sciences Laboratory, Pacific Northwest National Laboratory, Richland,Washington 99352, USA

Hui Yang, Peng Zhao and Sulin Zhang
Department of Engineering Science & Mechanics, Pennsylvania State University, University Park, Pennsylvania 16802, USA

Bing Li and Cunman Zhang
Clean Energy Automotive Engineering Center, Tongji University, Shanghai 201804, China

Yang Liu, Fang Liu and Yunfeng Lu
5 Department of Chemical and Biomolecular Engineering, The University of California, Los Angeles, California 90095, USA

Gao Liu
Environmental Energy Technologies Division, Lawrence Berkeley National Laboratory, Berkeley, California 94720, USA. These authors contributed equally to this work

Bo Wu, Taishan Wang, Yongqiang Feng, Zhuxia Zhang, Li Jiang and Chunru Wang
Key Laboratory of Molecular Nanostructure and Nanotechnology, Beijing National Laboratory for Molecular Sciences, Institute of Chemistry, Chinese Academy of Sciences, Beijing 100190, China

Yoann Prado, Aude Michel, Jérôme Fresnais and Vincent Dupuis
Sorbonne Universités, UPMC Univ Paris 06, UMR 8234, PHENIX, CNRS, F-75005 Paris, France

Marie-Anne Arrio
Institut de Minéralogie, de Physique des Matériaux et de Cosmochimie, UMR 7590, CNRS, UPMC, IRD, MNHN, F-75005 Paris, France

Niéli Daffé
Sorbonne Universités, UPMC Univ Paris 06, UMR 8234, PHENIX, CNRS, F-75005 Paris, France
Institut de Minéralogie, de Physique des Matériaux et de Cosmochimie, UMR 7590, CNRS, UPMC, IRD, MNHN, F-75005 Paris, France
Synchrotron SOLEIL, L'Orme des Merisiers, Saint-Aubin—BP 48, 91192 Gif-sur-Yvette, France.

Philippe Sainctavit
Institut de Minéralogie, de Physique des Matériaux et de Cosmochimie, UMR 7590, CNRS, UPMC, IRD, MNHN, F-75005 Paris, France
Synchrotron SOLEIL, L'Orme des Merisiers, Saint-Aubin—BP 48, 91192 Gif-sur-Yvette, France

Fadi Choueikani, Edwige Otero and Philippe Ohresser
Synchrotron SOLEIL, L'Orme des Merisiers, Saint-Aubin—BP 48, 91192 Gif-sur-Yvette, France

Thomas Georgelin
Sorbonne Universités, UPMC Univ Paris 06, UMR 7197, LRS, F-94200 Ivry-sur-Seine, France
CNRS, UMR 7197, Laboratoire de Réactivitéde Surface, F-94200 Ivry-sur-Seine, France

Nader Yaacoub and Jean-Marc Grenéche
Institut des Molécules et Matériaux du Mans CNRS UMR-6283, Universitédu Maine,
F-72085 Le Mans, France

Christophe Cartier-dit-Moulin, Benoit Fleury and Laurent Lisnard
Sorbonne Universités, UPMC Univ Paris 06, UMR 8232, IPCM, F-75005 Paris, France
CNRS, UMR 8232, Institut Parisien de Chimie Moléculaire, F-75005 Paris, France

Ritsuko Yaokawa and Tetsu Ohsuna
TOYOTA Central R&D Labs, Inc., 41-1, Yokomichi, Nagakute, Aichi 480-1192, Japan

Tetsuya Morishita
CD-FMat, National Institute of Advanced Industrial Science and Technology (AIST), Central 2, 1-1-1 Umezono, Tsukuba, Ibaraki 305-8568, Japan

Yuichiro Hayasaka
The Electron Microscopy Center, Tohoku University, Katahira 2-1-1, Aobaku, Sendai 980-8577, Japan

Michelle J.S. Spencer
School of Science, RMIT University, GPO Box 2476, Melbourne, Victoria 3001, Australia

Hideyuki Nakano
TOYOTA Central R&D Labs, Inc., 41-1, Yokomichi, Nagakute, Aichi 480-1192, Japan JST Presto, Kawaguchi 332-0012, Japan

Elise M. Miner, Tomohiro Fukushima, Dennis Sheberla, Lei Sun, Yogesh Surendranath and Mircea Dincă
Department of Chemistry, Massachusetts Institute of Technology, 77 Massachusetts Avenue, Cambridge, Massachusetts 02139, USA

Antonio Fernandez, Jesus Ferrando-Soria, Eufemio Moreno Pineda, Floriana Tuna, Iñigo J. Vitorica-Yrezabal, Christopher A. Muryn, Grigore A. Timco and Richard E.P. Winpenny
School of Chemistry and Photon Science Institute,The University of Manchester, Oxford Road, Manchester M13 9PL, UK

Christiane Knappke
Department of Chemistry, University of Oxford, Oxford OX1 3TA, UK

Jakub Ujma and Perdita E. Barran
School of Chemistry and Photon Science Institute, The University of Manchester, Oxford Road, Manchester M13 9PL, UK

The Michael Barber Centre for Collaborative Mass Spectrometry, Manchester Institute of Biotechnology, The University of Manchester, Oxford Road, Manchester M13 9PL, UK

Arzhang Ardavan
Department of Physics, Centre for Advanced Electron Spin Resonance, The Clarendon Laboratory, University of Oxford, Parks Road, Oxford OX1 3PU, UK

Debasis Banerjee and Praveen K. Thallapally
Physical and Computational Science Directorate, Pacific Northwest National Laboratory, Richland, Washington 99352, USA

Cory M. Simon
Department of Chemical and Biochemical Engineering, University of California, Berkley, Berkeley, California 94720, USA

Anna M. Plonka
Department of Geosciences, Stony Brook University, Stony Brook,
New York 11794, USA

Radha K. Motkuri and Jian Liu
Energy and Environmental Directorate, Pacific Northwest National Laboratory, Richland, Washington 99352, USA

Xianyin Chen
Department of Chemistry, Stony Brook University, Stony Brook, New York 11794, USA

Berend Smit
Department of Chemical and Biochemical Engineering, University of California, Berkley, Berkeley, California 94720, USA
Institut des Sciences et Ingénierie Chimiques, Valais, Ecole Polytechnique Fédérale de Lausanne (EPFL), Rue de l0Industrie 17, CH-1951 Sion, Switzerland

John B. Parise
Department of Geosciences, Stony Brook University, Stony Brook, New York 11794, USA
Department of Chemistry, Stony Brook University, Stony Brook, New York 11794, USA
Photon Sciences, Brookhaven National Laboratory, Upton, New York 11973, USA

Maciej Haranczyk
Computational Research Division, Lawrence Berkeley National Laboratory, Berkeley, California 94720, USA
IMDEA Materials Institute, C/Eric Kandel 2, 28906 Getafe, Madrid, Spain. Correspondence and requests for materials should be addressed to B.S.

Shiming Zhou, Xianbing Miao, Xu Zhao, Chao Ma, Jiyin Zhao, Lei Shi and Jie Zeng
Hefei National Laboratory for Physics Sciences at the Microscale, Hefei Science Center, University of Science and Technology of China, Hefei, Anhui 230026, China

Yuhao Qiu
School of Physics, Nankai University, Tianjin 300071, China

Zhenpeng Hu
School of Physics, Nankai University, Tianjin 300071, China
State Key Laboratory of Luminescent Materials and Devices, South China University of Technology, Guangzhou 510640, China

Wanbin Li, Congyang Zhang, Zehai Xu, Pengcheng Su, Zheng Fan, Lei Qin and Guoliang Zhang
Institute of Oceanic and Environmental Chemical Engineering, State Key Lab Breeding Base of Green Chemical Synthesis Technology and Collaborative Innovation Center of Membrane Separation and Water Treatment of Zhejiang Province, Zhejiang University of Technology, Chaowang Road 18#, Hangzhou 310014, China

Yufan Zhang
Department of Materials Science and Engineering, College of Engineering, University of California, Berkeley, California 94720, USA

Qin Meng and Chong Shen
Department of Chemical and Biological Engineering, College of Chemical and Biological Engineering, State Key Laboratory of Chemical Engineering, Zhejiang University, Hangzhou 310027, China

Qingbiao Li
Department of Chemical and Biochemical Engineering, College of Chemistry and Chemical Engineering, National Laboratory for Green Chemical Productions of Alcohols, Ethers and Esters, Key Lab for Chemical Biology of Fujian Province, Xiamen University, Xiamen 361005, China

Peter C. Ho, Patrick Szydlowski, Lucia Myongwon Lee, Hilary Jenkins, James F. Britten, Derek R. Morim and Ignacio Vargas-Baca
Department of Chemistry and Chemical Biology, McMaster University, 1280 Main Street West, Hamilton, Ontario, Canada L8S 4M1

Jocelyn Sinclair, Philip J.W. Elder, Joachim Kübel and Chris Gendy
Department of Chemistry, Dalhousie University, 6274 Coburg Road, PO Box 15000, Halifax, Nova Scotia, Canada B3H 4R2 (J.S.); ALS Environmental, 1435 Norjohn Ct #1, Burlington, Ontario, Canada L7L 0E6 (P.J.W.E.); Leibniz Institute of Photonic Technology e.V., Albert-Einstein-Stra_e 9, Jena 07745, Germany (J.K.); Department of Chemistry, University of Calgary, 2500 University Road, Calgary, Alberta, Canada T2N 1N4 (C.G.)

Baolin Wang, Hajime Hirao and Rei Kinjo
Division of Chemistry and Biological Chemistry, School of Physical and Mathematical Sciences, Nanyang Technological University, 21 Nanyang Link, Singapore 637371, Singapore

Yongxin Li and Rakesh Ganguly
NTU-SPMS-CBC Crystallography Facility, Nanyang Technological University, 21 Nanyang Link, Singapore 637371, Singapore

Zan Gao, Clifton Bumgardner, Ningning Song, Yunya Zhang and Xiaodong Li
Department of Mechanical and Aerospace Engineering, University of Virginia, 122 Engineer'sWay, Charlottesville, Virginia 22904-4746, USA

Jingjing Li
Department of Mechanical Engineering, University of Hawaii at Manoa, 2540 Dole Street, Honolulu, Hawaii 96822, USA

A. Gallagher, K.-W. Chen, C.M. Moir, D. Graf, S.C. Riggs, A. Shekhter and R.E. Baumbach
National High Magnetic Field Laboratory, Florida State University, Tallahassee, Florida 32310, USA

S.K. Cary and T.E. Albrecht-Schmitt
Department of Chemistry and Biochemistry, Florida State University, Tallahassee, Florida 32306, USA

F. Kametani
3 Applied Superconductivity Center, Florida State University, Tallahassee, Florida 32310, USA

N. Kikugawa
National High Magnetic Field Laboratory, Florida State University, Tallahassee, Florida 32310, USA
National Institute for Materials Science 3-13 Sakura, Tsukuba 305-0003, Japan

Yi-Yeoun Kim, Anna S. Schenk, Alexander N. Kulak and Fiona C. Meldrum
School of Chemistry, University of Leeds, Woodhouse Lane, Leeds LS2 9JT, UK

Johannes Ihli
School of Chemistry, University of Leeds, Woodhouse Lane, Leeds LS2 9JT, UK
Paul Scherrer Institute, 5232 Villigen PSI, Switzerland. (J.I.); London Centre for Nanotechnology, University College London, 17–19 Gordon Street, London WC1H 0AH, UK (A.S.C.)

Jesse N. Clark
Stanford PULSE Institute, SLAC National Accelerator Laboratory, 2575 Sand Hill Road, Menlo Park, California 94025, USA
Center for Free-Electron Laser Science (CFEL), Deutsches Elektronensynchrotron (DESY) Notkestrasse 85, 22607 Hamburg, Germany

Dorothy M. Duffy
Department of Physics and Astronomy, University College London, Gower Street, London WC1E 6BT, UK

Alexander S. Côté
Department of Physics and Astronomy, University College London, Gower Street, London WC1E 6BT, UK
Paul Scherrer Institute, 5232 Villigen PSI, Switzerland. (J.I.); London Centre for Nanotechnology, University College London, 17–19 Gordon Street, London WC1H 0AH, UK (A.S.C.)

Timothy P. Comyn
Institute for Materials Research, University of Leeds, Leeds LS2 9JT, UK

Oliver Chammas
School of Physics and Astronomy, University of Leeds, Leeds LS2 9JT, UK

Ross J. Harder
Advanced Photon Source, Argonne, Illinois 60439, USA

Ian K. Robinson
London Centre for Nanotechnology, University College London, 17–19 Gordon Street, London WC1H 0AH, UK. These authors contributed equally to this work

Index